A GEOGRAPHY OF RUSSIA AND ITS NEIGHBORS

Texts in Regional Geography

A GUILFORD SERIES

Edited by James L. Newman, Syracuse University

Latin America: Regions and People
Robert B. Kent

Africa South of the Sahara:
A Geographical Interpretation,
Second Edition
Robert Stock

Cuban Landscapes:
Heritage, Memory, and Place
Joseph L. Scarpaci and Armando H. Portela

A Geography of Russia and Its Neighbors
Mikhail S. Blinnikov

The Europeans:
A Geography of People, Culture, and Environment,
Second Edition
Robert C. Ostergren and Mathias LeBossé

A Geography of Russia and Its Neighbors

MIKHAIL S. BLINNIKOV

THE GUILFORD PRESS
New York London

A Division of Guilford Publications, Inc.
72 Spring Street, New York, NY 10012
www.guilford.com

Printed in the United States of America

This book is printed on acid-free paper.

Last digit is print number: 9 8 7 6 5 4 3 2 1

Library of Congress Cataloging-in-Publication Data

Blinnikov, Mikhail S.
A geography of Russia and its neighbors / by Mikhail S. Blinnikov.
p. cm. — (Texts in regional geography)
Includes bibliographical references and index.
ISBN 978-1-60623-933-9 (hardcover: alk. paper) — ISBN 978-1-60623-920-9 (pbk.: alk. paper)
1. Russia (Federation)—Geography. 2. Former Soviet republics—Geography. I. Title.
DK510.28.B55 2011
914.7—dc22

2010034117

Contents

List of Figures and Tables

Figures

Tables

The volume's photographs, maps, figures, and tables are available as PowerPoint slides on the book's page on The Guilford Press website (*www.guilford.com/p/blinnikov*).

Metric Units and Their Equivalents

1 kilometer (km) = 1,000 m, or about 0.621 miles

1 meter (m) = about 3.28 feet, or 1.09 yards

100 mm of precipitation = about 3.937 inches

1 hectare (ha) = 100 x 100 m, or about 2.471 acres

1 square meter = 100 x 100 cm, or about 10.76 square feet

1 metric tonne = 1,000 kg, or about 1.1 short tons

1 kilogram (kg) = 1,000 g, or about 2.2 pounds

1 liter (L) = 1,000 ml, or about 1.06 quarts or 0.26 gallons

CHAPTER 1

Introduction

Russia and Post-Soviet Northern Eurasia

Russia is a country unlike any other. It occupies much of the world's largest landmass, Eurasia; it stretches across 11 time zones and covers over 17 million km^2. Its average climate is the coldest of any country on earth. Its land is extremely varied, with large plains and bogs, forests and deserts, rivers and lakes. Underneath its soil are thousands of tons of precious and semi-precious metals; millions of pounds of iron ore, bauxite, and coal; billions of barrels of oil; and trillions of cubic meters of natural gas. Its peoples are numerous and diverse, speaking over 130 languages. Its main language, Russian, is among the world's 10 most common and has produced some of the greatest literary works. Russia is also home to world-class fine and performing arts. Its temples and museums display the precious heritage of countless generations, admired the world over. The two main religious traditions of its former empire—Orthodox Christianity and Islam—have had tremendous internal influence and are becoming more widespread in the rest of the world. Russia sent the first human-made object into space, as well as the first human to orbit the earth. In the 20th century it helped defeat fascism, but it also nearly destroyed itself in one of the bloodiest dictatorships ever known. This country remains an enigma to outsiders, and even to some people within its own borders. A full appreciation of Russia requires a firm grasp of geography. This book attempts to deliver a balanced presentation of the physical, historical/political, cultural/social, economic, and regional geography of Russia today. Although Russia is its main focus, the book also discusses other republics that were once part of the Soviet Union, so it should prove useful to a variety of courses on post-Soviet Eurasia.

What to Study: Russia or the Former Soviet Union?

Many teachers of college classes on post-Soviet geography face the question of whether to cover Russia only, or the entire former Soviet Union (FSU). In the United States during the Cold War period, courses on the region covered the U.S.S.R. as a whole. What do we do now, 20 years after the Soviet Union fell apart? Some professors no longer teach courses about the FSU. They may teach one course on Russia and another one on the emerging economies of Central Asia, for example. The Baltic states have joined the European Union (EU) and the North Atlantic Treaty Organization (NATO) and are now routinely treated as part of greater Europe, to which they rightfully belong. Ukraine is so large and complex that it might merit a textbook and a class of its own.

Nevertheless, although this book focuses mainly on Russia, it looks at all the FSU republics. All these republics were included for 50–70 years in one political entity that had a profound impact on them. Many of the processes that shaped these countries no longer exist, but the geographic patterns persist. There is still enough commonality among the countries in question to merit an overall discussion of what is going on in the FSU (which some believe may now be better referred to as Northern Eurasia). Besides the centrifugal tendencies that have forced these countries apart, there are also centripetal forces that have helped maintain some common identity for all 15 of them. One such force is the presence of numerous Russian speakers throughout the region. Another is heavy dependence on Russia for energy supplies, especially natural gas and electricity. Even the stubbornly independent Ukraine and Georgia are pragmatic enough to understand their reliance on their big neighbor. Economic patterns of production, once disrupted by the chaos of reforms, are likewise not all that different from the old Soviet ones. Kazakhstan, Ukraine, Belarus, and Russia remain particularly heavily interlinked with each other and are the most industrialized; the trans-Caucasian republics, Moldova, and the Central Asian states are more agricultural and less closely linked with either each other or the industrialized four, but remain somewhat interdependent.

In each discussion of a topic, this book addresses Russia first and in the greatest depth. Additional material on the other republics is included whenever this is necessary or appropriate. Part V of the book provides brief regional summaries about parts of Russia and various FSU republics (see Figure 1.1), and may be used as a quick reference or as a guide for more in-depth reading in advanced classes. But first let's discuss various terms referring to the region:

- **Rus** was the ancient state of the eastern Slavs, centered around what is today Kiev, Ukraine. It existed before Russians, Ukrainians, and Belarusians had become separate peoples, between ca. 800 and 1250 A.D. Gradually power shifted to the north, toward Moscow, where the Muscovy princedom evolved into a new and powerful state.
- The **Russian Empire** was the state centered on Moscow and St. Petersburg as its capitals; it existed from the 17th century until 1917.
- The **Soviet Union (U.S.S.R.)** existed between 1922 and 1991.
- The **former Soviet Union (FSU)** consists of the 15 republics that now make up this region. The adjective to describe these would be "post-Soviet."
- The **newly independent states (NIS)** refers to the same area. NIS is rarely used now (they are no longer "newly" independent).
- The **Commonwealth of Independent States (CIS)** is a loose alliance of 11 republics (12 until Georgia quit in 2008), excluding the Baltic states.
- **Russia and the Near Abroad** is an ambiguous term commonly used in Russia to describe Russia along with the other 14 republics (it is equivalent to the FSU), although geographically Finland or Mongolia could be added because they border Russia. Moreover, some FSU republics do not border Russia at all, so this term is best avoided.
- **Northern Eurasia** is a good physical definition of the region; it is now frequently used by biogeographers, ecologists, and other geoscientists. It is politically neutral and clearly describes the position of the region on the world's map. There is a problem with it, however: Few people who are not geography majors have any idea what or where it is.
- The **Russian Realm** may not be a bad title for a documentary, but it is too Russia centered to be of much use. On the one hand, the Russian sphere of influence in the world today extends into Israel or the United Kingdom, for example, but this does not make those countries part of the region in question. On the other hand, some countries in the region—for example, Armenia and Turkmenistan—have very few Russians left and have little to do with Russia proper.
- **Siberia** is a region within Russia, extending east of the Ural Mountains to the Lena River watershed. It is not a separate country. Everything west of the Urals is **European Russia**, while everything east of the Lena is the **Russian Far East** (or **Russian Pacific**).

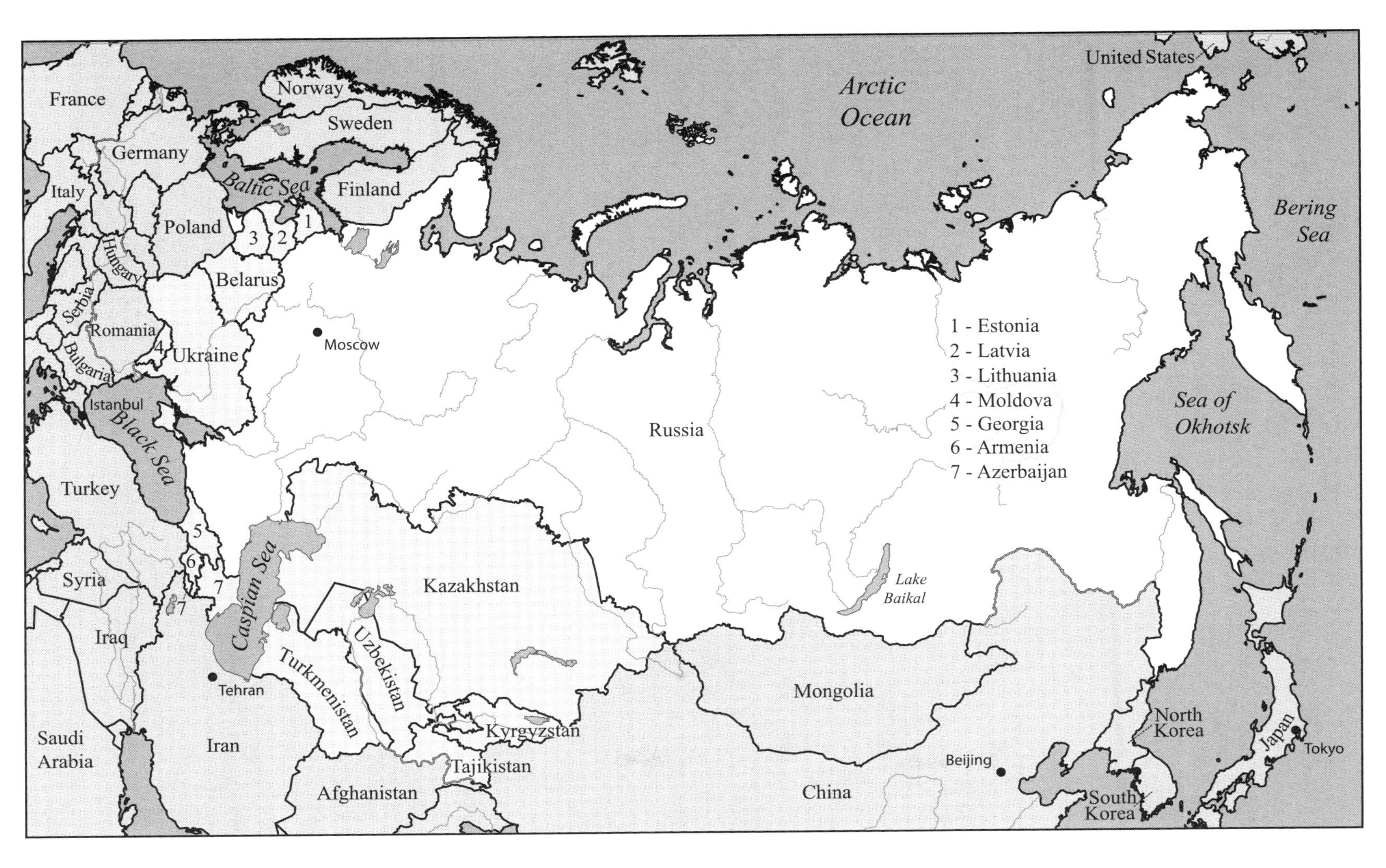

FIGURE 1.1. Russia and other Northern Eurasian republics.

The Organization of This Book

This book is organized into five parts, covering physical geography; history and politics; cultural and social geography; economics; and regional geography. Part I, the **physical geography** section, covers the natural environment. Issues of environmental degradation and conservation are addressed at the end of this section, because they are based on humans' interactions with nature and thus provide a link to Parts II–IV.

Part II briefly discusses **historical and current political events**, as a bridge between Part I and the remainder of the book. However, this book is not a history text, and students are encouraged to read additional sources on specific events in Russian and Soviet history, as needed. There are dozens of excellent books about the history of the region; some are included in the **Further Reading** lists at the ends of these chapters.

Part III, covering **cultural and social geography**, discusses population distribution; urban and rural patterns; social issues of income and health; cultures and languages; religions; and many other patterns. Whenever possible, examples are given from different republics.

Part IV, on **economics**, focuses on the current patterns of production in the FSU. One important statistic that I commonly use is gross domestic product (GDP) or gross regional product (GRP). When comparisons are made with other countries or regions, these are adjusted for purchasing power parity (PPP), based on the CIA World Factbook's methodology. The currencies of the countries discussed here are greatly undervalued in the world financial markets, so one must account for differences in prices between, say, the United States and Russia, to make a meaningful comparison. I have done this by using GRP PPP.

Part V, on **regional geography**, can serve as a handy reference. It is fairly concise, but its chapters provide brief descriptions of each main region of Russia and of all other republics. It complements the earlier thematic chapters well, but it can be skipped or incorporated into the study of specific topics. For example, the chapter on Central Asia (Chapter 31) complements the discussion of water problems in the Aral Sea in Chapter 5.

Each chapter in Parts I–IV has sections dealing with specific subtopics pertaining to Russia. Usually the last section of each chapter is devoted to a discussion of the other republics, to examine their similarities and contrasts with Russia. Classes that deal exclusively with Russia may skip that section. The Russia-centered parts of each chapter make some references to other republics, as appropriate.

Each chapter (except in Part V) ends with a set of **Review Questions** that can be answered as part of in-class discussion or homework. As a rule, these questions can be answered by using the textbook itself. **Exercises** are more involved tasks; they will typically require access to the Internet or a good library. Again, they may be completed either in class or at home. Some have been specifically designed as group projects. **Further Reading** lists can be used for additional study. The suggested **Websites** in most chapters in Parts I–IV are useful, but of course are subject to frequent change. **Vignettes** in some chapters contain case studies, personal stories, or technical notes.

A Note on Russian Names and the Metric System

I follow a modified version of the Library of Congress Russian-to-English transliteration system. In some cases, the accepted common spellings are used instead (e.g., Yeltsin, not El'tsin). I prefer to omit apostrophes that represent palatalized consonant sounds not found in English (e.g., Ob River, not Ob' River). Also, wherever possible and for the sake of consistency, I use the Russian names for place names in other republics—for instance, Kiev (Russian), not Kyiv (Ukrainian). The Russian alphabet is provided for reference in Chapter 13. Geographic names have been checked against *Merriam-Webster's Pocket Geographical Dictionary* (1999). Some names not found there have been transliterated to the best of my ability.

I use metric units throughout the book because these are the only ones used in the FSU. A list of these units and their U.S. equivalents is provided in the front of this book, after the table of contents.

Further Reading

These are either English- or Russian-language general sources on geography that can be consulted for additional information. Many are textbooks or monographs. This is by no means a comprehensive list; dozens of books on history and political science could be added. Specific topical readings, including some journal articles, are provided at the end of each subsequent chapter.

Sources in English

Bater, J. H. (1996). *Russia and the post-Soviet scene.* London: Arnold.

Berg, L. S. (1950). *Natural regions of the U.S.S.R.* New York: Macmillan.

Brunn, S. D., & Toops, S. W. (2010). *Atlas of Eurasia.* New York and London: Routledge.

Cole, J. P. (1967). *A geography of the U.S.S.R.* Harmondsworth, UK: Penguin Books.

Gilbert, M. (1972). *Russian history atlas.* New York: Macmillan.

Goldman, M. E. (Ed.). (2007). *Russia: The Eurasian republics and central/eastern Europe. Global studies* (annual editions). Dubuque, IA: McGraw-Hill.

Gregory, P., & Stuart, R. (2001). *Russian and Soviet economic performance structure.* Boston: Addison Wesley Longman.

Hill, F., & Gaddy, C. G. (2003). *Siberian curse: How Communist planners left Russia out in the cold.* Washington, DC: Brookings Institution Press.

Kaiser, R. (1991). *The geography of nationalism in Russia and the USSR.* Princeton, NJ: Princeton University Press.

Mathieson, R. S. (1975). *The Soviet Union: An economic geography.* New York: Barnes & Noble.

McKenzie, D., & Curran, M. W. (1986). *A history of the Soviet Union.* Chicago: Dorsey Press.

Pryde, P. R. (1991). *Environmental management in the Soviet Union.* Cambridge, UK: Cambridge University Press.

Shabad, T. (1950). *Geography of the U.S.S.R.* New York: Columbia University Press.

Shahgedanova, M. (Ed.). (2002). *The physical geography of Northern Eurasia.* Oxford, UK: Oxford University Press.

Shaw, D. J. B. (Ed.). (1995). *The post-Soviet republics: A systematic geography.* Harlow, UK: Longman.

Shaw, D. J. B. (1999). *Russia in the modern world: A new geography.* Oxford, UK: Blackwell.

Symons, L. (1992). *The Soviet Union: A systematic geography.* New York: Routledge.

Tikhomirov, V. (2000). *The political economy of post-Soviet Russia.* New York: St. Martin's Press.

Tomikel, J., & Henderson, B. (1966). *Russia and the near abroad.* Elgin, PA: Allegheny Press.

Trenin, D. (2002). *The end of Eurasia: Russia on the border between geopolitics and globalization.* Washington, DC: Carnegie Endowment for International Peace.

Turnock, D. (Ed.). (2001). *East Central Europe and the former Soviet Union: Environment and society.* London: Arnold.

Wixman, R. (1988). *The peoples of the U.S.S.R.: An ethnographic handbook.* Armonk, NY: Sharpe.

Recommended journals: *Eurasian Geography and Economics, Europe-Asia Studies, International Affairs, Post-Soviet Geography, Post-Soviet Geography and Economics, Soviet Geography, Soviet Studies.*

Recommended magazines: *Foreign Policy, National Geographic, Russian Life.*

Global statistical databases, including those of the Central Intelligence Agency (the CIA World Factbook); the Food and Drug Organization (FAOSTAT); the Population Reference Bureau; the United Nations Development Programme; the United Nations Environmental Programme; the United Nations Educational, Scientific, and Cultural Organization; the World Bank (the World Development Report); and the World Health Organization.

Sources in Russian, Including Both Classical Works and Modern Texts

Anuchin, V. A. (1972). *Teoreticheskiye osnovy geografii.* Moscow: Mysl.

Atlas Rossii: Design, informatsiya, kartografiya. Moscow: AST-Astrel.

Baranskiy, N. N. (1980). *Izbrannye trudy* (2 vols.). Moscow: Mysl.

Berg, L. S. (1947). *Geograficheskie zony Sovetskogo Soyuza* (2 vols.). Moscow: Geografgiz.

Dokuchaev, V. V. (1948). *Uchenie o zonakh prirody.* Moscow: Geografgiz.

Dronov, V. P., Barinova, I. I., Rom, V. Y., & Lobzhanidze, A. A. (2001 and other editions). *Geografiya Rossii* [8th–9th grades]. Moscow: Drofa.

Federalnaya Sluzhba Gosudarstvennoy Statistiki [Federal State Statistics Service]. (2006). *Regions of Russia statistics.* (Official edition two-CD set. Most statistics on Russia's regions are cited from this source.)

Gladkiy, Y. N., Dobroskok, V. A., & Semenov, S. P. (2000). *Sotsialno-ekonomichaskaya geografiya Rossii*. Moscow: Drofa.

Grigoryev, A. A. (1966). *Zakonomernosti stroeniyai razvitiya geograficheskoy sredy*. Moscow: Mysl.

Gvozdetsky, N. A. (1978). *Geograficheskie otkrytiya v SSSR*. Moscow: Prosveshchenie.

Isachenko, A. G. (1965). *Osnovy landshaftovedeniya i fiziko-geograficheskoe rayonirovanie*. Moscow: Vyshaya Shkola.

Khorev, B. S. (1981). *Terriorialnaya organizatsiya obshchestva (Aktualnye problem regionalnogo planirovaniya i upravleniya v SSSR)*. Moscow: Mysl.

Khruschev, A. T. (Ed.). (2006). *Ekonomichaskaya i sotsialnaya geografiya Rossii*. Moscow: Drofa.

Kolosov, V. A., & Mironenko, N. S. (2005). *Geopolitika i politicheskaya geografiya*. Moscow: Aspekt Press.

Kolosovky, N. N. (1969). *Teoriya ekonomicheskogo rayonirovaniya*. Moscow: Mysl.

Plisetsky, E. L. (2004). *Sotsialno-ekonomicheskaya geografiya Rossii. Spavochnoe posobie*. Moscow: Drofa.

Rodoman, B. B. (1999). *Territorialnye arealy i seti. Ocherki teoreticheskoy geografii*. Smolensk: Oikumena.

Rychagov, G. I. (Ed.). (1984). *Mir geografii*. Moscow: Mysl.

Saushkin, Y. G. (1973). *Ekonomicheskaya geografiya: istoriya, teoriya, metody, praktika*. Moscow: Mysl.

Tishkov, A. A. (2005). *Biosphernye funktsii prirodnyh ekosistem Rossii*. Moscow: Nauka.

PART I

Physical Geography

CHAPTER 2

Relief and Hydrography

The term "relief" refers to all the landforms on the surface of the earth. It is basically the same thing as "topography." "Hydrography" refers to the water features that produce some of the landforms. Every country has prominent features such as mountains, valleys, plateaus, and basins, which set the stage for climate types and biomes to develop, and these in turn determine to a large extent which human activities are possible. Surrounding every continent are peninsulas, islands, bays, gulfs, and seas. On land, lakes and rivers develop, depending on mountain systems and more local relief forms. The countries of the former Soviet Union (FSU) exhibit thousands of varied topographical and hydrographical features. Without knowing what and where they are, we cannot understand the region's climate types, biological communities, or human landscapes.

The Main Physical Features

The FSU (this term is used interchangeably with Northern Eurasia in this chapter) has numerous geographic features on a physical map. When you arrive in Moscow on an international flight, the land appears very flat. This is because Moscow is located in the middle of one of the largest plains on earth, the Eastern European Plain, stretching from Poland to the Urals. On the other hand, if you were to take the Trans-Siberian Railroad into Siberia, in a day's time you would be greeted by the Urals, and in less than 4 days by the Central Siberian Plateau and the mountains surrounding Lake Baikal.

Examine the map of Northern Eurasia (Figure 2.1) and the associated list of some important physical features (Table 2.1). The table is not an exhaustive list, but a good one to start with. Some features in this region are unique (biggest, deepest, highest, etc.). Here are some examples:

- Mt. Elbrus in the Caucasus is the tallest mountain in Europe and all of Russia, at 5,642 m (the famous Mt. Blanc in the French Alps is only 4,807 m).
- Ismail Samoni (formerly Peak Communism), in the Pamirs in Tajikistan, is the tallest mountain in the FSU (7,495 m). It is only 1,500 m shorter than Mt. Everest, but is considerably higher than any summits found in the two Americas.
- The lowest point in Russia is on the north shore of the Caspian Sea, at 28 m *below* sea level.
- Lake Baikal is the deepest lake on earth, at 1,620 m, and the biggest by freshwater volume (it contains 20% of the world's liquid freshwa-

FIGURE 2.1. Main physical features of Northern Eurasia.

TABLE 2.1. Main Physical Features to Know in Northern Eurasia

Seas and straits (from west to east)	Islands and peninsulas	Mountain ranges, plateaus, and lowlands	Rivers
• Baltic Sea	• Kola Peninsula	• Carpathians (Western Ukraine)	• Dnieper
• Barents Sea	• Crimean Peninsula	• Khibiny (on Kola Peninsula)	• Don
• White Sea	• Novaya Zemlya	• The Caucasus	• Volga (+ Oka and Kama)
• Kara Sea	• Yamal Peninsula	• The Urals	• Northern Dvina
• Laptev Sea	• Franz Joseph Land	• Eastern European Plain	• Pechora
• East Siberian Sea	• Severnaya Zemlya	• Western Siberian Lowland	• Syr Darya
• Bering Sea	• Taymyr Peninsula	• Central Siberian Plateau	• Amu Darya
• Sea of Okhotsk	• Novosibirskiy Islands	• The Pamirs (Tajikistan)	• Ili
• Sea of Japan	• Wrangel Island	• Tien Shan (Kyrgyzstan)	• Irtysh and Ob
• Bering Strait	• Chukchi Peninsula	• Kara Kum Desert (Turkmenistan)	• Angara and Yenisey
• Tatarsky Strait	• Commodore Islands	• Kyzyl Kum Desert (Uzbekistan)	• Lena
• Black Sea	• Kamchatka Peninsula	• The Altay	• Yana
• Sea of Azov	• Sakhalin Island	• The Sayans	• Indigirka
Lakes	• Kuril Islands	• Yablonovy range	• Kolyma
• Ladoga		• Stanovoy range	• Amur
• Onega		• Sikhote-Alin range	
• Aral Sea (Kazakhstan, Uzbekistan)		• Verkhoyansk range	
• Caspian Sea		• Chersky range	
• Balkhash (Kazakhstan)			
• Issyk-Kul (Kyrgyzstan)			
• Baikal			
• Khanka			

Note. Locate these geographical features on Figure 2.1 and additional atlas maps, and then label them on a blank map of the region from memory.

ter—the equivalent of all five Great Lakes in North America combined).

- The Caspian Sea is the world's largest saline lake. Its surface is four times greater than Lake Superior's.
- The Ob–Irtysh river system is the fifth longest worldwide, at 5,400 km (the Mississippi–Missouri system is fourth, at 6,019 km). Note that the Irtysh is the longer of the two rivers where they merge, but the Ob carries more water, so the combined river downstream retains the name Ob.
- Sakhalin Island is the biggest in Russia, with over 76,000 km^2. It is the 22nd biggest worldwide, about the same size as Hokkaido (Japan) and Hispaniola (in the Caribbean). Located in the Far East, it is over 900 km long, but only about 100 km wide.
- The Taymyr Peninsula is the biggest and northernmost in Russia. It ends at Chelyuskin Point (77°43'N), named after a famous Arctic explorer. In comparison, Alaska's northern shore is located at 72°N. The northernmost point of Russia on an island is Cape Fliegeli on Franz Joseph Land's Rudolf Island at 81°51'N, just 900 km south of the North Pole. The Soviet Union unilaterally claimed all the Arctic Ocean north of its shores all the way to the North Pole. The current Russian government is trying to get this claim recognized, but so far it has met with fierce resistance from Canada, the United States, and Norway.
- The southernmost point of Russia is Mt. Bazardyuzyu in Dagestan (41°10'N). For the remainder of the FSU, it is the city of Kushka in Turkmenistan (36°N).
- The westernmost point of Russia is on the border with Poland, on the Baltic Spit in Kaliningrad Oblast (19°38'E).
- The easternmost point of Russia is actually located in the Western Hemisphere! Dezhnev Point at 169°40'W, overlooking Alaska, is on the continent of Eurasia. Ratmanov Island in the Bering Strait is even closer to the United States, but it is not on the mainland (169°02'W).

Russia is enormous: It stretches for about 4,500 km from north to south, if the islands in the Arctic are included, and for 9,000 km from west to east. As noted in Chapter 1, it covers 11 time zones—definitely the world's record. (The entire country was placed 1 hour ahead of the true solar time by a decree of Lenin in 1918, thus effectively putting the whole country on daylight savings time. In the late 1980s, an additional hour of summer daylight savings time was introduced, beginning on the last Sunday of March and ending on the last Sunday of October.) If you are flying on a passenger jet from Moscow, it takes just 2 hours to reach Sochi or Murmansk; about 3½ hours to reach Paris or Tyumen; 4 to reach Novosibirsk; 7 to reach Khabarovsk; 8 to reach Magadan; and 9 to reach the Chukchi Peninsula. In comparison, nonstop flights from Moscow to New York City take about 10 hours.

Notice that whereas mountains in Northern Eurasia tend to run from east to west, the rivers mainly run from south to north, especially in Siberia. The Urals run from north to south; they divide Russia into its western (European) part and its eastern (Siberian) part, and separate Europe from Asia. The Volga flows mainly south and east into the Caspian Sea, and the Amur flows mainly east along the Chinese border into the Sea of Okhotsk.

The Geological History of Northern Eurasia

Older, Larger, More Stable Landforms

Like any other large landmass on our planet, Northern Eurasia has a long and complex geological history. However, the sheer size of Eurasia makes its geology particularly complex—unlike that of relatively simple and flat Australia, for example. The two largest "chunks," the Eastern European and Siberian platforms, are over 1,700 million years old, which is comparable to the age of the North American plate. They are two separate continental plates that were driven together by geological forces over long periods of time. About 550 million years ago, the two were still separate, drifting in the warm seas of the Southern Hemisphere. However, they came together about 500 million years ago, and the Urals formed between them about 220–280 million years ago. The Eastern European platform underlies much of what is European Russia and

Ukraine today. The Siberian platform is found east of the Yenisei River and west of the Lena. Parts of the Northern European plate are occupied by the Scandinavian and Baltic crystalline shields, which, like their Canadian counterpart, have some of the oldest rocks on earth (some over 2 billion years old) exposed at the surface. Other very old shields with rocks over 1 billion years of age are exposed in the northern part of the Siberian platform, called the Anabar Massif, and in the eastern part, the Aldan Plateau east of Lake Baikal. The oldest rocks here can be about 3 billion years old. Some of the famous gold and diamond deposits that formed in the Proterozoic period (about a billion years ago) are found in that area.

East of the Urals, the Western Siberia Lowland is covered with sea deposits from the Jurassic and Cretaceous periods (65–195 million years ago). This was a time of great warmth, supporting tropical plants and dinosaurs. This area can be compared geologically to parts of Colorado, Utah, and Wyoming in the United States, which were likewise submerged under the warm tropical sea at the same time and today have many dinosaur fossils. The vast oil and gas deposits of Russia date back to that time and are primarily concentrated in western Siberia.

Higher Mountains, Tectonic Movement, and Volcanoes

In contrast to these large and stable areas, many areas to the east and the south have a much more complex and recent history. In southern and eastern Siberia, some mountains south of Lake Baikal were formed by tectonic uplift in the Proterozoic era (over a billion years ago); the Altay and Sayans were similarly formed in the mid-Paleozoic (450 million years ago); the Sikhote-Alin and other Far Eastern ranges were thus formed in the Mesozoic (225 million years ago). The highest mountains are also the youngest: The Caucasus, the Pamirs, and the Tien Shan were formed primarily in the past 10–15 million years and are still exhibiting uplift today (Figure 2.2). They are part of the Alpine–Himalayan fold belt, which stretches from the Alps in Europe to the Zagros Mountains in Iran to the highest mountains on earth, the Himalayas in India and Nepal. This

FIGURE 2.2. The Caucasus Mountains have some of the youngest and tallest peaks in Northern Eurasia, formed just a few million years ago, as evidenced by the dramatic relief. More recently, glaciers carved deep U-shaped valleys. *Photo:* V. Onipchenko.

dramatic uplift began when the Indian subcontinent slammed into Eurasia from the south 40–50 million years ago. This same event apparently started the Baikal rift that produced Lake Baikal, the oldest lake on the planet, by about 25 million years ago.

The eastern and southern fringes of the FSU are mountainous, with active tectonic movement, frequent earthquakes, and (in the Russian Far East) active volcanism. Earthquakes reaching a magnitude of 7 on the Richter scale were recorded in the past in the Carpathians and the Caucasus, with magnitudes over 8 recorded in the Pamirs, the Tien Shan, the area east and north of Lake Baikal, and Kamchatka. Massive earthquakes devastated Ashgabat (1948, 100,000 casualties) and Tashkent (1966), two Soviet capitals in Central Asia. More recently, the Armenian earthquake of 1988 killed about 20,000 in Spitak, and the Sakhalin Island earthquake of 1995 caused about 3,000 fatalities in Neftegorsk. Most of these casualties were people trapped under poorly constructed concrete buildings, built in the Soviet period without regard to seismicity. Ninety percent of Northern Eurasia is earthquake-free, the chance of experiencing one in Moscow is close to zero. The greatest risk of earthquakes is in the mountainous belt in the south, especially in Moldova near the Romanian border; in Armenia and Georgia in the Caucasus; in Tajikistan; in the areas south and especially northwest of Lake Baikal; on Sakhalin Island; and, of course, in Kamchatka.

The Caucasus has a complex geological history, but essentially represents one long mountain wall trending from northwest to southeast, with associated smaller ranges extending north and south (average elevation 3,000 m). It is bigger, but less geologically complex, than the Alps. An extinct volcano, Mt. Elbrus (5,642 m), with two summits, sits to the north of the main range (Figure 2.3). The second highest point of the range in Georgia is Mt. Kazbek (Kazbegi; 5,033 m), to the southeast. Most of the Caucasus has granitic rocks, with a higher incidence of limestone farther east. Glaciers and perennial snowfields attract downhill skiers and mountaineers, to Dombai in Karachaevo-Cherkessia, Baksan in Kabardino-Balkaria, and Krasnaya Polyana near Sochi (the future home of the 2014 Winter Olympics). The north slope of the Caucasus has over 1,230 km^2 of glaciers, the most of any mountain range in Russia.

The highest mountains in the FSU are the Pamirs, which lie within Tajikistan and the Tien Shan ("Heavenly Mountains" in Chinese) in Kyrgyzstan and parts of Kazakhstan and China. Some peaks there rise above 7,000 m, higher than any summit in the Western Hemisphere (Figure 2.4). These ranges are the source of most river water and hydropower in Central Asia. They are also premier climbing and backpacking destinations.

The Altay and the Sayans in south central Siberia farther to the east are a bit lower than the Pamirs; they are comparable in height to the Caucasus or the Alps. They are complex mountain systems, with multiple ranges and substantial glaciers and snowfields. The Ob and the Yenisei originate in the Altay and the Sayans, respectively. More mountain ranges exist east of Lake Baikal (the Baikalsky, Barguzinsky, Yablonovy, and Stanovoy ranges) and in northeastern Russia (the Cherskogo and Verkhoyansky ranges). All of these are between 2,000 and 3,000 m in elevation, and have little glaciation despite being located in very cold places, because of the aridity so far inland. Along the Russian Pacific Coast runs the Sikhote-Alin range.

The volcanoes of the Kamchatka Peninsula and the Kuril Islands are legendary. About 28 active and 160 extinct volcanoes are found on Kamchatka, and 39 are active on the Kurils. The highest is the Klyuchevskaya Sopka, at 4,750 m in the central part of the peninsula. The skyline of the main seaport, Petropavlovsk-Kamchatsky, is dominated by the Avachinsky and Koryaksky volcanoes (3,500 m each). The central part of Kamchatka encloses a famous Geyser Valley, with 19 active geysers and 9 pulsing thermal springs, rivaling some Yellowstone and New Zealand counterparts. The Velikan ("Giant") geyser produces a pillar of boiling water 35 m high, with steam rising to an astonishing 250 m, which is the height of an average skyscraper in Seattle or Minneapolis. Massive eruptions are known to have occurred in Kamchatka in the late Pleistocene (20,000–30,000 years ago) and in the

FIGURE 2.3. Mt. Elbrus (in the background) is an extinct volcano in the Kabardino-Balkaria Republic of Russia and is the tallest peak in Europe at 5,642 m. *Photo:* V. Onipchenko.

FIGURE 2.4. The Tien Shan Mountains in Kyrgyzstan. *Photo:* L. Swanson.

mid-Holocene (7,500 years ago); some blasts produced enough ash to be found in substantial layers in Greenland's ice sheets, on the other side of the world! One of the most famous recent eruptions came without warning from Bezymyanny in 1953, with a powerful explosion comparable to that of Mt. St. Helens in Washington State in 1980. It did not kill any people, fortunately, because nobody lives in that area.

Ice Ages and Their Impact

As in North America, the Ice Ages of the Pleistocene made a profound impact on the landscape of Northern Eurasia, from 2.4 million years ago until approximately 10,000 years ago. Unlike in North America, however, there was no single giant ice sheet that covered the entire northern half of the continent. The biggest ice sheet covered all of Scandinavia and extended east as far as the eastern shore of the White Sea today. The Urals and parts of the Putorana Plateau in northern Siberia were also heavily glaciated. In between, however, and all the way to the Pacific Coast, only small areas of the highest terrain had much ice cover. The remainder was ice-free, but with hundreds of meters of permafrost extending deep into the soil. This may seem counterintuitive, but it can be understood if we remember that moisture available at cold temperatures is what makes ice and snow, not the cold temperatures themselves. Readers living east of the Great Lakes in the United States are no doubt familiar with the "lake effect" on snow formation: In a typical winter, parts of Ohio and upstate New York may get 10 feet of snow, while much colder North Dakota and northwestern Minnesota may get only a few inches. A similar effect operated in Eurasia during the Ice Ages. The area closest to the ice-free Atlantic Ocean, Scandinavia, received the most snow and consequently developed the most ice, while the colder parts farther inland received virtually no snow or ice.

Another impact of the Ice Ages was a worldwide lowering of the sea level by about 60–120 m, depending on the glacial stage, because much ocean water was frozen in the ice sheets on land. As a result, Eurasia was connected to North America via the Bering land bridge; Sakhalin Island was connected to Japan and the Eurasian mainland; and most Arctic islands were likewise connected to the Eurasian mainland. An amazingly rich fauna of large mammals existed in the ice-free cold areas in Siberia and the Russian Far East, with now extinct species (e.g., mammoth, woolly rhinoceros, camels, horses, saber-toothed tigers, and giant short-faced bears) mingling with some still-existing animals (e.g., musk oxen and bison). The abrupt end of the Ice Ages about 12,000 years ago, and the widespread arrival of human hunters in northern and eastern Siberia and in North America about 13,000 years ago, apparently led to the extinction of most of the 40 or so megafauna species. The last, albeit dwarf-sized, mammoths persisted until about 4,000 years ago on the lonely Wrangel Island of the northeastern Siberian coast—almost up to the time of the Egyptian pyramids!

The Ice Ages left numerous landforms in European Russia, including the morainal Valdai Hills and beautiful glacial lakes (Seliger, Ladoga, Onega, and hundreds of lakes in Karelia) north of Moscow (Figure 2.5). Large areas of drumlins, kames, eskers, and other glacial landforms familiar to Finns, Minnesotans, or Canadians are present in much of northern European Russia. The areas south of the ice sheets—in modern-day Ukraine; in the Bryansk, Kursk, and Voronezh regions of Russia; and in northern Kazakhstan and western Siberia—have extensive loess deposits consisting of fine wind-blown dust that came from the glaciers. The best chernozem soils producing the highest yields of grain in Ukraine and Russia owe their origin to these loessal deposits. The areas north and east of the Caspian and the Aral Seas have evidence of giant glacial outburst floods, like those in the Columbia Basin in Washington State. The rushing meltwater roared down from the ice fields of Siberia and the southern Urals toward the southwest and carved curious parallel channels, which are clearly visible from space today (e.g., use Google Earth and examine the areas north and northeast of the Aral Sea).

Originally, it was thought that only four major glaciations occurred, based on incomplete evidence from terrestrial records in Europe and North America. Deep drilling in the oceans since the 1970s has allowed scientists to conclude that

FIGURE 2.5. The Valaam Islands in Lake Ladoga. Scoured granite bedrock is exposed in low ridges. Thin, sandy soils develop in some areas. *Photo:* S. Blinnikov.

in the past 2 million years over 20 glaciations occurred worldwide, once every 100,000 years—each lasting about 80,000 years and separated by milder interglacial periods, like the one we are living in now. In European Russia, the most recent glacial stage is called the Valdai, after the Valdai Hills halfway between Moscow and St. Petersburg (a national park today). It corresponds to the Würm or Weichsel stages in Europe and the Wisconsinian stage in North America. The last interglacial period before the current one, Mikulino, happened about 120,000 years ago. Before that, the Dnieper glacial stage occurred in European Russia, corresponding to the Illinoian stage in North America between 120,000 and 200,000 years ago. As can be seen from its name, that ice sheet extended farther south than the Valdai, to the Dnieper River in modern-day Ukraine.

River Systems

Russia has over 120,000 rivers over 10 km long, which collectively create 2.3 million km of waterways. Fifty-four percent of their flow enters the Arctic Ocean, with only 15% entering the Pacific. Another 8% of water flows to the Atlantic Ocean via the Black and Baltic Seas, and 23% to the Aral–Caspian interior basin with no outlet to the ocean. Russian schoolchildren learn in the early grades that "the Volga flows to the Caspian Sea." This is interesting, because the biggest river in Europe does not even flow to the ocean! North America also has a few interior basins, the most famous being the Great Basin that includes the Great Salt Lake.

Northern Eurasia has a few of the world's largest rivers. Table 2.2 lists the top 11, and also some other large rivers around the world for comparison. The Volga is the biggest and longest river of Europe. Russians call it *Matushka*, meaning "Dear Mother," because their civilization developed around it (Figure 2.6). The basin occupies only 8% of the country, but is home to 40% of its population. Other important rivers in the European part of the FSU include the Northern Dvina and Pechora in the North; the Neva, flowing from Lake Ladoga to the Baltic Sea, with St. Petersburg at its mouth; and the Dniester, Dnieper, and Don in Moldova, Ukraine, and southern Russia, respectively. The "dn" root in the names of some rivers is not a coincidence; it probably comes from *dno*, meaning "bottom" or "low place" in the Slavic languages. The Volga, the Dnieper, and the Don are heavily tapped for hydropower, with many reservoirs behind dams. Dams slow the speed of water flow and increase evaporation off the reservoir surfaces, especially in the arid south. Irrigation and industrial and domestic consumption further reduce the flow. The Volga loses 7% of its annual flow to human consumption. Its flow has been reduced by about 20% in the last 100 years.

The Siberian rivers primarily flow north to the Arctic Ocean, with the exception of the Amur, which flows east into the Pacific. Four of the great rivers in Siberia are comparable to the Mississippi in length and flow (Table 2.2). The Yenisei and its tributaries, and to a lesser extent the Ob and the Irtysh, are tapped for hydropower. The Lena itself remains dam-free, with a few dams existing on its tributaries, and more dams on the Amur tributaries farther east. Because spring comes earlier in the south, north-flowing Siberian rivers are prone to catastrophic spring flooding, similar to the Red River of the North in North Dakota. While the spring meltwater is abundant in April

TABLE 2.2. Biggest 11 Rivers of Northern Eurasia Ranked by Runoff Compared to Other Biggest Rivers of the World

River	Annual runoff (km^3)	Length (km)	Basin size (× 1,000 km^2)
Northern Eurasia			
Yenisei–Angara	623	5,940	2,619
Lena	515	4,270	2,478
Ob–Irtysh	397	5,570	2,770
Amur	392	4,060	2,050
Volga	253	3,690	1,380
Pechora	130	1,790	327
Kolyma	123	2,600	665
Khatanga	121	1,510	422
Northern Dvina	110	1,310	360
Pyasina	84	680	178
Neva	82	74	281
World			
Amazon	5,509	6,400	6,915
Congo	1,229	4,700	3,820
Yangtze	687	6,300	1,826
Mississippi–Missouri	570	6,019	3,220
Nile	98	6,671	2,870
Danube	202	2,858	817

Note. The runoff shows how much water comes from the river in an average year. Northern Eurasia data from *The Physical Geography of Northern Eurasia* (Shahgedanova, 2002). World data recalculated from the *Rand McNally Atlas of World Geography* (2003).

in the Ob and Irtysh headwaters, the rivers are still solidly frozen in the far north. Thus a huge seasonal "pond" appears in the middle of western Siberia, creating great inconvenience for the residents.

Central Asia's main rivers are the Amu Darya and the Syr Darya; both now barely reach the Aral Sea because of irrigation diversions. The Kara Kum canal, dug in the 1950s to divert the Amu Darya water for cotton irrigation in Turkmenistan, was the longest in the country at 1,100 km. The total amount of diverted runoff in Soviet-era Central Asia approached the annual flow of the Dnieper, the largest river in Ukraine! Some short but powerful rivers flow from the Caucasus to the Black and Caspian Seas (the Kuban, Terek, Rioni, and Kura) and from the mountains of Central Asia (the Zerafshan and Vakhsh). These are tapped for irrigation and hydropower, but most are used for recreation and local water consumption.

Lakes

Lake Baikal is the oldest and deepest lake on the planet. It sits in a rift valley where the earth's crust spread apart about 25 million years ago (Figure 2.7). Baikal is almost 1 mile deep in places and covers 31,500 km^2. Some of its closest counterparts exist in East Africa (e.g., Lake Tanganyika, which is the second deepest lake in the world). Lake Baikal holds an astonishing 23,600 km^3 of freshwater, which is about one-fifth of the global liquid supplies of freshwater, as noted earlier in this chapter. The biggest lake of all, however, is the Caspian Sea. Its salinity is only about one-third that of the world's oceans. The Aral Sea and Lake Balkhash are also saline, but are much smaller. Lake Balkhash is famous for being fresh in its western half near the mouth of the Ili River, but saline in the eastern half. Lake Issyk-Kul in Kyrgyzstan is another great and famous lake of the region. It is fresh, relatively clean, and

FIGURE 2.6. The Volga River near its source north of Moscow. The statue represents the Volga's motherly aspect. *Photo:* S. Blinnikov.

FIGURE 2.7. Lake Baikal in winter. Photo: A. Osipenko.

extremely picturesque, with many resorts lining its mountainous shores. East of St. Petersburg, Lake Ladoga is the biggest in all of Europe (with 17,700 km^2 of surface), followed by Lake Onega (about half the size). Both are glacial in origin, like the North American Great Lakes.

Coastlines and Islands

The coastlines of the U.S.S.R. were among the longest on earth. Russia's current coastlines total about 37,000 km, third longest in the world after Canada's (202,000 km with all the Arctic islands) and Indonesia's (54,000 km). The U.S. coastlines are only 19,000 km by comparison. Most of Russia's longest coastline follows the Arctic Ocean coast. In Russian, the Arctic bears the name of "Northern Icy Ocean" for a good reason: For much of the year, ice comes right up to the shore. Therefore, although the coastline is long, sea travel there is very difficult. Russia has only one big year-round ice-free port in the European Arctic, Murmansk. St. Petersburg, much farther to the south, generally ices up, but Murmansk remains ice-free courtesy of the warm North Atlantic current. The second longest coast of Russia is along the Pacific Ocean, with Magadan, Petropavlovsk, Yuzhno-Sakhalinsk, Vladivostok, and Nakhodka as ports. Historically significant for the Russian Empire and later the U.S.S.R. were also ports on the Black Sea (Odessa, Sevastopol, Novorossiysk, Batumi) and the Baltic Sea (St. Petersburg/ Leningrad, Tallinn, Ventspils, Klaipeda, Liepaja, Kaliningrad). The internal ports of Astrakhan, Baku, Atyrau, and Aktau allow fishing and trade in the Caspian Sea basin.

Along the coast, a few physical features merit special mention. In the Black Sea, the prominent Crimea Peninsula in Ukraine is a famous resort with a rich history and well-preserved natural areas. The narrow Kerchinsky Strait allows ships access to the little gulf called the Sea of Azov, where the port of Taganrog is located. Access to the sea from the Mediterranean is controlled by Turkey.

In the Baltic Sea, the Curonian Spit is the longest sandbar feature in Europe. It is also an international nature park shared by Russia and Lithuania. The Gulf of Finland allows sea access to Europe from St. Petersburg—the main reason why Peter the Great built the city there after winning control over that territory from Sweden in the early 1700s. The port of St. Petersburg is now protected by an artificial dam stretching across the gulf 20 km offshore. It eases severe spring floods, but traps water pollutants.

The Kola Peninsula, in the Arctic portion of European Russia, contains important metal and phosphate deposits and separates the White Sea from the ocean. The Kanin Nos, Yamal, and Taymyr Peninsulas are prominent farther east. The Karskie Vorota Strait (33 km wide) in the eastern Barents Sea separates the southern island of Novaya Zemlya from the island of Vaigach. This is usually the impassable gate to the Arctic Ocean beyond, where ice melts only in July and August. When nuclear icebreakers are used, navigation through it is possible for about 4 months of the year. With global warming continuing to accelerate, it is likely that much of the so-called Great Northern Seaway Route will become navigable year-round by the end of the 21st century. The distance from Europe to Japan via the Suez Canal is about 12,000 miles, whereas it is only about 6,000 miles via the Northern Seaway.

Four main archipelagos exist in the Russian Arctic: Novaya (New) Zemlya and Franz Joseph Land in the European sector, and Severnaya (Northern) Zemlya and the Novosibirskie Islands in the Asian sector. The solitary Wrangel Island is an important wildlife area and a preserve in the easternmost corner of the Russian Arctic. The Bering Strait (90 km wide) separates Eurasia from North America, and Russia from the United States. Technically, the closest the two countries come together is between Ratmanov (Russia) and Kruzenstern (U.S.) in the Diomede Islands, a distance of just 4 km! There have been proposals to build an underwater railroad tunnel to connect the two continents. It would be about twice as long as the Channel Tunnel between England and France.

In the Russian Pacific, Chukchi and Kamchatka (peninsulas) and Sakhalin and the Kurils (islands) are important features. Kamchatka has the highest concentration of volcanoes in Russia, with over 30 being active. Chukotka, Sakhalin, and the Kurils (Figure 2.8) have strategic importance as fishing areas and for military reasons. About 20 large and 30 small Kuril Islands stretch for over 1,000 km from the tip of Kamchatka to Hokkaido. Japan still claims four of the southernmost Kurils as its own; they were taken over by the U.S.S.R. after World War II as a form of compensation for the damage caused by Japan as the aggressor. Although these islands themselves

FIGURE 2.8. The Kuril Islands in the Pacific. *Photo:* I. Smolyar, National Oceanic and Atmospheric Administration/National Oceanographic Data Center (*commons.wikimedia.org/wiki/Image:Kuril_Island.jpg—public domain).*

are not large or mineral-rich, the lucrative exclusive economic fishing zone of 200 miles around them and the opportunity of placing antimissile radar installations on them make the Kurils a prized possession for Russia, so it is highly unlikely that they will be handed back to Japan any time soon. An estimate from the Yeltsin period pegged their worth at $100 billion in U.S. dollars—a considerably heftier sum than the $7.2 million Russia wanted for Alaska in 1867, even after adjustment for inflation.

The Impact of Northern Eurasia's Relief on Humans

The overall impact of relief on human life in Northern Eurasia is not as significant as in many other parts of the world, because the region is flat in most places. The largest plains, the Eastern European Plain and the Western Siberian Lowland, allowed early settlers easy travel along meandering rivers, such as the Dnieper, the Don, the Volga, the Northern Dvina, and the Pechora in the European part, and the Ob–Irtysh system in western Siberia. In the central part of Siberia, despite the presence of a large elevated plateau, relatively easy travel along the Yenisei and Lena was likewise possible. Plenty of land has been available for human settlement on easily accessible, flat terrain (Figure 2.9).

FIGURE 2.9. Flat, gently undulating glacial relief covers much of central and northern European Russia, allowing easy travel and settlement. The area shown is in Tver Oblast, about 150 km north of Moscow. *Photo:* S. Blinnikov.

Only in the southern mountain belt does relief present some challenges to human travel and settlement. The jagged relief of the Caucasus and the Pamirs in particular, and the sheer size of these mountains, preclude easy travel across the ranges even today: there is only one year-round paved highway from Russia into Georgia across the main Caucasus range, for example. The most dangerous road in the U.S.S.R. as measured by accidents was the Khorog-Osh highway, in the remote parts of the Pamirs in eastern Tajikistan.

Relief may thus have played a role in producing cultures: Deep gorges separated by inaccessible mountain ranges made the Caucasus one of the most linguistically diverse areas on earth, as each group formed in relative isolation from others. Over 20 languages are recognized in just one part of the Caucasus, Dagestan. Furthermore, mountains provided a natural defense barrier against the invaders, and thus the Caucasus and mountainous Tajikistan were the last two areas added to the growing Russian Empire. The boundary between Tajikistan and Kyrgyzstan passes through some of the highest terrain on earth, and is therefore a natural as well as a political border.

REVIEW QUESTIONS

1. Name the main mountain systems of Northern Eurasia.
2. What are the two oldest, most stable platforms in Northern Eurasia? Where are they?
3. Where in the FSU is the danger of earthquakes highest?
4. What part of Russia is like Yellowstone in terms of geothermal features?
5. What role did the Bering land bridge play in the biogeographic history of North America?
6. Why was Siberia so poorly glaciated, compared to Scandinavia? Why was North America so well glaciated, compared to Eurasia?
7. What important coastal features can you mention?

EXERCISES

1. Develop a classroom presentation about the major topographical features of a particular mountain system (the Carpathians, Caucasus, Pamirs, Tien Shan, Altay, etc.). Try to find sufficient illustrations online that show different types of landforms and physical landscapes common to that mountain system.
2. Investigate where some of the glacial features can be found in Russia today (e.g., eskers, drumlins, kames). One good area to start is the Valdai National Park, but there are many others. Use Google Earth and Internet searches for specific types of glacial features.

Further Reading

Berg, L. S. (1950). *Natural regions of the U.S.S.R.* New York: Macmillan.

Cole, J. P. (1967). *A geography of the U.S.S.R.* Harmondsworth, UK: Penguin Books.

Hartshorne, J. (2004, September–October). Saving Baikal. *Russian Life*, pp. 22–29.

Horensma, P. (1991). *The Soviet Arctic.* London: Routledge.

Micklin, P. (1985, March). The vast diversion of Soviet rivers. *Soviet Life*, pp. 12–20, 40–45.

Newell, J. (2004). *The Russian Far East: A reference guide for conservation and development.* McKinleyville, CA: Daniel & Daniel.

Shahgedanova, M. (Ed.). (2002). *The physical geography of northern Eurasia.* Oxford, UK: Oxford University Press.

Website

earth.google.com (Google Earth)—Visualization in two and three dimensions of various prominent geomorphological features of Northern Eurasia mentioned in this chapter, including Lake Baikal, the Kamchatka and Kuril volcanoes, the Volga and Lena river deltas, the Crimean and Kola Peninsulas, and all the mountain ranges.

CHAPTER 3

Climate

"Climate" refers to the average weather conditions found over large territories. Climate is expressed in terms of daily, monthly, and annual values of air temperature and precipitation, as well as wind speed, moisture, seasonality, and other factors averaged over a standard period of observations, usually 30 years. Climates of the world are differentiated into five broad types, labeled with the letters A through E; this typology is known as the "Köppen system." A-type climates are tropical and are not found in the countries of the former Soviet Union (FSU). B-type climates are dry climates and are very common in much of Central Asia, Kazakhstan, southern Ukraine, and parts of Russia, just as they are in the western United States or the Middle East. C-type climates are mild, without much frost in winter. These gave rise to some of the earliest human civilizations and are generally considered pleasant (think of places like coastal California, Italy, or Japan). In Northern Eurasia, they are found only in small areas, mainly along the Black and the Caspian Sea.

The most common climate type in the FSU—covering much of Russia and good portions of Ukraine, Belarus, and Kazakhstan—is the D type. This is a microthermal climate of continental interiors. It features four seasons, including a distinctly cold winter; "cold" in this context requires the average monthly temperature to go below freezing. Some locations with this climate have average winter temperatures below –40°C in the coldest month, although a typical winter would be 3–5 months long with temperatures in the –10 to –15°C range. Most of Canada, Alaska, the upper Midwest in the United States, and Scandinavia have climates of this type. Can there be an even colder climate? Yes: The E type is the coldest, a true polar climate present on 10% of Russia's territory. Each of these broad climate types in turn has subtypes. For example, the climate of much of Moldova (or Peoria, Illinois) is the Dfa subtype, while the climate of Moscow (or Minneapolis, Minnesota) is the Dfb subtype. The main difference between them is how warm the summer gets—above or below +22°C on average, respectively. The letter f means that there is sufficient moisture year round.

What Factors Create a Particular Type of Climate?

Why are Moscow winters not like those in Baku? Why is much of Central Asia so dry? Why can people in Georgia grow tangerines, while people at exactly the same latitude in Vladivostok cannot? Why is northern European Russia fairly cold in winter, while eastern Siberia is mind-numbingly cold? Such questions arise when we try to understand the spatial patterns of climate distribution.

Climatologists generally consider the following factors important in producing a particular climate type:

- Latitude, or distance from the Equator. The farther a place is from the Equator, the less direct sunshine is available. All of Northern Eurasia lies far outside the tropics, north of 36°N; in comparison, southern Florida is at 25°N.
- Elevation above sea level. The higher this elevation is, the colder the climate gets. Some of the highest peaks in the FSU are over 7,000 m.
- Proximity to the ocean. Water cools down and heats up very slowly, thus reducing the differences between seasons in coastal locations; far inland, the seasonality is much greater. In Northern Eurasia, the inland effect is most pronounced in northeastern Siberia.
- Presence of ocean currents. Cold currents make coastal locations cool and dry; warm currents make them warm and wet.
- Prevalent wind direction. Over much of North America and Eurasia, the winds in the middle latitudes generally blow from the west, following the rotation of the earth.
- Position relative to a mountain range. Windward locations get orographic precipitation; leeward locations get almost no rain (the so-called rain shadow effect). Mountains may protect a city from cold northern winds, or expose it to dry and warm catabatic winds rushing down the slope.
- Cloud cover and dust. These may vary, depending on local natural or anthropogenic conditions, thus attenuating the climate.
- Human infrastructure. This may create a local "heat island" effect; the downtown areas of major cities are typically a few degrees warmer than the surrounding countryside.
- Global climate change. Increasingly, this is being driven by human-made emissions of greenhouse gases.

Two of the most striking things about Northern Eurasia in general, and Russia in particular, are how big and how northern this area generally is. Russia is located in the northern part of the biggest landmass on the planet, considerably north of the continental United States (Figure 3.1). The southernmost point of the region, Kushka in Turkmenistan at 36°N, still lies far north of the Tropic of Cancer (23.5°N). Thus we may expect winters to be generally very cold in the region, because of both its latitudinal position and a lack of moisture in much of the interior. Although Antarctica gets even colder, the cities of Oimyakon and Verkhoyansk in Yakutia hold the world record for the greatest temperature difference between summer and winter (55°C on average) and for the coldest spots in the Northern Hemisphere (–72°C vs. –65°C in parts of northwestern Canada).

Another prediction we may make is that because much of Russia is flat, the climate will not be greatly modified by mountains. Mountains, of course, do modify the climate of the Caucasus and Central Asia, but much of European Russia and Siberia have uniform climate conditions over large swaths of terrain. The climate zones pretty much run in parallel zones from west to east, in the following very predictable order from north to south: polar, tundra, subarctic, cold continental, semi-arid (steppe), and arid (desert), with a few pockets of subtropical climates in the extreme south. This phenomenon was noticed as early as the mid-19th century and was used by Vasily Dokuchaev, the founder of modern soil science, to predict the distribution of Northern Eurasian soil and vegetation zones in accordance with the "law of natural zonation."

Oceans play only a minimal role in forming the climates of Northern Eurasia, because they are too far away from most areas. The Arctic Ocean is frozen along most of the coast for about 6 months every year, thus climatically acting as a big snow field that gives no moisture to the interior. The Atlantic Ocean does have a strong moderating effect on the Kola Peninsula and the Baltic states (as it does on Europe), keeping them warmer than they should be, given their latitude. The Pacific Ocean has an influence on the extreme southeastern corner of Russia by bringing in monsoons and occasional typhoons, but during

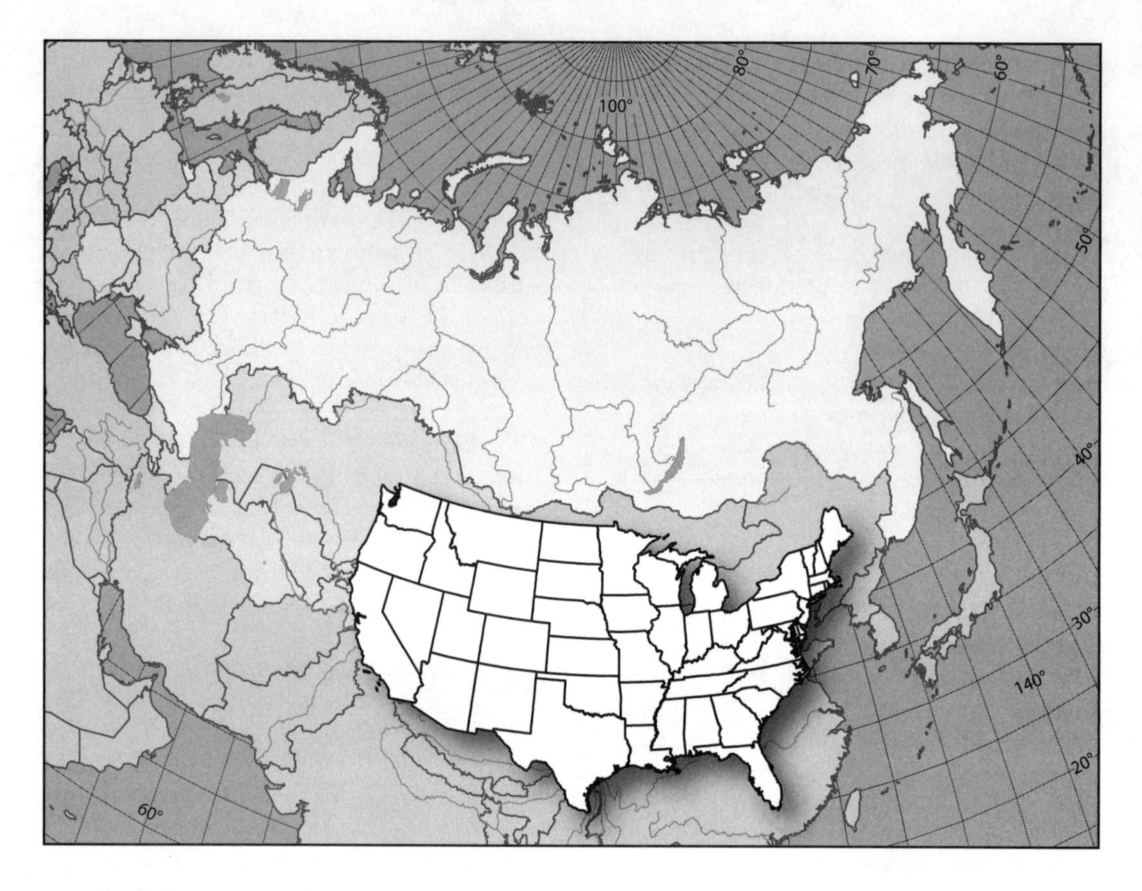

FIGURE 3.1. Russia is a huge northern country, more similar in its position to Canada than to the United States, and equal in size to both of them combined.

much of the year the winds in Siberia blow from the west (i.e., offshore), and again little moisture comes from the ocean to the land.

So, broadly speaking, Northern Eurasia has four major climate types. If we return to the Köppen classification system, these are as follows:

- Polar, or arctic climates of deserts and tundras (EF, ET).
- Subarctic climates of the boreal zone, where coniferous trees are common (Dc, Dw).
- Temperate climates, where either deciduous trees or steppes developed, depending on the availability of moisture (Dfa, Bs).
- Subtropical climates, where no freezing is observed in winter (Cs, Ca), or warm deserts (BW). There are no A-type tropical climates at all.

Climates at Different Destinations

To give us a clearer idea of what climates are like in different zones, let us take an imaginary trip to a few selected destinations in Northern Eurasia. We will visit places in each of the major climate types, learn what the climates are like there, and try to imagine what we would need to consider when packing for the trip.

To interpret the climate at each site, let us use climate diagrams (Figure 3.2). Such a diagram summarizes both average monthly temperature and precipitation in one easy-to-understand graph. The horizontal axis shows months, arranged from January to December. The vertical axis represents temperature, and the bars represent precipitation. Also shown are latitude, longitude, elevation above sea level, the mean an-

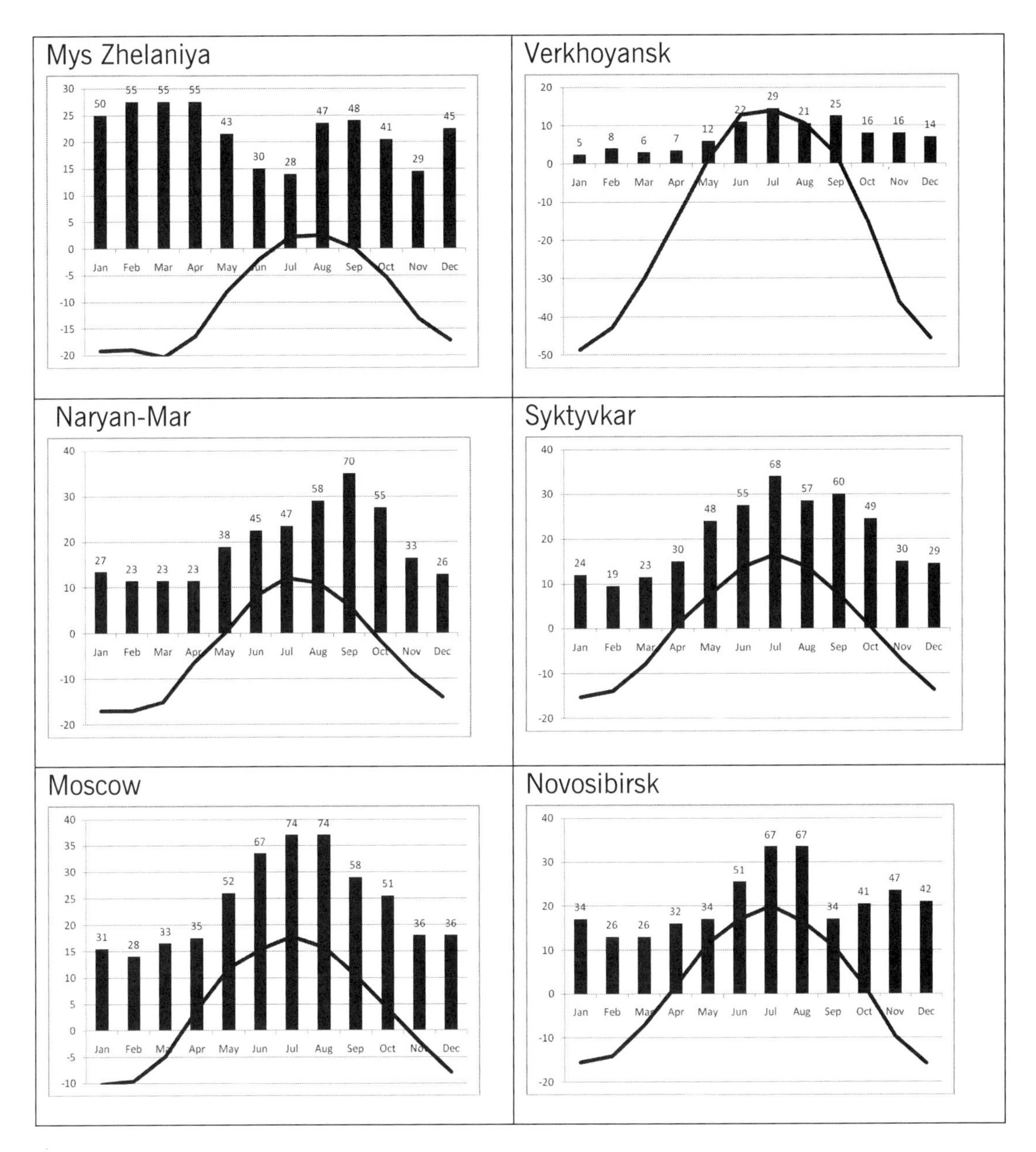

(cont.)

FIGURE 3.2. Climates of Northern Eurasia. For each climate diagram, the vertical axis represents mean monthly temperature (°C), while the bars represent mean monthly precipitation (mm). The map shows generalized Köppen climate types: ET, tundra; Dfc, subarctic; Dfb, continental cold winter; Dfa, continental warm winter; Dw, subarctic with very cold and dry winter; Ca, mesothermal; BSk, semi-arid; BW, arid. Data from *www.globalbioclimatics.org*, courtesy of S. Rivas-Martínez, Phytosociological Research Center, Spain.

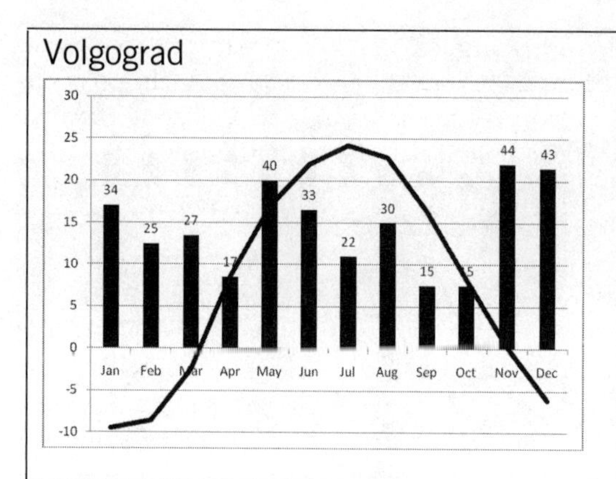

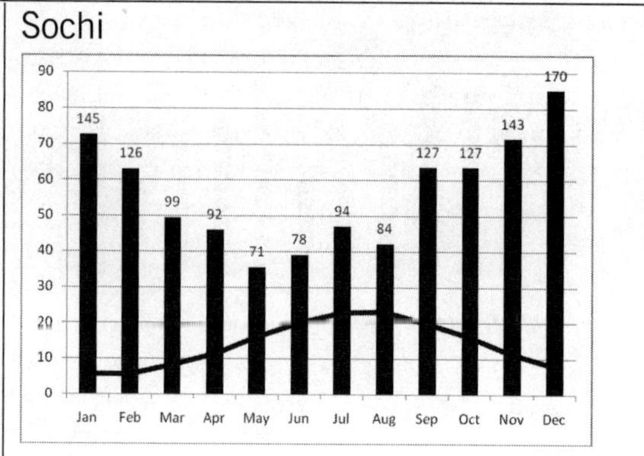

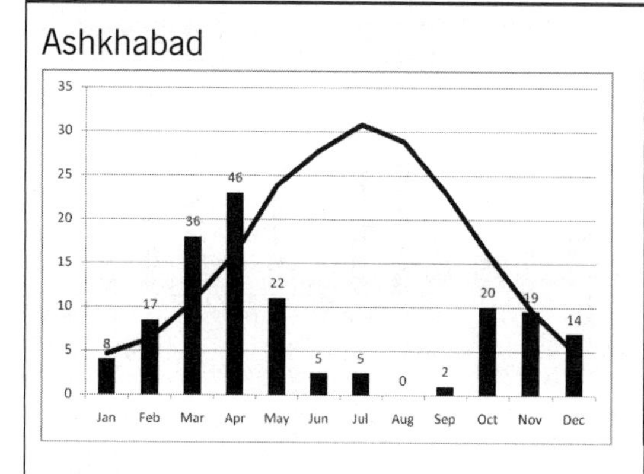

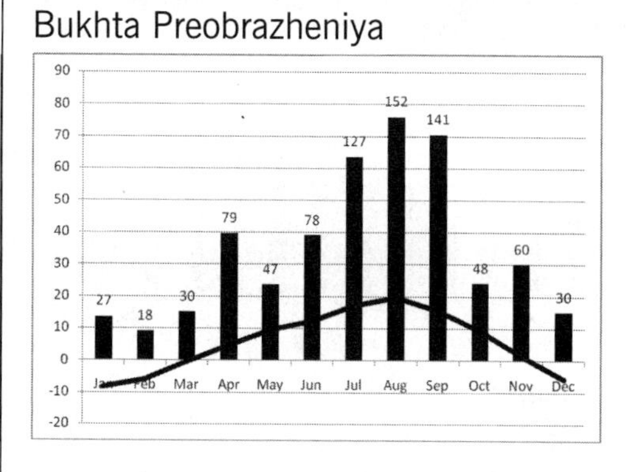

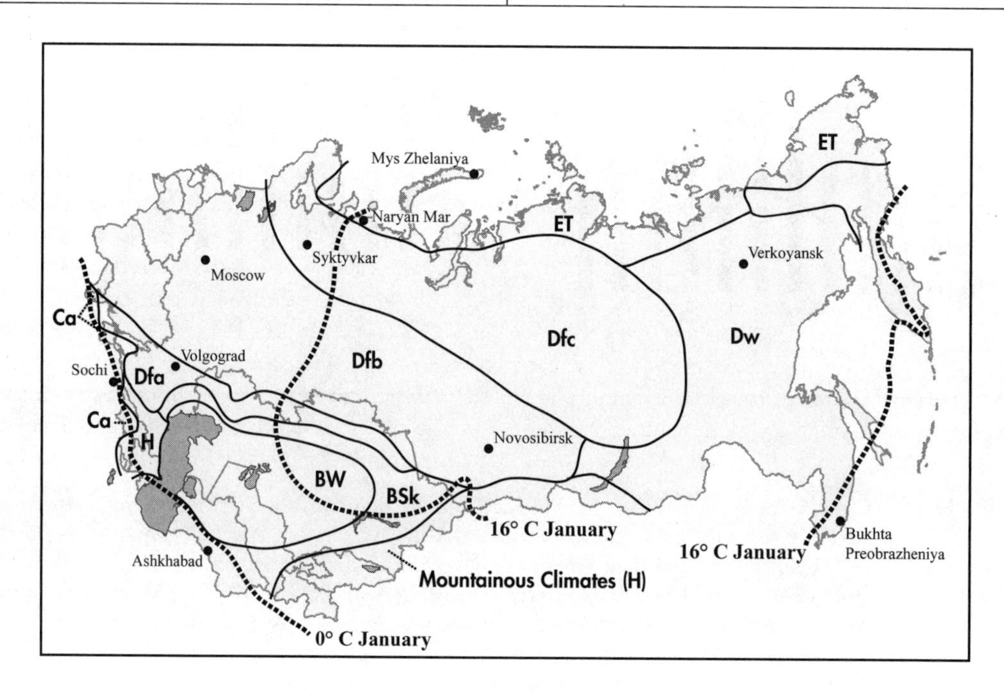

FIGURE 3.2. *(cont.)*

nual temperature (MAT), and the mean annual precipitation (MAP). The diagrams are scaled to have a bioclimatic meaning: Each 10°C gradation corresponds to 20 mm of monthly precipitation. At this scale, when the temperature curve rises above the precipitation curve, a moisture deficit is likely, and this will have a negative impact on plants.

Let's imagine traveling on a chartered plane, leaving Chicago and heading straight up north across the North Pole to the Novaya Zemlya islands in Russia. How long do you think it will take us to get there? 20 hours? In fact, the distance of 6,700 km can be covered in about 8 hours in a modern jet—less time than it takes to reach Paris! Our first stop is on the North Island of Novaya Zemlya, which is mainly covered with ice and snow. There is no permanent human settlement, and of course no big airport. Let's hope our imaginary plane can land on top of the ice cap near Mys Zhelaniya (the Cape of Desire). The climate here is similar to parts of Greenland or northern Iceland. It is a polar climate (E type), with temperatures near or below freezing all year (MAT= –9.7°C), and intermediate precipitation (MAP = 527 mm). Some parts of the eastern Arctic in this zone are much drier. For example, Wrangel Island in the East Siberian Sea gets only 300 mm of precipitation, almost as little as in a desert.

What matters the most to plants here is the length of the growing season, however, when temperatures rise above freezing: It is very short, just a few weeks in July and August. Only a handful of the hardiest species of plants (mainly lichens, mosses, and some Arctic grasses) can grow locally. No plant life exists on the ice cap itself. The North Island would be a tough place to spend even summer, let alone winter. Its analogues in North America include islands in the Canadian Arctic, although these tend to be drier than Novaya Zemlya (MAP = under 200 mm). You would need high-quality winter gear during most of the year. The presence of the ocean, however, modifies seasonality a bit; the coldest temperatures recorded at Mys Zhelaniya are "only" in the low –40°C, not –60°C as in Siberia. Even in July, though, temperatures do not rise above +10°C.

Our next stop, 1,100 km to the southwest, takes us to the tundra—still within the polar climate type (the subtype is ET). A good example would be the city of Naryan-Mar, Russia, where the Pechora River flows into the Barents Sea. The temperature here is a bit warmer (–3.5°C), but precipitation is about the same (468 mm). The growing season is longer, about 3.5 months. Winters are long and dark, because this area is still above the Arctic Circle. Snow stays on the ground for 220 days. Trees normally do not grow in the tundra, because they do not get enough warmth in the summer months to develop fully. Grasses, sedges, mosses, and small shrubs are best adapted for this climate type. You would still need a nice winter outfit during much of the year in Naryan-Mar. The coldest temperatures here are about –50°C, while the warmest may top +30°C in the summertime. More typical are cool summers (about 15°C in the middle of July). Nome, Alaska, has a pretty similar climate. Naryan-Mar is a fascinating place to visit, but not an easy place to stay over winter.

Our next stop will be in a D-type climate. D-type climates are the most widespread in Russia, covering over 80% of its territory. The air temperature in the coldest month is always below freezing, but the warmest month is generally above +10°C. Three distinct subtypes of the D climate type exist in Russia: subarctic Dfc (northern European Russia and western Siberia); subarctic with dry winter, or Dw (much of eastern Siberia); and the milder humid continental Dfb (central European Russia, including Moscow). Our subarctic stop in the European part will be in the city of Syktyvkar (MAT = +0.3°C, MAP = 492 mm). The growing season here is longer than in the tundra, between 5 and 6 months, with snow staying on the ground "only" 180 days. Trees can grow here. Most of these are pine, spruce, and fir—conifers whose needles are available year-round for photosynthesis, to compensate for the still relatively short growing season. The winters remain cold (–51°C is the record low), but summers can be surprisingly hot (+35°C is the record high). There is ample year-round precipitation. Dawson Creek, British Columbia, has a broadly similar climate, with a longer vegetative season of almost 8 months.

Minneapolis, Minnesota, and Moscow, Russia, can both be used as examples of the humid continental microthermal climate (subtype Dfb). This climate is warmer than the subarctic, but it still has a distinct, cold winter, with the average temperature below freezing. Summers are warm, but almost never hot. Moscow (MAT = +3.6°C, MAP = 575 mm) has moderately cold winters, with temperatures in January averaging about –10.3°C, and moderately warm summers, with July temperatures averaging +17.8°C (Figure 3.3). The coldest temperature ever recorded is –42°C, and the warmest temperature is +37°C. There are four distinct seasons, with winter lasting about 5 months. The Minneapolis climate is very similar (MAT = +6.6°C, MAP = 631 mm), with slightly warmer summers (+22.8°C average in July, –10.9°C in January). The primary difference between the two is the amount of available daylight in summer versus winter: Minneapolis is located much farther to the south (44°N vs. 56°N for Moscow), and thus has shorter days in summer, but longer days in winter. There are also more cloudy days in Moscow, in part because of its proximity to the Atlantic and in part because of the air pollution. Moscow's industries generate a lot of dust, which causes rain droplets to form. The city's actual temperatures are about 2–3°C higher in winter than in the surrounding countryside. When is the best time to visit Moscow? My personal recommendation is either the late spring (May), when flowers are in bloom and nightingales are singing in the city parks, or the midautumn (early October), when it is still relatively warm and all the leaves are at their peak color.

FIGURE 3.3. Moscow: Tsaritsyno Park in winter (a) and summer (b). *Photos*: (a) S. Blinnikov, (b) Author.

South of Moscow, we quickly enter dryer climates belonging to the B type. Notice that there is no C type between B and D. B-type climates are arid or semi-arid. Their exact classification is complex, but generally these climates have a moisture deficit at least part of the year. When there is not enough rain, but plenty of warmth, potential evaporation exceeds available precipitation, and a moisture deficit results. As noted earlier, we can see when that happens on the climate diagrams, whenever the temperature curve goes above the precipitation curve. Volgograd, the famous Stalingrad of World War II, is located in the semi-arid BSk climate (MAT= +7.7°C, MAP = 345 mm). An analogous climate in North America would be found near Pierre, South Dakota. For about 4 months in the summer, there is a moisture deficit. In midsummer in Volgograd, temperatures can be as high as +42°C (average about +24°C), while precipitation is scarce (22 mm per month, compared to 74 mm in Moscow). The plants best adapted to this climate are grasses and some long-rooted perennial forbs—in other words, prairie plants (Figure 3.4). In Eurasia, such grasslands are called "steppes." Steppes are semi-arid, meaning that the moisture deficit lasts only a portion of the entire year. Trees do not grow well in this type of climate. The winters can be still very cold (absolute minimum = –35°C) and windy, as the Nazi army fully experienced when it was trapped in November 1942 near Stalingrad. The snow stays on the ground for about 80 days a year.

FIGURE 3.4. A fragment of northern steppe in bloom near Pushchino, 100 km south of Moscow. *Photo:* Author.

True deserts are found in a small section of Russia next to the Caspian Sea in Kalmykia (this is the only desert in Europe, in fact), as well as in southern Kazakhstan (Figure 3.5), Uzbekistan, and Turkmenistan. The capital of Turkmenistan, Ashgabat (MAT = +16.9°C, MAP = 193 mm), has a typical desert climate (BW). Virtually no rain falls in summer, and, unlike in the U.S. Southwest, there is no August monsoonal rain. The peak of precipitation occurs in spring, when 35–45 mm of rain may fall per month, instantly turning the gray desert into a flowering garden. Winter temperatures on average do not drop below freezing (average January temperature = +4.7°C), and the summers are uncomfortably hot (+37°C is typical). Las Vegas, Nevada, has a similar climate, except that it is even drier (100 mm of precipitation per year vs. 193 mm in Ashgabat) and a bit warmer in winter.

If you live in the southeastern United States or in California, you may be wondering by now whether there are *any* climates in the FSU that would match yours. Specifically, such C-type climates are only found in Moldova; in the extreme southern part of Ukraine (Odessa), especially the southern portion of the Crimea near Yalta; and in narrow strips along the Black Sea in Russia and Georgia, and along the Caspian Sea. The warmest among these places is Batumi, a seaport in southwestern Georgia on the border with Turkey (Figure 3.6). This city is in a true subtropical climate (Ca), where many plants from Southeast Asia and Africa can survive winters. A famous

FIGURE 3.5. Semidesert near Kapshagai reservoir in southern Kazakhstan. Small trees with tiny leaves are the famous saxaul (*Haloxylon*). *Photo:* Author.

FIGURE 3.6. Batumi, Georgia, located on a narrow strip of land along the Black Sea, is in a subtropical climate and never experiences frost. Notice the evergreen Mediterranean-type vegetation. *Photo:* K. Van Assche.

Russian botanist, A. N. Krasnov (1862–1914), took advantage of this when he helped to establish a beautiful botanical garden in the city, full of exotic tropical trees and shrubs. In C climates, temperatures in the coldest month do not drop below freezing. This is extremely important to many plants (e.g., bananas or palms) that cannot tolerate even a short period of frost. The Crimea Peninsula and the Caspian Sea coast are relatively dry due to the mountain "rain shadow" effect, while the Black Sea coast is more humid. In a sense, the climate of the southern Crimea resembles that of the California coast, while areas near Sochi, Russia, feel more like the southeastern United States. However, Sochi's temperature and humidity levels are quite a bit below Florida's levels.

We have now completed our north-to-south transect. If we were to fly farther east (to Yakutsk and beyond), the climate would get on average much colder and dryer than in most of the European part of the FSU. The extreme Far East experiences monsoonal influence in later summer, and an occasional typhoon or two. Winters there are not as cold as in Siberia, but heavy wet snow is very common, while summers are moderately warm and muggy. Table 3.1 summarizes the climate extremes found in Northern Eurasia, and compares them to North American and world climate records.

Human Adaptations

Much has been written about the brutality of the Russian winters. Of course, the cultures of Russia developed in them and with them. The indigenous peoples of Siberia experience even colder average conditions than those of the Russian core. Parts of the Central Asian deserts may be very hot and dry in summer, but frigid in winter. Coastal St. Petersburg is foggy and cool year-round, and very dark in winter; it is located at the same latitude as Anchorage, Alaska, after all. Murmansk

TABLE 3.1. Extreme Climate Records for Northern Eurasia, North America, and the World

Extreme record	Northern Eurasia	North America (without Greenland)	World
Coldest temperature ever	–71°C (Oimyakon, Russia)	–63°C (Snag, Yukon)	–88°C (Vostok, Antarctica)
Warmest temperature ever	+46°C (Turkmenistan)	+57°C (Death Valley, CA)	+58°C (Al-Aziziya, Libya)
Most precipitation in a year	3,682 mm (Mt. Achishko, Caucasus)	6,500 mm (Henderson Lake, BC)	26,470 mm (Cherrapunji, India)
Least precipitation in a year	116 mm, Kosh-Agach (Altay, Russia)	30 mm (Bataques, Mexico)	0.8 mm (Arica, Chile)

is a city of 300,000 people located at the latitude of Barrow, Alaska (population 4,000). The sun does not rise above the horizon there for about 1 month each winter, so people often get depressed and have to be treated in sun rooms.

Obviously, all cultures of the FSU have had to learn to live with the climate, whatever it might be. Here just a few interesting cultural adaptations to climate are briefly mentioned.

- Traditional Russian peasant homes (*izba*) were one- to two-room log cabins, with a massive brick oven occupying about one-quarter of each home's interior space. The oven was stocked with wood. Peasants would not only cook in the oven, but sleep on its top.
- In northern Russia, farm animals would be kept indoors in a covered area adjacent to the main house, to save heat and to keep the animals warm (Figure 3.7).
- Much of the traditional dress is winter gear: *valenki* (felt boots with rubber bottoms), *tulup* (an overcoat made of sheepskin), and *ushanka* (a fur hat with ear flaps). Women have also made ample use of woolen scarves and shawls.
- Typical Russian food is heavy on fat and carbohydrates to provide much-needed calories in winter. However, two long fasts (one before Christmas and one before Easter) were also traditionally observed, when no animal products could be eaten. This reduced the amount of meat that had to be raised, but it also meant that the need for more fat and protein went unmet for lengthy periods.
- Only hot tea is drunk in northern Russia. Ice is never put in beverages.
- The calendar of feasts in the Russian Orthodox Church is busier in winter and freer in summer, to allow for ample time in the fields during the short growing season.
- Conversely, in the warmer climates of Central Asia, homes are constructed to keep the heat out, commonly with whitewashed walls, small windows, and good ventilation; people sit on low furniture or cushions spread on the floor to enjoy cooler air (Figure 3.8).
- In Central Asia, heads are always protected from the sun by a variety of creative headgear (e.g., *tyubeteika* hats for men and scarves for women).
- Central Asian cultures take a long midday break from work to avoid heat (similar to the Spanish *siesta*).

The traditional cultures of Northern Eurasia evolved many other unique adaptations to their particular environments. Two sets of these adaptations are described in greater detail in Vignettes 3.1 and 3.2). However, now all cultures are threatened by the increase in the rate of global climate change.

FIGURE 3.7. Typical houses in Malye Karely, Arkhangelsk Oblast, showing northern Russian architecture. Note the covered section that is lower than the rest of one house; this serves as a winter shelter for animals. *Photo:* A. Shanin.

FIGURE 3.8. Interior of a Kazakh house, taken at the Ethnography Museum at Ust-Kamenogorsk in eastern Kazakhstan. Notice the cushions, carpets, and low furniture designed to keep people close to the floor, where it is cooler in summer. *Photo:* Author.

Vignette 3.1. Living with Permafrost

"Permafrost" is perennially frozen soil and subsoil material that exists in climates below a certain temperature threshold. Usually it is found everywhere in tundra (ET) and subarctic (Dfc) climates. In North America, it is found in much of northern Canada and Alaska. In Russia, it occupies an astonishing two-thirds of the territory, primarily in the north and in central and eastern Siberia, where it extends all the way from the Arctic Ocean to the Chinese border near Chita. Isolated patches of it occur in many Siberian mountain ranges as far south and west as the Altay. The permafrost may extend hundreds of meters below the surface. The top layer of about 30–50 cm thaws in summer, turning the previously solid surface into liquid mud.

Russian scientists and engineers pioneered many studies of the permafrost. They also had to come up with ways of living with it. For example, houses in all northern Siberian towns have to be built on pylons above the ground, so that their undersides do not melt the permafrost. Oil and gas pipelines likewise must be propped up and suspended above ground. Roads and railroads crack and dip in summer, and must be frequently repaired. Even trees are affected: So-called drunken forests of larch cover much of Siberia, where permafrost conditions uproot the shallow roots of the trees and make them lean at odd angles.

Some spectacular paleontological finds have been made in the Siberian permafrost. Thousands of kilograms of mammoth bones were brought to world markets from Siberia in the 19th century. This "Russian ivory" was sold all over Europe. Some well-preserved remains of mammoths and other large wildlife are occasionally found along the big Siberian rivers, where they simply come to the surface from the lenses of ice and are exposed by the riparian erosion processes. In October 2007, the carcass of a female mammoth infant, nicknamed "Lyuba," was discovered on the Yamal Peninsula. She lived about 37,000 years ago and was about 1 year old when she died. The entire carcass was preserved, including eyes, trunk, and fur. In fact, for years now the possibility of extracting mammoths' DNA and cloning these animals has been discussed. Who knows, perhaps a Pleistocene Park may be possible in the near future, if not a Jurassic Park? Pending the arrival of the mammoth clones, S. Zimov of Magadan Zoological Institute of the Russian Academy of Sciences is working on creating a prototype wildlife park near the lower reaches of the Lena River, where all existing Siberian megafauna (musk oxen, bison, camels, horses, reindeer, saiga antelopes, bears, etc.) will be represented.

The Effects of Climate Change

Climate is always changing naturally. Seventy million years ago, there were no ice sheets anywhere in the world; palms were growing in Greenland, and dinosaurs roamed the earth. Conversely, just 20,000 years ago, the earth was in the grip of the last full Ice Age; ice sheets extended into Iowa in North America and Ukraine in Eurasia; and the woolly mammoth was the largest animal. In the past 150 years, however, the natural pace of change (mostly apparent as a warming trend) has greatly accelerated, due to human impact on the makeup of the atmosphere. The human role in global climate change is no longer contested in reputable scientific circles (although it may be by certain political groups). Al Gore's documentary *An Inconvenient Truth* won him a share of the Nobel Peace Prize in 2007. In the same year, the other winner, the Intergovernmental Panel on Climate Change (IPCC), released a new cache of global reports suggesting that the rest of the 21st century will see a much warmer climate. Not only is the climate warming up; it is virtually certain that it will continue to do so at increasing speed and with poorly anticipated consequences.

Generally speaking, Russia has relatively little cause for concern compared to its coastal European neighbors (especially the Netherlands and Denmark) or its southern Asian neighbors (Bangladesh, the Maldives). According to the IPCC, the two main impacts of the future climate change will be (1) rising sea levels and submergence of the coasts, especially if and when the western Greenland ice sheet melts; and (2) warm-

Vignette 3.2. Almaty, a City Designed with Climate in Mind

It is July in Almaty, the largest city of Kazakhstan and its former capital. The air is hot (it is 32°C in the shade), but the city feels cool. What's the secret? When you arrive at your hotel, you decide to leave the air-conditioned room behind and explore on foot. All streets are laid out in a classical grid pattern, with north–south avenues running uphill to the distant mountain peaks behind the city, and west–east streets running parallel to the slope. Lots of people are outside, going about their business.

Built by the Russians as Verny ("Faithful") in the 1850s, this city was later renamed Alma-Ata, meaning "Father-Apple" in incorrect Kazakh, and now is called simply "[City] of Apples," Almaty. Located in the heart of the Eurasian continent, as far from the ocean as one can possibly get, the city enjoys a fine climate despite its inland location. It also has spectacular scenery, not unlike that of Denver, Colorado. Right behind the last street, the jagged snow-capped peaks of the Zailiysky Alatau range soar to elevations of 4,000–5,000 m (Figure 1a). While not as huge as the Tean Shan further south in Kyrgyzstan, the Zailiysky range is an amazing unspoiled wilderness full of sublime beauty—a paradise for skiers and backpackers.

People began settling in the area in about 1000 B.C. In the Middle Ages, settlements in the Almatinka River valley served as stopover points on one of the few branches of the famed Silk Route from the Near East to China. When the Russians came in the 19th century, they seized the opportunity to build a grand, beautiful, modern city in a convenient location near water and well protected by mountains. Clearly, they wanted to establish a permanent Russian presence in Central Asia. In 1854, a small fort was built. In just 5 years, the population grew to 5,000 people; by 1913, it was 40,000. The city was the capital of the Kazakh Soviet Socialist Republic between 1936 and 1991. Today its population is about 1.4 million and very diverse, with Russians and Kazakhs evenly represented. There are also many residents now from other Central Asian FSU republics, China, Korea, and other countries.

The climate of Almaty is highly seasonal, but is milder than Siberia's, due to its more southern location at 43°N (the average temperature is –4.5°C in January, +23.6°C in July). The growing season is long, about 8 months, and there is little snow in winter. For 2 months in midsummer, there is a moisture deficit that affects vegetation, and temperatures may peak at 35–37°C in the afternoon (about as hot as it gets in Elko, Nevada).

The city planners designed Almaty with climate in mind. As you walk around, you notice a few features that allow for cooling in the scorching heat of summer. First, the streets are lined with huge, magnificent poplar or plane trees that provide ample shade. Second, right beside each sidewalk flows cool water in a concrete trough about 0.5 m across (Figure 1b). This water flow cools the surrounding air. Third, there are over 120 fountains in the city, many located in large parks. The parks themselves are everywhere, with beds of roses and other flowers, and beautiful deciduous and coniferous trees. Every city block has lots of additional vegetation, and many homes are built in a way to maximize ventilation in summer and to provide good views of the city. Some new commercial developments are being built underground, both on street corners in the pedestrian underpasses, and in the main downtown area. Cooler in summer and warmer in winter, these are popular gathering places for the city youth. Almaty is perhaps at its loveliest in late spring, when all the orchards around are in bloom; apple, peach, apricot, and cherry blossoms are truly spectacular.

(cont.)

er temperatures, especially in the Arctic and especially during winter nights, which may lead to moisture deficits in many areas because of less snow cover. On the first count, Russia has few seaports to worry about (see Chapter 2), and its capital and biggest city is far inland at a comfortable 156 m above sea level. Only a fraction of the Russian population (8%) lives near a seacoast. The main urban area that will be affected is St. Petersburg, which is right at sea level and is commonly flooded by the spring meltwater from the Neva. Compare this to the United States, where two-thirds of all people live within 200 km of a coast, *and* where the two biggest urban areas (the

(a)

(b)

FIGURE 1. (a) Almaty was built as a Russian frontier city in the mid-19th century, in the hot climate of the foothills of majestic Zailiysky Alatau. (b) Leafy plane-lined streets with water ditches next to the sidewalks keep the city cool, even in the hottest days of July. *Photos:* Author.

New York City and Los Angeles areas) are right at sea level.

The Central Asian states of the FSU have no oceanic coastline at all. The Caspian Sea is actually below sea level now, but is not expected to rise; it is just a big saline lake. Ukraine does have a few important seaports, but again most of its territory and population are far away from the sea. On the second count, Russian agriculture can greatly expand northward and eastward, especially in the currently undersettled Siberian and northeastern European parts of the country. So can we assume that all is rosy? Not so fast.

Among the seemingly inevitable consequences of global warming will be an increase in mid-continental droughts, floods, and other extreme weather events (Lynas, 2008). Much of Russia's grain is grown today in the "black soil" zone of the steppe, where precipitation is already scarce. Compared to the United States and Canada, Russia irrigates far fewer hectares of its crops; it mostly relies on the summer rainfall and winter snowpack, both of which are expected to become spotty in the future. In fact, in the most recent assessment from the IPCC, the amount of precipitation over Ukraine is expected to drop by almost 50% by 2070. Extreme hot spells in the middle of the growing season in the summer may decimate sensitive summer crops, like corn and soy. The loss of snowpack in winter may affect the growth of winter wheat, which is the staple grain produced in the region. Southern Ukraine, the Caucasus, and Central Asia will be even more severely harmed. The treeline is predicted to shift upward by a few hundred meters, and alpine ecosystems may disappear in the Carpathians, in much of the Caucasus, and even in some Central Asian mountains. Melting of the permafrost in Siberia is likely to cause major structural damage to the existing infrastructure there (see Vignette 3.1).

Furthermore, although global climate change scenarios differ in regard to the exact scope and magnitude of change in climate parameters, all agree that the change is likely to accelerate as the nonlinear feedbacks in the climate system begin to kick in (see below for a Russian example). We also need to begin preparing for the unexpected. For instance, an abrupt halt of thermohaline circulation in the North Atlantic may temporarily shut down the Gulf Stream and make Western Europe colder than it is today very quickly. This may lead to a frantic political scramble among European nations for more fossil fuels from Russia, with some unpredictable consequences. Also, a catastrophic melting of even a small portion of the western Greenland ice sheet may abruptly raise the oceans by a whopping 4 m in less than 30 years, which would wipe out not only New York City, Los Angeles, London, and Copenhagen, but also St. Petersburg, Murmansk, Odessa, and Vladivostok.

One of the fundamental feedbacks that seem to be speeding up the global rate of climate change is occurring right in Russia. In 2005, a group of American and Russian researchers discovered, with surprise and alarm, that methane was being released from thawing eastern Siberian bogs at a rate five times as fast as was previously estimated from observations in Alaska. Each molecule of methane escaping into the atmosphere equals in its impact 20 molecules of carbon dioxide. When the new rate of escape is plugged into climate models, they show a higher rate of global warming than previously believed (Walter et al., 2006).

Although carbon dioxide is responsible for roughly 65% of the enhanced global greenhouse effect, methane is already contributing 20%. Thus Russia, with the biggest tundras in the world, will be contributing an increasingly great share of this gas to the atmosphere; this is ironic, since Russia only just joined the Kyoto Protocol process in 2004.

QUESTIONS

1. Which major climate types are found in Northern Eurasia? Which are not found?
2. Explain in what direction climate gets warmer in Russia, and in which it gets dryer. Are these directions similar to or different from those in the country where you live? Why?
3. The famous Russian author Alexander Pushkin said that "nature waited and waited for winter, and finally the snow fell in January, on the third of the month, at night." What is the date of the latest start of winter snowfall in the area where you live (if you get any snow at all)?
4. Explain why Murmansk is an ice-free port, while Magadan (much farther to the south) freezes up in winter.

EXERCISES

1. Use the World Bioclimatics Website (in the lists of Websites below) to find climates analogous to those in Figure 3.2 for the country where you live. Which climates of Northern Eurasia do not seem to have good analogues in your country? Why do you think this is the case?
2. Use a few current books on global climate change to find out more about the predicted impacts of global warming on Eurasia. Stage a classroom debate about whether or not Russia (or any other FSU republic) should take measures to restrict its greenhouse gas emissions.

Further Reading

Bychkova-Jordan, B., & Jordan-Bychkov, T. G. (2001). *Siberian village: Land and life in the Sakha republic.* Minneapolis: University of Minnesota Press.

Hill, F., & Gaddy, C. G. (2003). *The Siberian curse: How Communist planners left Russia out in the cold.* Washington, DC: Brookings Institution Press.

Lydolph, P. E. (1977). *Climates of the Soviet Union.* Amsterdam: Elsevier.

Lynas, M. (2008). *Six degrees.* Washington, DC: National Geographic Magazine Press.

Pryde, P. (Ed.). (1995). *Environmental resources and constraints in the former Soviet Republics.* Boulder, CO: Westview Press.

Shahgedanova, M. (Ed.). (2002). *The physical geography of Northern Eurasia.* Oxford, UK: Oxford University Press.

Walter, K. M., Zimov, S. A., Chanton, J. P., Verbyla, D., & Chapin, F. S. III. (2006). Methane bubbling from Siberian thaw lakes as a positive feedback to climate warming. *Nature, 443,* 71–75.

Websites

www.globalbioclimatics.org—Access to thousands of climate diagrams for sites around the world, including a few hundred in Northern Eurasia.

www.gismeteo.ru—The main weather site for Russia. (In Russian only.)

www.ipcc.ch/ipccreports/ar4-wg2.htm—The IPCC 2007 assessment report focusing on the impacts, adaptations, and vulnerability of various regions around the world, including the Arctic and Northern Eurasia.

CHAPTER 4

Biomes

The nature of the former Soviet Union (FSU) is diverse and beautiful. It makes the most geographic sense to look at it from the perspective of "biomes," the largest ecosystem units. The biomes of Northern Eurasia are similar to those of Europe or North America: tundra in the north; taiga and deciduous forests in the middle; steppe and desert in the south. The extreme south has deserts or subtropical Mediterranean-like shrub vegetation. The boundaries of the biomes (see Figure 4.1) correspond closely to the major climate types (see Chapter 3, Figure 3.2).

For millions of years, Northern Eurasia and North America were connected to each other—mainly across the Bering Strait, but also sometimes via Greenland and Scandinavia. This resulted in an array of animals and plants that are shared by these two regions. In fact, much of the biota is so similar that biogeographers lump the two together into one "Holarctic" biogeographic realm. The flora and fauna of India (which is on the same continent as Russia), on the other hand, are completely dissimilar to Northern Eurasia's; they are more like Africa's. For example, North America and the FSU share many tree genera (e.g., pine, spruce, elm, maple, birch, aspen, and oak). Most tree species are different, but several look alike—so-called "vicariant" species. For example, the Siberian cedar pine (*Pinus sibirica*) generally resembles the North American white pine (*P. strobus*); the Norway pine (*P. sylvestris*) is very similar to the Minnesota red pine (*P. recinosa*). At the lower levels of the plant kingdom (e.g., among mosses), the similarity is even greater. Large swaths of Russia's and Canada's boreal forests have the same mosses (*Dicranum, Polytrichum, Pleurozium*) and *Cladonia* lichens, for example. There is a higher degree of difference among flowering forbs and grasses, but many Russian wildflowers are still instantly recognizable to visiting American botanists as a "buttercup," a "violet," a "lily of the valley," or a "lady's slipper," even if they do not know for sure what species they are looking at. The overall similarity is greatest between eastern Russia and Alaska, the former parts of the Bering land bridge (Hultén, 1937).

Many animal genera or even species are identical in North America and Northern Eurasia: Arctic and brown bears, gray wolves, red foxes, moose, elk, golden eagles, peregrine falcons, and black-capped chickadees, for example. If an exact match is missing, there is usually a pretty good substitute/vicariant species (e.g., American mink and Eurasian mink, otters, beavers, cranes, crows, etc.). The differences among songbirds are the greatest, because most of the migratory ones in North America originate in the neotropics, while

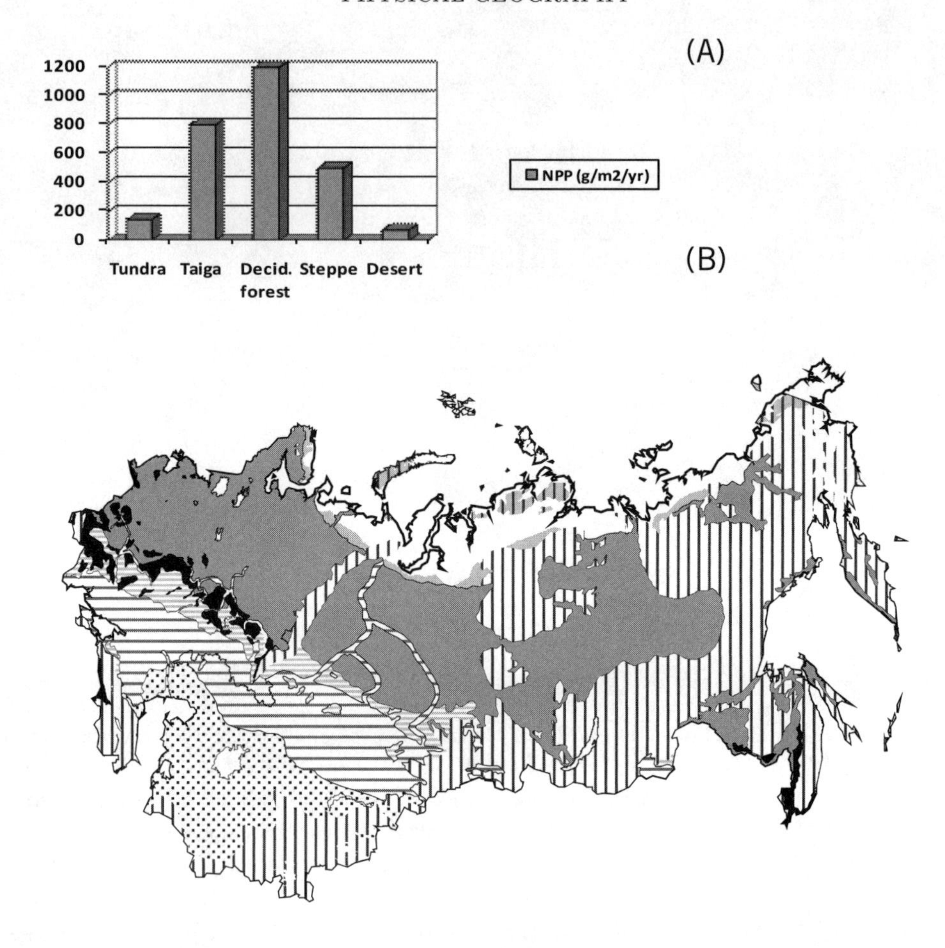

FIGURE 4.1. Biomes of Northern Eurasia. (a) Natural net productivity in grams per square meter per year. (b) Tundra is shown in white, taiga forest in gray, deciduous forest in black, steppe in horizontal hatching, desert in dots. Mountainous areas are shown in vertical hatching. Map data from ESRI ArcAtlas.

those in Northern Eurasia originate in Africa or South Asia. For example, Eurasian warblers or flycatchers are unrelated to the American birds of the same names, although they are similar in their ecology and behavior. Some apparently similar biomes also exhibit a higher degree of difference and endemicity. For instance, the Russian Far East shares some remarkable combinations of plants and animals with areas to either the south (China) or the north (Chukotka). No such forests exist in North America. Great uniqueness is observed in the ecosystems of coastal California and southern Florida instead, and there are no strong analogues for such ecosystems in Eurasia.

The main five biomes of the FSU (tundra, taiga, deciduous forest, steppe, and desert) are stretched across the Eurasian continent in wide belts from west to east. In between, there are transitional types (e.g., forest–tundra, mixed forest, and forest–steppe). Each biome or natural zone has a corresponding climate (see Chapter 3), a zonal soil type, and a characteristic set of plants and animals (Table 4.1). Some biomes are more extensive than others, depending on the climate pattern. Also, some are considerably better preserved than others. For example, whereas most of the taiga zone remains reasonably intact, with closed-canopy forests (even in areas with heavy logging), 99% of the virgin steppe has disappeared.

The overall diversity of the plants and animals in Russia is not great, because of its northern lo-

TABLE 4.1. Bioclimatic Characteristics and Biodiversity of the Main Biomes of the FSU

Biome	GDD	MAP	Plants	Amphibians	Reptiles	Birds	Mammals
Tundra	200–500	400–750	200–500	0	0	60	15
Taiga	500–2,000	500–900	700–1,000	3	5	160	40
Deciduous forest (W)	2,000–3,000	500–900	1,000–1,500	7	6	190	60
Deciduous forest (E)	2,000–3,500	700–1,500	1,500–2,500	8	7	200	70
Steppe	2,500–3,500	400–600	1,500–2,000	3	15	180	50
Desert	>3,500	100–300	1,500–2,000	0	40	140	40

Note. GDD, sum of growth-degree days above 10°C; MAP, mean annual precipitation (mm); Plants, number of vascular plant species in a local flora on 100 km^2; W, European part; E, Far East. Animal species numbers are those found in local faunas. Data from Zlotin (2002).

cation. For example, there are 11,000 species of vascular plants, 30 of amphibians, 75 of reptiles, 730 of birds, and 320 of mammals in the Russian Federation. By comparison, the United States (a more southern country half the size of Russia) has 19,000 species of vascular plants, 260 of amphibians, 360 of reptiles, 650 of birds, and 360 of mammals.

Tundra

Treeless tundra is found in the north of Russia, generally above the Arctic Circle. In European Russia, it occupies limited space on Kola Peninsula and in the Arkhangelsk and Komi regions along the coast. In Siberia, the most extensive tundra is found on Yamal, Taymyr, and Chukotka Peninsulas. In North America, tundra covers much of Alaska's North Slope, as well as about one-quarter of Canada. The word "tundra" comes from the Saami people and means "treeless." North of the tundra, the polar desert has virtually no life. Some hardy blue-green algae, and occasional mosses and lichens, are about all that can be found there. Nevertheless, even the northernmost islands of Russia, in Franz Joseph Land, have a flora of 57 flowering plants, 115 lichens, and 102 mosses. Polar bears, seals, and walruses are important mammals of the surrounding seas and ice. A few species of hardy Arctic birds—murres, puffins, gulls, and terns—live on inaccessible cliffs in "bird bazaars."

In contrast to the polar desert, the tundra has hundreds of species of plants and scores of birds and mammals. Although the precipitation in the tundra is low (usually under 300 mm per year), the evaporation rate is even lower, thus creating familiar soggy summer conditions. Soils are of the "tundra glei" type ("gelisols," in the U.S. classification), with a pronounced anaerobic zone. Underneath is permafrost, but the top 20–30 cm of soil near the surface can team with life in the summer months. These soils are subject to much frost churning, which pulls organic matter down the profile and brings rock fragments to the surface, creating spectacular patterned grounds (Figure 4.2).

The most common plants of the tundra are mosses and sedges. Dwarf shrubs, grasses, and forbs become more common in the southern tundra (Figure 4.3). Eventually, bigger shrubs and even small trees begin to appear as one travels

FIGURE 4.2. This patterned ground in the tundra is caused by frost, which churns up different-sized pieces of debris. *Photo:* V. Onipchenko.

FIGURE 4.3. The tundra is a treeless, unproductive community with a short vegetative season, little rainfall, and poorly drained soils. Plants must be short to take advantage of the warm boundary layer of air near the ground. *Photo:* Author.

farther south, giving way to forest–tundra (Figure 4.4). In European Russia this zone is located around the Arctic Circle (66°32'N); in Siberia it begins farther north, at about 70°N. In European Russia the treeline is formed by Scotch pine, spruce, or birch; in Siberia it is mainly larch. Climatically, the treeline corresponds to the point at which the mean July temperature goes above 10°C.

Typical animals of the tundra include Arctic foxes, reindeer, lemmings, gyrfalcons, swans, geese, ducks, various shorebirds, snowy owls, horned larks, redpolls, and buntings. Some are rare or endangered (e.g., Siberian red-breasted geese, Siberian cranes, and rosy gulls). There are many protected areas in the tundra biome: however, most of them are poorly accessible. The biggest three are the Great Arctic Zapovednik on the Taymyr Peninsula, the delta of the Lena River, and Wrangel Island.

Taiga

"Taiga" is a Siberian word; it has recently become better known through the efforts of the Taiga Rescue Network, doing important conservation work throughout the Northern Hemisphere. In North America, taiga is known as the "boreal coniferous forest," which is what covers much of Canada. Note that although the West Coast forests of British Columbia, Oregon, and Washington also have conifers, they have a much higher diversity of plants and much bigger trees, so they are not the true taiga. In Northern Eurasia, the taiga is a huge biome (covering over half of all Russia), but it is rather monotonous. In European Russia the main species are Scotch pine, Norway spruce, and European fir; in western Siberia they are Scotch pine, Siberian cedar pine, and Siberian fir and spruce; and in eastern Siberia they are two species of larch (Figure 4.5). Coniferous

FIGURE 4.4. Forest–tundra in the polar Urals. Most of the trees in the background are Scotch pines (*Pinus sylvestris*). *Photo:* A. Shanin.

FIGURE 4.5. The Siberian cedar pine (center) and larch are typical large trees in the southern taiga of the Altay Mountains, reaching 40–45 m in height. Note the people on the left for scale. *Photo:* Author.

but also deciduous, larch is the only tree that can survive the brutal cold of the Verkhoyansk area, which is the coldest in the Northern Hemisphere (see Chapter 3). Birch and aspen may be found as secondary-growth species on clearcuts and fire clearings. Low shrubs with berries of the *Vaccinium* group are very common, as are mosses and lichens. Interspersed with big trees are nutritionally poor bogs with peat mosses (*Sphagnum*), Labrador tea, cranberries, and carnivorous sundew (*Drosera*). The boreal forests of Eurasia make up about 21% of the world's total tree cover on 5.3 million km^2; this area is twice the size of Argentina!

From north to south, three subzones can be distinguished in the taiga: northern, middle, and southern. The biodiversity and the productivity are highest in the southern taiga, which extends south to an imaginary line from Moscow to Yekaterinburg to Krasnoyarsk. Over 2,500 species of flowering plants occur in the taiga. Some, especially orchids, can be beautiful, but are very rare. Mosses, lichens, ferns, and mushrooms thrive under the canopy of the coniferous trees. Soils of the taiga are poor in nutrients and acidic; the most typical are called "podzols," or "spodosols" in the U.S. classification. Consequently, few crops can be grown in the taiga zone. The main crops are the hardiest grains, like barley and rye, which are raised on small clearings of land near the rivers. Meadows in the floodplains can produce good hay, and berries and mushrooms from the forest complement the diet.

Typical taiga mammals include the symbol of Russia, the brown bear (the same as the North American grizzly, albeit a different subspecies). They also include gray wolves, lynxes, red foxes, Siberian sables, minks, wolverines, moose, elk, shrews, red squirrels, flying squirrels, chipmunks, and mice. The local mammal fauna ranges from 30 to 50 species. Over 160 species of taiga birds include black storks, various raptors, eagle owls, capercaillie (turkey-sized black forest chickens), grouse, black woodpeckers, waxwings, and many finches (crossbills, hawfinches, siskins, etc.). Some of the same species occur in North America.

The best places to visit taiga in European Russia (Figure 4.6) include the Darwinsky, Tsentralno-Lesnoy, Kivach, and Kostomuksha Zapovedniks and the Paanayarvi, Vodlozerski, Kenozerski, and Valdaiski National Parks. (A *zapovednik* is a protected nature preserve; see Chapter 5.) For the ultimate in European taiga, the virgin forests of the Komi Yugyd Va area

FIGURE 4.6. Taiga (boreal forest) in winter in Arkhangelsk Oblast, in the northern part of European Russia. The biome is dominated by conifers, especially spruce, pine, and fir, which are adapted to long, cold winters. *Photo:* A. Shanin.

near the polar Urals are worth a visit; this is the largest remaining fragment of original forests that covered northern Russia and Scandinavia, and it is a World Heritage Site. In Siberia, most of the taiga can be observed directly along the Trans-Siberian Railroad, although much of it is secondary growth. More pristine landscapes include Visimsky Zapovednik in the central Urals and Yuganski Zapovednik in the Tyumen region. Lake Baikal is surrounded by three zapovedniks and two national parks, and is mainly in the taiga zone. East of Lake Baikal, Zeisky and Bureinsky are two relatively new zapovedniks protecting the true wilderness of the eastern taiga. If you are only visiting Moscow and St. Petersburg, several of the forests near these cities are southern taiga as well; there are many local nature parks and wildlife sanctuaries, including Losiny Ostrov National Park, partially within the Moscow city limits! The park's name literally means "Moose Island," and it used to be the hunting preserve of the tsars.

Mixed and Deciduous Forests

South of the taiga zone, a narrow wedge of mixed and deciduous forests stretches from the Baltic republics to the Urals and beyond, to Novosibirsk and the Mongolian border. This zone is smaller than the taiga, but it has a warmer and generally wetter climate. Moscow is located in the middle of it, with pine and spruce being more common to the north of Moscow, and oak, maple, and linden being more common to the south. The exact mixtures vary, depending on previous logging, fire history, plantings, and bedrock. The majority of secondary forests in this zone are pure birch stands, very popular among the Russian landscape artists. (Vignette 4.1 describes the influence of nature on Russian artists in greater detail.) Deciduous forests can grow faster and utilize resources better than conifers, provided that the weather is not too cold. When autumn comes, they shed their leaves and become dormant for winter to avoid death by desiccation. Broad leaves are efficient water evaporation machines; if they are left on the trees in winter, all the water will escape the trunk. In North America the same zone is found throughout much of the mid-Atlantic region, parts of New England, Ohio, Ontario, and central parts of Wisconsin and Minnesota. Much of Western Europe likewise is in this zone.

The soils of the deciduous forest zone are gray forest soils ("alfisols"). These are richer and less acidic than the spodosols of the taiga, but are only modestly better for agricultural purposes. The main feature of these soils is a very quick turnover of nutrients. Wheat and rye are commonly grown in this zone. Forests have a well-developed layered-canopy structure, with tall trees like oaks, lindens (basswoods), or maples dominating the top layer (Figure 4.7). The second layer of smaller trees and tall shrubs (chokecherries, mountain ash, hazelnuts) give way to small shrubs and herbaceous layers, and finally a layer of moss on the ground. Mushrooms are plentiful, as well as wild berries.

The deciduous forest zone is warm enough for some amphibian and reptile species as well; toads, frogs, vipers, and lizards are common. The typical mammals of the zone include many of the taiga species mentioned above. In addition, hedgehogs, martens, European roe deer, beavers, and dormice are common. Endangered European wood bison can be seen in a few preserves, such as the Prioksko-Terrasny Zapovednik, about 2 hours south of Moscow. The secretive Russian desman is an endemic of the Soviet Union; it looks like an oversized water shrew and spends most of its life in clean, slow-flowing forest rivers. It is a threatened species. The birds are very diverse, with a few hundred species present in the forests surrounding Moscow, for example. Not all of them are true forest species, but every May the forests ring with dozens of different voices. Typical forest birds include falcons, eagles, owls, woodpeckers, nuthatches, titmice, and thrushes. The famous nightingales sing majestically in early May through late June in much of European Russia. These secretive, drab olive birds with rusty tails do not look at all remarkable and are hard to see; their song, however, has 12 different parts and is remarkably rich and beautiful. They are even more common in the forest–steppe, where the legendary Kursk nightingales were

Vignette 4.1. Traveling through Biomes and Seasons with the Russian Painters

Russia has two outstanding museums of Russian art: the Tretyakov Gallery in Moscow (established in 1856) and the Russian Museum in St. Petersburg (established in 1895). Both collections include numerous exquisite depictions of Russian landscapes, biomes, and seasons by the nation's leading artists. The artists whose work is best suited for our purposes would be the realist painters of the 19th century. As Europe was becoming preoccupied with the Impressionists, a few Russian artists stubbornly persisted in depicting nature in the traditional way, usually with a social theme. Although Impressionist and modernist paintings can also be found in many Russian museums, the realist school provides the most accurate depictions of Russian nature. Many artists worked in the taiga and especially the deciduous zones of the country, where the four seasons are sharply distinct. In particular, an estate called Abramtsevo (Figure 1), northeast of Moscow, was depicted by many painters of the period (Repin, Surikov, Serov, Vrubel, Korovin). The owner of Abramtsevo, S. Aksakov, was an art connoisseur and a Slavophile writer; he provided room and board to many distinguished artists. Another area frequently depicted near Moscow was the Oka River basin, where a number of prominent painters lived in summer (e.g., Polenov and Ivanov-Mussatov lived near Tarusa). The Russian North in the taiga zone was another favorite: With its quaint villages, cozy wooden churches, and quiet life amid the harsh natural conditions, it provided many subjects for paintings.

FIGURE 1. The estate of Abramtsevo, near Moscow, was a site where many classical works of landscape art were painted in the late 19th century. This river valley, painted in different seasons, appears in the work of several famous artists (Repin, Surikov, Serov, Korovin) of the period. *Photo:* Author.

(cont.)

If you visit the Tretyakov Gallery either in Moscow or online (*www.tretyakov.ru/en*), the following paintings are worth exploring:

- Winter: I. I. Levitan's *March* (1895), F. A. Vasilev's *Thaw* (1871), and B. M. Kustodiev's *Maslenitsa* (1916) all depict the frosty landscape of Central Russia at the beginning of the Great Lent Fast (see Chapter 14).
- Spring: A. K. Savrasov's *The Rooks Are Back* (1871) shows a late March scene in Central Russia, with rooks and birch trees amidst snow melt. A. G. Venetsianov's *Plowing Fields* (1820s) and L. L. Kamenev's *Spring* (1866) are other fine views of this season.
- Summer: I. I. Levitan's *Above the Eternal Peace* (1890) depicts a midsummer landscape, with a big river and a wooden church, in the taiga biome. A. I. Kuindzhi's *Birch Grove* (1879) is a midsummer view of birch trees in the deciduous forest biome, and Kuindzhi's *Dnieper in the Morning* (1886) is a wonderful depiction of wild steppe vegetation on the banks of the largest river in Ukraine. I. I. Shishkin's *Rye* (1878) and A. G. Venetsianov's *Summer Harvest* (1820s) are both views of Central Russian fields in late summer.
- Autumn: I. I. Levitan's *Golden Autumn* (1895) shows birch trees at their fall color peak. V. D. Polenov's *Golden Autumn* depicts birch and aspen trees on a river bank. M. V. Nesterov's *Vision of Youth: Bartholomew* (1889–1890) shows native flowers in early fall in Russia near Radonezh.

FIGURE 4.7. Deciduous and mixed forest biomes dominate the central part of European Russia. Typical trees include European oak and birch. The oak trees (*Quercus robur*) in Kolomenskoe are the oldest in Moscow, 600 years old, with trunks 1 m in diameter. *Photo:* Author.

greatly admired by 19th-century Russian writers and poets.

The best places to see the deciduous zone in European Russia include the Prioksko-Terrasny Zapovednik, mentioned above; the Oksky Zapovednik in the Ryazan region, about 4 hours east of Moscow; and the Kaluzhskie Zaseki in the Kaluga region, about 4 hours southwest of Moscow. In Belarus on the border with Poland, the famous Belovezhskaya Puscha is home to one of the last herds of European bison. A few national parks in this zone also exist in the Baltic republics.

Forest–Steppe and Steppe

South of Moscow, the forest gradually gives way to the steppe. Across the Oka River, the first patches of steppe begin to appear. The Tula and Orel regions have forest–steppe, while the Kursk and Belgorod regions are primarily in the true steppe zone. The steppe stretches across much of Ukraine to the lower Volga, to northern and central Kazakhstan, and to the foothills of the Altay.

Steppe forms in areas with moisture deficit that precludes tree growth. Although steppes are on average warmer than most of the forested biomes to the north, it is really the lack of water that determines the tree boundary. In North America, crossing from eastern to western North Dakota or from Iowa into Nebraska takes you across this climate boundary. In Europe, the most extensive steppes exist in Hungary. Although Eurasian steppes are warmer than the taiga zone, they can be brutally cold in winter with temperatures dropping to –40°C (plus massive wind chill). Snowfall is highly variable, and some winters see very little snow. The mean annual temperature may range from +9°C in Moldova to –6°C in the Tyva Republic.

The classic Eurasian steppe is treeless (Figure 4.8). The main plants are perennial grasses and forbs with deep root systems. They can resist droughts, fire, and cold extremely well. The two most widespread grasses are sheep fescue (*Festuca ovina*) and species of feathergrass (*Stipa*). Unlike in North America, there is no tallgrass prairie in Eurasia; its closest analogue is the northernmost and the wettest type of steppe, the meadow–steppe. One square meter of meadow–steppe can support over 50 species of flowering plants! Some shrubs (e.g., wild plum) and diverse wildflowers are common, especially members of the rose, legume, and sunflower families.

The soils underneath the Eurasian steppe are the legendary "chernozems" (literally "black earths"). They were extensively studied by Vasily Dokuchaev (see Vignette 4.2) and are similar to the "mollisols" of the United States. The topsoil may exceed 1 m in depth, and is a rich black color due to a high proportion of organic matter (10–15%). Calcium carbonate accretions occur deeper in the profile. Salinization is a common problem in the drier areas, where so-called chestnut soils become dominant. The productivity of virgin chernozem is several times greater than that of the gray forest soils or podzols, allowing a bountiful harvest with minimal fertilization. Over many years of farming, however, even the best chernozems will be depleted. There is a considerable need for irrigation, especially when spring wheat or other summer crops are grown. Soil erosion due to plowing is common. Even 5% of tree cover in the form of windbreaks may dramatically reduce erosion, and many such windbreaks were planted in southern Russia, Ukraine, and Kazakhstan in the 1950s.

The typical mammals of the Eurasian steppe include steppe foxes, ferrets, wild steppe cats, saiga antelopes, field hares, ground squirrels, gerbils, jerboas, and marmots. The typical birds

FIGURE 4.8. The steppe of Eurasia is dominated by bunchgrasses with occasional shrubs, but no trees except in the floodplains. Shown here is the Barabinsk steppe, west of Novosibirsk, Russia. *Photo:* Author.

Vignette 4.2. Vasily Dokuchaev: The Founder of Soil Science

In front of the Moscow State University building overlooking the Moscow River, one can see two rows of solemn busts depicting men of great fame. All were scientists who lived and worked in the late 19th and early 20th centuries to make Russia great. One of these scientists was Vasily Vasilyevich Dokuchaev (1846–1903); see Figure 1). The world knows him as the founder of modern soil science. Although many Russian scientists of his time made important contributions, he was one of the very few who achieved truly global fame and founded an entire new branch of science. He was a geographer and an ecologist as well. His main contribution was the development of the genetic method of soil classification, which still forms the backbone of the Russian and several other systems of soil classification. In this method, a scientist must evaluate all physical and biological factors responsible for the production of each type of soil before assigning a definite classification label. Although U.S.-based scientists no longer use this method and rely on a more formally prescribed taxonomy of soil types instead, Dokuchaev's name is still mentioned first in any American soil science textbook. Many of his terms are still in common international usage (e.g., "glei," "podzol," and "chernozem").

FIGURE 1. Vasily Dokuchaev was the founder of soil science, a geographer, and an ecologist. He was one of the most famous Russian scientists of the 19th century. *Photo:* Author.

Dokuchaev believed that nature is a united, complex system, not a collection of disjointed parts. In this sense his works foreshadowed the writings of the American naturalist Aldo Leopold. Working out of St. Petersburg, he did much fieldwork in the steppes of Russia and Ukraine, trying to understand the factors that guided soil development there. He coined the word "chernozem" to describe the most productive soils on earth, found in the steppes. He followed Alexander von Humboldt in describing ordered natural zones dependent on climate, but went much further in proposing precise scientific explanations for their distribution. His laws of natural zonation explain the reasons behind the orderly succession of Northern Eurasian biomes from north to south. Although in North America several of the same zones follow a meridianal pattern, in Northern Eurasia they change strictly with latitude. Dokuchaev conducted a number of pioneering scientific experiments and published many papers. His contributions helped develop Russian intensive agriculture in the 20th century. His two most famous pupils were G. F. Morozov, the founder of Russia's forestry school, and V. I. Vernadsky, who presciently wrote about the biosphere as the world's largest ecosystem.

include demoiselle cranes, bustards, eagles, harriers, kestrels, stilts, avocets, quails, hoopoes, bee-eaters, rollers, larks, and magpies. There are also a few dozen species of reptiles, including snakes and lizards.

There are few places where virgin steppe can still be seen. As in North America, over 99% of this biome in Eurasia was plowed under in the 19th and 20th centuries. There are very few restored steppe patches. However, small preserves provide glimpses of the steppe's original vegetation. The best examples in Russia include the Galichya Gora (Lipetsk), Kursky (Kursk), and Voronezhsky and Khopersky (Voronezh) Zapovedniks, as well as the Orlovskoe Poleye (Orel), Ugra (Kaluga), and Samarskaya Luka (Samara) National Parks. In Ukraine, the most famous preserve is Askaniya Nova in the Kherson region near the Black Sea. This unique territory was established by a visionary German landowner, F. Falts-Fein, in 1886. Today it is one of a handful of virgin steppe fragments left in Eastern Europe. The early history of the preserve included acclimatization experiments with exotic fauna; ostriches, zebras, antelopes, and llamas roamed the first Ukrainian safari park. Today, the descendants of many of these animals can still be seen in large enclosures. The remainder of the Askaniya Nova steppe is home to the native fauna.

Desert

With its spacious, rainless interior, Eurasia is home to the northernmost deserts in the world. Located entirely outside the tropics, the deserts of Central Asia have all the usual desert features, including sand dunes, desert pavement, rock formations, small saline lakes and playas, and very little vegetation. However, the northern, boreal elements of their flora and fauna are unique. The main deserts in North America are found at latitudes between 25° and 35°N, whereas in Eurasia they occur between 38° and 44°N. The four main deserts of Central Asia are the Kara Kum in Turkmenistan, south of the Aral Sea; the Kyzyl Kum in Uzbekistan, southeast of the Aral Sea; the Moyynqum in Kazakhstan, east of the Aral Sea; and the Saryesik Atyrau, south of Lake Balkhash. There is also a small desert north of Makhachkala and west of the Caspian Sea in Russia, in Kalmykia (the only true desert in Europe). Altogether, the Central Asian deserts occupy 3.5 million km^2—an area as large as Saudi Arabia and Iran combined.

Deserts generally form in areas with potential evaporation exceeding precipitation by a factor of 10 or more. In temperate deserts, the average rainfall is <250 mm per year. The sandy desert is the most common type, with large dune fields of various shapes. The most famous dune form is the crescent-shaped "barkhan," with horns pointing downwind. Barkhans form in areas with little vegetation. Parabolic dunes, star dunes, and longitudinal dunes are also common. Some dunes may be 30–40 m high. Most of the Kara-Kum is sandy desert ("black sand"). East of the Caspian Sea is the gravelly Ustyurt desert. There are also stony and salty deserts in Central Asia. When soils are present, they are of the desert type ("aridisols"). In the United States, such soils are common in parts of the western Great Plains and much of the Southwest.

Plants of the deserts are "xerophytic," which means they are adapted to very dry conditions. Typically they lack leaves and have extensive but shallow root systems, capable of catching whatever moisture may be available on short notice. There are no cacti, because those are native only to the Americas. Instead, *Artemisia* forbs and small shrubs (*Atriplex, Salsola, Tamarix*, and *Anabasis*) are widespread. One genus, saxaul (*Haloxylon*), grows into a small-sized tree. Unique communities develop on saline flats that are flooded during the rain period, the so-called *takyrs* (similar to the playas of North America). Many desert plants are adapted to tolerate severe salinity. Along the seasonal watercourses, gallery forests or *tugai* develop, with poplar and willow species. Reeds develop around isolated saline desert lakes.

The fauna of the deserts can be surprisingly diverse, but elusive. Animals spend most of the day underground, avoiding heat; at night they are everywhere. Unfortunately, some of the most spectacular representatives of large desert mammals are now extinct (wild tarpan horses and tigers), while others are endangered (Asiatic wild donkeys, Przhevalsky horses, saiga antelopes, Persian gazelles) and are confined to a few preserves or zoos. The most common mammals are rodents (22 species); also common are insectivores, including long-eared hedgehogs, and carnivores, including weasels and wildcats. Birds are represented by eagles, Asian pheasants, sand grouse, pratincoles, desert jays, crested larks, and desert wheatears. Reptiles thrive in this biome, with monitor lizards, agamas, skinks, epha vipers, cobras, and others. There are some spectacular butterflies, beetles, cicada, and spiders in the deserts as well. Gerald and Lee Durrell (1986) provide some excellent descriptions of the ones they found in the Repetek preserve of Turkmenistan.

Other Biomes

Besides the main five biomes of Northern Eurasia, there are some rarer types, of which four merit mention here: mountainous ecosystems; the subtropical vegetation of the Black and Caspian Sea coasts; the unique forests of the Russian Pacific; and the azonal communities of the floodplains and marine coasts.

All mountain ranges have their own zonation of ecosystems from bottom to top. For example, in the Karachaevo-Cherkessia Republic of Russia in the northern Caucasus, the following ecosystems are found: true steppes (200–500 m above sea level); oak–hornbeam forests (500–1,300 m); beech forests (1,300–1,500 m); fir–spruce forests (1,500–1,700 m); pine forests (1,700–2,100 m); subalpine tall-grass vegetation (2,000–2,500 m); and alpine short-grass vegetation (2,500–3,200 m). Snow and glaciers extend above the highest alpine vegetation (Figure 4.9). The lower timberline is determined by moisture availability, and the upper by temperature during the vegetative season. The timberline at about 2,100–2,500 m is formed by a *krummholz* of crooked pines, beeches, birches, aspens, and other trees that grow only as tall as shrubs. In the subalpine belt, rhododendrons, tall forbs from the rose and sunflower families, and some tall grasses play an important role. In the alpine zone, graminoids (grasses, sedges, rushes) and forbs (roses, pinks, primroses, and sunflowers) predominate. The exact sequence and elevation of the vegetation belts are determined by the direction of the slope (north-facing slopes

FIGURE 4.9. Vertical zonation of the Caucasus in the vicinity of Mt. Elbrus. The ridge in front is treeless steppe; the foothills of the ridge behind are mixed forest. Subalpine meadows start in the middle of the ridge behind, with only alpine tundra extending to the summits. Small perennial snowfields are found on northern slopes. *Photo:* V. Onipchenko.

are always colder and have a lower treeline) and by local climatic and biological factors. The treeline, for example, occurs at 300 m in the polar Urals and the Khibins in the Kola Peninsula in the Arctic, but at 2,000 m in the Carpathian mountains, 2,500 m in the Caucasus, and above 3,000 m in much of Central Asia (which is considerably warmer and drier). The main species at the treeline will also differ among mountain ranges. In much of Siberia it is Siberian cedar pine shrub (*Pinus pumila*), while in the Caucasus it may be birch, beech, or Scotch pine.

Subtropical vegetation can be found at the southern tip of the Crimea Peninsula; in a narrow strip along the Black Sea coast of Russia and Georgia (see Chapter 3, Figure 3.6); and in the southeastern corner of Azerbaijan (Lenkoran) along the Caspian Sea coast. These areas all have a subtropical C-type climate, where frosts do not occur even in January. Protected by the mountains from the cold northern wind, these sheltered areas can support Mediterranean-like vegetation. Of the three areas, the Crimean Peninsula is the driest; its native communities consist predominantly of sclerophyllous scrub, but cork oaks, junipers, wild madroños, pistachio trees, and other unusual plants are well represented. Visually, it bears a striking resemblance to the vegetation of Italy and Greece, much farther south. Much of this native ecosystem has been replaced with fruit orchards, vineyards, and parks full of introduced Mediterranean trees and shrubs (e.g., cypresses, cedars of Lebanon, Italian pines, and palms). Massandra and Livadia Parks, and Nikita Botanical Garden at Gurzuf, have particularly famous arboreta. The Black Sea coast has lush vegetation forming under wetter conditions. Native plants include many evergreen shrubs or small trees (boxwood, laurel, yew, etc.). Many of these are relics of the much warmer Tertiary period, 2–65 million years ago. Lianas and epiphytes are common in the forests. Tea and tangerines can be planted and survive winters here. In Russia, the Great Caucasus Zapovednik near Sochi, including a famous box–yew grove, can be visited for the best representative look at the whole Black Sea coast ecosystem. In Georgia, a few preserves and arboreta existed in the Soviet period (e.g., the Pitsundo-Mussersky Zapovednik south of Gagry and the Sukhumi Botanical Garden); however, many of these are now in the separatist province of Abkhazia, and their status and ease of access are thus uncertain. The Lenkoran region of southeastern Azerbaijan is covered with humid subtropical forests with many Tertiary relics well represented (ironwood, chestnut oak, Hyrcanian box tree, Lenkoran acacia, and others). The Hirkan National Park protects over 150 rare and endemic plant species along with many native bird and mammal species with limited distribution.

The unique mixed and deciduous forests of the Russian Pacific combine northern elements from Siberia with southern elements from Manchuria, and have no analogues in North America or elsewhere. This is the only area of the FSU influenced by summer monsoons; 60% of all rain falls between July and September. The summers are warm (the average temperature of Vladivostok in July is +17°C), but winters are very cold (the mean January temperature is -15°C, colder than Moscow's). These forests have the greatest tree diversity in the FSU, with over 70 species. By comparison, the mixed forest in the Moscow region has at most 15 species. (Parts of New England, on the other hand, have over 70 species of trees, and the Great Smoky Mountains National Park has over 130!) Korean pine, two firs, two spruces, four lindens, and a few oak and maple species are the dominant trees in the north, along the Amur River. In the south near Vladivostok, walnuts, elms, and other southern species with Chinese affinities become more prominent. *Actinidia* is a common large vine, and Siberian ginseng and lemon-scented *Schisandra* are common in the understory. On Sakhalin and the Kurils, even bamboo can grow among the fir and spruce trees! The Amur tigers, of course, are the flagship animal species of the Far Eastern forests, numbering in the low 400s. Other interesting mammals include brown and Himalayan black bears, Far Eastern leopards, elk, wolverines, sables, lynxes, and giant shrews. Many rare bird species with limited distribution are found here (Blackiston's fish owls, Mandarin wood ducks, and blue and green magpies). There are 20 species of reptiles and amphibians here; although the state of Virginia (at a comparable location, but farther south) has 67 species of reptiles, these numbers are the highest for Russia. Turkmenistan deserts have over 40 species of reptiles.

The azonal communities of the floodplains, lakeshores, and marine coasts occur everywhere near water, regardless of the natural zone they are in. The river floodplains and lake shores have tall meadows and emergent marshes composed of a few dozen widely distributed species (e.g., cattails, reeds, bullrushes, sedges, grasses, and other wetland plants). Likewise, marine coasts have a rather uniform set of species in the tidal zone (brown and green algae, barnacles, sea urchins, sea anemones, and a few flowering plants), all adapted to saline water and fluctuating tides. In the FSU, the most diverse marine life and the highest productivity of marine life are found along the Barents and White Sea coasts and in the Pacific. The Black Sea, in contrast, is a species-poor basin, because of the extensive anaerobic hydrogen sulfide zone in its depths and a high degree of local water pollution. The Baltic Sea has an intermediate degree of productivity and is the freshest of the major sea basins of Russia.

REVIEW QUESTIONS

1. Name the main five biomes of Northern Eurasia. What are the biggest distinctions among them? Where are those found? Are there corresponding biomes in Central and Western Europe? In North America?
2. Which biomes are the most productive? Which are the least productive? In each case, what reasons can you give?
3. Which biomes have the highest and lowest biodiversity? How can you explain that in each case?
4. If you were to protect one biome in the FSU through the creation of a network of protected areas (parks), which one would you pick? Where within that biome would you place those territories? Explain your rationale.
5. Why do you think both the Caucasus and the Far Eastern coastal forests are unique?
6. What are the main factors that determine biome distribution?

EXERCISES

1. Using a physical geography atlas, analyze the distribution of typical temperature and precipitation values for each of the biomes in the FSU. Then compare the values to those commonly observed in your area. Which biome would be the closest match to the area where you live?
2. Search for and watch a documentary film on any endangered species in Russia or any other FSU republic (to find an international list of endangered species by country, go to *www.redlist.org*).

Further Reading

Aksenov, D., Dobrynin, D., Dubinin, M., Egorov, A., Isaev, A., Karpachevskiy, M., Laestadius, L., Potapov, P., Purekhovskiy, A., Turubanova, S., & Yaroshenko, A. (2002). *Atlas of Russia's intact forest landscapes*. Moscow: Global Forest Watch.

Amirkhanov, A. M. (1997). *Biodiversity conservation in Russia*. Moscow: State Committee of the Russian Federation on Environmental Protection/GEF Project Biodiversity Conservation.

Davydova, M., & Koshevoi, V. (1989). *Nature reserves in the U.S.S.R.* Moscow: Progress.

Durrell, G., & Durrell, L. (1986). *Gerald and Lee Durrell in Russia*. New York: Simon & Schuster.

Hultén, E. (1937). *Outline of the history of Arctic and boreal biota during the Quaternary period*. Stockholm: Cramer.

Knystautas, A., & Flint, V. (1987). *The natural history of the U.S.S.R.* New York: McGraw-Hill.

Onipchenko, V. G. (Ed.). (2004). *Alpine ecosystems in the northwest Caucasus*. Dordrecht, The Netherlands: Kluwer.

Tishkov, A. A. (2005). *Biosphernye funktsii prirodnyh ekosistem Rossii*. Moscow: Nauka.

Zlotin, R. (2002). Biodiversity and productivity of ecosystems. In M. Shahgedanova (Ed.), *The physical geography of Northern Eurasia* (pp. 169–190). Oxford, UK: Oxford University Press.

Websites

www.biodiversity.ru.eng—Biodiversity Conservation Center, Moscow.

www.forest.ru/eng—Global Forest Watch Russia (with links to other resources about Russian forests).

www.rbcu.ru—Birdlife International Russia. (In Russian only.)

www.taigarescue.org—Taiga Rescue Network.

www.tretyakov.ru/eng—Tretyakov Gallery.

www.wild-russia.org—Center for Russian Nature Conservation; presents protected areas of Russia.

CHAPTER 5

Environmental Degradation and Conservation

The Soviet Union was commonly perceived as one of the most polluted places on earth. A list of the major environmental disasters of the 20th century includes many that happened in the U.S.S.R.: the Chernobyl disaster in Ukraine; the less publicized Kyshtym nuclear accident near Chelyabinsk in the Urals; the Aral Sea water loss; the Semipalatinsk and Novaya Zemlya nuclear bombing fallout; and the industrial pollution of rivers, air, cities, and entire regions. One book about the late Soviet period published in the West was even entitled *Ecocide in the U.S.S.R.* (Feshbach & Friendly, 1992). It claimed that in the U.S.S.R. the water was toxic, the land was polluted, and the air was unbreathable. A much more balanced treatment was provided by Pryde (1991).

At the same time, one cannot help wondering just how much impact all these disasters really had over such a large territory. Because the region is so large, there had to be unpolluted areas of considerable size. The perception of pollution is subjective, and much of this perception depends on the spatial scale involved. For example, Moscow does have relatively polluted air. In fact, the first thing you notice upon arrival at one of its three international airports is the pervasive smell of car exhaust and cigarette smoke outside the terminal. Nevertheless, 15 km away you can be in the summer cottage country, relaxing near one of the many lakes and inhaling impeccably clean pine forest air while fishing for carp. In addition, some of the largest and cleanest streams on the planet are in the vast Siberian taiga forests. The wilderness ranges of the Altay are famous for their pristine beauty. Most of Siberia and the Russian Far East are unspoiled by humanity. And, outside Russia proper, the Central Asian deserts, steppes, and mountains are almost beyond compare, with few tourists and even fewer roads. Remember that population density in the former Soviet Union (FSU) is less than a quarter of the U.S. level and less than one-eighth of the European.

As we discuss environmental issues in the countries of Northern Eurasia, let us keep in mind that while some areas were heavily affected by pollution and the like, many remain pristine. This chapter describes both environmental degradation (air and water pollution, as well as nuclear and toxic waste issues) and biodiversity conservation.

Air Pollution

Air pollution is common everywhere in the industrialized world. The U.S.S.R. was one of the largest polluters of air on the planet, and Russia still is today. The difference is primarily in the total amounts: Whereas the U.S.S.R. was a polluting monster, releasing over 60 million metric tonnes (mmt) of pollutants per year from stationary sources, Russia today releases 25 mmt or so. The United States released 145 mmt in 2005, of which slightly less than half (or about 60 mmt) was from stationary sources. Table 5.1 provides a more detailed comparison of emissions. Russia is of course a smaller country than the U.S.S.R., so logically it would produce less pollution. Also, its industrial output dropped about 50% between 1991 and 1998. Although there has been some increase in production since 2000, Russia generally pollutes less today than it did 20 years ago. However, a major new contributor to air pollution is car exhaust. Moscow, for example, had only 500,000 automobiles in the late 1980s. Today there are about 4 million cars and trucks in the city, only about half of which comply with modern emission control standards. Russia's total carbon monoxide emissions are higher than those of the entire U.S.S.R. Although the general trend of U.S. air pollution has been steadily downward, because of the improved pollution control devices required by the Clean Air Act, Russia is actually beginning to produce more pollution now that its industry is recovering (Figure 5.1).

Pollution from industry (e.g., coal-fired electricity plants, metal smelters, and chemical factories) remains a significant concern in at least four countries of the FSU: Russia, Belarus, Ukraine, and Kazakhstan. A few hundred cities in Russia alone, such as Norilsk, Cherepovets, or Magnitogorsk, were built around a single huge enterprise. In cases like these, several hundred thousand people in each city are breathing the air polluted by the industrial monster. In the biggest cities, like Moscow or Yekaterinburg, there are dozens of smaller factories. Although some of these were shut down during the 1990s, many are still operating today, and only a handful have been upgraded enough to reduce their emissions substantially. About 40 cities are on the national watch list of the most polluted (out of about 200). Some of the most notorious ones include these:

- In European Russia, the cities of Cherepovets (a major steel factory); Ryazan, Vladimir, Saratov, and Volgograd (machine building and chemical plants); and Naberezhnye Chelny (petrochemicals and the KAMAZ truck plant).
- In the Urals, the cities of Yekaterinburg, Chelyabinsk, Magnitogorsk, Pervouralsk, Nizhniy Tagil, and Ufa (all major centers of heavy industry, such as production of weapons and/or chemicals).

TABLE 5.1. Emissions of Major Atmospheric Pollutants (in Millions of Metric Tonnes per Year)

	CO	NOx	Hydrocarbons	SO_2
U.S.S.R. (1988)[a]	14.9	4.5	8.5	17.6
United States (1985)[b]	170	26	27	23
Russia (2004)[c]	17.3	3.1	3.1	5
United States (2005)[d]	93	19	18	15

Note. The Russia (2004) data only include official data on industrial and automobile emissions; the actual totals are probably about 20% higher because of underreporting. CO, carbon monoxide; NOx, nitrous oxide; SO_2, sulfur dioxide.

[a]Data from Pryde (1991).

[b]Data from "Progress in Reducing National Air Pollutant Emissions 1970–2015," by the Foundation for Clean Air Progress (*www.cleanairprogress.org*), and from the U.S. Environmental Protection Agency.

[c]Data from "Annual Report on the Status of the Russian Environment," by the Ministry of Natural Resources of Russia, 2004 (*www.mnr.gov.ru*).

[d]Data from the U.S. Environmental Protection Agency.

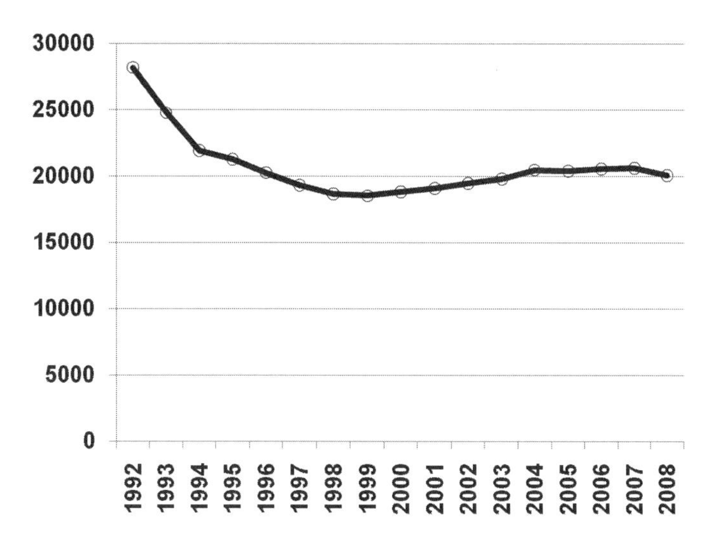

FIGURE 5.1. Total industrial air pollution from stationary sources in Russia, in thousands of metric tonnes per year. Data from Federal Service of State Statistics, Russian Federation (*www.gks.ru*).

- In Siberia and the Far East, the cities of Norilsk (nickel and copper processing); Angarsk and Bratsk (aluminum smelters); Novokuznetsk and Kemerovo (coal processing, chemical industries); and Omsk, Chita, Blagoveshchensk, Yuzhno-Sakhalinsk, and Magadan (all centers of heavy industry, including military factories).

European Russia suffers from transboundary air pollution as well, from factories in Poland, Ukraine, Belarus, Romania, Sweden, and even Germany. All of these countries have a lot of heavy industry and many coal-fired plants, and all are located west (downwind) of Russia. Similarly, industry in Moscow pollutes the Volga region, which in turn pollutes the Urals, and finally the Urals pollute western Siberia. Some of the air pollution from eastern Kazakhstan reaches the Russian Altay. It should also be stressed that these regions are not merely experiencing heavy levels of conventional pollution; they are dealing with increasing levels of toxic pollutants, such as benzene, aniline, formaldehyde, hydrochloric acid, hydrogen sulfide, lead, methanethiol, and the like. Just a small amount of these in the air will make people seriously sick. Chronic lung diseases are very widespread in Russia, although these are also often due to a high rate of smoking (over 60% among adults vs. 18% in the United States), not to industrial pollution. At the same time, some cities where factories either closed or were reprofiled in recent years now have much cleaner air.

As stated above, car exhaust is a major problem in all large cities. Moscow's traffic jams are now worse than those in most U.S. cities, including Seattle, Minneapolis, San Francisco, and even New York City, if measured by the amount of time spent sitting in traffic. Commutes of 2–3 hours across Moscow are no longer unusual, with the average one-way commute being about 1 hour. This is a dramatic increase from the Soviet period, when it would have taken merely 40 minutes. To make matters worse, only about half of the car fleet is equipped with catalytic converters. No Soviet/Russian car models had pollution control devices until just a few years ago, and many of the older, polluting Ladas and Volgas are still running. Although there has been an upsurge in imported models and in the assembly of Western-quality vehicles inside Russia in recent years (see Chapter 18), the overall increase in car ownership has more than offset any reduction in pollution caused by better controls on cars or by cleaner central power production (see Figure 5.2). Between 2000 and 2005, an average big city in Russia saw a 30% increase in air pollutants. In 2007, Russia as a whole had 195 passenger cars per 1,000 people, and Moscow had 261. The cor-

FIGURE 5.2. The smokestacks of the Yuzhnaya power station in Moscow, as seen from the beltway. Twenty such plants surround the city and provide both electricity and hot water to millions of customers. Increasing car traffic more than offsets any gains produced by cleaner central power production, however. *Photo:* Author.

responding number in the United States was 453, or 783 if light trucks and SUVs were included. In the late Soviet period, Russia had only 50 cars per 1,000 people.

Russia reluctantly ratified the Kyoto Protocol for greenhouse gas reduction in 2004, after deliberating for 6 years. Russia emitted about 1 billion tons less of greenhouse gases in 1990 than in 2004, and thus it was in a position to benefit from the lucrative trade in emissions permits. However, as Russia continues to expand its economy, it is likely that by 2012 it will cease to be a net seller of credits and will have to start buying them instead.

Because Russia's economy is less energy-intensive than the U.S. economy, its per capita carbon dioxide production is only moderately high. It was in the 16th place worldwide in 2005, with the United States being in the 5th spot. Qatar was at the top of the list, while China was only the 80th. However, in total carbon dioxide emissions, Russia trails only the United States and China and is ahead of India and Japan. If Russia develops more postindustrial, high-tech industries, its emissions are likely to fall in the future. However, the presence of large gas, coal, and oil reserves precludes serious changes in the interim period.

Water Pollution

Clean, fresh water is in limited supply on our planet and is likely to become the top environmental concern of this century (Gleick, 2009). Five of Russia's rivers are in the top 25 worldwide by water volume (the Yenisei, Lena, Ob, Amur, and Volga, in descending order). Of these five, the Yenisei carries about as much water as the Mississippi (without the Missouri); the Volga carries more water than the Yukon or the Indus, and about twice as much as the Nile. Although the Volga is heavily polluted, the Siberian rivers are relatively pollution-free, and the Lena and the Amur remain dam-free. In addition, Lake Baikal contains approximately 20% of the liquid freshwater on our planet, as much as all five North American Great Lakes, and is relatively unpolluted. At the same time, some smaller lakes and rivers in the European part of the FSU and the Urals are notoriously polluted. Some of the greatest environmental catastrophes involving water happened in the FSU (the Techa River nuclear waste dumping in the Urals in the 1960s, and the Aral Sea destruction in the 1970s).

What is happening with water in Russia today? As in the rest of the developed world, much of it is diverted for the cooling of coal-fired power plants, as well as for other industrial purposes (59% vs. 53% in the United States), irrigation (13% vs. 34%), and household consumption (21% vs. 12%). The Soviet factories were notoriously inefficient water users. Note, however, that less water is used for irrigation (in both relative and absolute numbers) or for household consumption in Russia than in the United States. Why? First, many of Russia's cultivated crops have traditionally been grown without much irrigation, except for those in southern Ukraine and Central Asia. The Soviet Union developed relatively few grand irrigation schemes (the Kara Kum canal in Turkmenistan was an exception). In contrast, the farmers of central California and much of the American West could not possibly grow crops without irrigation. Second, until very recently few Russians owned homes that had lawns (or cars that required washing). Lawn sprinklers are the leading consumers of water in U.S. households, but not yet in Russia.

The most polluted rivers and streams include those of the Kola Peninsula (with copper, nickel, and phosphate mining nearby); the Northern Dvina River (with paper and pulp industry in its basin); the Volga (with many industries nearby, especially machinery building, chemicals, and petrochemicals); the Don in the south (with much agricultural runoff); and the Ob–Irtysh system (with pollution from the Urals, Krasnoyarsk, and Novosibirsk, as well as from the petroleum and gas industries in the midbasin) (Figure 5.3). The Angara River receives major pollution from Bratsk. The Lena is relatively clean, but the Amur has been seriously polluted in recent years by both China and Russia. Typical types of water pollutants include (but are not limited to) petrochemicals, lead and other heavy metals, complex organics, phosphates, and nitrates. Fecal matter in river water is common, as well as many parasitic diseases.

Lake Baikal remains mildly polluted, despite all the media hype, but this is because it has rather limited development in its basin (primarily the paper and pulp mill in Baikalsk in the extreme south); the polluted area of the lake is

FIGURE 5.3. The Obskoe reservoir on the Ob River in Novosibirsk attracts swimmers during the short summer. Despite heavy industry, the Ob is only moderately polluted here, given its enormous size and the availability of pollution control devices at most factories. *Photo:* P. Safonov.

about 20 km^2. The lake itself is so huge that this pollution fortunately has little overall impact, which is not to say that it is in any way desirable. A plan to locate a new oil pipeline to China north of the lake was met with tremendous public opposition nationwide, and was modified by then-President Putin to be routed outside the lake basin and over 100 km to the north.

The water pollution in European Russia is spotty. It is possible, for example, to swim safely in most small rivers and lakes even close to Moscow, as long as there is no major chemical plant upstream. Compared to North America, few feedlots exist in Russia, and pesticide/herbicide applications to the fields have been drastically reduced in recent years through economic restructuring of the agricultural sector. At the same time, one cannot guarantee that someone is not washing an SUV upstream from where you are swimming, because local enforcement of water pollution laws is lax and the culture is permissive. In addition, someone may dump broken glass, rubber, plastics, or household chemicals into the river at any time. In any event, it is not advisable to drink from any open water source without filtering the water first, even in a wilderness. Bottled water is widely available throughout Russia today. In most municipalities, tap water is purified, although not necessarily to the average U.S. standards. Every spring, Moscow faucets run with brownish-tinged water smelling faintly of manure; it enters the Moscow water supply system from agricultural fields upstream. Since most Russians routinely drink only boiled tea, bottled water, juices, or alcoholic beverages, it does not hurt them much. (Visitors should not consume tap water, if possible.)

Much has been written about the destruction of the Aral Sea (see, e.g., Micklin, 2006), so it is only discussed briefly here. The famous desert lake of Central Asia lost much of its water because the two main rivers feeding it, the Amu Darya and the Syr Darya, were diverted for cotton irrigation in Uzbekistan and Turkmenistan in the late 1960s. The lake straddles two countries, Uzbekistan in the south and Kazakhstan in the north; it is actually no longer a lake at all, but a combination of two unconnected evaporation ponds (see Figure 5.4 for how it looks from space). The situation remains pretty grim. The

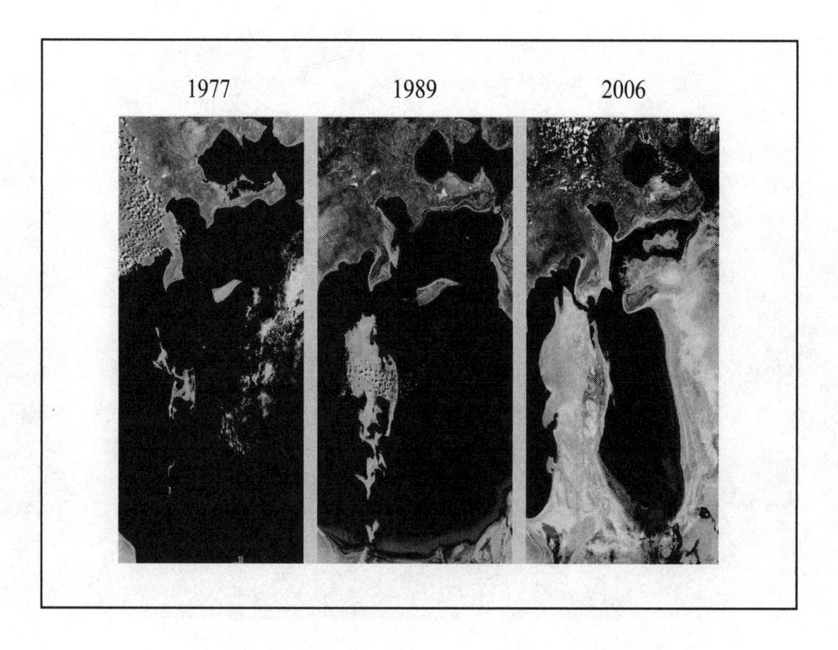

FIGURE 5.4. The Aral Sea on June 4, 1977, September 17, 1989, and May 27, 2006. Landsat imagery courtesy of NASA Goddard Space Flight Center and U.S. Geological Survey (public domain).

steadily receding lake, formerly with a surface area of 67,500 km^2 (1960), had split into two parts and shrunk to 17,380 km^2 by 2006. Only about 26% of the surface area and 10% of its former volume remain. Some water is allowed to reach the smaller northern fragment in Kazakhstan from the Syr Darya. However, the larger southern fragment does not receive any water and is likely to disappear completely by 2015. The lake's salinity levels have risen from 1% to over 8% (for reference, the salinity of normal ocean water is 3.5%, that of the Great Salt Lake in Utah is between 15% and 28%, and that of the Dead Sea in Israel is about 30%).

More than 30 fish and 200 invertebrate species have completely disappeared from the Aral Sea, including three endemic sturgeon species and one salmon, even though some of these may still remain in the river deltas and in the small northern fragment. Of particular concern are the health effects of salt on the human population in the basin. The desert winds whip up salt storms and blow them into towns. Since the mid-1970s, satellite images have revealed major salt–dust plumes extending from 200 to more than 500 km downwind; these drop dust and salt over a considerable area adjacent to the sea in Uzbekistan, Kazakhstan, and (to a lesser degree) Turkmenistan. The incidence of lung disease in Karakalpakistan is three times the normal rate. Tens of thousands of fishing jobs were lost because fish could no longer be caught. For a few years now, the cannery at Aralsk has been surviving on fish brought in by train from the Far East.

A proposal currently exists to replenish the Aral, as a revival of a water transfer scheme invented in the late 1960s. The plan calls for diverting about 10% of the Irtysh River south, in an aqueduct. Although this may seem far-fetched, it certainly is not without precedents. The Central Arizona Project of the 1970s in the United States, and the current south–north (Chang Jiang to Huang He) water transfer project in China, have had technological challenges and financial costs similar to those proposed for the Aral project. The price tag is expected to exceed $10 billion, but in Putin and Medvedev's Russia it may still happen, despite the vocal protests that the environmental community is bound to make.

Nuclear and Toxic Waste

If there is one environmental topic that concerns all those visiting or living in Russia, it is certainly the topic of nuclear and toxic waste. The U.S.S.R. was the second country in the world after the United States to develop an atomic bomb, in 1949. It was also the second to develop the considerably more powerful thermonuclear (hydrogen) bomb, in 1955. Eventually the U.S.S.R. developed and tested the largest thermonuclear bomb in the world, a "tsar" bomb code-named "Ivan" (about 50 megatons, although a 100-megaton bomb was initially proposed). "Ivan" was blown up over Novaya Zemlya on October 30, 1961. The bomb exploded at about 4 km above the surface, forming a fireball about 8 km in diameter. It could be seen and heard from a distance of 1,000 km. The mushroom cloud reached 64 km into the atmosphere. An eyewitness told my father, a physicist, that the ocean would open up to the bottom as a result of such a blast. Many smaller bombs were tested in Semey, Kazakhstan. Until the late 1960s, the Soviet Union and the United States continued testing these powerful weapons in the earth's atmosphere. Fortunately, both nations signed the partial testing ban treaty in 1963, which stopped any future atmospheric tests (although France stubbornly carried on nuclear explosions over its Pacific atolls for over two decades afterward).

Besides building nuclear weapons, the U.S.S.R. was also at the forefront of peaceful nuclear research. The nuclear power station in Obninsk, Kaluga Oblast, started operating in 1954. It was the first plant in the world to generate electricity by using nuclear power. Soviet engineers also equipped military submarines and civilian icebreakers with nuclear reactors, giving them the power necessary to reach the North Pole. Initially, nuclear bombs were thought to be good for major earth-moving projects like diverting rivers. Luckily, this civilian use of nuclear weapons was never fully realized, although a number of tests were in fact conducted. At the end of the Soviet period, the U.S.S.R. boasted over 40 reactors at 15 sites (today Russia has 31 reactors at 10 operating plants), not counting a few dozen small research reactors at scientific institutes. By comparison, the United States has slightly over 100 commercial reactors, Japan has 63, and France has 59. The total energy production from nuclear power in the United States is 97,000 megawatts (MW), as compared to only 23,000 MW in Russia. This number does not include the Soviet-built reactors in Ukraine, Armenia, or Lithuania that continue to produce electricity.

Nuclear pollution may result from the following:

- Uranium enrichment, and production of plutonium and other fissile materials.
- Atmospheric and underground nuclear testing.
- Nuclear accidents at power plants (such as Chernobyl).
- Nuclear fuel transportation and storage.
- Nuclear waste storage, either at power plants, underground, or at sea.

Concerns exist about all of these. The most infamous nuclear accident in history was, of course, the explosion of Chernobyl reactor #4 in the town of Pripyat, Ukraine, in 1986. We still do not know what exactly happened there. Although the official version is that some hydrogen gas was released from water steam and exploded during the emergency shutdown procedure in an experiment that went wrong, another explanation suggests that a low-power nuclear explosion actually took place instead; other theories exist as well. It is pretty clear, however, that both the reactor's construction flaws and the faulty experimental design were to blame for the blast. (See Chapter 8 for a more detailed discussion.) What is also undeniable is that the total amount of radioactive fallout was immense—as much as 14×10^{18} Bq, comparable to the fallout expected from a 1-megaton thermonuclear bomb. (The becquerel, or Bq, is a very small radioactivity unit equaling 1 fission per second.) About 200,000 km^2 of land, including dozens of villages and prime farmland, were seriously contaminated with long-lasting nuclides (especially ^{137}Cs and ^{90}Sr, both with half-lives of about 30 years). Sixty percent of the radiation fell on Belarus, and about 20% each on Ukraine and on Bryansk Oblast in Russia. Today, people still should not spend any significant amount of time in the 30-km security zone around the reactor. Many areas to the north near Grodno, Belarus, and Bryansk, Russia, 100–300 km away, have been seriously affected. About 600,000 "liquidators" (persons responsible for dealing with the various consequences of the explosion) received high doses of radiation, with an additional 300,000 residents affected in the vicinity of the station.

However, many less-publicized nuclear accidents happened earlier. For instance, a number of accidents occurred at the Mayak facility in Kyshtym, Chelyabinsk Oblast (a plutonium production, storage, and reprocessing facility in the Urals), as well as several others throughout the FSU (Medvedev, 1979). Nuclear pollution is unevenly concentrated in the FSU, and much of the information about former accidents is still classified. However, it is certain that the highest levels of such pollution are found in and around Chernobyl (northern Ukraine, southeastern Belarus, and southwestern Russia); in the Novaya Zemlya islands and Semey, Kazakhstan; and at the production facilities in Sarov, Kyshtym, and a few cities near Krasnoyarsk. Furthermore, there are several submarine staging areas where offshore dumping of nuclear waste took place in the Far East and off the Kola Peninsula. Beyond these areas, there are a smattering of sites polluted by radiation—for example, in European Russia in Ivanovo and Perm Oblasts close to Moscow, as well as in the Komi Republic, where small underground tests were conducted in the 1960s and 1970s. Generally, however, the level of background radiation in the vast majority of places in Russia is no different than in the United States and presents no danger to a visitor.

A major international concern of the 21st century is the possibility that organized terrorist groups may smuggle nuclear materials across national borders. Although no major incidents have been reported at the time of this writing, several potential target sites exist in Russia and Ukraine today—sites where a person with proper connections could conceivably obtain at least some radioactive material for a "dirty bomb," if not for a real nuclear weapon.

Another concern is toxic waste, particularly industrial and chemical waste similar to that found at the U.S. Superfund sites. As in the United States, much of this waste is a by-product of the Cold War. Unlike in the United States, information on the actual location of such sites in Russia or other post-Soviet states is not readily available. There is no online EnviroMapper for the FSU, at least not yet. These sites number in the hundreds, if not in the thousands—and they are difficult to find. Only a few cities can be identified that were known to produce highly toxic materials for the Soviet weapons program (see Chapter 18). The ironically named Vozrozhdeniya (Restoration) Island in the middle of the Aral Sea is now a peninsula connected to the mainland. It is known to contain caches of biological, and possibly chemical, weapons. Another notoriously polluted chemical dump is located near Dzerzhinsk in Nizhny Novgorod Oblast. This area has a much higher rate of birth defects than Russia's average.

Biodiversity Conservation

Despite its large size, Russia's biological diversity as measured by the number of species is relatively limited. This has to do primarily with climate. Like Canada, the majority of Russia is suitable only for tundra or taiga species, although there are also some deciduous forest, steppe, and desert species. It does not have any rainforests. Its zone of subtropical vegetation along the Black Sea coast is diverse, but tiny. The highest diversity of plants, birds, and mammals is found in the south, especially in the Caucasus Mountains, the Altay in Siberia, and the Far East along the Pacific Coast. The Central Asian republics have a high diversity of desert and mountain species.

Russian conservation efforts have a long history, dating back to the late 19th century, when a number of game preserves and zoological gardens were created (Weiner, 1988). Some of the finest Russian zoologists, botanists, geographers, and ecologists were at the forefront of conservation efforts in the early 20th century (Boreiko, 2001). Boreiko lists over 150 names, including biogeographers Vasily Alekhin, Vladimir Sukachev, Andrei Veniamin and Semenov-Tian-Shansky; zoologists Georgy Kozhevnikov, Sergei Buturlin, and Vladimir Stanchinsky; forester Georgy Morozov; and many others. The main difference between the conservation approaches of these people and of famous American conservationists of the same time period—people like John Muir, Robert Marshall, Sigurd Olson, and Gifford Pinchot—was the Russians' emphasis on the ecological integrity of landscapes, rather than on aesthetic preservation or utilitarian conservation. The closest American in spirit to the Russians was Aldo Leopold, who understood the need for protecting representative large and wild ecosystems as early as 1924.

Another important component of biodiversity conservation has been education. Many schoolchildren in the Soviet period were members of clubs for young naturalists, learning the basics of nature conservation in after-school programs throughout the country. Such clubs and other efforts to educate youth about environmental issues remain popular today (Vignette 5.1). In addition, Russia now has many local, regional, and national environmental groups, such as the Socio-Ecological Union, the Biodiversity Conservation Center, Greenpeace Russia, and World Wildlife Fund Russia. Some are domestic groups stemming from the student movement of nature conservation started in the 1960s; others have recently arrived from the West; but all use local staff and resources. However, the overall level of environmental awareness in Russia continues to be lower than in Western countries, especially among older people and state bureaucrats.

Protection of Species

Russia does not have an Endangered Species Act like the United States. Instead, it relies on the Red Data Book, which lists threatened and endangered species in a colorful volume with detailed descriptions, range maps, and pictures. In theory, the book should assist land managers in making the appropriate decisions about conserving these species. However, as respectable as the book is, it is not legally enforced as the Endangered Species Act is in the United States. Few people, if any, are ever fined or imprisoned by the authorities for taking one of the listed species from the wild. The book does convey im-

Vignette 5.1. Saving Nature . . . by Teaching Kids

My trip to Siberia in the summer of 2006 started with a long bus ride from the international Tolmachevo Airport in Novosibirsk. After about 2 hours of bumpy road on the national Trans-Siberian Highway (which in places resembles a local access road somewhere in Montana), I was relieved to get off on a curve somewhere in Bolotniki district and to see a four-wheel drive UAZ waiting for me. A friend picked me up to get through 15 km of barely passable jeep trails to a forestry camp on the banks of the Ob River, near the village of Novobibeevo. This innovative summer project, sponsored by the SibEcocenter of Novosibirsk, attracted students and teachers from seven villages in the vast Novosibirsk Oblast. The region around is heavily forested, mostly Scotch pine and birch planted after World War II. Much of the original forest was cut down during the war, but today the 60-year-old timber stands are impressive in their unbroken natural beauty. However, logging has increased recently because of growing timber demands in China.

The camp we were heading to was held in the forest on a scenic tributary of the Ob River. Supported by the World Resources Institute forestry initiative and some local funding, the students, their schoolteachers, and college-age instructors from SibEcocenter spent 7 days living together in tents and sharing meals, sports, swimming, music, and dancing, in addition to being exposed to a vast array of forestry-related disciplines (Figure 1). These disciplines included plant ecology, geography, geographic information systems (GIS), field orientation with the global positioning system (GPS), cardiopulmonary resuscitation, wilderness survival, local customs and folklore, timber cruising and management, and even (with my humble input) U.S. conservation policies. Participants were schoolchildren from 8th to 10th grade. Some students came from the local village, while others came from 100–200 km away. During the school year, the students would keep in touch with each other by mail and phone (and, on two occasions, personal meetings at the follow-up winter camps in Novosibirsk). The project attracted regional TV attention and a visit from the head of the local government, who pledged support for organizing removal of the litter collected by students during the program. Most of this litter had been left by careless hunters and tourists, and now these schoolchildren had shown the adults what it means to take care of the forest.

FIGURE 1. The Novobibeevo forestry camp for middle and high school village children in Siberia, organized by SibEcocenter, Novosibirsk, in 2006. The campers spent 1 week of training in forestry, ecology, and sustainability. They collected plastic trash from nearby woods and practiced minimal-impact camping with leave-no-trace techniques. *Photo:* Author.

TABLE 5.2. Selected Examples of Endangered Wildlife Species from the Red Data Book of Russia

Species	How many remain?	Where in Russia?	Main threat(s)?
Mammals			
Baltic nerpa seal	A few thousand	Baltic Sea coast	Poaching, sea pollution
Russian desman	50,000	European rivers	Habitat alteration, pollution
Dahurian hedgehog	Unknown	Steppe, Far East	Habitat alteration
Snow leopard	A few hundred	Altay, Sayans	Poaching
Birds			
Steller's sea eagle	2,000–3,000	Pacific, Kamchatka	Hunting, tourism
Black stork	Unknown	Throughout taiga	Deforestation, natural rarity
Blackiston's fish owl	About 500	Southern part of Far East	Loss of old-growth forests
Short-toed creeper	unknown	Caucasus	Natural rarity

portant information to decision makers and the public, and it helps them assess overall strategies for species' recovery. Since the mid-1990s, some laws have been passed in Russia that attempt to manage rare species and protected areas in more explicit manner. Among the most protected and rare species (see Table 5.2) are several that exist only in Russia. Such endemics include the Russian desman (which resembles an oversized water shrew), the red-breasted goose, the Siberian crane, and the Blackiston's fish owl. Some endangered species also live in other FSU republics, such as the snow leopard in Kyrgyzstan and the wild donkey in Turkmenistan.

Protection of Natural Areas

Russia was one of the first countries in the world to start establishing scientific nature preserves, called *zapovedniks*, as early as 1916. Compared to the U.S. National Parks, they are primarily wilderness areas without roads, allowing very limited human recreation. Numbering about 100, they contain representative samples of naturally functioning ecosystems. Some are very large (such as the Great Arctic Preserve in Taymyr, with over 4 million ha), while others are small (such as Prioksko-Terrasny in Moscow Oblast, with fewer than 5,000 ha). Most now have limited ecotourism programs and have established scientific monitoring stations. Some are also listed as internationally recognized Biosphere Preserves and/or World Heritage Sites (see *www.wild-russia.org* for a complete list). The closest zapovednik to Moscow is Prioksko-Terrasny, about a 2-hour drive south of the city. As described in Chapter 4, it houses a thriving population of European wood bison and many other typical deciduous forest species.

Since the late 1980s, Russia has also created about 30 national parks; these are usually less scenic than their U.S. counterparts, but are nevertheless popular. Unlike the zapovedniks, they primarily emphasize nature tourism, and resemble U.S. state parks more than they do the national parks like Yellowstone or Yosemite—primarily because they are smaller and less well known than the zapovedniks. One of these parks, Losiny Ostrov, is partially inside the city of Moscow. Another fine example is Ugra National Park in Kaluga Oblast, about 4 hours' drive southwest of Moscow. Since the 1990s, the annual March for Parks program has attracted thousands of local residents, especially schoolchildren, in spring rallies around individual parks in every region of the country and in some other FSU republics.

In addition to its zapovedniks and national parks, Russia has *zakazniks* (wildlife refuges), small natural monuments, and a variety of both regional nature parks and historical–natural parks (Colwell et al., 1997). All of these protect unique natural and/or cultural landscapes, but they are typically poorly staffed. However, they do provide another important form of protection, because development in and around such areas is quite limited by law.

Other FSU countries have similar systems of zapovedniks and/or national parks. Some of

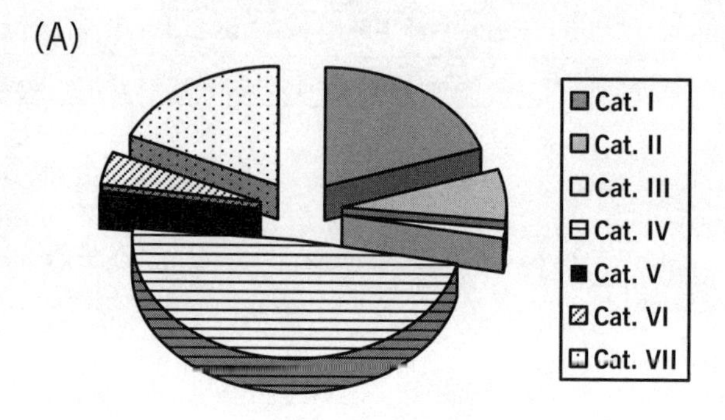

Total area protected = 1,816,735 km^2 (8.22% of all land area)

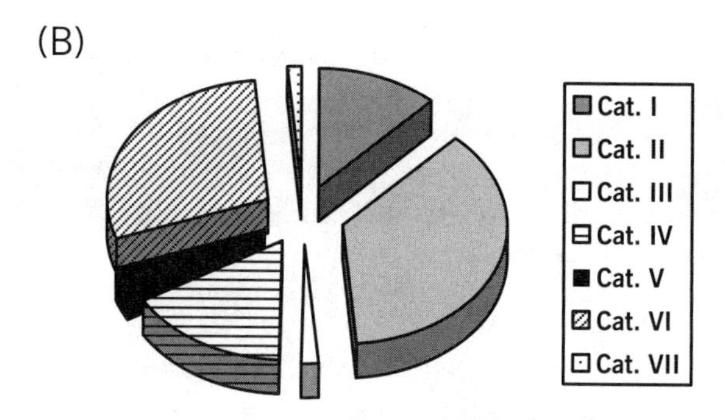

Total area protected = 4,552,905 km^2 (20.79% of all land area)

FIGURE 5.5. International Union for the Conservation of Nature categories of federally protected natural areas in (a) Northern Eurasia and (b) North America: I—strictly defined nature reserves (e.g., zapovedniks) in a and wilderness areas in b; II—national parks; III—natural monuments; IV—habitat and species management areas (e.g., zakazniks in a and wildlife refuges in b); V—protected landscapes; VI—managed resource protected areas (e.g., national forests in b); VII—all other areas. North America has more protected land, mainly because of its national forests and a huge preserve in Greenland. Data from the United Nations List of Protected Areas (2003).

these countries tragically lost many of their former protected natural areas because of political chaos and financial collapse during their early years of independence in the 1990s. Of particular concern is the situation in Georgia, where the government has lost control over parts of its own territory, and in Turkmenistan, where a closed-off autocratic regime makes independent environmental monitoring impossible. The charts in Figure 5.5 compare the status of protected areas in Northern Eurasia (the FSU, except the Baltic states) and North America (the United States and Canada), based on United Nations data.

As a final sobering reminder of the importance of conservation in this region, Figure 5.6 highlights some of the areas with the highest amount of environmental degradation in Northern Eurasia. Note that the distribution is not uniform and is generally correlated with areas with high population density and industrial development.

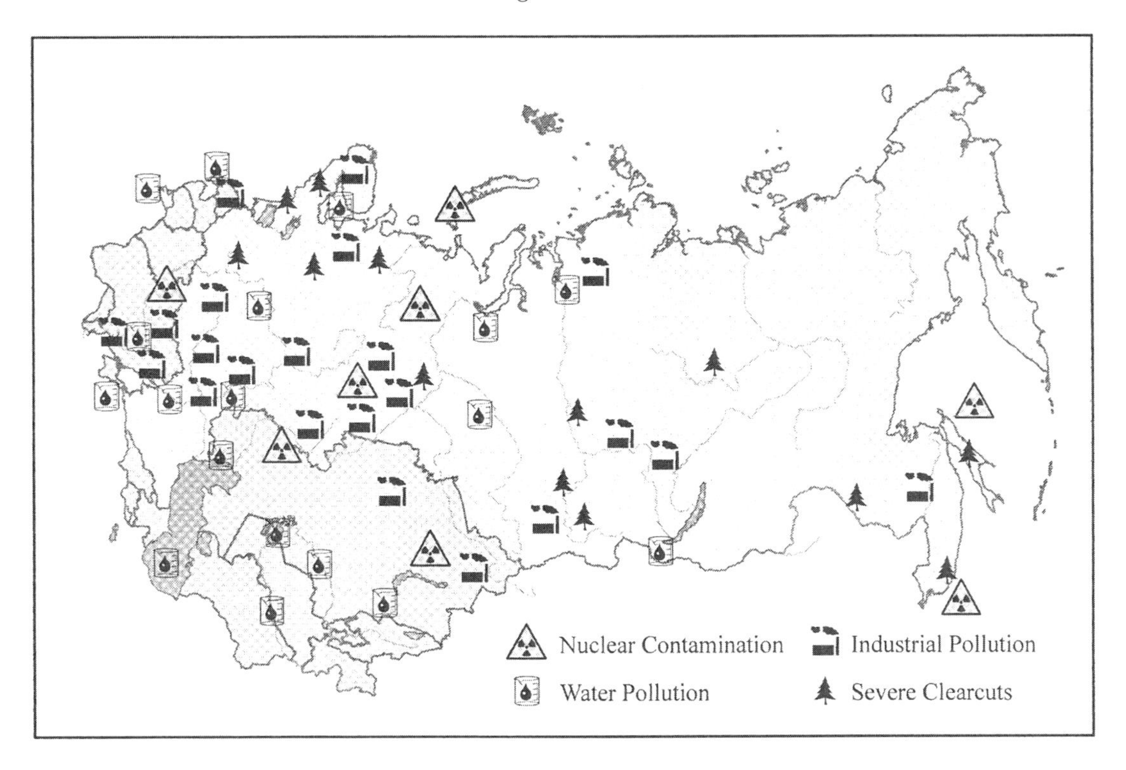

FIGURE 5.6. Environmental pollution in Northern Eurasia. Map: J. Torguson.

REVIEW QUESTIONS

1. What seems to be the environmental issue of greatest importance in Russia today? Why?
2. How does the air pollution situation differ between Russia and the United States? Why?
3. If you choose to go on a vacation in Russia, where would be good places to go to avoid exposure to high levels of air or water pollution?
4. What lines of evidence support the idea that Russia is an environmental dump? A green paradise?
5. Compare the pie charts in Figure 5.5. Which region has more protected land by total acreage? What are the differences in types of protected areas between the two regions? What does this suggest about conservation priorities and policies in each region?

EXERCISES

1. Find online data on air pollution in some cities in China, Mexico, or Brazil today. Compare them with data available for any Russian cities. Which ones are worse?
2. Study the series of images of the Aral Sea in Figure 5.4. During what time period did the lake surface decline the most? What was, or were, the most likely reason(s)? Based on current observations, how soon would you estimate that the lake will completely disappear?
3. Choose one zapovednik (see *www.wild-russia.org* for a complete list). Create a 5-minute slideshow highlighting the preserve to a prospective ecotourist from a country in the West. Highlight opportunities for viewing wildlife and landscapes. Also, describe how this visitor would travel to and from the particular zapovednik. Try to find any existing commercial tours in this part of Russia that include the preserve. Present your slideshow to the class.
4. Prepare a report about one endangered or threatened species in the FSU. Consult *www.redlist.org* or one of the Websites listed below, or conduct additional online searches. Explain what is currently being done to protect this species, and what, in your opinion, should be done. Share your findings as part of an in-class activity.

Further Reading

Boreiko, V. (2001). *Slovar deyatelej ohrany prirody*. Kiev: Kiev Ecologo-Cultural Center.

Chebakova, I. V. (1997). *National parks of Russia: A guidebook*. Moscow: Biodiversity Conservation Center of Russia.

Colwell, M. A., Dubynin, A. V., Koroliuk, A. Y., & Sobolev, N. A. (1997). Russian nature reserves and conservation of biological diversity. *Natural Areas Journal, 17*(1), 56–68.

Feshbach, M., & Friendly, A., Jr. (1992). *Ecocide in the U.S.S.R.* New York: Basic Books.

Gleick, P. H. (Ed.). (2009). *The world's water, 2008–2009 report*. Washington, DC: Island Press.

Global Environmental Facility Project. (1997). *Biodiversity conservation in Russia*. Moscow: Author.

Medvedev, Z. A. (1979). *Nuclear disaster in the Urals*. New York: Vintage Books.

Micklin, P. (2006). The Aral Sea crisis and its future: An assessment in 2006. *Eurasian Geography and Economics, 47*(5), 546–567.

Mould, R. F. (2000). *Chernobyl record: The definitive history of the Chernobyl catastrophe*. London: Taylor & Francis.

Pryde, P. R. (1991). *Environmental management in the Soviet Union*. Cambridge, UK: Cambridge University Press.

Weiner, D. R. (1988). *Models of nature: Ecology, conservation, and cultural revolution in Soviet Russia*. Pittsburgh, PA: University of Pittsburgh Press.

Yanitsky, O. (1993). *Russian environmentalism: Leading figures, facts, opinions*. Moscow: Mezhdunarodnye Otnosheniya.

Websites

www.biodiversity.ru/eng—Biodiversity Conservation Center of Russia (Moscow).

www.greenpeace.org/russia/en—Greenpeace Russia.

www.mnr.gov.ru—Ministry of Natural Resources of Russia. (In Russian only.)

www.wild-russia.org—Database on federally protected natural areas of Russia.

www.rosenergoatom.ru/eng—Information on all nuclear plants in Russia.

www.wwf.ru/eng—World Wildlife Fund Russia.

PART II

HISTORY AND POLITICS

CHAPTER 6

Formation of the Russian State

The Russian state has a long history, encompassing over 11 centuries (see Table 6.1 for a brief timeline). Archeological work in Ukraine points to the existence of settlements north of the Black Sea in the Paleolithic period, placing human presence in the Dnieper basin well over 10,000 years ago. The ancient Slavic tribes that gave rise to the Russian, Ukrainian, and Belarusian people originated in the Dnieper basin shortly before the time of Christ, probably by the 4th century B.C. The Greeks and Romans came in touch with these people as their cultural spheres of influence intersected north of the Black Sea more than 2,000 years ago. However, little is known about them prior to the late 9th century A.D. The *Primary Chronicle* (also called the *Tale*

TABLE 6.1. Brief Timeline of Russia's History

Dates (A.D.)	Main events
880	Oleg establishes Kiev as the capital of Kievan Rus. Wars with the Pechenegs, Khazars, and other nomadic invaders from south and east.
988	Prince Vladimir of Kiev converts to Orthodox Christianity and baptizes the people of Rus.
Early 1000s	Yaroslav the Wise compiles the first legal code. The St. Sophia Cathedral is built in Kiev. The Kiev Caves Monastery is established by Sts. Anthony and Theodosius near Kiev.
1147	Moscow is founded in the Vladimir-Suzdal region by Yuri Dolgoruky.
1219–1240	Mongolian conquest of Rus (Genghis Khan, Batu, etc.). The period of the Tatar–Mongol Yoke begins.
1242	Prince Alexander Nevsky defeats the Teutonic knights on Lake Chudskoe.
1288–1340	Ivan I ("Kalita") strengthens the principality of Moscow.
1380	Prince Dmitry Donskoy of Moscow scores a victory over the Tatars at Kulikovo.
1392	St. Sergius of Radonezh (founder of Holy Trinity Monastery) dies. Andrei Rublev paints his famous religious icons at about this time. Flowering of Russian Orthodox spirituality.
1480	Ivan III of Moscow calls himself the first tsar. Building of the white-stone Kremlin in Moscow. The Tatar–Mongol Yoke is finally broken; Novgorod, Vyatka, Pskov, and Tver are subordinated to Moscow.

(cont.)

TABLE 6.1. *(cont.)*

Dates (A.D.)	Main events
1533–1584	Reign of Ivan IV ("the Terrible"). Kazan and Astrakhan are conquered. Yermak crosses the Urals into western Siberia (Tobolsk is founded in 1587).
1598–1613	Death of Tsar Feodor ends the Rurik dynasty. Reign of Boris Godunov. Two "false Dimitrys" on the throne. "Time of Troubles" begins with Polish invasion, ends with crowning of Mikhail Romanov (1613).
1645–1680s	Reign of Alexei Mikhailovich Romanov. Rapid eastern expansion into Siberia (many cities are founded—Yeniseisk in 1619, Yakutsk in 1632, Anadyr 1649, Irkutsk 1652).
1666	Patriarch Nikon's church reforms lead to the Great Schism between "Old Believers" and the newly reformed church.
1712	St. Petersburg becomes the new capital. Peter I (the Great) brings in Western customs, creates first Russian Navy, greatly diminishes church power by abolishing patriarchate. Two expeditions to discover Alaska (1728, 1741).
1762–1796	Reign of Catherine the Great (originally from Germany). "Russian Baroque" period. Westward expansion into Lithuania, Belarus, and the Crimea. Annexation of Poland. Buildings in Italian style are constructed in St. Petersburg and Moscow.
1812–1814	War with Napoleon during the reign of Alexander I. Battle of Borodino. Moscow is burned down. The French are eventually expelled, and the Russians invade Paris.
1825	Nicholas I becomes tsar. Decembrists' Revolt. Reactionary period. Alexander Pushkin is writing his famous works at this time.
1850s	Crimean War against Turkey. Russian expansion into the Caucasus.
1861–1882	Serfdom is abolished (1861). Alexander II is murdered by anarchists (1882). Dostoevsky and Tolstoy write their great novels.
1860–1875	Manchuria is annexed from China (1860), Sakhalin Island from Japan (1875).
1904–1905	Russo-Japanese War. "First Russian Revolution"; Duma legislature is established by Nicholas II as a concession.
1905–1914	Prime Minister Stolypin implements agricultural reforms, but is assassinated (1911). Rapid industrialization. World War I begins (1914).
1917	Nicholas II abdicates the throne in February. Interim government is formed. Bolsheviks seize power on October 25 (November 7 on the Gregorian calendar).
1917–1922	Civil War; White Army loses. Hunger. New Economic Policy (NEP) is instituted by Lenin. U.S.S.R. is formed in 1922. First labor camps are founded.
1924	Lenin dies. Struggle for succession between Joseph Stalin and other followers of Lenin, most notably Leo Trotsky.
1928–1953	Stalin's period. Collectivization, industrialization, cultural revolution. *Kulaks* (more prosperous peasant farmers) exiled into Siberia (early 1930s). Mass terror beginning in 1935, especially 1937–1938. The GULAG system matures.
1941–1945	The U.S.S.R. is invaded by Germany. World War II. Key battles: Moscow (autumn, 1941), Stalingrad (winter 1942–1943), Kursk (summer 1943), siege of Leningrad (1941–1943), Germany's defeat (1944–1945).
1953–1962	Nikita Khrushchev initiates reforms. Cold War begins. Sputnik is launched (1957); Yuri Gagarin becomes the first man in space (1961). Cuban missile crisis (1962). Khrushchev is replaced by Leonid Brezhnev.
1963–1985	"Stagnation" or late Soviet period. Dissidents' movement arises. Increasing economic problems. Afghanistan is invaded (1979).
1985–1991	Gorbachev's *perestroika*. Failed coup in Moscow and the end of Communist government (August 1991). The U.S.S.R. ceases to exist, and Yeltsin rises to power (December 1991).
1991–2008	Economic reforms under Yeltsin. Two Chechen wars. Vladimir Putin becomes president (2000) and then prime minister (2008). Dmitry Medvedev becomes president (2008).

of Bygone Years, compiled in Kiev ca. 1113 A.D.) and other historical documents begin their narrative at about 850 A.D., the time when the Slavs were beginning to realize their collective identity as a people united by language and culture. Ironically, they invited foreigners—Varangians from Scandinavia—to rule them at the time. A Varangian prince, Rurik, first came to Novgorod in the north. He was selected as a common ruler by several Slavic and Finno-Ugric tribes in about 860 A.D., before moving south and extending his authority to Kiev. The *Primary Chronicle* cites him as the progenitor of the Rurik Dynasty. It is possible that the word *Rus* comes from the typical red color of the Varangians' hair.

Early History (850–1480 A.D.)

Geographically, the old Kievan Rus was centered on the city of Kiev (see Figure 6.1, left side). Located on the right (west) bank of the Dnieper, just above the rapids, Kiev (Figure 6.2) provided a convenient, highly visible, and defensible outpost, well suited for control over the southern reaches of the big river. The Dnieper originates not far from Smolensk, and people could easily travel from the Baltic to the Black Sea via the Neva and Volkhov Rivers into the Dnieper, with minimal portaging near the headwaters; this was the famous route used by the Varangians to trade with the Greeks. Kiev's location along this major north–south thoroughfare of medieval Eastern Europe facilitated its quick rise to prominence. Also noteworthy was its location at the "ecotone" (transition area) between the deciduous forests to the north and the open steppe to the south. Each biome provided some unique products to the nascent nation. For example, timber and furs came from the forest, while many agricultural crops could be grown in the steppe.

The Slavs were historically people of forested floodplains; they avoided large expanses of open grassland, which were harder to defend against hostile tribes. Other important cities of the period, such as Chernigov, Novgorod, Pskov, and

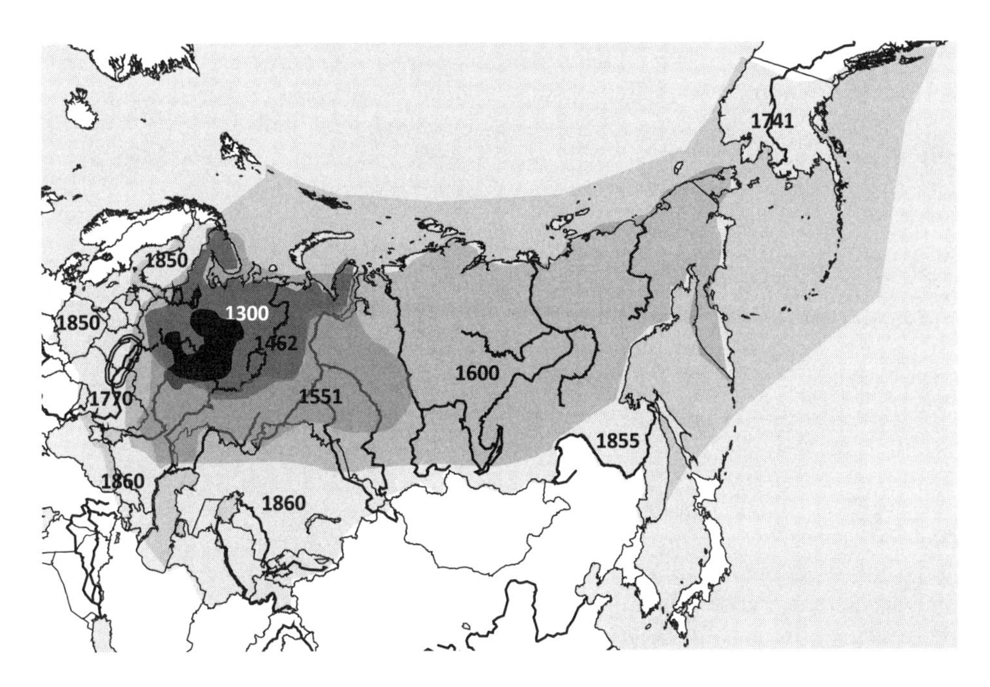

FIGURE 6.1. Map of territorial expansion of Muscovy/Russia. The original position of Kievan Rus (ca. 1000 A.D.) is indicated by the double-outlined oval at left. Alaska was sold to the United States in 1867; while Poland and Finland were lost in 1914. Parts of western Ukraine, Belarus, and Moldova have been repeatedly claimed by Germany, the Austro-Hungarian Empire, Poland, and Russia over the past 300 years. These boundary claims are too complex to be shown here.

FIGURE 6.2. Kiev, Ukraine—the birthplace of Rus. Visible are churches and the bell tower of the Kiev Caves Monastery (the oldest monastery in the Russian Orthodox Church, established ca. 1050 A.D.) *Photo:* J. Lindsey.

Smolensk, were located farther north. The first few centuries of this early state were filled with numerous battles between various Slavic princes for the control of Kiev, and more substantial fights against the invading Asiatic nomads from the eastern steppe: the Khazars, the Pechenegs, the Polovtsians, and finally the Tatars, all of whom were eager to sack and loot Kievan Rus. During the years from 1054 to 1224, no fewer than 64 principalities existed; about 300 princes put forth succession claims, and their disputes led to a few dozen local wars. In this sense, the Eastern Slavs were no different from most Western European tribes of the period (Gauls, Franks, Anglo-Saxons, and others).

The early Slavs were animists (Vignette 6.1). The conversion of Prince Vladimir to Orthodox Christianity in 988 A.D. was a significant event, in that it allowed a powerful alliance between the Greek-based Byzantine Empire (the surviving eastern half of the original Roman Empire) and the Slavic people. This opened up possibilities for mutual defense, cultural enrichment, and improved trade. Vladimir's successors remained in Kiev for about two more centuries (until the mid-1300s), but eventually the relentless nomadic attacks from the southeastern steppes forced a geographic resettlement much farther to the north, toward present-day Vladimir, Suzdal, and Yaroslavl, along the Volga River. The Volga basin provided a convenient forested retreat away from the less defensible Kiev.

The eventual rise of Moscow to the preeminent position among Russian cities had to do with some pure luck and the political talents of the early princes there, but it also owed a good deal to geography: Originally an insignificant wooden fort (established in 1147), it was located at a perfect midpoint between the sources of the Dnieper and the Volga. It was situated on a tributary (the Moscow) of a tributary (the Oka) of the Volga—not on the main water artery, but close enough to Smolensk (100 km to the west in the Dnieper basin) that the Dnieper headwaters could be easily reached. In the age before highways, all transportation of goods took place by rivers. The forests of the area were mixed pine, spruce, basswood, maple, and oak, providing a sheltered existence and plenty of timber. The agricultural potential was lower than in the south because of the colder climate, but barley, oats, rye, and even wheat could be grown, along with

Vignette 6.1. Slavic Gods

Before the Eastern Slavs were converted to Christianity, they were animists. "Animism" is a belief in spirits as expressed in forces of nature. The ancient Slavs believed in a number of gods, both male and female. Each tribe had one most important god and a variety of others. Wooden totemic statues (idols) were commonly erected at prominent sacrificial sites. Many were located in sacred groves, near springs, or on promontories between two rivers. The gods included Perun, the god of thunder; Dazbog, the god of fertility and sunshine; Svarog, the blacksmith god; Khors, the god of the sun; Mokosh, the goddess of fate; Lada, the goddess of spring; and many others. Some deities had clear parallels with Greek and Roman mythological characters, whereas others were unique. In addition, the Slavs believed in various supernatural creatures who lived in the forest (*leshy*), in the water (*vodyanoy* and *kikimora*), in houses (*domovoy*), and so on. Some of these resembled the dwarves, elves, and leprechauns of the western Celtic and Germanic peoples. They were not spirits, but may have had some supernatural powers.

The open worship of the ancient gods came to an end with Prince Vladimir's official baptism of the people of Rus in 988 A.D., although many folk traditions continued to be retold in tales and legends for many centuries thereafter. The sacred geography of ancient Rus is poorly studied. V. Boreiko from the Kiev Ecological–Cultural Center has published a few books (in Russian) that elucidate some of these landscape connections for the early Ukrainians and Russians.

a variety of common vegetables (beets, turnips, carrots, and cabbage) along the floodplains. Hunting for wild boar, bear, moose, European deer, wood bison, and wild cow provided enough meat for the growing population.

Moscow's real rise started with Prince Daniel in the early 14th century. It was situated on a high pine forest hill (*bor*) above the Moscow River at its confluence with the smaller Neglinnaya—an extremely defensible site. In the middle of the 14th century, the head of the Russian church moved his see from Vladimir to Moscow, thus making the latter not only a political but also a spiritual center. At the heart of the city was the *Kremlin*, meaning "stronghold" in Russian—a large white-stone (later red brick) fortress, with its oldest cathedrals dating back to the early 15th century. It occupies about 30 ha today and is triangular in shape: Its south side runs along the Moscow River, its western side along the now-buried Neglinnaya River, and its eastern side where the Red Square was formerly protected by a moat (Figures 6.3 and 6.4). Incidentally, the name "Red" means "beautiful" in Russian, and has nothing to do with either bloody history or Communism. Although the Moscow Kremlin is the most famous one, many older Russian cities have kremlins as well: Novgorod, Pskov, Yaroslavl, Vladimir, and Suzdal, for example. Typically these settlements were located in similar spots, on hills high above the confluence of two rivers in the generally flat Russian plain.

Between 1230 and 1480, Russia was under the foreign rule of the Tatars and Mongols. The invasions started during the rule of Genghis Khan and continued for more than two centuries. The

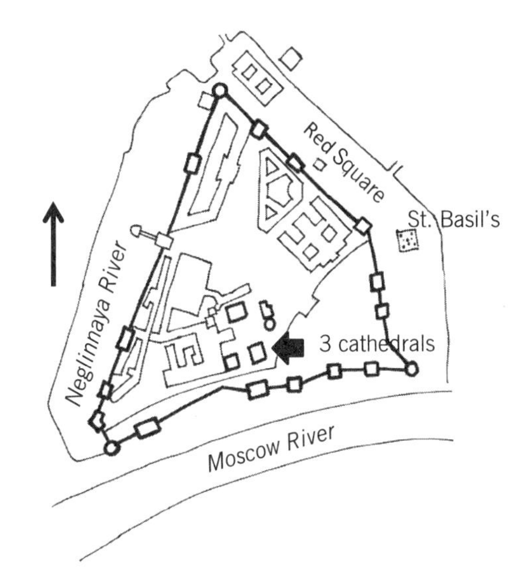

FIGURE 6.3. Map of the Kremlin of Moscow as it exists today. *Drawing:* I. Blinnikova.

FIGURE 6.4. The Moscow Kremlin (view from the Moscow River). *Photo:* P. Safonov.

Mongols ruled from a distance, requiring Russian princes to pay tribute and sometimes extorting contributions of slaves as well. The Tatars forged alliances with the Mongols and were their main foot soldiers; as a result, this period came to be called the time of the "Tatar–Mongol Yoke." Although self-ruling Slavic princedoms persisted, few were powerful enough to challenge the Mongols directly, except Muscovy.

Maturity and the Great Tsars (1480–1917)

By 1480, the new Slavic state of Muscovy was firmly centered on the city of Moscow and extended out to the north and east for about 800 km into the Volga River basin. Through forging alliances with some states and through conquering others, the great princes of Moscow managed to extend their reach into the territory of Novgorod (a city as old as Kiev, and traditionally very independent in spirit) by the time of Ivan III (1480). Ivan married the daughter of the last Byzantine emperor and claimed that Russia was to be the successor of the rapidly vanishing empire of his in-laws. Accordingly, he was the first to be crowned as a "tsar" (Caesar) of All Rus, and undertook a series of aggressive building projects to enhance Moscow's power and prestige. He enlarged the stone-walled Kremlin and invited the best Italian architects to complete magnificent cathedrals in the early 1400s. Two of these cathedrals, honoring the Assumption of the Virgin Mary (Figure 6.5) and the Archangel Michael, are especially famous achievements from this era. By the birth of his grandson, Ivan IV ("the Terrible," which is better translated into modern English as "the Majestic"), in 1530, Moscow's geographic reach extended all the way to Arkhangelsk on the White Sea. The emerging state would not, however, gain access to the Baltic Sea for another two centuries, or to the Black Sea for over two and a half.

Ivan IV conquered Kazan and the Astrakhan khanates of the Volga Tatars in the mid-1550s, thus ending the period of the Tatar–Mongol Yoke and opening up vast expanses of the lower Volga and the Urals to Russian settlement. Many of the settlers were frontiersmen, called Cossacks, who form an ethnic subgroup within the Russian people today. The Cossacks are a mixed group with both Slavic and Tatar cultural traits. The Caspian Sea and western Siberia were now within the reach of Moscow. The first capital of Russian Siberia was established in Tobolsk (on the Tobol

FIGURE 6.5. The main cathedral of the Moscow Kremlin is dedicated to the Dormition of the Theotokos (i.e., the Assumption of the Virgin Mary). It was built between 1475 and 1479 by the Italian architect A. Fioravanti. *Photo:* Author.

River, which is part of the Ob–Irtysh system) in 1587. Tyumen, Yeniseisk, Irkutsk, Yakutsk, and other Siberian cities followed shortly afterward.

The main exploratory push and the expansion of the Russian frontier across Siberia came in the mid-17th century with the new Romanov dynasty (see Figure 6.1). After a time of troubles and a major war with Poland in the early 17th century, the period of Rurik rule ended, and a time of relative peace and prosperity came. The lure of Siberian furs, gold, and timber, coupled with a relatively small and not very hostile native population, encouraged rapid Russian expansion into Siberia. Astonishingly, in less than *one century* (from 1580 to 1650), the Russian state was extended from Tyumen in western Siberia all the way to Okhotsk on the Pacific Coast! Of course, this vast area was not fully settled by any means, but about two dozen forts were built at strategic locations. Typically these forts were located along major rivers at convenient confluence points, because the exploration proceeded primarily along the great waterways by boat in summer and by sleigh on ice in winter. Every major Siberian city that was established during this period is situated on a big river.

The movement was somewhat analogous to the opening of the American West, except that it was driven less by farmers and more by fur traders (similar in lifestyle to the French trappers of Canada), and that the direction of movement was of course from west to east (not the other way around). The early settlers were a highly mobile force, not interested in farming or other sedentary pursuits. Virtually all of central and eastern Siberia is underlain by permafrost, which makes farming almost impossible in any case. Still, it took only 70 years for the state to quadruple its size—a feat probably unmatched in human history. In comparison, the movement to the west, north, and south was much slower, because more developed states and tribes there made rapid expansion impossible. To the west and north were the Swedes, Germans, and Poles. To the south were the Crimean Tatars and the Turks, as well as Central Asian and Caucasian tribes.

Under Peter the Great, the Baltic Sea became accessible through the creation of the new seaport of St. Petersburg. Built on the coastal swamp at the mouth of the Neva River at the cost of a few thousand lives, it became known as Russia's "Window to Europe." The great project began in 1703, and the capital was moved from Moscow to St. Petersburg in 1712. Catherine the Great pushed the Russian frontier to the Black Sea by defeating the Crimean Tatars and their Turkish allies. This was accomplished by capturing a few strategic fortresses along the Azov Sea in the second half of the 18th century. In the mid- to late 19th century, the Russian Empire expanded into Central Asia to the present-day border with Afghanistan, and into the Caucasus and Manchuria on the Pacific Coast. Although these land acquisitions into the Russian Empire were by no means small, they were still dwarfed by the giant Siberian expansion. Further advances in the south were halted by very high mountains (the Pamirs and the Tien Shan) and strong, hostile groups of people in Persia, Afghanistan, Turkey, and China. Japan finally stopped the Russian advance into northeast China with its victory in the Russo-Japanese War in 1904, when the Russian colony of Port Arthur (now Lushunku) fell at the southern tip of Liaodong Peninsula.

At its peak, the Russian Empire occupied over 22 million km^2 (i.e., it was equal in size to all of North America and made up 15% of the world's landmass). In 1913, it was second in the world by area after the British Empire, third by population after China and the British Empire, and fifth in terms of gross domestic product (GDP) (Treivish, 2005). Great Britain controlled almost 25% of the world's landmass (including Canada, Australia, and India); in the mid- to late 19th century, it clashed with the Russian Empire repeatedly along geographic fracture zones in the Black Sea basin, in Persia, and in Afghanistan. By the start of World War I in 1914, the Russian Empire included most of Ukraine, Belarus, and Moldova (Bessarabia); Finland, Armenia, Azerbaijan, and Georgia; the Central Asian states (Russian Turkestan); Lithuania, Latvia, and Estonia; significant portions of Poland; and some Turkish cities in the Balkans. Only about 45% of its population consisted of ethnic Russians. The total population was 125 million in 1897, the time of the first Russian census.

Alaska was sold in 1867 to the United States for $7.2 million, or merely $100 million in today's dollars—an astonishingly cheap price, although back then Secretary of State Seward was asked in Congress why *so much* money was spent on the acquisition of "rocks and ice." The Russian government wished to sell off this territory, largely because of the expenses it had incurred while fighting the Crimean War with Turkey and Britain in the Black Sea basin. Russia had lost this war in 1856; that same year, British and allied French warships attacked and took the town of Petropavlovsk in Kamchatka. This latter attack raised a question about the security of Russian America. If Russia could not successfully protect even Petropavlovsk, would it be able to protect Sitka or Kodiak across the Bering Sea? There was also concern about losing Alaska as a result of British invasion from Canada, and advisors to the Romanovs advocated making the sale to a third party (the United States) while they were still in a position to negotiate a fair price. Russian settlements in Alaska had always been sparse (Figure 6.6); the total of fewer than 800 Russian settlers included a few businessmen, some government officials, and some Orthodox missionaries who worked to convert the Aleuts to Christianity, although it did not include several thousand inhabitants of mixed Russian and Aleutian descent. The southernmost point in North America to which Russian influence extended was Fort Ross (just north of Santa Rosa, California), which is now a state historical park.

In a landmark political decision, Alexander II abolished the serfdom of the Russian peasants in 1861, allowing millions to begin life as free people and not subject to the rule of their landlords. Virtually no land was provided to the serfs, however, except in distant Siberia and southern Ukraine; as a result, many freed serfs emigrated to the United States and Canada toward the end

FIGURE 6.6. Baranov Museum (Erskine house) in Kodiak, Alaska, is the oldest surviving Russian structure in North America (built ca. 1790) and is a National Historic Landmark. *Photo:* Author.

of the 19th century. Among those who remained, discontent with the lack of land sparked dissent and riots in the early 20th century. In the urban settings, Jews were subject to many pogroms at this time (especially in Ukraine), because they were perceived as economically savvy but unfair merchants, and of course as culturally and ethnically distinct.

The Romanov Empire came to a bitter end in 1917, as two successful revolutions shook the country. The capitalist "February Revolution" removed the last Romanov emperor, Nicholas II, and installed a provisional bourgeois government, which in turn was overthrown by the Bolsheviks (the early Communists) in October of that year. The reasons for the "October Revolution," as it became known, are complex. The disastrous Russian involvement in World War I, growing political dissent among the non-Russian peoples within the empire, a lack of rapid reforms in agriculture, and rapid industrial growth all played major roles. Some researchers also point out the direct involvement of the British and German intelligence services in tacitly helping the Bolsheviks to assume power, because a strong Russia was not in Western European interests. As a result of the civil war, many of Russia's western territories—including Poland, about half of Ukraine and Belarus, and the Baltic states—were lost over the ensuing few years.

The Soviet Period (1917–1991)

After a bitter civil war between the Bolsheviks (known as the "Reds") and the anti-Bolsheviks (known as the "Whites") in 1917–1922, the Soviet state renamed itself the Soviet Union—or, officially, the Union of Soviet Socialist Republics (U.S.S.R.). It reconstituted itself within the former borders of the Russian Empire, with the exceptions of Finland, Poland, the Baltic states, much of western Ukraine and Belarus, and Moldova. This may be explained by not only political and cultural but also geographic factors. As suggested by Harold Mackinder in his famous series of papers on the world's "Heartland" (see Cohen, 2009), northern Eurasia forms a large, easily-defensible area bounded by some of the highest mountains in the world on the south, by the frozen Arctic Ocean on the north, and by the Pacific Ocean on the east. It is much more open and vulnerable in the west, and this is precisely where all the major wars were fought.

Once these boundaries were reclaimed by the Soviets in the 1920s, there was relatively little change for 70 years. Following the defeat of the Nazis in 1945, the Kaliningrad region was added (carved out of what had been Prussia), as well as the Baltic states, Moldova, and western Ukraine. Some islands in the Far East were gained from Japan. This produced the instantly recognizable shape of the U.S.S.R. that dominated the tops of world maps for about 50 years, until its collapse in 1991. The period from 1917 to 1991 is discussed in greater detail in Chapter 7.

The creation of the Soviet Union's internal borders was of geographic importance, too. Most of these were drafted in the 1920s by early Soviet leaders, including Lenin, Trotsky, Zinoviev, Bukharin, and Stalin. Some of the boundaries were well designed to account for certain national groupings within the U.S.S.R. (e.g., Georgia, Uzbekistan), but others were drawn more in line with the economic or political needs of the moment. For example, there was no compelling reason to place the border between Ukraine and Russia, or that between Kazakhstan and Russia, exactly where it exists today. These nations have genuine transition zones between largely Russian and non-Russian speakers that stretch for hundreds of kilometers; these have no clearly defined boundaries, however, but rather are overlapping cultural, ethnic, and linguistic zones. Where the borders were drawn in these and similar cases had more to do with the Soviet economic rationale than with politics. Some examples in particular reflect the whimsical politics of the moment: The Crimea Peninsula, a mainly Russian-speaking area, was abruptly turned over to the Ukrainian Soviet Socialist Republic in 1954 as a gift to a political friend there from Nikita Khrushchev, who was himself from Ukraine. Armenians did not get the predominantly Armenian-populated Nagorno-Karabakh region, while Azerbaijan was given the Azeri-speaking Naxicevan region inside Armenia; these decisions reflected the personal tastes of Stalin, who, himself from Georgia, particularly disliked Armenians. Fergana, the most fertile valley in Central Asia, was carved into a maddening jigsaw puzzle of borders in an attempt to accommodate Tajiks, Uzbeks, and Kyrgyz living in the area. Most of the territorial conflicts of the post-Soviet period (Table 6.2) can be traced back to these ill-fated policies of the Stalin and Khrushchev periods.

The Post-Soviet Period (1991 to the Present)

After the U.S.S.R. was dissolved by mutual agreement of the Russian, Ukrainian, and Belarusian presidents in December 1991, many internal borders became external (Vignette 6.2). Numerous conflicts started, some with thousands of casualties. A few started even before 1991, during Gorbachev's awkward *perestroika* attempts (see Table 6.2); these and other events of post-Soviet history are discussed in more detail in Chapter 8. To the credit of the people and leaders of the region, the situation did not come to resemble the horrific Yugoslavian scenario. Most conflicts remained localized, and the boundaries of the 15 republics today are essentially unchanged. Some areas, such as Abkhazia, South Ossetiya, and Chechnya, do see persistent military conflict; other areas experience occasional tensions, but without bloodshed. In addition, some self-proclaimed "republics" that have not been officially recognized by the United Nations or any individual nations do exist. They are greatly emboldened now by the recognition of Kosovo's independence by some European Union (EU) and North Atlantic Treaty Organization (NATO) members. Of particular note is the recent recognition of South Ossetiya and Abkhazia by Russia in the wake of the Ossetian–Georgian conflict in August 2008.

It is important to understand that the Russian Federation today is not merely a smaller U.S.S.R. It is qualitatively different from either the Russian Empire or the U.S.S.R. The latter two had fewer than 50% ethnic Russians and had external borders with nations of very different cultures (e.g., Hungary, Turkey, Iran, Afghanistan), whereas Russia is over 80% ethnically Russian and mainly borders other Russian-speaking territories in Ukraine, Belarus, or Kazakhstan (see Vignette 6.2). Although Russia remains the biggest state in the world by area, it is half of its original size and is now only 9th in terms of population and 6th in terms of GDP adjusted by purchasing power parity (PPP). It has also lost its status as one of the world's two superpowers. Indeed, in terms of overall trade and economic strength it is now part of the world's semiperiphery, more

TABLE 6.2. Main Territorial Conflicts or Disputes of the Post-Soviet Period

Conflict	Parties in conflict	Interethnic? Violent?
Estonia claims in Peipus region	Estonia vs. Russia	Yes; no
Latvian and Estonian ethnic issues	Russian-speaking minorities vs. Estonia and Latvia	Yes; no
Crimea	Crimean Tatars vs. Russian majority	Yes; no
Trans-Dniester Republic (Moldova)	Moldova vs. Trans-Dniester industrial region	Partially ethnic; violent in the early 1990s
North Ossetia (Russia)	Ossetians vs. Ingushs	Yes; yes
Chechnya (Russia)	Chechens vs. Russian Federation	Yes; yes
Dagestan (Russia)	Chechens and Dagestanis vs. Russian Federation	Partially ethnic; yes
Abkhazia (Georgia)	Abkhazs vs. Georgians	Yes; yes
South Ossetiya (Georgia)	Ossetians vs. Georgians	Yes; yes
Nagorno-Karabakh	Armenians vs. Azerbaijanis	Yes; yes
Fergana Valley	Uzbekistan vs. Tajikistan vs. Kyrgyzstan	Yes; somewhat violent
Tajikistan	Clans within Tajik society	No; yes

Vignette 6.2. Current Boundaries of Russia

Russia occupies 11.3% of the world's landmass. The total length of the land border is 20,097 km. The countries Russia borders, and the length of the border with each country, are as follows: Norway, 196 km; Finland, 1,340 km; Estonia, 294 km; Latvia, 217 km; Lithuania (Kaliningrad Oblast), 280.5 km; Poland (Kaliningrad Oblast), 232 km; Belarus, 959 km; Ukraine, 1,576 km; Georgia, 723 km; Azerbaijan, 284 km; Kazakhstan, 6,846 km; China (south), 40 km; Mongolia, 3,485 km; China (southeast), 3,605 km; and North Korea, 19 km. The total coastline is 37,653 km. The Soviet Union claimed all of the Arctic Ocean to the North Pole, approximately along the 32°E meridian to the west and the 169°W meridian to the east. These claims have not been universally supported.

Here are the extreme points of Russia's territory today:

- In the north, Cape Fliegeli on Franz Joseph Archipelago (81°49'N).
- In the continental north, Cape Chelyuskin (77°43'N).
- In the south, Bazardyuzyu Mountain (41°11'N).
- In the west, a spit in the Gulf of Gdansk (19°38'E).
- In the continental east, Cape Dezhnev (169°40' W).
- In the east, Ratmanov Island in the Bering Strait (169°02'W).

comparable to Brazil or South Africa than to the United States, China, Germany, or Japan. Politically, too, it is relatively isolated; it has lost most of its influence over Eastern Europe, including even the traditional friends Bulgaria and Serbia, as well as over countries in Asia, Africa, and Latin America that were tightly aligned with the U.S.S.R. Russia is also embroiled in a number of conflicts, either on its own territory (Chechnya, Ingushetia) or in close proximity to its borders (Abkhazia, South Ossetia, the self-proclaimed Trans-Dniester Republic in Moldova). Although Russia and China have successfully settled their disputes along the Amur River border, Japan still expects Russia to return the annexed four southern Kuril Islands, although there is no indication from the Russian side that this will be forthcoming.

Most of the independent non-Russian republics have strong, if not enthusiastic, economic ties to Russia. However, they have relatively few continuing political connections with Russia, at least among the elites. A good case in point is Georgia—a country culturally similar to Russia and with a long history of mutual connection and even admiration, but now politically alienated from Russia both by its own pro-Western ambitions, and by the uncompromising stance of Russia on Abkhazia and South Ossetiya.

Thus Eurasia's heartland is no longer strong and is rather divided. It is also shrinking in population size. Among the signs of the times is the rise in Russian nationalism evident everywhere in the new post-Soviet Russia—from newspaper headlines and political pronouncements to ultraright demonstrations and even pogroms of Caucasian ethnic minorities in some peripheral Russian cities. The increasing cost of travel across the vast territory raises a possibility of farther devolution, especially in the Russian Far East; this extremely remote part of the country is 8–10 time zones away from Moscow and has a growing Chinese and Japanese presence and influence.

REVIEW QUESTIONS

1. What are some geographic advantages and disadvantages of Kiev's location along the Dnieper River, between the forest and the steppe?
2. What are the reasons why Moscow was found to be a better location for a capital city during the time of the Tatar conquests?
3. At the height of World War II, Harold Mackinder wrote that "the Heartland (i.e., the U.S.S.R.) is the greatest natural fortress on earth. For the first time in history it is manned by a garrison sufficient both in number and quality" (quoted in Cohen, 2009, pp. 16; 252). What did he mean by

"a fortress"? What has happened to this "garrison" since the fall of the Soviet Union in 1991?

EXERCISES

1. Using online research, compile a list of the Russian cities that have kremlins. What are the dates of their foundation? After what year were kremlins no longer needed? Why?
2. On a single page, make two lists: on the left, a list of countries that the U.S.S.R. bordered, and on the right, a list of countries that Russia borders today (see Vignette 6.2). Compare the lists and discuss the possible implications for national security in the past and now.
3. Write an essay comparing and contrasting the expansion of the American frontier from east to west and the Russian frontier in the opposite direction. Estimate the amount of area that was absorbed into each country per century (for the United States, start with the year 1600; for Russia, start with the year 1400).
4. Use a world gazetteer (this is a list of place names, either published or online) to explore the "language gradient" across a segment of the Russian border with Kazakhstan today. That is, how many kilometers on average does it take to get to the point where more than half of the names are Kazakh? What does this suggest about the placement of the actual border?
5. Use a blank map showing the rivers of Siberia. Locate about 20 major cities from the list below and, with a pencil, draw the shortest routes to connect them all; try to maximize the use of rivers and to minimize portages. A similar exercise can be done with the aid of a geographic information system (GIS). Cities to locate: Yekaterinburg, Tobolsk, Tyumen, Omsk, Berezov, Turukhansk, Narym, Yeniseisk, Novonikolaevsk (Novosibirsk), Tomsk, Krasnoyarsk, Bratsk, Irkutsk, Yakutsk, Chita, Nerchinsk, Okhotsk, Verkhoyansk, Khabarovsk, Verkhnekolymsk, Nikolaevsk, and Anadyr (more could be added).

Further Reading

Baddeley, J. F. (1969). *The Russian conquest of the Caucasus.* New York: Russell & Russell.

Bassin, M. (1999). *Imperial visions: Nationalist imagination and geographical expansion in the Russian Far East, 1840–1865.* New York: Cambridge University Press.

Black, L. (2004). *Russians in Alaska.* Fairbanks: University of Alaska Press.

Bobrick, B. (1992). *East of the sun: The epic conquest and tragic history of Siberia.* New York: Poseidon Press.

Christian, D. (1998). *A history of Russia, Central Asia and Mongolia.* Oxford, UK: Blackwell.

Cohen, S. B. (2009). *Geopolitics: The geography of international relations* (2nd ed.). Lanham, MD: Rowman & Littlefield.

Gilbert, M. (1972). *Russian history atlas.* New York: Macmillan.

Grousset, R. (2002). *The empire of the steppes: A history of Central Asia* (8th ed.). New Brunswick, NJ: Rutgers University Press.

King, C. (2008). *The ghost of freedom: The history of the Caucasus.* New York: Oxford University Press.

Lane, G. (2004). *Genghis Khan and Mongol rule.* Westport, CT: Greenwood Press.

Lincoln, B. (1994). *The conquest of a continent: Siberia and the Russians.* New York: Random House.

Mackinder, H. J. (1904). The geographical pivot of history. *The Geographical Journal, 23*, 421–437.

Magosci, P. R. (2007). *Ukraine: An illustrated history.* Seattle: University of Washington Press.

March, J. P. (1996). *Eastern destiny: Russia in Asia and the north Pacific.* Westport, CT: Praeger.

Obolensky, D. (1994). *Byzantium and the Slavs.* Crestwood, NY: St. Vladimir Seminary Press.

Pipes, R. (1995). *A concise history of the Russian revolution.* New York: Knopf.

Shubin, D. H. (2004). *A history of Russian Christianity.* New York: Algora.

Treivish, A. (2005). A new Russian heartland: The demographic and economic dimension. *Eurasian Geography and Economics, 46*(2), 123–155.

Websites

www.bucknell.edu/x20136.xml—Russian history: chronology.

www.pbs.org/weta/faceofrussia—*The Face of Russia* PBS series.

www.feefhs.org/maplibrary.html—Historical maps of the Russian Empire.

CHAPTER 7

The Soviet Legacy

The Soviet period started in October 1917, with the victory of Vladimir Ilyich Lenin's Bolshevik party over the bourgeois Provisional Government in the political revolt later referred to as the "October Revolution." It ended with the Communist hardliners' coup against Mikhail Gorbachev in August 1991. Thus the period covers 74 years of Russia's recent history. The word "Soviet" means "council" in Russian, and as such refers to an idealistic concept of a government of peasants and workers ruling through local, regional, and national councils of people's representatives. Such a system was in fact put in place in 1917, *before* the Bolsheviks hijacked it for their own purposes. As the Communist Party of the Soviet Union (C.P.S.U.) matured, the lower-level Soviets became completely subordinate to one-party rule and in the later Soviet period they did little more than give a nod of approval to all of the party's decisions. Nevertheless, the entire country became known as the Union of Soviet Socialist Republics (U.S.S.R.), or the Soviet Union. The Soviet Union was not fully formed until 1922, because it took the Communists about 5 years to defeat the White Army in a civil war. Even after the Communist Red Army's victory over the Whites, there were still significant territorial losses in comparison with the former Russian Empire. Finland, Poland, the Baltic states, Bessarabia (in contemporary Moldova), and much of western Ukraine did not become part of the U.S.S.R. for about 20 years. Because of the Molotov–Ribbentrop pact, the Soviet Union would regain most of these territories just before World War II. Finland fought back and successfully defended its independence in 1939, while Poland was allowed to regain its sovereignty (albeit under socialist rule) in 1945.

Politically, the U.S.S.R. not only had a hierarchical one-party government, but permitted no freedom of political expression and held merely token single-candidate elections. Ordinary party members, numbering about 17 million in a country of 280 million, had only token membership and played almost no role in formal decision making, while a small group at the top made all political decisions. Nevertheless, the small group at the top (the so-called *nomenklatura*, discussed later) would recruit its new members from the large party base.

Economically, the Soviet Union was a socialist state running as a command economy on 5-year plans without the aid of the free market. Although making the transition to a communist economy was the nominal goal, Lenin and his followers quickly discovered that its implementation as envisioned by Marx, Engels, and their philosophical followers of the 19th century did not work in practice. Marx envisioned communism as an egalitarian society in which production is voluntary and abundant, while coercion (taxes, police, prisons, etc.) is unnecessary. Idealistic (usually religious) leaders over the course of human history have managed to create communes reflective of the Marxist ideal on a small scale. Creating a national-level communist economy, however, proved impossible in Russia or anywhere else.

The Communist regime of Lenin in Russia failed to create anything like a utopian social system. After 5 years of bloody civil war and the draconian measures of so-called war Communism, when even staple foods were forcibly taken from the peasants by bands of armed soldiers to feed hungry cities, Russia had to find an alternative. Lenin shrewdly replaced the dream of Marxist communism with the reality of Marxist–Leninist socialism. Socialism was supposedly a temporary fix—an economy not based on the Communist slogan of "from each one according to abilities, to each one according to need," but rather on "from each one according to talent, to each one according to labor." Thus money, courts, an army, prisons, and taxes could be retained, and people would still have a strong incentive to work. In return, many state benefits would be provided free of charge or at a nominal fee (e.g., housing, health care, education, and guaranteed employment).

The early socialism retained some free-market elements under the so-called New Economic Policy (NEP) of Lenin, which was successful at producing surplus food. The NEP allowed small artisan cooperatives and private farms. Stalin later abolished the NEP and changed the system into a top-heavy state socialism, where even the remaining small pockets of coops and private owners completely disappeared.

When Lenin died in 1924 after a few years of illness, he left no designated successor. Instead, Trotsky, Zinoviev, Kamenev, Bukharin, Stalin, and other Communist leaders were pitted against each other in a vicious behind-the-scenes fight to control the party and the country. By the early 1930s, Stalin had emerged as the victor, having dispatched his enemies one by one through cleverly playing them off against each other. All his comrades of the 1920s were eventually either executed in the U.S.S.R. (Bukharin, Zinoviev, Kamenev) or killed in exile by Stalin's agents (Trotsky). Stalin did not have any personal friends, only subordinates who lived in constant fear for their lives. Even the wives of some of his closest associates, such as Khrushchev and Kalinin, were arrested and imprisoned in GULAG camps to ensure the associates' loyalty. Joseph Stalin belongs to the group of infamous bloody dictators of the 20th century, along with Hitler of Germany, Mussolini of Italy, Mao of China, and Pol Pot of Cambodia. Tens of millions of lives were lost in the famines, executions, prisons, and labor camps of the Stalin period—so many from so many sectors of society that it is impossible to quantify the death toll accurately.

Geographically, Stalin's Soviet Union after World War II corresponded almost exactly to the boundaries of the Russian Empire, without Poland and Finland. The 15 constituent republics of the postwar U.S.S.R. had all been, at one time or another, parts of the Russian Empire of the 18th and 19th centuries. Therefore, although it is technically incorrect to refer to the Soviet Union as "Soviet Russia," it was a common name given to the country in the United States at the time. Russian political émigrés in Europe refused to call the country anything else but Russia, as a matter of principle.

Lenin strategically moved the capital of the country from the coastal and vulnerable St. Petersburg/Petrograd (renamed Leningrad in 1918) to the much more defensible inland Moscow. Lenin correctly felt the imminent threat posed by Germany and other Western countries to the new socialist state. When a socialist revolution in Germany failed in 1918, Lenin rightly concluded that sooner or later the two countries would be on a collision course again. His decision to move the capital proved critically important in the fall of 1941.

Territorial Administrative Structure

Figure 1.1 in Chapter 1 shows the 15 post-Soviet republics. (Some of the present-day Central Asian and Caucasian republics were integrated before World War II into the Soviet republics of Turkestan and Trans-Caucasus.) Each of the Soviet republics had its own flag, coat of arms, legislature, and ruling committee of the Communist Party.

In theory, the republics were equal units joined into a voluntary federation, like the United States. The actual decision making, however, was very top-down and unitary in nature, not federal. Each of the republics got to send 32 delegates to the Council of Nationalities at the federal level, for example, but those delegates had no power over what would actually happen back home. Their role instead was to approve party decisions in a cheerful unanimous show of hands broadcast on state TV. Each republic was headed by a Communist leader who was a member of that republic's principal ethnic group, with a Russian vice-secretary as the second in command. Such a system ensured Moscow's control over the nationalist agenda in each republic.

Given the fact that the Soviet Union included close to 200 nationalities, you may ask why only these 15 republics were officially recognized. Three general criteria had to be met for a republic to be formed:

1. The unit in question had to have over 1 million ethnically non-Russian people. Thus the smaller ethnic groups of the Caucasus or Siberia did not qualify, while Estonia just barely qualified.
2. The unit had to have a border with the outside world, so that its constitutional right to secede could be exercised, albeit only in theory. Thus the large internal region of Tatarstan, with 3 million Tatars, did not qualify.
3. Over 50% of the non-Russian population had to be of the main, or "titular," ethnicity. Thus Armenia, with 90% ethnic Armenians, qualified easily. Kazakhstan, with only 40% Kazakhs, should not have qualified under this rule, but an exception was made because of its enormous territory and the importance of the Kazakh culture in the cultural life of Central Asia. Latvia and Kyrgyzstan had about 50% of ethnic Latvians and Kyrgyz, respectively, but exceptions were also made for them.

Note that Moldavia, Armenia, and the Central Asian states had no internal border with Russia. The capital of each republic was typically its largest city, in most republics including at least 10% of the republic's population and fitting the definition of the "primate city." The best schools, universities, hospitals, museums, theaters, and research centers, and of course the republic's governmental structures, were located in the capital. The capital city was therefore the most desirable place to live in each republic.

The Russian Federation, then called the Russian Soviet Federated Socialist Republic (R.S.F.S.R.), was by far the largest and most complex unit. It had about half of the country's population. It also had the most diverse array of internal regions, including predominantly Russian *oblasts* and *krays*, as well as more ethnically diverse autonomous republics and autonomous *oblasts* and *okrugs*. The logic behind these various regions was that many ethnic groups that did not qualify for a full-fledged Soviet republic could at least have their own autonomous units within the R.S.F.S.R. Some of the most populous of these republics were Tatarstan (Tataria), Bashkortostan (Bashkiria), Yakutia, Karelia, Chuvashia, and Checheno-Ingushetiya. Most of these territorial units had an ethnic Russian majority (exceptions included Tataria, Checheno-Ingushetiya, and Tyva), but all had sizable ethnic minorities (e.g., the Komi Republic had 23% ethnic Komi people). In Dagestan, dozens of minorities were packed into one territorial unit. In other republics of the northern Caucasus, two unrelated ethnic groups were forced into one unit (e.g., Kabardino-Balkaria and Karachaevo-Cherkessiya). This was done deliberately as a form of "divide-and-conquer" policy. Politically, each autonomous republic could send 11 delegates to the Council of Nationalities.

Autonomous republics and/or autonomous oblasts or okrugs also existed in Georgia (Abkhazia, South Ossetiya, Adjaria), Azerbaijan (Nagorny-Karabakh, Nakhichevan), Uzbekistan (Karakalpakia), and Tajikistan (Gorno-Badakhshan Autonomous Oblast). In the Soviet Union as a whole, there were 20 autonomous republics, 8 autonomous oblasts, and 10 autonomous okrugs.

The autonomous okrugs and oblasts differed from the autonomous republics, in that they included only very small minorities of the mostly indigenous, tribal peoples of Siberia and the north. Many of the titular ethnicities in those only numbered a few thousands, living among much larger Russian populations. For example, in Yamalo-Nenets Autonomous Okrug, which had half a million people, the indigenous Nenets and Khanty made up only 5% of the population. The rest were ethnic Russian and Ukrainian settlers, mainly oil and gas workers from the European part of the country, who had moved to the okrug for work. Whereas autonomous oblasts could send five delegates to the Council of Nationalities, autonomous okrugs, given their small population size, could send only one.

Although no independence from the party's political line was allowed, many ethnic units of the U.S.S.R. enjoyed significant cultural autonomy with respect to using their local languages in education (especially at the primary level), in the arts, and in local administrative affairs.

Political Structure

Politically, it is helpful to think of the Soviet Union as a pyramid of power with one man (the Secretary General of the Communist Party) at the top (Figure 7.1). During the late Soviet period, the same man would also assume the title of Chairman of the Supreme Soviet of People's Deputies of the U.S.S.R., thus making himself into the leader of both the party and the government. The top decisions were made by this person in consultation with a small circle of close allies, called the Politburo of the Central Committee of the C.P.S.U. This oligarchy had about 15 members, typically all men. The broader Central Committee would have slightly over 60 members, with maybe 5 or 6 women among them, and would be supplied with an apparatus of about 5,000 technical workers (*apparatchiks*) organized into 23 departments (Theen, 1980). The regional and local party committees would exist at every level—including republics, smaller regions (oblasts, okrugs, or krays), and districts or municipalities—as well as at every large state enterprise. Each party chapter was headed by a secretary, who was the real leader, not a clerk. This odd usage of the word was introduced by Stalin, who indeed was a secretary under Lenin, but later refused to change the familiar title when he became an absolute ruler.

Although the Communist Party was in charge of making all actual decisions, the rubber-stamping legislatures of the Soviets likewise existed at every level, from the Supreme Soviet to the republican, smaller regional, and local levels. These legislatures consisted of party-picked loyal representatives of workers and farmers, who would simply "sign on the dotted line" and raise their hands in unison without any debate. The lower-level Soviets met infrequently, usually when a new party program was announced and had to be formally approved. Typically, these Soviet members were card-carrying members of the

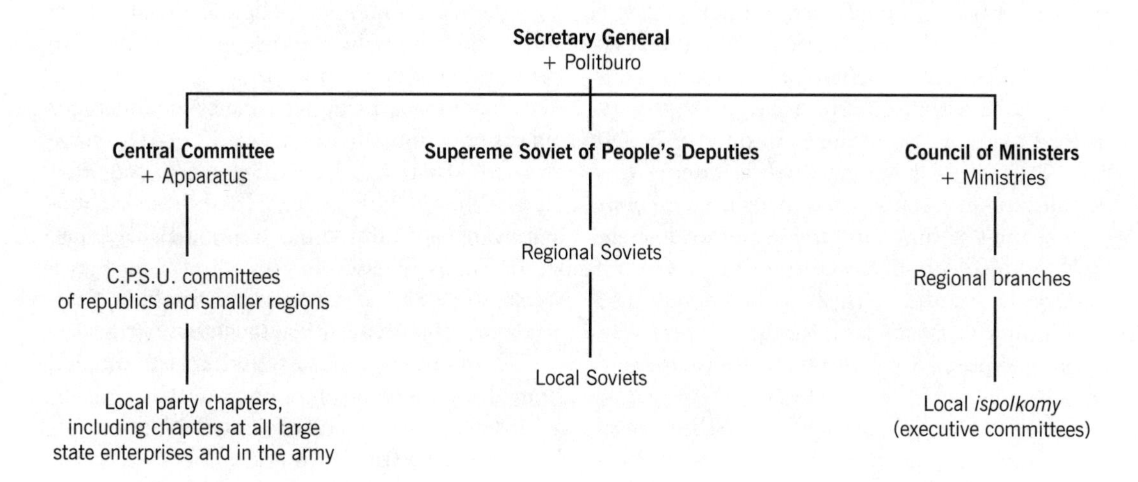

FIGURE 7.1. The general structure of the Soviet governmental system. The left side represents Communist party structures; the center represents the legislature; and the right side represents the executive branch.

Communist Party themselves, so of course they would not disobey their own leadership. Persons who were not party members (*bespartijnye*) could be theoretically elected as well, but in practice rarely were.

The third component of the system was the executive branch of the Soviets, called the *ispolkomy*. These would be put in charge of the government's actual daily operations; they would respond first to the party bosses, and then also to the Soviets at each level. Many of the actual economic decisions were made by national ministries—about 50 in all, each responsible for a sector of the economy (iron and steel, nonferrous metals, oil and gas, agriculture, railroads, etc.). Each ministry had regional branches and was run very much like a large state-owned corporation, with factories, construction bureaus, research institutes, schools, sanatoria, clinics, and even entire cities under its control. Nonindustrial sectors had ministries, too; the Ministry of Culture, for example, had directorates for theaters, music, art, and museums.

Missing from the diagram of Soviet governmental structure in Figure 7.1 is the all-pervasive secret police (KGB, literally translated as the "Committee on State Security")—loyal to the party, but with an independent leadership in charge of spying on party members and the common people. At the national level, the KGB was a state committee, not a full ministry, but it actually had more power than any ministry. The KGB head was always a member of the Politburo. During the Stalin period, some of the worst atrocities were perpetrated by the KGB (then known as the NKVD or MGB), with the tacit approval of Stalin himself. Every factory and institute in the country had the infamous "First Department" unit, whose members (KGB plainclothes agents) would ensure that the leadership and workers did not get out of line. Intimidation was a common tactic, and of course during the Stalinist period (from about 1930 to 1953) millions were arrested, sent to prison, and sometimes tortured and shot for very minor offenses, or frequently for no offense at all. For example, a 15-minute unauthorized break from work could lead to an imprisonment. Many party members were arrested simply to scare others into complete submission.

Such a system ensured strict compliance out of fear. No true thoughts could be expressed in public (cf. the "doublespeak" of George Orwell's *1984*). To be sure, the KGB attracted a lot of bright young people to its ranks with high pay, perks, and status. It is highly symbolic that the first president of free Russia (Boris Yeltsin) was mistrustful of the KGB during his tenure, but had no choice but to appoint a representative of this organization (Vladimir Putin) as his successor.

The country had a planned command economy, as noted earlier: Every enterprise was state-owned, and everything from paper clips to intercontinental ballistic missiles (ICBMs) was produced according to a 5-year plan. Gosplan was the agency in charge of economic planning. No private enterprises of any sort were allowed, except that some traditional craftspeople, piano or language tutors, or domestic servants would work in the informal economy for cash. There was also, of course, a black market—a dangerous, illegal, and highly profitable enterprise.

All salaries were fixed on a countrywide schedule, and there was little difference in pay among various levels. For example, in the 1980s a lowly lab assistant at an institute had a salary of 80 rubles a month, while the director would be paid about 600 rubles, with the majority of workers making between 100 and 250 rubles almost regardless of qualifications. What did vary tremendously were the perks that came with various jobs. A really good state farm worker could hope to go on a free state-paid trip to a Black Sea resort once or twice in a lifetime; an advanced party member at a state committee enjoyed more than one such trip a year, plus free use of a large city flat, a nice summer cottage, access to a limousine with a chauffeur, weekly deliveries of delicatessen food not available from regular stores, and privileged seats at theaters and concerts, all paid for by the state. Some of the most trusted party members were even allowed to travel abroad (usually to the socialist states of Eastern Europe or to Cuba), and a selected few even to the capitalist countries. Some of the latter defected, and this was how the rest of the world learned how the system in fact worked (Voslensky, 1984).

Since few goods were available in regular state stores, money per se meant little, compared to the status that came with a job. The inner circle of the Communist Party (about 10% of its membership, or roughly 2 million people, according to Voslensky's estimates) would enjoy the most privileges. These people were called *nomenklatura*, a word derived from the card file that was kept by the party on each of these members. Although you did not have to be a Communist to work at a factory, you had to be one to get promoted to a manager, and you could not be a director of a large plant without being entered into the nomenklatura's ranks. The nomenklatura was the secret ruling class of the Soviet society, concealed in censuses under innocuous-sounding names such as "servants of the people" and "senior executive managers" (Voslensky, 1984).

The Impact of Collectivization and Industrialization

The Soviet period left a profound impact on the national geography. Let's consider the city of Moscow, for example. Prior to the revolution, it was the historical capital of the nation, with the Kremlin, famous churches, palaces, squares, museums, theaters, shops, and parks. It had some factories as well, but the overall character of the city was oriented toward consumption, not production. By contrast, in the 1980s Moscow had hundreds of factories, including a huge truck plant, a large automobile plant, and scores of secret military research labs. In addition, hundreds of new power plants, warehouses, railroad stations, and industrial complexes were built throughout the city during the Soviet period. Across the nation, numerous large-scale construction projects (dams, coal mines, oil fields, metallurgy plants, railroads, etc.) were initiated. Dozens of new cities were built in the Arctic, in Siberia, and in Central Asia (Hill & Gaddy, 2003).

In the late 1920s, Stalin sensed that a great leap forward was needed to protect the "socialist revolution" from the enemies around the Soviet Union. The traditional potential enemies at that time were the British and the Germans, and more distantly the United States. Although tsarist Russia had been the fifth largest economy in the world and had developed particularly fast in 1910–1914, World War I and the subsequent civil war greatly diminished the country's industrial strength over the next decade, and the period of small-scale cooperative development known as the NEP in the 1920s only allowed for limited development of large enterprises. Innovation was stymied as hundreds of the best scientists and engineers left the country during the civil war, mostly for the United States. Sikorsky (father of the U.S. helicopter industry) and Zworykin (inventor of modern TV and certain types of bombs) were both brilliant Russia-educated engineers, but ended up in America.

To turn things around, Stalin proposed three things in his ambitious program presented to the 15th Party Congress in 1927:

- *Industrialization.* The goal was to create large-scale mines and industrial factories in order to double the gross domestic product (GDP) in less than 8 years, so that the U.S.S.R. could compete against the German, British, and American military machines.
- *Collectivization.* The primary goal was to create large state farms to supply food. As discussed later, another goal was to ensure that independent peasants would be destroyed, as their way of life posed a threat to Stalin.
- *Cultural revolution.* The goal was to provide for rapid education and subsequent indoctrination of the masses, and eventually to forge one Soviet nation out of the many ethnicities of the Russian Empire.

Industrialization and collectivization are considered in this section; the cultural revolution is discussed in the next section.

The most important geographic legacy of industrialization lies in the creation of large state-funded enterprises, often in very distant areas of Siberia and the north. These projects were accomplished with much heroic effort by all involved, but especially with the aid of political prisoners. Entire new cities would be built to accommodate the new coal mines, metal smelters, steel combines, wood and pulp mills, tractor and textile factories, and of course the GULAG camps themselves. Figure 7.2 is a map showing the locations of the main projects undertaken during this period.

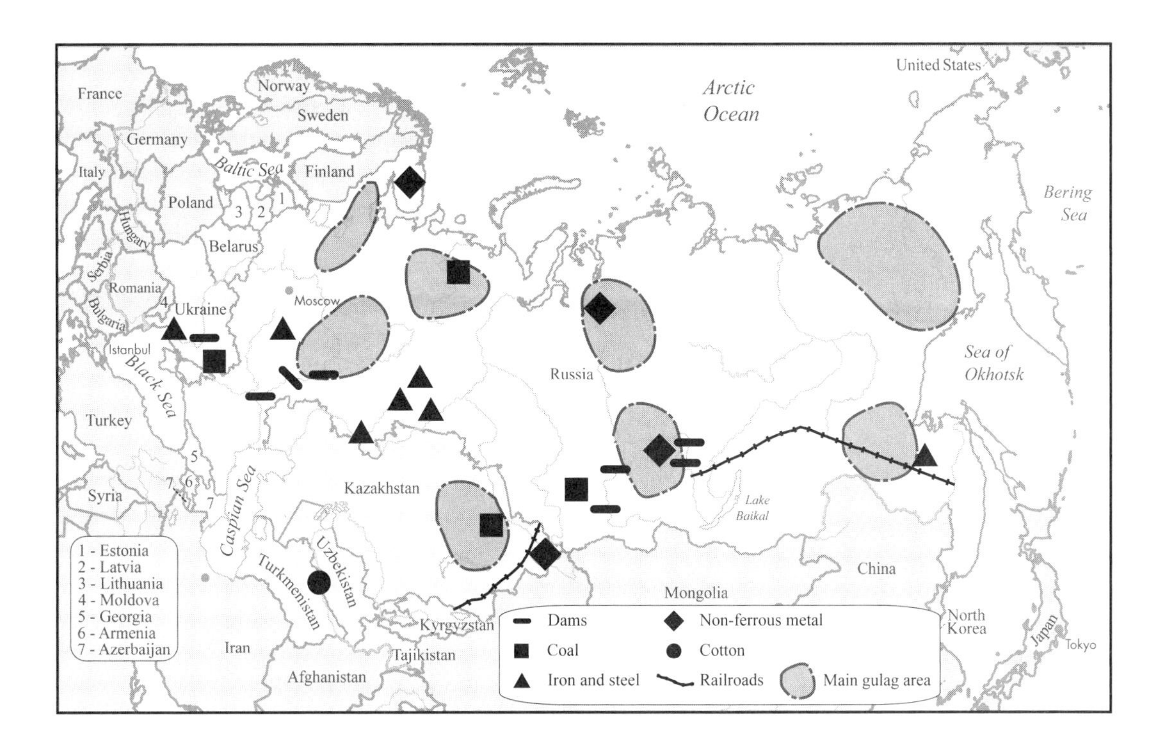

FIGURE 7.2. Major projects of Stalinism (1930–1953). The main GULAG areas are circled. Map: J. Torguson.

- DneproGES was built on the Dnieper in Ukraine in 1932. It was the first large hydropower installation in the U.S.S.R., with a capacity of about 650 megawatts (MW) (Hoover Dam, built on the Colorado River at about that time, has a capacity of about 2,000 MW). After World War II, numerous large dams were built on the Volga and in Siberia.
- The Belomorcanal, a canal 227 km long, was built in less than 2 years and connected the White Sea to the Baltic Sea and to the Moscow–Volga canal systems.
- Development of the Donbass, Vorkuta, Kuzbass, and Karaganda coal-mining basins allowed production of the coke necessary for making steel, and provided fuel for other factories and power plants.
- The central Urals (Chelyabinsk, Nizhny Tagil, Sverdlovsk) and southern Urals (Magnitogorsk) saw the creation of some of the largest steelmaking combines in the world.
- Norilsk and Kola were tapped for deposits of copper, molybdenum, nickel, and rare metals (e.g., platinum and palladium).
- The encircled areas in Figure 7.2 are areas where GULAG labor camps were located: the Karelian and Komi camps in the north; the Mordovia camps east of Moscow; the West Siberian, Norilsk, and Karaganda camps east of the Urals; and the Far Eastern and Kolyma camps on the Russian Pacific side. Of these, the most infamous and deadly were the extremely cold and remote Kolyma camps, where between 500,000 and 2 million people perished.

The results of industrialization were profound. In 1929, for example, the country made only 1,800 tractors; in 1937 it made over 66,500. In 1929 only 35 million metric tonnes (mmt) of coal were produced; by 1937 over 128 mmt were mined. The Soviet economy grew between 10% and 15% per year in the mid-1930s. Some Western journalists were flown in and shown the new

FIGURE 7.3. Soviet-era nonferrous metallurgy plants were huge, like this one in Oskemen, East Kazakhstan. The smokestacks are at least 100 m high. *Photo:* Author.

great spectacle of Communism—usually only the best examples, of course (Figure 7.3). In reality, the feverish growth was partially fueled by constant fear of long prison sentences and partially by many workers' genuine enthusiasm about building something big and new.

Collectivization involved the forcible creation of huge farms called *kolkhozy* (i.e., collective farms). By creating these huge factory-like farms, the state accomplished two things: (1) Independent farmers lost their private land holdings and therefore could not possibly ever stage a revolt; and (2) a more efficient system of mass food production and distribution was supposedly created. Effectively, the system not only did away with private farming, but returned the country to the period of serfdom, when peasants could not own any land themselves. The main targets of collectivization were the so-called *kulaks* ("fists")—basically, any peasants with means, such as a few horses or cows. Massive expropriations of the kulaks' property started in 1932. By the end of 1935, over 2 million kulak families were sent into exile to Siberia or to Kazakhstan, to languish under unbearable conditions in the cold, merciless taiga or empty steppe. Those who attempted to resist were promptly shot. Because the most productive peasants were the first victims, enormous hunger (*golodomor*) ensued, especially in the breadbasket regions of Ukraine and the lower Volga. Unknown numbers simply died of hunger in one of the darkest chapters of Stalinist history.

In less than 5 years, however, the entire agricultural sector was moved to the new system of collective farm production. A typical Soviet collective farm (kolkhoz) consisted of a few thousand agricultural workers who lived in a few villages within a radius of perhaps 20 km around a central town (Figure 7.4). The fields, barns, seed, fertilizer, pesticides, fuel, tractors, and harvesting equipment were all farm-owned and shared. Each kolkhoz would usually have a school, a club, a common cafeteria, and a medical clinic, which were available at no charge to the workers. Much production shifted from diverse local crops to monocultures. The most emphasized were staple grains: wheat, rye, barley, and (since the 1950s) corn. Regional versions included farms specializing in orchard crops, vegetables, milk farming, or beef ranching, depending on the region. Fish and forestry farms also existed. Despite the collective farms' large size and supposedly efficient planning and management, Soviet agricultural productivity lagged far behind North American or European yields. In 1990, one U.S. farmer fed about 80 people, one in Canada fed 55, one in Spain fed 25, and one in the U.S.S.R. only fed about 13. About half of the difference could be attributed to the harsher climate of Russia: Even

FIGURE 7.4. State collective farms would encompass tens of thousands of hectares of land with thousands of workers in a few large villages. These villagers are still working in one, now called "an agricultural enterprise," in Altaysky Kray, Russia. *Photo:* A. Fristad.

with the best techniques, a field of potatoes in France will produce double the yield of the same-size field in Russia, because of the much longer and warmer season in Western Europe. However, the other half of the effect was entirely due to the inefficiency of Soviet production.

To understand why, imagine that you were the director of a kolkhoz in about the year 1980. Your payment from the government would not be determined by how much food you could grow; it would be pretty much fixed. You would also have to meet rather arbitrary annual targets of production (e.g., "Produce 1,000 tons of apples by October 1"). Although these targets were not completely unfounded, the unpredictable weather patterns or local demand on workers to do other tasks could interfere with meeting them. If you missed the target or were late with the harvest, you might be chastised by the local party officials, or the kolkhoz (not yourself) would have to pay a nominal fine. In the worst-case scenario, you could be put in prison, although this was unlikely after 1960. Even if you grossly missed the target, you could usually still explain it away as something due to bad weather, pests, or lax workers' discipline, which you could not improve despite your best efforts. Now if you met your target ahead of schedule, you would be patted on the back, given a token prize, sent for a nice vacation, or maybe even promoted in the party ranks. Thus the incentives were largely nonmonetary, and not really worth much. The majority of farms simply grossly overstated their real harvests—lied to their bosses, in other words—and got away with it. Many would choose to have prizes in the office over a good harvest in the barn. The resulting chronic shortage of even basic food staples in the state stores became so widespread by the 1980s that even the notoriously senile Brezhnev's government had to tackle it with the so-called national food program, which did little to change the situation.

The 1950s saw widespread irrigation projects in Central Asia and in southern parts of European Russia and Ukraine. A very ambitious program of land development called Virgin Lands was launched in northern Kazakhstan by Nikita Khrushchev in 1953. Over 330,000 km^2 of virgin semi-arid steppe were plowed under and planted with wheat there. Many farmers needed to be brought in from all over the U.S.S.R. to work this new agricultural land, but the time of the GULAG was almost over, so the Komsomol (the Soviet organization for youth) was charged with recruiting them. Over 300,000 people, mostly Russians and Ukrainians, arrived in the Virgin Lands to begin working on large state farms. Perhaps an additional million came as soldiers, students, mechanics, and other service workers, as well as members of their families. By the end of the mass immigration to the Virgin Lands, Slavs outnumbered Kazakhs in many areas in the north—a trend that is now reversing itself. The main town was renamed Tselinograd, or "Virgin Lands City." It is the capital of today's Kazakhstan, renamed Astana. Although production per hectare in this marginal habitat was only one-quarter to one-half of the American yields in comparable areas in North Dakota, the scale was completely unprecedented for Central Asia. After Khrushchev's visit to the United States in 1959 (the first such visit by a Soviet leader), he was so impressed with American achievements in farming that he decided to greatly increase cotton, corn, soy, and hog production. Some progress was made, although overzealous party officials tried promoting the growing of corn even in the far north of Russia, where it failed miserably for climatic reasons. By the 1970s, however, the chronic inefficiency of the agricultural sector forced the country to begin massive imports of grain from the United States and Canada, sugar from Cuba, and some processed food from Europe, in exchange for Soviet petroleum and natural gas. Today this is happening all over again: Post-Soviet Russia needs to import over 40% of all its food, despite some recent improvements in private farms and food processing.

Cultural Sovietization

One of the great Soviet myths was that the diversity of cultures in Northern Eurasia could eventually be fused into one great Soviet nation. Thus, it was argued, there would be no more Russians, Kazakhs, Jews, or Estonians; instead a new nation would be made. There was a problem with this myth: Culture is stubbornly resistant to change, and governments, even very repressive ones, can do little to change it. Although Soviet society was unquestionably founded on the idea of the internationalism of all workers, Russians, Ukrainians, and a few other large groups had an undeniable edge in getting promoted to the top jobs. Some of the early Politburo leaders were Jewish (Trotsky, Zinoviev, Kamenev) or Georgian (Stalin, Ordzhonikidze) by nationality. However, the late Soviet Politburo included mainly ethnic Russians and Ukrainians. To be sure, each Soviet republic was always headed by a Communist secretary of local ethnicity, as noted earlier—but the second in charge was a native Russian vice-secretary, whose job was primarily to spy on the secretary and to ensure local compliance with Moscow's decisions.

The process of Sovietization promoted the Russian language as a common form of communication, or Soviet lingua franca. Starting in preschool and continuing throughout life, Russian was taught along with, or in place of, the local language (Figure 7.5). Although primary and middle schools would use both the local language and Russian for instruction, in high school and especially in college almost all instruction would be done in Russian, and virtually all textbooks were available only in Russian (some Ukrainian texts were available in Ukraine). One needs to bear in mind that for some languages, especially in the Caucasus and in parts of Asia, no written form existed even in the early 20th century. Therefore, it is easy to understand why Russian had to be used. A firm command of Russian was required to enter the Communist Party ranks and to have a good career. Newspapers, radio, TV programs, and books were available in the native languages, however.

Another powerful tool of Sovietization was mandatory military service. Starting at age 18, every man had to serve for 2 years (or 3 in the navy). Exemptions from the draft were made for those enrolled full-time at a few dozen of the most prestigious universities, and also for medical reasons. Once in the military, a young man was typically sent very far away from home—from Moscow to the trans-Baikal region of Chita, or from Azerbaijan to the Kola Peninsula in northern Russia, for example (Figure 7.6). This was done deliberately, for several reasons: to prevent soldiers from running back home, to homogenize the military, and to instill a common culture. Ethnic groups would still naturally form supportive communities (*zemlyachestva*)

FIGURE 7.5. The Kazakh (left) and Russian (right) languages coexist on this sign at the entrance to the young naturalists' station in Almaty. Kazakhstan is 40% Russian-speaking, but Russian is no longer an official state language, merely a "language of cultural communication." *Photo:* Author.

FIGURE 7.6. Russian Army soldiers doing a drill in the distant Chita region. *Photo*: P. Safonov.

based on their shared home region or language. However, all military instruction was conducted in the Russian language. Furthermore, in an effort to destroy all non-Soviet nationalism, the commanders instilled a view of the Soviet Union as the Motherland. Upon returning home at 20 or 21, a young man would no longer fit in with his familiar domestic environment. He would become in a sense orphaned, because for 2 or 3 years his life had been among a very different set of people. Whatever he had learned in high school or college often had to be learned anew.

Yet another form of Sovietization was shared interest in and support of arts and sports (Chapter 15). The arts were heavily promoted by the Soviet government (Table 7.1). Some art forms of distinct ethnic heritage were supported (e.g., embroidery or the production of carved wooden toys). At the same time, many artists, actors, writers, and sculptors from the ethnic regions of the U.S.S.R. would study and work in the best central locations—most importantly Moscow and Leningrad, but also in the republican capitals, where their works would become known to many.

Sports were also heavily promoted by the state. In fact, all the "amateur" teams in hockey, soccer, volleyball, and other team sports were actually heavily subsidized professional clubs, whose members were on the state payroll. The successes of the Soviet Olympic teams are legendary. The Soviet Union first participated in the 1952 games. In 1972 in Munich, Germany, the U.S.S.R. won 50 gold medals, 27 silver, and 22 bronze; the United States ran a distant second, with 33 gold, 31 silver, and 30 bronze medals. The U.S.S.R. was the foremost winner of medals seven out of nine times in both the Summer and the Winter Olympics. Of the summer sports represented, the highest gold medal counts for the Soviet team over its history were earned in gymnastics, athletics (i.e., track and field events), wrestling, weightlifting, canoeing, fencing, shooting, boxing, and swimming (in descending order). In winter sports, the most Soviet medals were won in cross-country skiing, speed skating, figure skating, biathlon, and ice hockey. Table 7.2 lists all Soviet gold medal holders from the 1976 Summer Olympics in Montreal. This is the best example from the late Soviet period, because the American team boycotted the Moscow Summer Olympics in 1980, and the Soviet Union boycotted the Los Angeles Summer Olympics in 1984 (both citing political reasons), thus skewing the

TABLE 7.1. Some Great Cultural Figures of the Soviet Period

Name	Occupation	A major accomplishment
Sergei Eisenstein	Film director	*The Battleship Potemkin* (movie)
Kazimir Malevich	Painter	*Black Square* (painting)
Boris Pasternak	Poet and writer	*Doctor Zhivago* (novel)
Andrei Platonov	Writer	*Foundation Pit* and *Chevengur* (novels)
Serge Prokofiev	Composer	*Piano Concerto #2*
Mikhail Sholokhov	Writer	*And Quiet Flows the Don* (novel)
Dmitry Shostakovich	Composer	*Seventh Symphony*
Konstantin Stanislavsky	Theater producer	Productions of the *Seagull* and other plays by Anton Chekhov
Andrei Tarkovsky	Film director	*Andrei Rublev* (movie)
Marina Tsvetaeva	Poet	Many great poems
Agrippina Vaganova	Ballet dancer and teacher	The Kirov Ballet School in Leningrad
Vladimir Vysotsky	Actor and poet	Hundreds of songs

Note. This table includes some persons who remained in the U.S.S.R. and others who left and then came back. All were active before 1991.

TABLE 7.2. Gold Medals Won by the U.S.S.R. in the Montreal 1976 Summer Olympics

Discipline	Events	Name/Team
Artistic gymnastics	Men's floor exercises	Andrianov, Nikolay
Artistic gymnastics	Women's floor exercises	Kim, Nelli
Artistic gymnastics	Men's individual all-around	Andrianov, Nikolay
Athletics	Women's 1,500 m and 800 m	Kazankina, Tatiana
Athletics	Men's hammer throw	Sedykh, Yuri
Athletics	Men's triple jump	Saneev, Viktor
Basketball	Women's basketball	U.S.S.R.
Canoe/kayak, flatwater	Men's 500 m single canoe	Rogov, Aleksandr
Canoe/kayak, flatwater	Men's 500 m double canoe	Petrenko, Sergei/Vinogradov, Aleksandr
Cycling, road	Men's team time trial	U.S.S.R.
Diving	Women's 10-m platform	Vaytsekhovskaya, Elena
Fencing	Women's foil, team	U.S.S.R.
Fencing	Men's sabre, individual	Krovopuskov, Viktor
Fencing	Men's sabre, team	U.S.S.R.
Handball	Men's handball	U.S.S.R.
Handball	Women's handball	U.S.S.R.
Judo	Men's +93 kg (heavyweight)	Novikov, Sergei
Judo	Men's 63–70 kg (half-middleweight)	Nevzorov, Vladimir
Rowing	Men's four-oared shell with coxswain	U.S.S.R.
Shooting	Mixed 50-m running target (30 + 30 shots)	Gazov, Aleksandr
Swimming	Women's 200-m breaststroke	Koshevaya, Marina
Weightlifting	Men's +110 kg, total (super-heavyweight)	Alekseyev, Vasily
Weightlifting	Men's –52 kg, total (flyweight)	Voronin, Aleksandr
Weightlifting	Men's 56–60 kg, total (featherweight)	Kolesnikov, Nikolai
Weightlifting	Men's 60–67.5 kg, total (lightweight)	Korol, Pyotr
Weightlifting	Men's 75–82.5 kg, total (light-heavyweight)	Shary, Valeri
Weightlifting	Men's 82.5–90 kg, total (middle-heavyweight)	Rigert, David
Weightlifting	Men's 91–110 kg, total (heavyweight)	Zaitsev, Yuri
Wrestling freestyle	Men's +100 kg (super-heavyweight)	Andiev, Soslan
Wrestling freestyle	Men's 52–57 kg (bantamweight)	Yumin, Vladimir
Wrestling freestyle	Men's 62–68 kg (lightweight)	Pinigin, Pavel
Wrestling freestyle	Men's 82–90 kg (light-heavyweight)	Tediashvili, Levan
Wrestling freestyle	Men's 90–100 kg (heavyweight)	Yarygin, Ivan
Wrestling, Greco-Roman	Men's +100 kg (super-heavyweight)	Kolchinsky, Aleksandr
Wrestling, Greco-Roman	Men's –48 kg (light-flyweight)	Shumakov, Aleksei
Wrestling, Greco-Roman	Men's 48–52 kg (flyweight)	Konstantinov, Vitali
Wrestling, Greco-Roman	Men's 62–68 kg (lightweight)	Nalbandyan, Suren
Wrestling, Greco-Roman	Men's 68–74 kg (welterweight)	Bykov, Anatoli
Wrestling, Greco-Roman	Men's 82–90 kg (light-heavyweight)	Rezantsev, Valeri
Wrestling, Greco-Roman	Men's 90–100 kg (heavyweight)	Balboshin, Nikolai

Note. Data from *www.olympic.org*.

picture for those two sets of games. Since typical Russian names end only with "-ov" or "-in," it is clear that other ethnicities besides Russian are represented (the list includes at least one Armenian, one Georgian, one Korean, one Pole, and one German—all raised in the U.S.S.R.).

Achievements and Problems of the Late Soviet Period

Although it is reasonable to expect that without a Communist government Russia could have achieved similar or even better development over the course of the 20th century, it is undeniable that by the end of World War II the U.S.S.R. emerged as the world's second-largest superpower, able to openly challenge the United States. The fact that the Allies won World War II at all, despite the extremely heavy human toll (officially, over 20 million Soviet people died in the conflict, as compared to about 9 million Germans and slightly over 500,000 Americans), is a testimony to the tremendous resilience and sacrifice of the Soviet people.

Many Soviet achievements of the 1950s and 1960s were in the social and economic spheres, as well as in military might:

- Universal education was achieved, with a corresponding 100% literacy rate among adults. School attendance was made compulsory through the 8th grade (later the 10th grade). In addition, free education was available at the university level for a selection of the best students; many new universities were founded, and existing ones were expanded (Vignette 7.1).
- Free, comprehensive health care was available, including access to world-class surgery procedures, pioneering diagnostic techniques, and domestically developed and produced medical drugs.
- Maternity benefits were among the best in the world (3 years' leave of absence at close to full pay!), and free child care was available for preschoolers.
- Mortality rates were low (though slightly above those in Western Europe or North America), and birth rates were moderately high.
- The Soviets launched the first artificial satellite, Sputnik, in 1957, and put the first man in space in 1961. An ambitious program of permanent orbital space stations (Salyut and Mir) was developed in the 1970s and 1980s.
- Nuclear parity with the United States was achieved, with over 10,000 nuclear warheads on each side. Soviet-built ICBMs were capable of carrying multiple warheads and reaching anywhere in the world in less than 20 minutes.
- Many new and superior conventional weapons were developed (e.g., the MIG and Su-series jet fighters; T-70, -80, and -90 tanks; and S-200, -300, and -400 antiaircraft mobile missile launchers).
- Large-scale production of passenger jets included the Tu-144 supersonic jet and the Il- and Tu-series long-range passenger jets of domestic design.
- Large-scale production of certain types of consumer goods began, although these were rarely comparable to Western goods in quality; TVs, stereos, washing machines, and refrigerators were all domestically made.
- Excellent transit systems, including subways, were built in about 10 cities (including Moscow, Leningrad, Novosibirsk, and the biggest republican capitals).
- Several Nobel Prizes were won in physics, chemistry, medicine, and literature.
- World-class resorts were built on the Black and Baltic Seas.

More details on some of these accomplishments are presented in Chapters 13–16. By the late 1970s, however, despite continuing homage to Lenin and other Soviet heroes (Figure 7.7), it was becoming clear that the system was showing signs of major problems.

Much has been written about the political and economic challenges of the late Soviet period. Rosefielde (2007) provides a robust theoretical framework for economic analysis of the failure of the Soviet system, based on the application of the Pareto–Arrow–Bergson (PAB) model. The PAB model allows analysts to directly compare and contrast the outputs of two very different systems: command and market economies. The challenge, among other things, is to compare

Vignette 7.1. The Moscow State University Building: One of the Projects of Stalinism

If you have been to Moscow, you probably have seen the main building of Moscow State University, with a spire soaring to 250 m (Figure 1). The building was completed in 1954, the year after Stalin's death. It looks like a wedding cake, and its neo-Empire design is rather similar to that of some New York skyscrapers of the 1920s. The spire is topped with a massive five-pointed star, which is almost 9 m across! The original plans called for a statue of Stalin to be placed on the top, but this plan had to be scrapped because of the danger that the wind would topple it. The building has 33 floors and houses the schools of mathematics, geology, and geography, as well as numerous dormitories, about 150 apartments for professors, a few cafeterias, a radio station, and a few large assembly halls. The building not only goes above ground; it goes below the ground surface for about seven floors and has a massive bomb shelter at its base. The entire complex is almost 300 m wide at ground level; one would have to walk for over 20 minutes to get around it.

The Moscow State University building was constructed with prison labor. Thousands of workers toiled for about 5 years to complete the project. (A few escaped their misery by jumping off the walls to a certain death.) Inside the building, massive oak panels cover the walls, and the floors are marble and granite. According to one estimate, it took almost one-third of the entire country's hardwood production in 1952–1953 to produce enough wood for the paneling. Although we may disagree about its aesthetics, the mere fact that Moscow's tallest building in the past was not a bank or even a palace of Soviet delegates, but a university, testifies to the Soviet emphasis on science and education—an emphasis that many observers see as lacking in post-Soviet Russia.

FIGURE 1. Moscow State University's main building was built between 1949 and 1954. It is about 250 m high and was the tallest building in the city for over 50 years; it is now surpassed by a few office skyscrapers. *Photo:* Author.

FIGURE 7.7. Lenin's statue still graces the square in front of the old Communist Party city headquarters building in Biysk, Altaysky Kray. *Photo:* Author.

the amounts of products and services, or the prices, produced by two different mechanisms. The Soviet system had fixed prices that did not reflect actual supply or demand, and the ruble was not directly exchangeable with any foreign currency, so year-to-year comparisons with the West are not immediately possible. Moreover, the official Soviet statistics were notoriously and deliberately misleading. Among the other main inefficiencies of the period, Rosefielde (2007, p. 136) cites these: (1) State demand controlled all aspects of production, so that there were no free agents available to counterbalance the state's monopoly; (2) coercion was substituted for monetary incentives for workers, so that the supply of workers was not reflective of what was actually needed, resulting in oversupply of some items and chronic undersupply of others; and (3) no market equilibration was possible because of the state-fixed prices.

According to the best CIA estimates and other common studies of the late Soviet period, the growth of the Soviet GDP slowed from a robust 4.5–6% per year (1961–1965) to an anemic 0.5–2% two decades later (1981–1985). The only branch of the economy that kept growing in the late 1970s was the military. This was paid for in part by hidden inflation and in part by oil and gas sales from the newly developed fields in the West Siberia economic region. On a per capita basis, the Soviet GDP as measured by the CIA in 1990 was about 30% of the U.S. GDP, whereas the Soviet estimates put it at 60%. In an independent assessment with the PAB model, Rosefielde puts the per capita GDP at only 20%.

Anyone who visited the late Soviet Union from the West was uniformly struck by how poorly the people lived, in comparison either to what was imaginable from the official Soviet propaganda, or to the lives of their counterparts in the West—all the free services notwithstanding. The area where the contrast was most apparent was housing: The average Soviet citizen had less than 20% of the square footage available to the average American, and perhaps about 40% of the level available to the average European. In addition, over half of the country's population had no access to indoor plumbing.

The U.S.S.R. in the late 1980s still seemed to be a superpower, but increasingly this was only a facade. It could not feed itself without imported food; workers' productivity lagged far behind that in the West; stealing from employers was commonplace; there were long lines to buy anything of value (such buying was essentially a form of hidden inflation); and the growing international military competition with the West was not easing up. A new paradigm was urgently needed to allow the country to respond to the increased internal and external challenges. To do this would require changing the system, but who could do that?

REVIEW QUESTIONS

1. Name the three main components of Stalin's plan for moving the U.S.S.R. forward.
2. Explain the roles of the Communist Party, the Soviets, and the executive committees in the Soviet Union. Compare and contrast the Soviet system with the U.S. system of three branches of government.
3. Why was industrialization necessary in the 1930s?
4. What parts of Russia were most affected by industrialization?
5. Why do you think famine occurred in the early 1930s in the U.S.S.R.?
6. Name any two nationalities in the U.S.S.R. that had no Soviet republic of their own. Why do you think this might have been the case?
7. Choose any five major accomplishments of the late Soviet Union from the list near the end of this chapter. Compare and contrast those with what was accomplished in the United States (or Western Europe) in the same period of time.
8. Argue that fixed prices may have some benefits, along with disadvantages. Try to convince another person in class that having one set price for the same product nationwide has some advantages over the free-market system, in which price is set in accordance to local supply and demand.
9. How would you go about figuring out the Soviet economic performance level, based on indicators other than prices? Can you think of some objective parameters that could be measured by an independent observer?

EXERCISES

1. Use the Olympic Committee Website (*www.olympic.org*) to research in-depth changes in the numbers of Soviet gold, silver, and bronze medals won at all Olympic Games from the end of World War II until 1988 (excluding the 1984 Summer Olympics in Los Angeles). What sports seem to be consistently lacking Soviet Olympic champions? What sports have seen the best Soviet results? Can you explain why?
2. Prepare a 5-page report on any of the large projects of Stalinism (use the map in Figure 7.2 for ideas). Describe what was built, when, where, by whom, and why. Does it still exist today? Add any pictures and/or descriptions of it that you like. Do you know any similar projects in the country where you live? If so, when and where was each one built? What are the similarities and differences between them?
3. Interview an older relative who may have lived outside the U.S.S.R. during the late Soviet period. Ask this person to describe the perception of the U.S.S.R. that he or she had as an outsider. Think about how what you now know about that period is different from what your relative describes.

Further Reading

Applebaum, A. (2003). *Gulag: A history.* New York: Doubleday.

Bahry, D. (1987). *Outside Moscow: Power, politics, and budgetary policy in the Soviet republics.* New York: Columbia University Press.

Baykov, A. (1954). The economic development of Russia. *Economic History Review, Second Series,* 7(2), 137–149.

Cole, J. P. (1967). *A geography of the U.S.S.R.* Harmondsworth, UK: Penguin Books.

Conquest, R. (1986). *The harvest of sorrow: Soviet collectivization and the terror-famine.* New York: Oxford University Press.

Conquest, R. (1990). *The great terror: A reassessment.* New York: Oxford University Press.

Davis, H., & Scase, R. (1985). *Western capitalism and state socialism.* Oxford, UK: Blackwell.

Dobrenko, E., & Naiman, E. (Eds.). (2003). *The landscape of Stalinism: The art and ideology of Soviet space.* Seattle: University of Washington Press.

Douglas Jackson, W. A. (1965). *Soviet Union.* Grand Rapids, MI: Fideler.

Ellman, M. (2007). Stalin and the Soviet famine of 1932–33 revisited. *Europe–Asia Studies,* 59(4), 663–693.

Frankel, B., Horan, A., Manners, J., & Thompson, D.

S. (1985). *The Soviet Union.* Alexandria, VA: Time–Life Books.

Hill, F., & Gaddy, C. (2003). *The Siberian curse: How Communist planners left Russia out in the cold.* Washington, DC: Brookings Institution Press.

Hooson, D. J. M. (1964). *A new Soviet heartland.* Princeton, NJ: Van Nostrand.

Hosking, G. (2006). *Rulers and victims: The Russians in the Soviet Union.* Cambridge, MA: Harvard University Press.

MacKenzie, D., & Curran, M. W. (1986). *A history of the Soviet Union.* Chicago: Dorsey Press.

Rosefielde, S. (2007). *The Russian economy: From Lenin to Putin.* Malden, MA: Blackwell.

Service, R. (2000). *Lenin: A biography.* Cambridge, MA: Harvard University Press.

Service, R. (2005). *Stalin: A biography.* Cambridge, MA: Harvard University Press.

Shearer, D. R. (1996). *Industry, state, and society in Stalin's Russia, 1926–1934.* Ithaca, NY: Cornell University Press.

Smith, G. B. (Ed.). (1980). *Public policy and administration in the Soviet Union.* New York: Praeger.

Solzhenitsyn, A. I. (1974). *The Gulag archipelago 1918–1956: An experiment in literary investigation.* New York: Harper & Row.

Theen, R. H. W. (1980). Party and bureaucracy. In G. B. Smith (Ed.), *Public policy and administration in the Soviet Union.* New York: Praeger.

Voslensky, M. S. (1984). *Nomenklatura: The Soviet ruling class.* Garden City, NY: Doubleday.

Websites

www.ibiblio.org/expo/soviet.exhibit/entrance.html#tour—The Library of Congress's Soviet Archives online exhibit tour; an excellent and concise overview of the internal workings of the Soviet system, covering everything from the KGB to Chernobyl.

www.cnn.com/SPECIALS/cold.war/episodes/01/—CNN special on the Cold War.

www.lib.utexas.edu/maps/commonwealth.html—Collection of CIA-produced maps of the Soviet Union; useful for many class assignments.

sovietart.com—Paintings and sculpture from the Soviet era that can be analyzed for hidden meanings of what it meant to be Soviet.

CHAPTER 8

Post-Soviet Reforms

The 1985 election of Mikhail Gorbachev as a new leader of the Communist Party of the Soviet Union (C.P.S.U.) ushered in a new era. (Table 8.1 summarizes the political characteristics of this era to date, and Table 8.2 provides a brief general timeline of it.) The stagnation of the Brezhnev period had ended with his death in 1982. After two successors to Brezhnev died in rapid succession, the Communist elite wanted someone younger and healthier in the lead. Gorbachev was apparently chosen because of his relative youth and unassuming demeanor. He was a good compromise: A peasant boy from the grain-rich Stavropol region, he seemed provincial enough to present little danger of despotism. He was also well educated and was supported by some of the most forward-looking members of the Central Committee of the C.P.S.U.

Gorbachev's Perestroika

As discussed in Chapter 7, Gorbachev inherited a deeply entrenched, but increasingly dysfunctional, totalitarian political system and a sickly state-run economy. On the one hand, even the party elite was getting tired of the old-fashioned, inefficient command economy and other methods of running the country. On the other, the economy stopped growing. Much of the country's foreign earnings came from exports of petroleum from the west Siberia economic region. Unfortunately for the Soviets, Saudi Arabia and other Organization of the Petroleum Exporting Countries (OPEC) members greatly expanded their oil production in the early 1980s, to counterbalance the price shocks in the aftermath of the Iranian revolution. Global oil prices went from $75 to less than $20 per barrel—roughly the break-even point for the Russian oil producers. Much of the hard currency earned by the Soviet Union from the oil sales had to be spent on purchases of imported food and basic consumer goods in any case. In short, the economic picture was not pretty. There is evidence that Gorbachev, even when he tried, could not obtain reliable in-country statistics on how bad things truly were (Åslund, 2007).

In the late 1980s, over 60% of the Soviet Union's industrial output was in the form of heavy machinery (tractors, turbines, engines, etc.), thought to be necessary for the production of better goods and weapons. Less than 30% was accounted for by consumer goods. The persistent problems were these:

TABLE 8.1. Basic Political Characteristics of the Brezhnev, Gorbachev, Yeltsin, and Putin Periods

	Brezhnev (in 1975)	Gorbachev (in 1989)	Yeltsin (in 1995)	Putin (in 2004)
Head of the country	Secretary-general	President of the U.S.S.R.	President of the Russian Federation	President of the Russian Federation
Head of government	Chairman of the Supreme Soviet	Prime minister	Prime minister	Prime minister
Parliament	Supreme Soviet	Congress of People's Deputies	Federation Council and Duma	Federation Council and Duma
Number of parties in parliament	One	One (later a few)	Five or six	Three or four
Regional governors	First secretary (appointed)	Appointed (later elected)	Elected governors	Appointed governors
Freedom of press	No	Limited freedom	Free	Limited freedom
Independent TV	No	No	Yes	No
Freedom of religion	No	Limited	Yes	Yes, except for some new religious movements
Wars/conflicts	Afghanistan	End of Afghanistan, Karabakh, Trans-Dniester Republic	Chechnya I	Chechnya II
Defense alliances	Warsaw Pact	CIS	CIS + NATO partnership	CIS + NATO partnership
Private economy	0%	5%	20%	75%

TABLE 8.2. General Timeline of the Post-Soviet Reforms in Russia

Dates	Main events
1985	Gorbachev elected secretary-general of C.P.S.U.
1985–1986	His ill-fated antialcohol campaign.
April 1986	Chernobyl nuclear disaster in northern Ukraine.
1987	Beginning of perestroika and glasnost.
Dec. 1988	First multicandidate elections to the Soviet Parliament.
1988–1990	Rising nationalism in the Baltics, Georgia, Ukraine, and Moldova.
June 1991	Ex-Communist Yeltsin elected first president of the R.S.F.S.R.
Aug. 1991	Hardliners' 3-day coup.
Dec. 1991	U.S.S.R. dissolved; Gorbachev resigns; Commonwealth of Independent States (CIS) formed with 12 of the 15 former republics as members (the Baltics do not join).
Jan. 1992	Liberalization of prices; inflation close to 1,000% by year's end.
Sept. 1992	First voucher auction.
Dec. 1992	Reformer Gaidar resigns as the prime minister; "gas man" Chernomyrdin takes office.
Oct. 1993	Parliamentary crisis in Moscow; Yeltsin sends in tanks.
Dec. 1993	New constitution gives the president sweeping powers; Duma elected.
July 1994	Voucher investment scam collapses; millions lose savings.
Dec. 1994	First war in Chechnya begins.

(cont.)

TABLE 8.2. *(cont.)*

Dates	Main events
Mar. 1995	Loans-for-shares scheme proposed by Potanin, Khodorkovsky, Smolensky.
Dec. 1995	Communists do very well in Duma elections.
Feb. 1996	Oligarchs meet in Davos with members of Yeltsin's circle; they promise political support before upcoming elections.
May 1996	Chechen rebels take hostages at the Budenovsk hospital; a cease-fire is declared between Chechen and Russian forces.
June 1996	Yeltsin wins first round of presidential election; he sacks his long-time bodyguard and friend, Korzhakov, at the oligarchs' instigation.
July 1996	Yeltsin suffers a massive heart attack, but defeats Zyuganov in the second round of presidential elections.
Fall 1996	Yeltsin undergoes open-heart surgery; some oligarchs occupy various government positions.
1997	Russian emergent economy is rattled by the spreading Asian currency crisis; inflation runs about 20% per year.
Mar. 1998	Chernomyrdin is sacked as prime minister and replaced by young, inexperienced Kiriyenko.
May 1998	Russian stock market crashes; Chubais and others plead for help from the International Monetary Fund (IMF).
July 1998	IMF approves a $22 billion loan for Russia as a bailout; $4.8 billion is disbursed.
Aug. 1998	Partial default: Ruble is devalued; default on GKO bond payments; temporary moratorium on foreign debts of Russian companies is announced; Kiriyenko is sacked.
Fall 1998	Primakov comes in as new prime minister, stabilizes situation, and scares oligarchs with promises to put many in jail.
May 1999	Primakov is dismissed; Stepashin is appointed as transitional prime minister; search for a successor for Yeltsin quietly goes on.
Aug. 1999	Putin appointed as prime minister and declared heir apparent by the media.
Sept. 1999	Bombs explode in a few Russian cities; Chechens are blamed (although some evidence indicates that the Federal Security Service is at least complicit), and a new round of war in Chechnya begins.
Dec. 1999	Yeltsin steps down; Putin becomes acting president.
Mar. 2000	Putin elected second president of Russian Federation.
2000	Kasyanov is appointed prime minister; members of Yeltsin's government are being gradually replaced with personal acquaintances of Putin.
2001–2002	Growing state control over media: NTV and ORT TV channels are turned over to companies loyal to the Kremlin; their owners, Gusinsky and Berezovsky, flee the country.
2001–2002	Tax code is streamlined, and a flat tax of 13% is introduced. Seven federal districts are proposed for the country, with each having a personal presidential representative (vertical structure of power).
Oct. 2003	Richest man in Russia, Khodorkovsky, is put in jail on corruption charges.
Dec. 2003	Pro-Putin "United Russia" party wins an overwhelming majority of seats in Duma.
Mar. 2004	Putin easily wins reelection; Fradkov is appointed prime minister.
Mar. 2008	Medvedev is elected president; Putin becomes prime minister.

- Lack of variety. Only a few basic designs in each category were available.
- Lack of quantity. Some regions had more than others; planners routinely overplanned or underplanned production, which was inevitable, given the lack of a free market.
- Lack of quality. There was no incentive to produce better goods, because there was no competition among the factories; some quality control was in place, but it was rarely adequate to ensure durability, consistency, freshness, and so on.

The productivity per worker was only a fraction of that in the West (see the discussion of per capita gross domestic product [GDP] at the end of Chapter 7). Ministries duplicated some of their functions: One would be busy shipping coal 4,000 km from Kuzbass in central Siberia to Rostov-on-Don near the Black Sea, while another would ship local Donbass coal from Rostov-on-Don to Krasnoyarsk, bypassing Kuzbass on the way. Stealing among workers was common, as people tried to improve their lives by stocking up on goods that were not available from the half-empty Soviet stores. Special warehouses for the *nomenklatura* (see Chapter 7) would distribute Western-made consumer goods and luxury items to the privileged party members. In the biggest cities, including Moscow and the republican capitals, a higher diversity of goods and services was available to all. For instance, one could buy beef sausage at a Moscow grocery store at almost any time in the 1970s, albeit sometimes after a long wait in a line. Meat products were simply not available in state shops in most of the rest of the country. Most people survived by growing their own food on small *dacha* plots, by stealing whatever was available through work, by bartering rare Western goods, and by getting some exotic food items a few times a year through their employers.

The economy of the Soviet Union was not only struggling to provide for itself. It also was supporting millions in the developing world: Cuba, Nicaragua, Angola, Mozambique, Ethiopia, Vietnam, North Korea, six countries in Eastern Europe, and many others were all directly dependent on supplies from the U.S.S.R. Moreover, the mounting military costs of the Cold War were beginning to take a toll on the country's ability to protect itself. Finally, few party members seriously believed in the coming bliss of Communism any more, and even fewer wanted the return of a Stalinist level of repression to make people work harder. Gorbachev realized that if things were allowed to continue in the old ways, the Soviet system would quickly collapse under pressures from both within and without. Still, it seemed impossible to dissolve the party or to abolish socialist ideals overnight. Gorbachev felt a need to reform the system slowly and after much deliberation.

An early reform idea was to require state enterprises to become more accountable. This *khozrasschet* system was intended to ensure that every enterprise kept a running inventory of all supplies and products, and to provide regular reports to the planning authorities as feedback. No enterprise was supposed to run at a deficit. Of course, such a system was utopian from the onset, for what Soviet directors would want to report bad things about their enterprises? Or the government could try to replicate Lenin's New Economic Policy (NEP), which had been sidelined by decades of Stalinism, with its emphasis on gargantuan factories and massive farms. A return to the NEP in the 1980s would not be impossible, but would certainly be difficult. For example, in a city like Cherepovets—with a massive steel combine employing 50,000 workers at a loss, and no other factories around—what could possibly be done? Open small barber shops? Gorbachev felt that perhaps something more realistic was needed.

To add farther urgency to the situation, the Chernobyl nuclear disaster happened in April 1986. After denying the rumors about the incident for 76 hours, the state news agency finally had to admit that something went terribly wrong, after Swedish scientists began picking up increased radioactivity over northern Europe and complained. The Gorbachev government's first serious failed test was its inability to effectively confront the disaster, mobilize resources, and ask for foreign help, all in the matter of a few critical days. Over the summer of 1986, hundreds of thousands of people had to be relocated; the destroyed reactor had to be sealed; and the hard questions about how it all happened needed to be answered. The accident was not just preventable, but was absolutely avoidable: It resulted from a very poorly conceived idea of fooling around with the cooling system in the absence of an external source of power. (It was a little bit like trying to drive your car after disconnecting the alternator and draining all the oil.) The Chernobyl reactor was also of an obsolete graphite-controlled type. When the core got too hot during the planned experimental shutdown, the rods could not go back in and preclude the meltdown. All U.S. and many Soviet reactors at the time used pressurized water, not graphite, and those would be much safer to tinker with. Incredibly, the chief designers of the Chernobyl reactor were not even consulted before the experiment.

Although we still do not know every detail about the accident, it was one of the final nails in the coffin of the Soviet system. Too many people felt that the government had failed them on too many counts: The closed society could not adequately protect its citizens, or adequately explain to them what had happened and why. Some people in the regions began to demand more political openness. Environmentalists were at the forefront of this movement. Besides the environmental vulnerability to large-scale disasters, and the political powerlessness of the masses, the accident also highlighted the poor communication between the center and the periphery of both the government and the whole nation. The strict top-down hierarchical chain of command, common under leaders from Stalin to Brezhnev, was beginning to fall apart. Chernobyl was in Ukraine, a separate republic from Russia, and it had its own branch of the ministry of atomic energy carrying out the experiment without proper consultation with Moscow. Also, local police, firefighters, and political leaders had to depend on some decisions being made for them in Kiev and other decisions in Moscow.

To sum up, three factors played a role in moving Gorbachev toward the reforms: (1) the ineffective, stagnating economy; (2) political pressures from abroad, coupled with growing dissent at home; and (3) the environmental fiasco of Chernobyl. Early in 1987, Gorbachev addressed the party and the nation by proposing a three-pronged approach to reforms. He was very cautious; in no uncertain terms, he explained that this was to be an evolution, not a revolution, of the Soviet economic structure. The three aspects he announced were these:

- *Perestroika*, or restructuring of the worst elements of the Soviet planning system.
- *Glasnost*, or political openness, including freedom of the press and real elections.
- *Uskorenie*, which means "acceleration" (i.e., not simply rebuilding industries, but producing more, better, and faster to catch up with the West in the production of high-tech and consumer goods).

In the end, the only success was glasnost. Gorbachev's major accomplishments here were releasing political prisoners; abolishing the one-party policy; lifting most restrictions on the mass media; and allowing multiple-candidate local and federal elections, freedom of meetings and demonstrations, freedom of association, freedom of religion, and (toward the end of his tenure) freedom to travel abroad.

With respect to perestroika (i.e., actual economic reforms), little progress was made. The main problem was Gorbachev's inability to go beyond mere cosmetic changes. Not remodeling, but whole-scale demolition and rebuilding was needed. Gorbachev was a smart man, but his main fault seemed his inability to realize the ultimate futility of the socialist system of production, at least in its late Soviet form. He seemed to be willing to allow a few new types of semi-private or private ownership, but on a very small scale of cooperatives: a toy shop here, a barber shop there. Most of all, he was afraid to lose the Soviet Union, the Communist Party, and Russia's central place in both, and of course this was precisely what happened in 1991 anyway. On the one hand, he started promoting a reformist agenda; on the other, his hands were tied by his connections to many of the still-powerful Communists who were not at all convinced that his reforms were needed. Gorbachev did manage to assemble a strong reformist team of political advisors and economists, some of whom continued to work with Boris Yeltsin on the much more drastic reforms of the 1990s.

Some of the political reforms implemented in 1987–1989 included curtailment of the power of central administrators to control agricultural and industrial production; greater autonomy of Soviet directors to decide on what to produce, when, and with whom; expansion of workers' rights; encouragement of some small-scale private enterprises in food production and services; a focus on the production of critically needed consumer goods; and pursuit of joint ventures with foreign capital. Ventures of this last type were not entirely new: Pepsico had been present in the U.S.S.R. since the early 1970s, for example.

The C.P.S.U. monopoly was broken in 1988, and the Central Congress of People's Deputies was transformed from a merely rubber-stamping body into a real parliament, with the deputies introducing diverse legislative proposals. The first true multiparty elections took place on March 26, 1989. Gorbachev also had to control the military, which was a hard task, especially with the rise in nationalism in Georgia, Armenia, Azerbaijan, Moldova, and the Baltic states at the time. The state's weakening grip on power was correctly interpreted by the various oppressed social groups in the Soviet republics as an indication that the time to act was now. Some examples of the rising nationalism included violent protests in Baku, in Tbilisi, and in Vilnius in 1988–1990, resulting in casualties after Soviet tanks moved in. In 1988 pogroms took place in Baku against the Armenians, and in Yerevan against the Azerbaijanis, as two republics were preparing to commence a real war over control of the disputed Nagorno-Karabakh area. Gorbachev chose not to intervene.

The End of the Soviet Union

It is sometimes stated that the Soviet "empire" collapsed in 1991. Although the U.S.S.R. was a multiethnic entity, it was not an "empire" in the same sense as the British or French colonial holdings were. The Soviet Union's dissolution was a result of a deliberate political act by a few republican leaders, not of a popular revolt by the oppressed indigenous masses. The dramatic events of August 1991 took place primarily in Moscow, as those in the periphery waited quietly. In fact, in the spring of 1991, the majority of Soviet citizens (75%) had expressed their desire to keep the U.S.S.R. intact in an open referendum. With the exception of the Baltic states, which clearly wanted out at any cost, all the other republics actually could have stayed together, because there were many advantages to it. However, the Communist coup of August 1991 and the resulting power grab by Boris Yeltsin made preservation of the Soviet Union all but impossible. Just a few months after the referendum, over 75% of the voters in countries like Ukraine approved their leadership's decision to pull out of the now forever compromised U.S.S.R.

The events leading up to that point were dramatic. On August 19, 1991, the country and the world woke up to a stunning announcement by Gennady Yanaev, the vice-president, on Soviet state TV: His boss, Gorbachev, had been arrested while vacationing in Foros, Crimea, and Yanaev and five other men from the Politburo were taking full responsibility for the country. Radio stations were pulled off the air, creating an information vacuum (remember that the Internet was not yet commonly used). At the time, Yeltsin was the newly elected president of the Russian Federation—a post below Gorbachev's, but nevertheless sanctioned by the people. Yeltsin was a proven independent leader who, unlike Gorbachev, had officially quit the Communist Party a year earlier. He announced that Russia would not follow the coup leaders back to Communism. Tanks were ordered to the capital. The country seemed to be descending into a lockdown, if not an outright civil war.

The outcome of the standoff was decided in less than 3 days; the hardliners, after all, were not hardened criminals and did not have a resolve to use brutal force. Significantly, they could not manage to arrest Yeltsin or the popular and independent-minded mayors of Moscow and St. Petersburg. Even the elite units of the army were not prepared to use lethal force. The ordinary people poured out into the streets in Moscow to talk to the bewildered soldiers perched on tanks, and to erect barricades around the seat of the Russian Federation's government. Most Soviet republics' leaders had not issued any definite statements, but were waiting on the sidelines. The conflict ended on August 22, 1991—remarkably peacefully, with three young men dying in a street clash, but no major shootouts.

Gorbachev was soon back in Moscow, but was quickly sidelined by Yeltsin, who emerged as the real leader of the new Russia. The white, blue, and red flag of the Romanovs flew atop the Kremlin again, and all of a sudden everyone was a "democrat." (A small but important detail: Although Yeltsin officially gave up his C.P.S.U. membership in 1990, he never endorsed any one particular party. He was nevertheless supported by a broad range of anti-Communist forces, including many democratic and nationalistic ones.) On December 8, 1991, the presidents of Russia, Ukraine, and Belarus signed an agreement that formally dissolved the Soviet Union. They chose symbolically to meet near the Polish (i.e., European) border. After this, each of the remaining republics was officially free to pursue its own independent way. The three leaders deserve credit for avoiding the worst possible scenario—the one that played out with massive bloodshed and horror in the former Yugoslavia during the 1990s (Åslund, 2007).

The important geographic outcome of 1991 was that a single, unitary state, the U.S.S.R., with its capital in Moscow, was replaced on the world maps by 15 newly independent states (NIS), each with its own capital, president, parliament, and so on. Twelve of these would soon form the Commonwealth of Independent States (CIS), a military and economic alliance; three others, the Baltics, would be admitted to the North Atlantic Treaty Organization (NATO) and the European Union (EU) in 2004. From 1991 on, the political and economic changes in each NIS were decoupled to a large extent from those in others, and proceeded along individualized trajectories. There were very rapid reforms in the Baltic states, almost no reforms in Uzbekistan and Belarus, and intermediate levels of reforms in others.

Some important geographic realities, however, remained unchanged. The U.S.S.R. had uniform control over its external, but not internal, borders. Now every republic in the Former Soviet Union (FSU) would have to design its own security border system, where previously there were none. The U.S.S.R. also had a uniform electric grid; a national network of gas and petroleum pipelines; a centralized postal, telegraph, and telephone system; a unified railroad network; a centralized airspace control system; and so on. All of these would of course continue to operate, but now each country was free to replace some of the old elements with the new or to quit the common system altogether. The Soviet Army was still present in every FSU republic. It largely withdrew from the Baltics in 1992–1993, but remained present to some extent in all other republics. Border patrol units, for example, remained positioned along the borders between Afghanistan and Tajikistan, and between Armenia and Iran, as well as in Moldova, Georgia, Kyrgyzstan, Uzbekistan, and Turkmenistan. Ukraine and Kazakhstan promptly nationalized their armed forces, but had to give up their nuclear arsenals to Russia, upon the insistence of the United States and the European Community (the predecessor of the EU).

Economically, many of the republics remained interdependent. Tractors or radio sets assembled in Minsk, Belarus, for example, had parts made mainly in Ukraine and Russia; Ukrainian coal was powering factories in the Urals; Uzbekistan's cotton was made into fabric in the Ivanovo region of Central Russia; and so on. In short, an abrupt termination of the state covering one-sixth of the earth's land surface was going to be very painful for all.

Yeltsin: "Painful, but Quick" Reforms?

Yeltsin called Gorbachev's ambiguity irrelevant and dangerous, and promised that real political and economic reforms would be made quickly. It was clear, he stated in the fall of 1991, that the country had to move toward a democratic state and a free-market economy. He had overwhelming public support for this at first. Ordinary people were tired of the long lines, absence of products, and waffling political statements of the Gorbachev period. The Communist political elite had largely prepared itself for the major property grab that would soon follow (see "Privatization and the Rise of the Oligarchs," below). Foreign policy makers were eager to loan a lot of advice and a little money to the new, ostensibly no longer Communist, government. Yeltsin's chief economic advisors, largely recruited from abroad, were eager to extol the virtues of unrestrained capitalist production and consumption (Sachs & Lipton, 1993). Some of the same economic advisors who had helped transform the Polish economy just a few years earlier declared that the reforms in Russia would be "painful, but quick." A common estimate was that after about 3 years of a downturn, the economy of Russia would rebound once the necessary restructuring, adjustment, and privatization were completed. Similar predictions were made for most of the other FSU republics.

This was not to be the case. No country has ever attempted such an ambitious and sweeping program of reforms in so little time as Yeltsin wanted to try. Poland's example was not a good analogy. Poland and other Eastern European countries were not at all as deeply socialized as the U.S.S.R. was, and of course they were much smaller. For example, Soviet-style collective farms were never implemented on a large scale in Poland, so most of the agricultural land there had already been in small private holdings during socialist times; privatization of those required only paper shuffling without much physical restructuring. Similarly, industry and retail in Poland were much more consumer-goods-oriented, because Poland's tanks, planes, missiles, power plant boilers, and so on were built for it in the U.S.S.R. So privatizing small factories and shops in Poland was an easier task (Dunn, 2004).

Another recent model of free-market transition is that in China. Market reforms there began in the late 1970s, but the Chinese planners were very cautious. They chose to make changes to the economic framework slowly, carefully, and gradually, without much concession to any democratization. In a way, the country had no glasnost, only perestroika: China remains politically dominated by the Communist Party to this day, with a drastically different economy (Lai, 2006). Russia went the opposite way—almost too much political freedom very quickly, and not enough state control, especially in the first few years of Yeltsin's reign. In hindsight, Chinese-style reforms might have been better for Russia; however, they were simply never an option, given the Soviet people's overwhelming desire for freedom. A very important difference between Russia and China was the attitude of the rulers. The first priority that members of the nomenklatura in Russia set for themselves as early as 1991 was to get rich as quickly as possible in a privatization grab, regardless of the cost to society at large. Waiting years while gradually changing laws sounded foolish to them.

On October 28, 1991, Yeltsin outlined his proposed reforms to the Russian Congress of People's Deputies. On November 6, symbolically on the eve of the anniversary of the Great October Revolution—he appointed the economist Yegor Gaidar as the prime minister (later the deputy prime minister in charge of economic reforms). Gaidar was a grandson of a famous revolutionary writer, and he got the very best education the Soviet system could provide. He liked the works of classic contemporary American and European economists. In fact, he was a good example of a new generation of the Soviet elite: modern, civilized, and Western, with a keen perception of the complexity of the world's real economy, and quite unfettered by communist dogma.

The two main ideas presented in Yeltsin's reform proposal were these:

- To liberalize prices, so that each vendor could set whatever price the market could bear.
- To start privatization by allowing pieces of state property to be auctioned off.

On January 2, 1992, the prices in all state stores were allowed to float. Within a few weeks, the shelves were full of goods, including even some items that had been largely absent from the old Soviet stores; however, prices were shockingly high (Figure 8.1). In most regions, there were monopolist suppliers. Competition could not appear overnight. In the absence of competition and with chronic underproduction, not enough goods were available, so prices went through the roof as dictated by the market. In the first few weeks of 1992, prices doubled, then tripled, and then quadrupled. By the end of the year, the inflation was approaching 1,000%—something the Soviet

FIGURE 8.1. A store in a Siberian village today looks still much the same as it did during the late Soviet era, 20 years ago. Prices are now much higher, but there are many more goods on the shelves. The scale on the right is still the old Soviet model. *Photo:* A. Fristad.

people had previously only read about in novels about Weimar, Germany, in the 1920s. Riots did not break out, because a handful of staples continued to be provided at subsidized prices via coupons, and also because many people were able to grow some of their food themselves.

Despite the abrupt release of prices, the second step was slow in coming. Privatization required more preparation. According to the plans drafted by the Ministry of Privatization under the reformer Anatoly Chubais, every citizen of Russia, young and old, was to receive a voucher with a face value of 10,000 rubles. These rubles then could be invested in some state property, either directly at an auction or through an investment fund. However, when the vouchers became available in the spring of 1992, no auctions had yet been set up; thus their value rapidly plummeted. A few enterprising individuals started collecting them, hoping to invest them later, when the auctions would eventually begin. The going rate of one voucher rapidly went down from the price roughly comparable to that of a new Soviet-built car to the price of a pair of shoes, or even two bottles of vodka. Lots of people simply cashed the vouchers in by selling them to unscrupulous sharks on street corners. A few people managed to hold off until the autumn, by which time a handful of auctions did open.

Chubais, the man in charge of privatization in Gaidar's government, has traditionally been made a scapegoat for the failure of the voucher-based privatization effort (Brady, 1999; Freeland, 2000). Effectively, the charge goes, he deliberately waited an unacceptably long time to begin the auctions, until most people had lost faith in the vouchers. The truth is more complex than that. On the one hand, every single auction had to be planned months in advance. Only some enterprises were attractive enough to be auctioned off quickly. The directors had to be coached, the trade unions persuaded, the prospective buyers found. On the other hand, there was a lot of opposition among the Congress of People's Deputies—and among the Communist-era directors, regional governors, and workers themselves—to the very idea of simply giving away pieces of state property to some unknown figures with vouchers in hand. The possibilities that organized crime or foreign capitalists might take over were particularly feared. The most important (and probably deliberate) failure of Chubais, Gaidar, and Yeltsin was that only a tiny proportion of the total Soviet state assets got auctioned off at all. The idea of a fair distribution of wealth implied by the vouchers consequently went out the window. By September 1992, only a handful of marginal factories had been auctioned off. Correspondingly, few people were able to obtain a piece of the state pie.

The majority of state enterprises were eventually privatized in 1994–1996 through a few very different schemes. One of the main alternatives to voucher-based privatization was the simple reorganization of an enterprise into a stock venture (corporation) or limited-liability partnership. The former Soviet director typically retained a controlling packet of stock, and workers were given a number of shares as well to appease them. This suited most directors just fine, because they wanted to make sure that they would not be deprived of property in the new Russia. To the workers, it also seemed like a better deal; at least they had some shares and knew the director well. Some of the best and most profitable enterprises, such as the Norilsk nickel smelter and many key oil fields and refineries, were turned over to private owners later, in 1996, in a very different loan-for-shares scheme (described below). Yet another scheme was employed by the Moscow city government under Mayor Yuri Luzhkov and in some regions: The regional elite would simply convert real estate, construction companies, or municipal organizations into private or semiprivate ventures, sometimes with very little federal or public involvement, to be directly controlled by shadowy offshore structures ultimately accountable to the governors themselves. This would be a bit like Governor Arnold Schwarzenegger privatizing the Golden Gate Bridge in San Francisco and beginning to collect tolls not for the California state treasury, but for his private venture registered in the Bahamas.

Chubais went on record as saying that he hoped to have a quick privatization, not a fair one. In other words, his main goal was to create a class of owners very rapidly, without ever hoping to please everyone. Ideally, this would lead to competition among the newly created private ventures to produce more and better goods. It required over 10 years for this to become a reality in Russia, and even longer in some other FSU republics. A few natural monopolies were not privatized at all, including the railroad system; the postal service; the military; the unified energy system; the Russian Academy of Sciences; parts of the state telecommunications industry; the oil pipeline monopoly; and most hospitals, universities, and schools. Others, such as the giant Gazprom monopoly, the strategically important TV Channel One, and Sberbank (the largest consumer savings bank in the country), were only partially privatized (i.e., the federal government retained over 50% of stock).

Rosefielde (2007) argues that the privatization process in Russia was inefficient and unfair—an exercise in murky politics rather than economics. Åslund (2007) takes a more positive view of the process, explaining that few other options were realistically available and that the results were quick and impressive. By early 2000, over 80% of the Russian economy was in private hands.

A New Political Structure: The Russian Federation

Besides economic reforms, a great deal had to be done politically by Yeltsin's government. The Soviet constitution no longer worked and had to be replaced. The roles of the president of Russia, the Congress of People's Deputies, and the executive branch had to be redefined. An independent system of courts had to be established. Virtually all Soviet laws—including the civil and penal codes, as well as regulations of land, property, natural resources, labor, and taxes—had to be overhauled or adjusted. New political parties were mushrooming, in the absence of clear constituents or goals. (My favorite one was the Party of the Lovers of Beer.) Hundreds of nongovernmental organizations were being established monthly in every imaginable field (from cultural to environmental to political) and had to be regulated. New businesses were starting up, growing, breaking up, failing, and disappearing. Millions of state employees (professionals and blue-collar workers alike) were chronically paid their wages late by the partially privatized payroll system: the banks were making money on interest and were in no hurry to pay people state wages on time. Frequently workers would stage protests over unpaid wages—sometimes very dramatic ones, as in the case of gold miners in Siberia blockading the Trans-Siberian Railroad on a few occasions. In short, a new political structure and new laws were badly needed.

Thanks to the resilience of the population and some clever maneuvering by regional governors, widespread starvation and riots were avoided. One of the key ways common people survived was through trade. Yeltsin allowed anyone to be a trader, so lines of *babushkas* (grandmothers) selling food and cigarettes at bus stops to commuters on their way to and from work became common in every major city. The grandmas had spare time to wait in lines and buy goods at a lower price during the day, to resell them quickly in the evening at a small profit. Another major form of private enterprising was called "shuttle trading." Over 3 million people took to it. It became possible, and very profitable, to travel to Turkey, Poland, or Cyprus and come back loaded with Western goods (jeans, VCRs, coffeemakers, etc.) to be sold on street corners or in hastily constructed city markets. Many early entrepreneurs of the new Russia made their first million rubles this way. Some of the most successful shuttle traders were middle-aged, aggressive people with university degrees, and over half were women. They were not afraid to bargain hard; to learn a few words in Turkish, Greek, or Polish; and to use some of their higher education in math to make good money.

Again, however, the overall situation was dicey. By the fall of 1992, Prime Minister Gaidar had accomplished the key steps of his ultraliberal economic agenda and could be conveniently dismissed by Yeltsin, to be replaced by a high-ranking apparatchik from the Soviet period, Viktor Chernomyrdin. Chernomyrdin's main training was in the gas industry. He was effectively the chief lobbyist for the state Gazprom monopoly, and as such he remained very useful to Yeltsin for the following 5 years. A national referendum on April 25, 1993, unexpectedly expressed high confidence in the course of Yeltsin's reforms. However, the question that was asked was essentially "Would you support reforms or go back to Communism?" Since few wanted to go back, the majority said "yes" to the reforms. This was, of course, not an unqualified endorsement. Later in 1993 voucher-based privatization would get into full swing, and Gaidar would come back as a deputy prime minister under Chernomyrdin.

Then a political disaster happened. The Congress of People's Deputies was composed of a variety of political parties and forces, having been elected under Gorbachev. Many deputies were supportive of Yeltsin; however, even more were critical of him and openly hostile to his government. In the absence of a new constitution, Yeltsin's hands were tied with respect to what he could and could not do with the reforms. His attempts to ram some key privatization bills through the Congress repeatedly failed. A standoff was brewing. Then on September 21, in a bold move, Yeltsin dissolved the Congress by a decree—something that he had no clear constitutional authority to do. He appealed directly to the people, as he had done in August 1991, to let him lead the nation out of the political impasse to a better and richer future. Overall public opinion would support Yeltsin, not the Communist-leaning deputies in the Congress.

The Congress refused to comply. Yeltsin's own vice-president, a charismatic ex-general and Afghan war hero named Alexander Rutskoi, was chosen as Yeltsin's replacement. The deputies, ensconced in the Russian parliament building (known, like the U.S. president's house, as the White House), prepared for a brutal standoff. Details of those fateful days can be found elsewhere (Brady, 1999; Åslund, 2007). With the tacit support of the Group of Seven (G7) governments, Yeltsin felt that the time to act was at hand, and that no one would dare to question his tough and undemocratic measures. On October 4, tanks summoned by Yeltsin into the capital shelled the White House, and riot police in full gear stormed the Ostankino TV tower, where supporters of the Congress were hiding. Over 150 casualties resulted from this massacre—the first major bloodshed of the supposedly democratic period, and a much larger toll than that of August 1991. Yeltsin won, and the hardliners were put in jail.

On December 12, 1993, new parliamentary elections and a constitutional referendum took place. The system of power in Russia from this point on was much different from the previous model. Much more power became concentrated in the president's tsar-like hands. The Congress was transformed into a bicameral legislature: The upper house was the Federation Council (Senate), composed of regional governors and their representatives elected in their regions (two for each of the 89 regions); and the lower house was the Duma (House of Representatives), with 450 deputies elected every 4 years, either by a direct vote or by party lists.

Although the new Russian model superficially resembled that of the United States, the president in Yeltsin's Russia played a much bigger role than the U.S. president, while the parliament had a much smaller role than the U.S. Congress. One of the key differences was the ability of the Russian president to propose new bills. In the United States, the power of introducing a bill rests solely with the members of Congress. Also, the president in Russia was given the power of appointing the prime minister over the will of the Parliament. The president's administration grew to be a huge body of several thousand bureaucrats, much like the old apparatus of the Central Committee of the C.P.S.U. A national security council was established under the president to respond to pressing threats. It was composed of the heads of the power ministries (the police, KGB, the army, etc.).

One of the big outcomes of 1992–1993 was a geographic transformation of the country's federal administrative structure. As described in Chapter 7, the old R.S.F.S.R. was a federation (at least on paper) of many diverse units, called oblasts, krays, autonomous republics, autonomous oblasts, and autonomous okrugs. Many of these were retained, but their names and roles were modified (Figure 8.2). In his initial push to appease as many regional elites as possible, a jubilant Yeltsin proclaimed that local autonomous ethnic units should feel free to grab "as much sovereignty as they could swallow." The more powerful political units, such as Tatarstan, Sakha (Yakutia), and Chechnya, took this slogan seriously and began procedures to become "self-governing nations" within the larger body of the Russian Federation. In one case, Chechnya, this process culminated in a full-blown war for secession that is not quite over yet (Chapter 25).

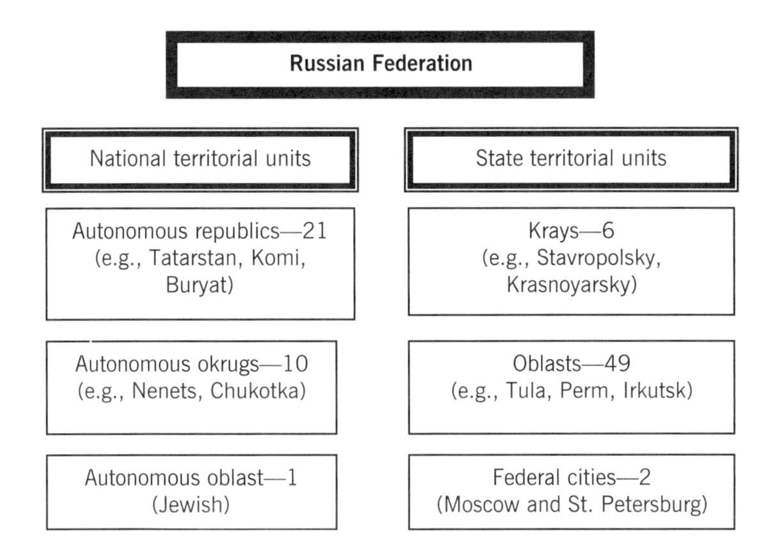

FIGURE 8.2. Russian Federation administrative units according to the first post-Soviet (1993) constitution (89 units). Since 2000, a few autonomous okrugs have been merged with nearby oblasts or krays (see Vignette 8.2).

In contrast to the autonomous republics, the autonomous okrugs received less power than they could have hoped for. Some were recently merged with neighboring oblasts (e.g., Komi-Permyak Autonomous Okrug was merged with Perm Oblast to form Permsky Kray). This made economical and political sense, because some of the smallest okrugs had very few people in a huge territory. At this writing, there are 83 units in the Russian Federation (Figure 8.3), including 21 republics and 4 autonomous okrugs. More mergers are being planned. According to some proposals, the optimal number of units would be about 50, as in the United States.

Privatization and the Rise of the Oligarchs

One of the notorious results of privatization à la Chubais was the emergence of new wealthy private owners, dubbed "oligarchs." In Greek, *oligos* means "few" and *archon* means "power." Basically, then, an oligarchy is a system in which a few people control a lot, and an oligarch is one of these people. A typical oligarch of the mid-Yeltsin period was a man in his mid-30s to mid-40s with a Soviet background (e.g., a Komsomol leader or son of a well-heeled party bureaucrat); he usually also had an engineering degree, personal connections with Yeltsin's family, and a few hundred million dollars in a bank (Hoffman, 2003). Some oligarchs had been members of the Communist elite in the past, but the majority were either children of the nomenklatura bosses or obscure engineers who emerged due to their entrepreneurial spirit, lack of scruples, and uncanny business sense. Some were economists or mathematicians, others came from the petroleum and metallurgy industries, and still others were former managers of state factories or cities. Contrary to the common belief, few had criminal backgrounds; however, more than a few used the services of shadowy protection bureaus.

How did these people become so wealthy so fast, in a country with inflation in double digits and an average salary of less than $100 per month? Well, all had some key "insider" connection that enabled them to get in on the grand privatization early. Some of the earliest fortunes, not surprisingly, were made by cashing in the wealth accumulated by the C.P.S.U. from both domestic and foreign sources (real estate, gold, jewelry, Swiss bank accounts, etc.). Privatizing the Soviet state treasury was the goal of the late Soviet apparatchiks who supported Gorbachev's

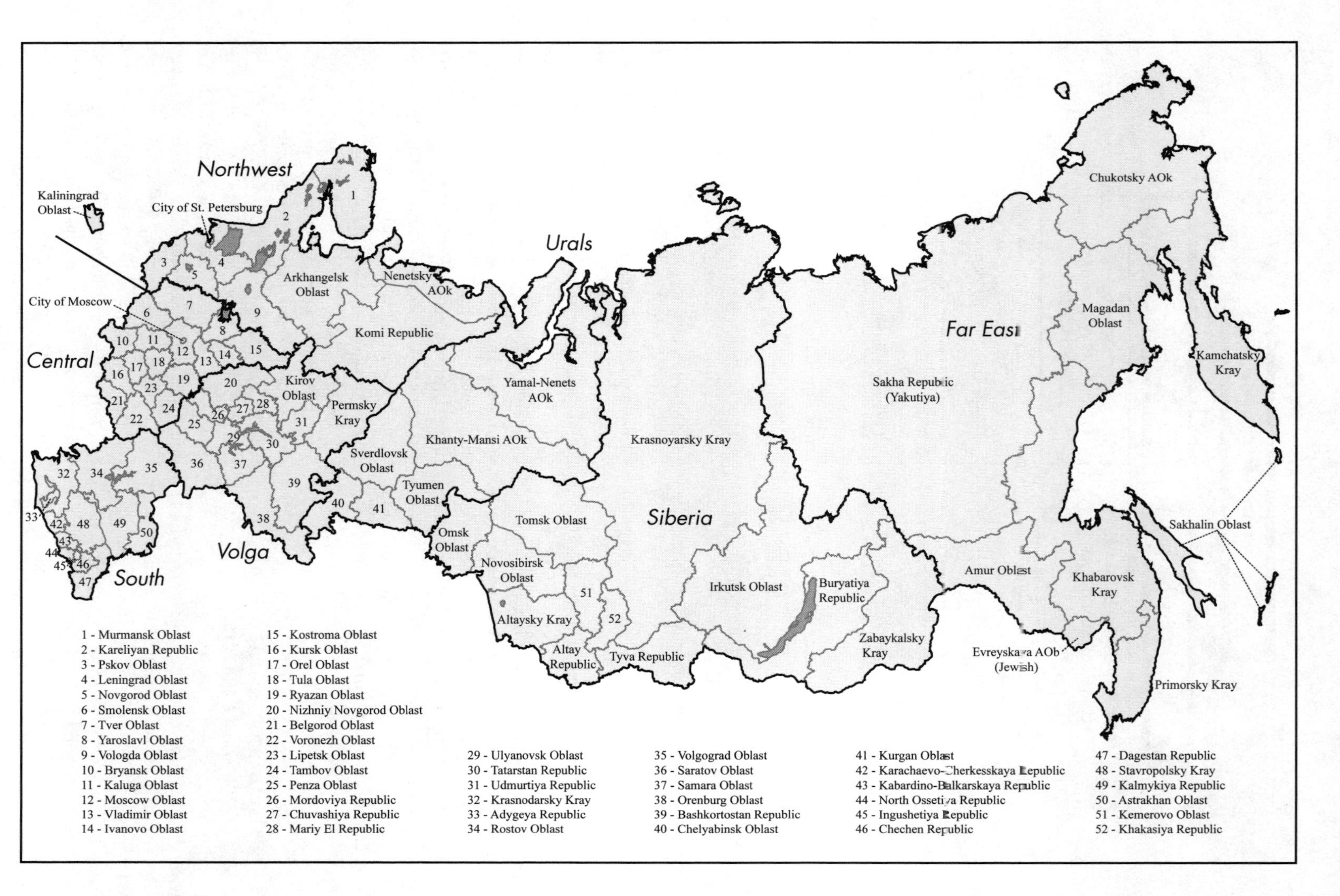

FIGURE 8.3. The 83 “subjects of federation” (internal units) in Russia in 2010. AOb, autonomous oblast; AOk, autonomous okrug.

reforms. One of the early oligarchs, Vladimir Potanin, was a son of the Soviet chairman of the state foreign exchange bank. Potanin was able to set up one of the first private banks under Gorbachev. The initial capital clearly had to come from a state (party) source. A few other oligarchs were somehow known to one of Gorbachev's top aides through their party or Komsomol connections (e.g., Mikhail Khodorkovsky) and were likewise allowed to set up commercial banks early. In this initial period, the banks were little more than cash machines designed to convert state noncash accounts into real rubles, and increasingly into dollars. A few oligarchs who had risen seemingly out of nothing (a toy coop entrepreneur, a physicist, etc.) turned out either to have married someone close to the president, or to be personally trusted by Yeltsin and his close family.

Once a few banks started out, they were able to make money through a variety of creative "get rich quick" schemes. Importing all sorts of Western goods duty-free because of bribes or permissions to bypass customs; cashing in noncash factory accounts; withholding interest on state workers' wages for a few weeks; directly looting the state treasury via fake invoices; and many other creative schemes generated millions of dollars very quickly for those few who knew how to work the system. Besides the oligarchs themselves, a few other new categories of wealthy Russians emerged, usually collectively known as "new Russians" (*novye russkie*):

- Small private entrepreneurs, many of whom got wealthy early by either importing Western goods or privatizing bakeries, barber shops, shoe repair businesses, and the like, and who gradually grew to become owners of larger firms.
- Professional voucher traders and commodity traders.
- Stockbrokers and investment bankers.
- Many Soviet-era factory directors who simply pocketed their entire factories without paying a dime for them.
- Former Soviet mob bosses (*vory v zakone*) with criminal connections and black-market cash, who had been released by Gorbachev's government from the overflowing state prisons.
- Local, regional, and federal politicians, as well as army, police, and KGB bosses, who were able to convert their relational capital into real cash. For example, many ex-KGB agents started their own protective services by using their connections in the local underworld.
- Anyone with solid foreign connections, including some emigrants who came back, or enterprising and bold citizens of Western countries.
- A few particularly lucky individuals who happened to be in the right place at the right time.

The key characteristic of all these individuals was the desire to take very high risks to make a lot of money quickly (Tikhomirov, 2000). Many of them paid with their lives, particularly in 1993–1995, when full-blown gangster wars erupted over the key state assets that were up for grabs (e.g., aluminum smelters in Krasnoyarsk). By 1996 some of the most dangerous criminals had exterminated each other, and from then on business contract killings became less useful, as the legal and economic system evolved.

In 1996 Yeltsin came up for reelection. Given the hardships endured by most people because of his reforms, his approval rating was less than 5%, much lower than that of his main opponent (a Communist, Gennady Zyuganov). In one of the most fateful stories of the reform period, a group of seven oligarchs controlling a little less than 50% of all privatized assets of the entire country (according to them) came to the president and proposed a Faustian bargain: They would use the power and money of their new private media empires to rally public support, if Yeltsin would agree to let them keep shares of some of the most lucrative, yet still unprivatized, enterprises. This loans-for-shares program was originally conceived by Potanin, the owner of the Interros business empire. However, it was not until the 1996 elections that the oligarchs received this unprecedented leverage. The program would let the oligarchs loan some money to the state and keep state enterprise shares as a collateral for a while, but in reality everybody understood that the state government would default, and so those assets would forever be transferred to the oligarchs (for details, see Freeland, 2000; Hoffman, 2003; and Klebnikov, 2000). In return, the oligarchs promised that they would support Yelstin's bid for reelection with all the means at their disposal. Some of the best assets were privatized very cheaply under this scheme, including the Norilsk nickel combine (worth billions) for $170 million, and many oil fields and refineries for a fraction of their cost.

Yeltsin agreed to the deal. With the private NTV and ORT television channels bombarding the public with the images of an apparently reenergized Yeltsin dancing on stage; with the relentless private newspaper coverage of the sinister plots to restore Communism, should Zyuganov come to power; and with massive financial backing from the oligarchs as well as some Western funding, Yeltsin's victory was assured. However, he suffered a massive heart attack and almost died just a few days before the second round of elections in the summer of 1996. He did win the round, but few people understood just how sick he was then. It is unclear how much electoral fraud was perpetrated by Yeltsin's electoral commission during the vote counts, but he won by only a slim margin; nevertheless, he stayed in power for another 4 years. After the summer of 1996, Boris Berezovsky (the leader of the so-called gang of seven oligarchs) began wielding an ominous influence behind the scenes at the Kremlin, largely through free access to Yeltsin's two daughters and some of his key staff (Klebnikov, 2000). Every oligarch on the team had received very lucrative rewards: Potanin got hold of the Norilsk nickel combine, Khodorkovsky got the Yugansk oil field and some key refineries in the Volga region, and so on. The independent press critical of the Kremlin concluded that oligarchic capitalism not only had taken root in Russia, but had grown to become the main trunk of the economic tree.

Hitting Bottom: The Default of 1998

The period between 1996 and 1999 was characterized by continued privatization, growth in the big private companies, some political maneuvering over the passing of new legislation, a few high-profile assassinations, and (very importantly) the rapid growth of government short-term bonds. The so-called GKO bonds were issued for a few months each, and typically had an interest rate just ahead of the inflation rate, to keep the public interested. Because the situation appeared to be under control, many people—including many oligarchs, key government officials, and even some representatives of reputable Western investment funds—were attracted to the GKOs in large numbers. The economic realities, however, were not at all as rosy as they seemed, and after a few months of skyrocketing yields (to attract ever more investors in what was essentially a Ponzi scheme), the Russian government started suspecting a looming default. In the absence of real economic growth, it basically had to keep raising the interest rates to attract new buyers to pay off the old bonds to bring in new cash. After replacing the stalwart Chernomyrdin with the young, inexperienced Kirienko, Yeltsin seemed to do nothing at all to deal with the situation. The default finally occurred in August 1998, partially in response to the widening financial crisis in the emerging markets in Asia. The Russian state defaulted on its ruble GKOs, and some foreign loans in hard currency. The investors fled; many banks collapsed; and the ruble plunged from about 6 to 31 against the U.S. dollar in less than 2 months.

It took about 2 years and a new Russian president, Vladimir Putin, to restore some confidence in the Russian market. To be fair, many countries with much older market economies and in better economic shape than Russia (e.g., South Korea, Brazil, Mexico, and Argentina) experienced similar problems at about the same time; even the mighty United States itself became embroiled in a major financial collapse in 2008. The default of 1998, however, made it abundantly clear that Russia's economy had indeed hit bottom. By the fall of 1998, the Russian GDP was about half of what it was in 1990—an unprecedented decline in any country in the absence of war.

When citing macroeconomic statistics, however, we need to consider the unofficial sectors of the Russian economy (thought to account for perhaps 20% of it), so the real situation was actually better than the official numbers alone portray. Also, much of the decline in the GDP was in heavy industrial production; Russian tractors, boilers, combines, and so on were of inferior quality and greatly overproduced anyway. Growth in retail trade, services, construction, and a few other consumer-oriented sectors was occurring at the same time as the big industries were faltering, and this growth partially offset the gloomy statistics. Nevertheless, particularly hard hit were some of the most consumer-oriented enterprises, such as the textile and shoe industries, where production in 1998 was a mere 20% of the 1990 level.

Putin Rising: The Beginning of a New Order

After the 1998 default, a new prime minister was brought in to restore some credibility to the country's image: a former Middle East career diplomat and spymaster, Yevgeny Primakov. He managed to stabilize the situation within a few months, but was abruptly dismissed by Yeltsin in the spring of 1999, when it was discovered that Primakov had formed an alliance with the powerful mayor of Moscow to run for parliamentary elections in opposition to Yeltsin's allies. Primakov was replaced for a few months by Sergei Stepashin, who was a transitional figure. Yeltsin and his family (his two daughters and a few loyal oligarchs) were quietly seeking a permanent replacement for the aging leader, and finally they settled on Vladimir Putin, who replaced Stepashin as the prime minister in August 1999.

Vladimir Putin was, like his two predecessors, a KGB man; he was a career foreign intelligence officer who had spent 8 years supervising the recruiting of agents in East Germany. Unlike any Russian leader since Stalin, however, he never became a nomenklatura member. Many accounts of how and why Putin was introduced to Yeltsin as a possible successor have been given. The primary factor was probably Putin's demonstration of his loyalty to his former employer—the mayor of St. Petersburg, Anatoly Sobchak—in the fateful days of the August 1991 coup. Putin was the vice-mayor of St. Petersburg then and helped to protect Sobchak, who risked everything by firmly standing with Yeltsin against the coup leaders. Putin was also a complete unknown to the country at large, which had its advantages.

Putin came to power at a difficult time in the south, where an incursion of armed guerrillas from Chechnya under Shamil Basaev was beginning to destabilize volatile Dagestan. Less than a month after Putin's installation as prime minister, a series of apartment building explosions rocked Moscow, Volgodonsk, and Buynaksk, killing over 300 people. Chechen terrorists were promptly blamed. Within a few weeks Putin authorized a new major operation in Chechnya, reigniting hostilities there that had been dormant since 1996.

On December 31, 1999, Yeltsin made his final gift to the nation by announcing his unexpected resignation. He said that he trusted that people would like Putin enough to elect him as the new president. Putin easily won the March 2000 elections: After 8 years of having a sickly, frequently drunk leader, Russia finally had a young, energetic man in charge. The alternatives to Putin were mainly recycled from the 1996 elections: the Communist, Zyuganov; a bombastic maverick nationalist, Zhirinovsky; a mumbling pro-Western intellectual, Yavlinsky; and a handful of others, none of whom were a match for Putin. Clearly, the new century required a new leader, and Putin happened to be available at the right place and the right time.

The economic situation also began to improve in 1999. This had less to do with Putin than with the aftermath of the 1998 default. Although a lot of Western investors fled for a while when the ruble lost value, all Russia-made goods became very competitive on the world markets while the imports became too expensive, so Russian production picked up. Also, while some wealthier people lost their savings in rubles, most had few savings left, and usually those who did wisely kept them in dollars or some European currency. Thus the stock market crash and the GKO default had a minimal impact on the general population; they primarily hurt the rich and the "new Russians." Also, and perhaps more significantly, privatization finally began producing real results. Production was finally up after years of decline. Some private owners managed to make investments in better equipment and revitalized their aging Soviet factories. Others started completely new companies from scratch (e.g., retail supermarket chains, book publishers, food-processing factories, computer and software retail and manufacturing, or furniture factories). Many changes began to appear in urban neighborhoods (Vignette 8.1).

Several important pieces of legislation were approved in 2000–2001 that paved the way toward greater stability and slightly greater transparency in business transactions. The tax code was streamlined. A flat personal income tax of 13% was introduced, to encourage people to pay taxes legally. Although a progressive tax would have been a better long-term strategy, the flat rate did help bring billions of black-market rubles into the "white zone." Salaries were raised for some categories of state workers, especially soldiers, police officers, physicians, and teachers. Pensions for state workers were also increased and actually began to be disbursed on time, which had rarely happened under Yeltsin. A few corrupt regional officials (some at very high levels) were tracked down, sacked, fined for bribes, or even imprisoned on corruption charges. The oligarchs Berezovsky and Gusinsky fled Russia to avoid prosecution for tax evasion, money laundering, and fraud. It is widely believed that the real reason for their departure was the need for Putin to control the private media empires that they had built, including the all-important NTV and ORT television channels and a number of newspapers. Khodorkovsky, the main owner of Yukos (an oil company) and MENATEP (a bank), was offered the same chance to flee the country, but he courageously refused. He was arrested, tried, and sentenced to 8 years in prison for tax evasion and fraud, in a case widely believed to be politically motivated. Khodorkovsky had refused to give the Kremlin a share in a planned merger between Yukos and Sibneft; he had also spent a lot of money on supporting Putin's political opponents; and he had even publicly questioned the involvement of personal friends of Putin in murky privatizations (Baker & Glasser, 2005; Åslund, 2007).

Several other pieces of important federal legislation were passed in 2002–2003, including a new penal code, a civil code, a labor code, a forest code, a water code, and a land code. As the rules became better known and observed, further economic growth followed. The ruble strengthened, and Russia's stock market doubled in value in less than 2 years. Little improvement, however, was achieved in some of the most persistent problem areas: crime, corruption, and inflation. Although rates of crime (particularly the most violent types) dropped slightly in the early 2000s because of improved policing and prosecution, they still remained at much higher levels than those common during the Soviet period. For example, homicides went down from 35,000 per year in 2001 to 29,000 in 2005; the Soviet level, however, was about 30% lower.

When Putin came into office, addressing state corruption was ostensibly one of his top priorities, but he quickly announced that little could be done: The situation was too pervasive, too entrenched. All public officials' salaries were greatly increased under Putin, without a corresponding drop in bribe taking or an increase of quality of

Vignette 8.1. The Evolution of Retail Establishments on a Typical Moscow Street

To give you a better sense of the pace of post-Soviet reforms, we could take a walk through time on any Moscow street and look for clues. One such street, Borisovsky Proezd in southeastern Moscow, is near the flat where I grew up. The district of Orekhovo-Borisovo was founded in 1974, and 10 years later it had a population of 300,000 people located in 16 *microrayons* (microdistricts for living quarters, covering about 10–80 ha each). Borisovsky Proezd, named after the former village of Borisovo (the birthplace of Boris Godunov), is a typical Moscow street near the city periphery. It starts at Kashirskoe Shosse, one of the main city arteries running from the areas near the center to the ring road on the city periphery. The stretch of Borisovsky from Kashirskoe to the intersection with Shipilovskaya street is about 1 km long. The street curves almost 90° at its midpoint, so that you travel first east and then south along it.

(cont.)

In the late 1980s, the following typical Soviet retail establishments were located on Borisovsky, or immediately off it in one of the nearby microrayons: a vegetable/fruit store, a radioelectronics store, a pharmacy, a post office/telephone office, a bakery, a small factory manufacturing school lunches for a few nearby schools, and a general grocery store. A bit farther away along one of the nearby streets were also a hardware store, an office of the only bank in the U.S.S.R. that worked with the general public (called Sberbank), and a dairy grocery store. Each store was housed in a standard one-story concrete building with big windows and a flat roof. Besides the big stores, you would see a few small kiosks selling newspapers, tobacco, and ice cream near each bus stop (there were four bus stops along this stretch). Each shop was thus rather specialized, and together they provided the most essential services to the nearby microrayon, with a population of about 15,000.

By the mid-1990s, a new microrayon along the northern side of Borisovsky was completed. Most of the local stores got privatized, some through immediate leadership and others by being sold to outside investors. Making the same trip, you would notice the following changes:

- A major new supermarket was built at the intersection between Borisovsky and Kashirskoe Shosse.
- The radioelectronics store closed and was turned into a general grocery store. A new electronics shop opened up across the street from the old one.
- The vegetable/fruit store just across the street was turned into another general store, selling everything from meat to dairy to bread to medical supplies to perfume.
- The hardware store likewise was turned into a general store, selling everything except food.
- Three new cafes and a bar appeared on the street, which previously had none.
- An optical store appeared.
- Two specialized meat stores appeared, one selling specific sausages from a reputable Moscow factory.
- The bakery became yet another general store.
- The former general grocery store, ironically, went bankrupt and closed its doors.
- Lots of street vendors appeared, selling stuff from the backs of pickup trucks, or standing on the sidewalk near every bus stop.
- The post office and the bank continued to operate.

So there was some diversification of retail forms, along with, curiously, generalization of some previously specialized stores. (Think why this might have been the case).

Let's "fast-forward" to 2007 now. Most of the stores described above for the mid-1990s are still here. However, the large supermarket is now an entertainment megacenter, complete with a discotheque, a spa, a tanning salon, a casino, and a bar. It still sells groceries as well, but now there is a lot more competition. The store that was the 1990s radioelectronics store went through two name changes and now is a discount minimarket. Just next door to it is a bigger and more expensive new supermarket. Inside, it is very slick, almost indistinguishable from its American counterparts. The street vendors are all tightly regulated now; most work from semipermanent kiosks, rather than directly on the streets. The former hardware store is completely gone: It went up in flames one night 2 years ago (probably due to arson), and has been sitting on the corner as an empty, blackened shell ever since. This is, by the way, rather uncommon in contemporary Moscow. Most importantly, a huge new shopping center, Ramstor (part of a Turkish chain), opened up in what used to be a local apple orchard. Inside it has a huge supermarket, as well as a few dozen shops selling everything from jewelry to cosmetics to Gap outerwear or Victoria's Secret underwear. It also has a cinema multiplex, together with a few fast-food and moderately priced sit-down restaurants. In short, it looks a lot like an American enclosed mall. What has not changed much yet is the housing: These are the same 9- to 22-story buildings of the late 1970s or 1980s. However, the retail options provided to the new inhabitants are radically different from those of the past. Welcome to yet another example of Russian capitalism.

life for the masses. Virtually anything one needs from the state in Russia today requires an "incentive" or a kickback—ranging from a few dollars for a traffic cop to avoid receiving a ticket, to a few thousand for an army official to dodge the draft, to millions for members of the top administration to receive a particularly lucrative federal contract. In fact, corruption in the government today is estimated to be unprecedented (Åslund, 2007). There are dozens of articles on the issue in any Russian newspaper. Although the state macroeconomy has somewhat improved, and the official budget has been running solidly in the black (given the exorbitant revenues from petroleum sales and tight monetary policies), very little has been done to abate inflation (which has held steady at 10–12% a year), to help small business owners and regular citizens to make ends meet, or to improve people's personal safety. In fact, billions of state surplus money has been stashed away in foreign bank accounts in "stabilization funds" for some hypothetical time in the future, instead of being invested right now in the decaying infrastructure or used to stimulate industrial growth.

A few political changes during Putin's early administration received a generally negative response in the Western press. These included abolition of democratic elections for governors in the regions; rebuilding of the vertical structure of federal power through the creation of seven presidential envoys; an increase in the number of votes needed to secure a presence in the Duma, which led to the elimination of most parties (Figure 8.4); abolition of non-party-affiliated independent candidates; greater pressure on the courts; and restoration of the Soviet anthem music and the Red Army symbols on banners.

Geographically, the most significant change was the introduction of seven federal districts nationwide, into which the 83 units of the Russian Federation are now grouped (Vignette 8.2 and Table 8.3): Central, Volga, Northwest, South, Urals, Siberia, and Far East. These districts are important units to know, because many government economic statistics are reported by districts now, as well as by the individual units. The old Soviet system had 11 economic regions, some of which are now merged in the present system

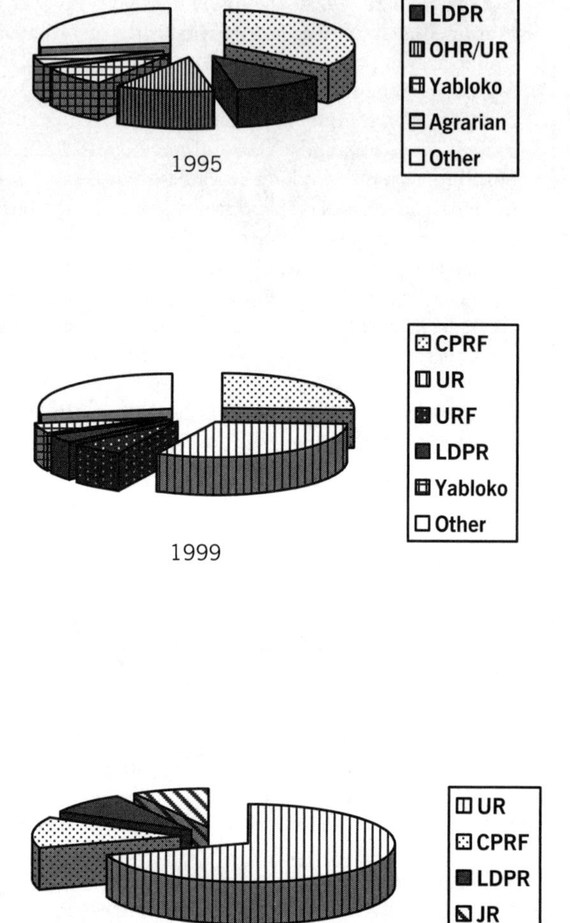

FIGURE 8.4. Composition of the Duma of the Russian Federation after parliamentary elections in 1995, 1999, and 2007. CPRF, Communist Party of the Russian Federation; LDPR, Liberal Democratic Party of Russia (nationalist leanings); OHR/UR, Our House Russia/United Russia (Our House Russia was a pro-Yeltsin party, which was later merged with others to create United Russia, the present-day pro-Putin party); Yabloko, a democratic, pro-Western party popular with the intelligentsia; URF, Union of Right Forces; JR, Just Russia (another pro-Kremlin party that was formed in 2006 to present a more socialist-leaning alternative to United Russia).

Vignette 8.2. The New and Old Regional Units of the Russian Federation

Russia, like the United States, is a federation of regional units. They are called "subjects of federation," not "states," but the idea behind them is similar: Each has its own governor, legislature, flag and seal, borders, and so on. Overall there were 89 subjects of federation in 2000, but only 83 in 2010, including 21 autonomous republics, 4 autonomous okrugs, 46 oblasts, 9 krays, 1 autonomous oblast, and 2 federal cities (Moscow and St. Petersburg). Each region has proportional representation in the federal Duma and two delegates each in the Federation Council. Regional governors were mostly elected by a popular vote until 2004, when they began to be appointed by the Kremlin.

From the Putin administration's standpoint, the situation with the regions was untenable. The idea of rebuilding the vertical structure of power began to take shape when Putin created a system of seven federal districts into which all 89 units were grouped in 2000. Each district received a personal envoy appointed by the president. Each envoy was given hundreds of staff members, a generous budget, and an imperative to promote the presidential agenda in the regions and to serve as a liaison between the Kremlin and the regional elites. The bloody siege of the school in Beslan on September 1, 2004, was used as a pretext to move farther toward abolishing governors' elections in all regions; eliminating independent delegates and permitting only party-affiliated delegates to run in the parliamentary elections; and removing some obstinate governors from their posts. The political map underwent some changes as well. Specifically, in 2005–2007 three autonomous okrugs (Komi-Permyak, Koryaksky, and Aga-Buryat) were merged with nearby oblasts (Perm, Kamchatsky, and Irkutsk, respectively). A year later, two more okrugs were merged into Krasnoyarsky Kray, and another okrug into Chitinskaya Oblast. More such mergers are planned in the future. Pre-Communist Russia was much larger, and it had only 30 regions. Larger regions are deemed more efficient and are easier to control from Moscow.

The map in Figure 8.3 depicts the regions as they exist in 2010. Table 8.3 details which oblasts and republics are included in which federal districts, as well as the 11 economic regions used for reporting during the Soviet and Yeltsin periods. Those were purely statistical units used for reporting aggregate economic data. The new seven districts are political units, but many aggregate data are now reported by these districts instead of by the old 11 units. Note that while there is a lot of overlap, these are not at all identical lists. One thing that has not yet happened is the actual redrawing of any internal or external borders of the subjects of federation. When two subjects merge, their shared border disappears, but no changes are made to the external borders. At least something stays the same!

(e.g., North and Northwest are now simply called Northwest). At the same time, other old regions are split between new districts: For instance, the Povolzhye (Lower Volga) region is now partially included in the Volga district and partially in the South district, which also now includes the northern Caucasus. In addition, the oil-rich Tyumen Oblast in western Siberia is now included in the Urals, not in Siberia, as one would expect. When I discuss economics in Part IV, I refer to both the new and the old units when necessary, but I generally emphasize the new federal districts. Each such district has an appointed presidential envoy representing the Kremlin administration's interests. In 2007, five of these seven envoys had strong ties to the former KGB.

Early in 2010, President Medvedev announced that a new eighth federal district would be created in the Russian North Caucasus region to strengthen the political and economic control of Moscow over this volatile territory.

The Kremlin Corporation and Putin Forever?

In December 2007, *The Wall Street Journal* published a story suggesting that Putin's personal wealth, if measured by the value of the assets that he is believed to control personally, may approach $40 billion. This would have been about double the net worth of the officially richest Rus-

TABLE 8.3. Internal Units of Russian Federation During the Times of Yeltsin and Putin

Unit	Economic region	Federal presidential district (2000)
Belgorod Oblast	Chernozemny	Central
Bryansk Oblast	Central	Central
Vladimir Oblast	Central	Central
Voronezh Oblast	Chernozemny	Central
Ivanovo Oblast	Central	Central
Kaluga Oblast	Central	Central
Kostroma Oblast	Central	Central
Kursk Oblast	Chernozemny	Central
Lipetsk Oblast	Chernozemny	Central
Moscow Oblast	Central	Central
Orel Oblast	Central	Central
Ryazan Oblast	Central	Central
Smolensk Oblast	Central	Central
Tambov Oblast	Chernozemny	Central
Tver Oblast	Central	Central
Tula Oblast	Central	Central
Yaroslavl Oblast	Central	Central
City of Moscow	Central	Central
Kareliyan Republic	North	Northwest
Komi Republic	North	Northwest
Arkhangelsk Oblast	North	Northwest
Nenetsky Autonomous Okrug	North	Northwest
Vologda Oblast	North	Northwest
Kaliningrad Oblast	Northwest	Northwest
Leningrad Oblast	Northwest	Northwest
Murmansk Oblast	North	Northwest
Novgorod Oblast	Northwest	Northwest
Pskov Oblast	Northwest	Northwest
City of St. Petersburg	Northwest	Northwest
Adygeya Republic	Caucasus	South
Dagestan Republic	Caucasus	South
Ingushetiya Republic	Caucasus	South
Kabardino-Balkariya Republic	Caucasus	South
Kalmykiya Republic	Povolzhye	South
Karachaevo-Cherkessiya Republic	Caucasus	South
North Ossetiya Republic	Caucasus	South
Chechen Republic	Caucasus	South
Krasnodarsky Kray	Caucasus	South
Stavropolsky Kray	Caucasus	South
Astrakhan Oblast	Povolzhye	South
Volgograd Oblast	Povolzhye	South
Rostov Oblast	Caucasus	South

(cont.)

TABLE 8.3. *(cont.)*

Unit	Economic region	Federal presidential district (2000)
Bashkortostan Republic	Urals	Volga
Mariy El Republic	Volga-Vyatka	Volga
Mordoviya Republic	Volga-Vyatka	Volga
Tatarstan Republic	Povolzhye	Volga
Udmurtiya Republic	Urals	Volga
Chuvashiya Republic	Volga-Vyatka	Volga
Permsky Kray	Urals	Volga
Kirov Oblast	Volga-Vyatka	Volga
Nizhniy Novgorod Oblast	Volga-Vyatka	Volga
Orenburg Oblast	Urals	Volga
Penza Oblast	Povolzhye	Volga
Samara Oblast	Povolzhye	Volga
Saratov Oblast	Povolzhye	Volga
Ulyanovsk Oblast	Povolzhye	Volga
Kurgan Oblast	Urals	Urals
Sverdlovsk Oblast	Urals	Urals
Tyumen Oblast	West Siberia	Urals
Khanty-Mansi Autonomous Okrug	West Siberia	Urals
Yamal-Nenets Autonomous Okrug	West Siberia	Urals
Chelyabinsk Oblast	Urals	Urals
Altay Republic	West Siberia	Siberia
Buryatiya Republic	Central Siberia	Siberia
Tyva Republic	Central Siberia	Siberia
Khakasiya Republic	Central Siberia	Siberia
Altaysky Kray	West Siberia	Siberia
Krasnoyarsky Kray	Central Siberia	Siberia
Irkutsk Oblast	Central Siberia	Siberia
Kemerovo Oblast	West Siberia	Siberia
Novosibirsk Oblast	West Siberia	Siberia
Omsk Oblast	West Siberia	Siberia
Tomsk Oblast	West Siberia	Siberia
Zabaykalsky Kray	Central Siberia	Siberia
Sakha (Yakutiya) Republic	Far East	Far East
Primorsky Kray	Far East	Far East
Khabarovsk Kray	Far East	Far East
Amur Oblast	Far East	Far East
Kamchatsky Kray	Far East	Far East
Magadan Oblast	Far East	Far East
Sakhalin Oblast	Far East	Far East
Evreyskaya Autonomous Oblast	Far East	Far East
Chukotsky Autonomous Okrug	Far East	Far East

sian at the time, Roman Abramovich. This may be either an overstatement or an understatement. Personal wealth is a sensitive issue, and little is known in present-day Russia about who owns exactly what. To be sure, the president should be the last person to be poor, the way the cards are stacked. Most big companies now have shareholders, many of whom are registered under fictitious names or are represented by murky offshore firms. Who else but the president would know who all these people are?

Putin systematically appointed his most trusted friends from St. Petersburg (KGB buddies, or colleagues from his former job as vice-mayor there) to the top positions in his administration. Many of these people also ended up controlling key government ministries or regions. A few have been chosen to sit on boards of the wealthiest semiprivate or state corporations. It has never been known what proportion of the privately issued stock of these companies these people control, but, more importantly, they also control the state's packets of shares. If the Kremlin shareholders consisting of Putin's closest friends constituted a corporation, they would control 20–30% of the country's GDP, as estimated by *Novaya Gazeta* experts in a series of articles on corruption. This is a smaller proportion than that claimed by the seven top oligarchs in 1996, but back then only a handful of enterprises had been privatized, while today over 75% have been. The GDP itself had also returned to its 1990 level by 2007.

Further changes began in early 2008, when Dimitry Medvedev won the presidential elections with 70% of the votes, against 18% for the main Communist contender. In the absence of any real opportunity for other candidates to campaign, the result was predictable. Putin remained in power, however, by becoming the head of the majority party (United Russia) and agreeing to become a national "leader" and the prime minister. This allowed him not only to save face and to avoid changing the constitution, but also to keep an eye on Medvedev. The truth is that the Kremlin Corporation, in all likelihood, is very unlikely ever to step down voluntarily. Unlike Ukraine and Moldova, which have seen post-Soviet swings from one political party to another and have real competition within their elites, Russia has been dominated by the blue color (United Russia) for fear of the red (the Communist Party). The ironic thing is that the same former nomenklatura (or KGB) members are still sitting in power, but under different colors. The future of Russia—at least for now, while petroleum prices are high—seems to be in the hands of intelligence men turned oilmen.

The first two years of Dmitry Medvedev's presidency have not changed the overall situation dramatically, despite some early hopes. The control of the government, and in reality, much of the country, seems to remain largely in the hands of now-Prime Minister Vladimir Putin, who is still considered to be the most influential politician based on expert polls. Medvedev did announce a number of important initiatives in modernizing Russia's domestic economy and in deeper engagement with the European Union and the United States. Medvedev has been frequently portrayed in the media as a "moderate" and even a "liberal" in contrast to Mr. Putin. In reality, however, the two men share much in common, and even closest sympathizers of Medvedev have little doubt that the so-called tandem in power is little more than a facade covering up the increasingly autocratic and extremely corrupt top of the Russian oil-and-gas driven bureaucracy headed by Vladimir Putin.

Other FSU Republics

The pattern of economic transition in other FSU republics followed broadly the same path as Russia's. The Baltic states were the earliest adopters of the Western free market, with large proportions of their economies privatized by the mid-1990s. By the turn of the century they were already well off enough to be considered for membership in the EU, which they successfully joined in 2004. They have multiparty systems; nationalistic free-market parties won elections easily to begin with, but more recently they have lost to the more left-leaning or liberal parties.

Ukraine and Moldova have followed a slower pace of reform than either the Baltics or Russia, but have essentially had the same periods as Russia of rapid inflation, unfair privatization, and battles for key assets. Unlike Russia, however, both have experienced political shifts back and forth from one clan to another—something that Russia has not yet seen. The president of Moldova in 2008, Vladimir Voronin, was a Communist, although his predecessor was not. Belarus went into an early period of political isolation because of the authoritarian antics of Alexander Lukashenko, its autocratic president. Nevertheless, Belarus is economically better off than neighboring Ukraine, at least in terms of its GDP. A large part of the reason for this is the strong interdependence between the Belarusian and Russian economies, which has remained largely unchanged since Soviet times. The two countries have managed to preserve a much higher degree of economic integration than the rest of the FSU. Many products assembled in Belarus factories are made from Russia-made parts. Many Russian goods travel to Europe via Belarus, and many imports to Russia travel in the opposite direction; each time, Belarus takes a cut in the proceeds. Also, Russia provides natural gas to Belarus at a much lower cost than for any other country in the FSU. The two countries are negotiating an even tighter integration into a union of sorts, as you can already see in Russian passport control lines at the airports, where Belarusian citizens are the only ones allowed passage alongside Russians.

The trans-Caucasian republics and the Central Asian states have all managed to go through reforms. Kazakhstan has arguably been the most successful, with many years of positive GDP growth, heavy privatization, and a high share of foreign participation in its new industries. Turkmenistan represents the opposite case of very inward-looking development, with its authoritarian leader managing to destroy much of the economy and trade in the process of self-adoring nation building. The economies of Kyrgyzstan, Uzbekistan, Tajikistan, Georgia, and Armenia have all done fairly poorly lately. Azerbaijan's economic revival is tied to high petroleum prices at the moment, but even there the situation remains rather desperate for the majority of its population. Regional conflicts, political unrest, and/or autocratic regimes in these countries make reforms go slowly. Additional information on each of these countries is provided in Chapters 25 and 29–31.

REVIEW QUESTIONS

1. Explain the main internal and external rationales for Gorbachev's reforms.
2. What is the difference among perestroika, uskorenie, and glasnost? Which one(s) was/were successfully implemented by Gorbachev's team?
3. Why did the August 1991 coup happen? Who was behind it? What was its outcome?
4. Which republics of the FSU were the ones most eager to leave the union? Which ones were the most reluctant? In each case, why?
5. Theorize what would happen if Russia had chosen to make Chinese-style reforms under Gorbachev (i.e., not much political openness, but slow and gradual economic change).
6. What was the low economic point of the post-1991 reforms? How are things different now?
7. Is Russia's economy stronger or weaker now than in the late Soviet period? Than in the mid-Yeltsin period (1996)? What are the most pressing issues that must be addressed by the government of President Medvedev?

EXERCISES

1. Compare and contrast Gorbachev's and Yeltsin's styles of leadership, based on the descriptions in this chapter and on additional readings.
2. Stage a role-playing game in your class. Imagine that you are trying to privatize a state factory. The following roles may be used: head of the state privatization committee, local hoodlums, a Russian coop owner who made some money during the late 1980s, a foreign investor from Europe (or the United States), a foreign investor from Turkey (or India), a Communist hardliner politician, concerned factory workers, and the current factory director. Who do you think will get the factory in the end?
3. Research and debate the Yukos 2003–2007 story: how the largest oil company and the wealthiest Russian tycoon, Mikhail Khodorkovsky, were criminalized and bankrupted by Putin's prosecutors. What could have been done to avoid the showdown? How did the Russian economic and political landscape change as a result of this?
4. Suppose you have an extra $10,000 in the bank. Research realistic options available to you for investing the money in Russian (or other post-Soviet) markets. Can you do it without leaving your country? Your state? Your house? Will you feel safe and se-

cure about the investment? Would you go through a Western intermediary firm, or invest directly with Russian stockbrokers or companies? Track the stock price of one Russian company listed on an exchange for 2–3 months. If you had invested the money in reality, would you have made any money?

Further Reading

Åslund, A. (2007). *Russia's capitalist revolution: Why market reform succeeded and democracy failed.* Washington, DC: Peterson Institute for International Economics.

Baker, P., & Glasser, S. (2005). *Kremlin rising: Vladimir Putin's Russia and the end of revolution.* New York: Scribner.

Brady, R. (1999). *Kapitalizm: Russia's struggle to free its economy.* New Haven, CT: Yale University Press.

Dunn, E. C. (2004). *Privatizing Poland: Baby food, big business, and the remaking of labor.* Ithaca, NY: Cornell University Press.

Freeland, C. (2000). *Sale of the century: Russia's wild ride from communism to capitalism.* New York: Crown Business.

Gregory, P., & Stuart, R. (2001). *Russian and Soviet economic performance structure.* Boston: Addison Wesley Longman.

Hoffman, D. (2003). *The oligarchs: Wealth and power in the new Russia.* New York: Public Affairs.

Klebnikov, P. (2000). *Godfather of the Kremlin: Boris Berezovsky and the looting of Russia.* New York: Harcourt.

Lai, H. (2006). *Reform and the non-state economy in China: The political economy of liberalization strategies.* New York: Palgrave Macmillan.

Remnick, D. (1993). *Lenin's tomb: The last days of the Soviet empire.* New York: Random House.

Rosefielde, S. (2007). *The Russian economy: From Lenin to Putin.* Malden, MA: Blackwell.

Sachs, J. D., & Lipton, D. A. (1993). Remaining steps to a market-based monetary system. In A. Åslund & R. Layard (Eds.), *Changing the economic system in Russia.* New York: St. Martin's Press.

Smith, H. (1993). *The new Russians.* New York: Avon Books.

Tikhomirov, V. (2000). *The political economy of post-Soviet Russia.* New York: St. Martin's Press.

Volkov, V. (1999). Violent entrepreneurship in post-Communist Russia. *Europe–Asia Studies, 51,* 741–754.

Websites

www.economy.gov.ru/minec/main—Ministry of Economic Development of Russian Federation. (In Russian only.)

go.worldbank.org/6HBRBYQ7G0—Russia Economic Reports (2001–2010) from the World Bank.

www.wilsoncenter.org/index.cfm?fuseaction=topics.home&topic_id=1424—The Kennan Institute, the oldest program of the Woodrow Wilson Center, brings scholars and governmental specialists together to discuss political, social, and economic issues affecting Russia and other FSU republics.

CHAPTER 9

The Geopolitical Position of Russia in the World

Now that we have considered the main economic and political reforms of the last 20 years, it makes sense to look at the Russian Federation and the other countries of the former Soviet Union (FSU) with respect to their geopolitical position. Although Russia is a successor to the Soviet Union, it only has half of the U.S.S.R.'s population and 70% of its territory; it is much more ethnically homogenous; and it is far less influential in global affairs.

"Geopolitics" may be defined as "the analysis of interactions between . . . geographic settings and . . . political processes" (Cohen, 2009, p. 5). The early geopolitical studies of Ratzel, Mackinder, Mahan, Bowman, and Kjellen sought to elucidate the general principles of the global world order in the periods before and between the two great wars of the 20th century. Particularly salient for us is Harold Mackinder's (1904) notion of the "Heartland" (i.e., continental Eurasia, more or less coterminous with the Russian Empire) as a pivotal world region that theoretically is destined to control the rest of the world. Mackinder's "Heartland" can be contrasted with Nicholas Spykman's (1944) "Rimland" (i.e., the coastal areas of Europe, Asia, and North America). The Heartland has a strategic advantage over the Rimland in having more natural resources and less vulnerability when attacked inland by conventional weapons (tanks, artillery). The Rimland, however, has a strategic advantage in shipping and is able to leverage its coastal positions in any warfare that involves aircraft carriers and submarines. Although the developments of the last 20 years have given much greater prominence to the Asia–Pacific and North Atlantic Rimland, the Heartland theory did receive some validation when the Soviet Union developed to rival the United States in the Cold War, and it is still an interesting starting point for discussions about the present and the future of Northern Eurasia.

The Russian Empire reached its zenith at the time of the Crimean War in the 1850s, when the country stretched from Poland in the west to Alaska in the east. By that time, it already included much of trans-Caucasia and Central Asia, and was posed to enter into several prolonged battles: with Turkey and Britain over the Balkans; with Persia over the entire Caspian Sea basin and the Caucasus; and with Japan and China over Manchuria (Figure 9.1). The only empire in recent history that was physically bigger was the British Empire, which controlled about 25% of the world's surface, whereas the Russian Empire controlled about 17%. The British Empire accounted for 13.6% of the world's gross domestic product (GDP) in 1913, while Russia's

FIGURE 9.1. A Russian church in Harbin, northeastern China—currently a historical museum—testifies to the strong Russian presence in northeast China between 1880 and 1940. *Photo:* K. Wong.

accounted for 8.3%. The U.S.S.R. was a smaller entity than the Russian Empire, because it did not include Alaska, Finland, or Poland. It did expand farther into Central Asia and the Caucasus, however. After World War II, the Soviet Union came to dominate the affairs of Eastern Europe, Cuba, and parts of Southeast Asia and Africa by setting up Communist governments there.

As one of the victorious powers in World War II, the U.S.S.R. became a dominant force in global affairs, along with its allies (the United States, Britain, and France). The four countries established themselves as permanent members of the U.N. Security Council, with veto powers (China was added in the late 1960s). They thus greatly influenced the composition and decision making of the entire United Nations and the postwar world order in general. With its socialist satellites, the Soviet Union controlled close to one-quarter of all U.N. votes. Nuclear parity with the United States was largely achieved by the mid-1960s. Although the Soviet Union was trailing the United States in developing atomic and hydrogen bombs in the early 1950s, it was the first to develop intercontinental ballistic missiles (ICBMs) by the late 1950s, and the first to put a man in space in 1961. The development of nuclear weapons and space research ensured that the Soviet Union began to be taken seriously everywhere in the world. It was the only country besides the United States capable of destroying the entire planet in a nuclear war—a true superpower.

How is Russia today different geopolitically from the U.S.S.R.? First, it is much smaller. Although Russia did retain the bulk of the richest extractive and manufacturing zones and about 70% of Soviet manufacturing capacity, it lost access to about half of the productive agricultural areas in Ukraine, Kazakhstan, Georgia, and Uzbekistan; some essential mining areas (chromium and uranium ores in Kazakhstan, manganese ores in Georgia); and most of the coastline along the Black and Baltic Seas. A lot of high-tech manufacturing and final assembly of machinery and equipment used to take place in Ukraine, Belarus, and the Baltic states. Much of the infrastructure built in the Soviet period with nationwide efforts (e.g., hydropower plants in Tajikistan and Georgia, or nuclear stations in Armenia, Lithuania, and Ukraine) is divided now among the successor states. The Russian military had to pull out of most republics, notably the Baltic states, Georgia, Ukraine, and Kazakhstan. The nuclear warheads and missile ingredients that were deployed in Ukraine and Kazakhstan were dismantled and moved to Russia, in accordance with international agreements with the United States and Europe. However, much of the civilian infrastructure (radiolocation and generation equipment, military bases, etc.) has been given over to the respective national governments, with no compensation to Moscow. One can of course argue that this is only fair, because the entire U.S.S.R. participated in the production of those. Nevertheless, Russia's share in constructing these was greater than its proportion of the population. Moscow did retain some control over a few of these assets within the FSU (e.g., the Sevastopol naval base in the Crimea, Ukraine; an early-warning radar station in Gabala, Azerbaijan; the Baikonur space launching pad in Kazakhstan). However, given the skewed distribution of production in the Soviet period, it is safe to say that Russia did not benefit from the collapse of the U.S.S.R. as much as the newly independent periphery did.

Second, Yeltsin's agreement with the presidents of Belarus and Ukraine in December 1991 essentially accepted the Soviet internal boundaries as the new international ones: The FSU republics' outlines today are the same as they were in the Soviet period. This was probably the easiest choice, and it helped to prevent a major conflict developing along Yugoslavian lines. However, those internal boundaries only loosely conformed to where the respective ethnic groups actually lived in the U.S.S.R., and they were never intended to become permanent international borders. They were physically unmarked, had no checkpoints, and frequently did not follow any physical landmarks. Locals used to cross them routinely on the way from home to work, just as people in the two Kansas Cities do when they travel between Missouri and Kansas every day. The borders were internal matters of administrative convenience for the Communist planners in the 1920s through the 1950s, not matters of international politics.

Today, however, each new country has its borders recognized by the international community as if they were indeed national borders that had been carefully delineated by some impartial committee. Unfortunately, they were not. Large Russian minorities (totaling about 25 million in 1991) lived in Estonia and Latvia; in eastern Ukraine; on the Crimea Peninsula and much of Ukraine's Black Sea coast; in Moldova; in northern and eastern Kazakhstan; in parts of Kyrgyzstan; and elsewhere. Russians had only moved to some of these places during the last 60 years or so, but they had lived in others ever since permanent settlements of any kind were established by the expanding empire. (The special case of Kaliningrad Oblast—an "exclave" of Russia that is now completely surrounded by other FSU republics—is described in Vignette 9.1.) Similarly, millions of Ukrainians lived throughout Siberia, Kazakhstan, and the Russian Far East. Ossetians found themselves divided between Russia and Georgia. The Abkhazy people in Georgia, who are closely related to the Cherkesy and Adygi people of the Russian northern Caucasus, were now part of independent Georgia—a country with a very different predominant ethnicity and a strongly nationalistic government. Many Armenians, Georgians, Azerbaijanis, Uzbeks, Tajiks, Estonians, and members of other ethnic groups lived in large numbers in most big Russian and Ukrainian cities, in villages along the Black Sea coast, in the Caucasus, and so forth. All of these people were suddenly thrust into dealing with the increasingly nationalistic governments of the new states. Many chose to move, but many others stayed and had to adapt to the new realities. A few are still living as unrecognized citizens of the now extinct country, without passports or even a path toward full citizenship.

Third, Russia lost much of its international influence outside the former Soviet borders. The Soviet Army withdrew from central Europe (in particular, East Germany) and from Afghanistan in 1989. It also left dozens of allied countries in the developing world (e.g., Cuba, Angola, and Vietnam) without critical economic assistance. Gorbachev's decision not to oppose unification in Germany led to a hasty withdrawal of the Soviet troops, with virtually no compensation from the North Atlantic Treaty Organization (NATO). In fact, Gorbachev made an extremely generous gift to the West: Not only did he not request any financial support for troop withdrawal and resettlement; he did not even ask for a firm political guarantee from NATO that it would not

Vignette 9.1. Strategic Kaliningrad

If you look at a map of present-day Russia, you may wonder why a triangular piece of its territory is isolated between Poland and Lithuania, right on the Baltic Sea coast. Historically, this was part of the now extinct country Prussia, populated by the Baltic people of the same name. However, the ethnic Prussians were absorbed over several centuries by the Polish, Germanic, and Slavic inhabitants of this region. The German Teutonic knights made this area one of their Baltic strongholds and brought Roman Catholicism here in the 1300s. Later Prussia became the first country in the world to adopt Lutheranism as its state religion. Under a post-World War II arrangement, the Soviet Union claimed the territory for itself, to gain a strategic foothold in Central Europe and to help cover the enormous costs of postwar reconstruction. The territory is small (slightly under 15,000 km^2), but it is strategically important for Russia. The total population is just under 1 million.

The city of Kaliningrad was formerly known as Koenigsberg, "the city of kings." It is known as the birthplace of Immanuel Kant, a famous German philosopher who lived and is buried there. The city's architecture and layout show strong German influences. It is a big seaport. Manufacturing in the region includes ships, railroad cars, automobiles, and TVs. Kaliningrad Oblast is also one of the leading areas of amber production and has thriving fisheries. More significantly for Russia, its ports serve as a gateway to Europe. Since 2004, the oblast has been surrounded by EU territory from all sides except the sea. Its residents must have visas to visit Lithuania or Poland. Without visas, they cannot travel to Russia except via direct airplane flights or an express train that crosses Lithuania without stopping. There is also an unfinished highway to Berlin, which ends at the Polish border and bypasses most inhabited areas.

The strategic importance of this exclave lies in its geographic position close to Europe and in the southern part of the Baltic Sea. The city of Kaliningrad is the closest port in Russia to Europe. Because of its southerly location, it is also the only Baltic Sea port that does not freeze in winter. About 12 million metric tonnes of goods pass through the port per year. The oblast has a special economic zone status with favorable tax rates for foreign investors, to stimulate local industry. It is also one of the few areas where Russia can locate its early-warning radiolocation stations to keep an eye on possible NATO expansion and can stage its antiaircraft missile complexes and fighter jets. Finally, the region has high tourism potential because of its dunes and beaches.

expand its borders toward the U.S.S.R. (or later Russia). Gorbachev did ask for and receive plenty of financial loans from the International Monetary Fund (IMF), the World Bank, and various Western governments (which Russia is now repaying with interest), but he obtained little free assistance. Billions of rubles of assets in Poland, the Czech Republic, Slovakia, Hungary, and East Germany were simply left behind. Putin's final task as an official of the KGB in East Germany was to personally oversee the destruction of KGB archives there, as well as to dispose of Soviet assets in a last-minute "fire sale." The Soviet troops' withdrawal from Afghanistan in that same year led to a creation of a power vacuum there, which eventually was filled by the Taliban movement. By 1990 the Baltics were de facto free, and the collapse of the Soviet regime in 1991 left each country of the FSU pursuing divergent goals in a new geopolitical space.

Russia's Neighbors

Table 9.1 illustrates the position of Russia vis-à-vis other nations in the world today. It remains an important player worldwide: It is still the biggest country by size, with plenty of natural resources, one of the largest military complexes on the planet, thousands of nuclear warheads, and brisk arms sales to other countries. It is far less significant in cultural and "soft" economic endeavors. For example, lots of Russian movies are being made, but they are little known outside the country; Russian computer software is generally of low quality and, with the exception

TABLE 9.1. Selected Rankings of Russia in Relation to Other Countries, 2009

Characteristic	Ranking
Area	1st
Land border length	2nd (after China)
Population size	9th (smaller than Pakistan, Bangladesh, or Nigeria)
Armed forces personnel	4th (after China, the United States, and India)
Number of nuclear warheads	1st
Conventional arms sales	2nd (after the United States)
GDP purchase parity (total)	6th
GDP purchase parity (per capita)	73th
Coal production	5th (after China, the United States, India, and Australia)
Oil production	2nd (after Saudi Arabia)
Nickel production	1st
Natural gas production	1st (about one-quarter of the world's total)
Hydropower production	5th
Potassium fertilizer production	1st
Diamond production	3rd (after Democratic Republic of the Congo and Australia)
Motor vehicles production	12th
Electricity production	4th
Arable land	4th (after China, the United States, and India)
Timber production	8th (1st in amount of standing timber)
Full-length movies produced	10th
Number of tourists sent abroad	9th

Note. Data from many sources, including the U.S. Geological Survey, *nationmaster.com*, the CIA World Factbook, and others.

of the Kaspersky Internet Security suite, is virtually absent from Western stores; Russian furniture cannot compete with Italian or Swedish furniture; and so on.

Russia is located on the largest continent, Eurasia, with 15 direct neighbors (see below) and lots of other countries it does business with. Only China has as many neighbors. It is convenient to divide Russia's neighbors and other related countries into four tiers: immediate neighbors (Tier I); second-degree neighbors (Tier II); more distant countries with which Russia has strong past and/or present ties (Tier III); and the rest of the world (Tier IV).

Immediate Neighbors (Tier I)

Tier I includes Norway, Finland, Estonia, Latvia, Lithuania, Poland, Belarus, Ukraine, Georgia, Azerbaijan, Kazakhstan, Mongolia, China, North Korea, Japan, and the United States (via Alaska). Of these, Finland and the three Baltic countries are European Union (EU) members. The Baltics are NATO members and staunch U.S. allies; they have an ambivalent relationship with their big eastern neighbor. On the one hand, they have deep suspicions about a possible resurgence of the Kremlin's imperial ambitions. On the other, pragmatically speaking, these countries greatly benefit from transshipment of Russia's oil, gas, metals, and timber, as well as from Russian tourism and investment opportunities. The stickiest points from Russia's perspective are the lack of full citizenship rights for Russian-speaking minorities in Estonia and Latvia; the sometimes uncivil behavior of Baltic politicians with respect to the past (e.g., the rise of neo-Nazis in Latvia, with tacit approval or even encouragement from the nationalist politicians, as well as the desecration of Soviet war memorials there); and arguments over portions of the common border between Estonia and Russia near Lake Peipus/Chudskoe.

Because of its autocratic president, Belarus is the most marginalized country in Europe right now. However, as described at the end of Chapter 8, it is a critical partner of Russia in two areas: shipping goods to and from Europe (Belarus ships more freight to and from Russia than any other country), and shared manufacturing ventures. Thus Belarus is one of Russia's strongest allies; it is even negotiating a formal union between the two nations, with shared borders, currency, armed forces, and tax system planned for some point in the future.

Ukraine, Azerbaijan, Kazakhstan, and Georgia have all gone through a gradual process of dissociation from Russia, to a greater or lesser degree. Ukraine is perpetually torn between its nationalistic but economically underdeveloped western half on the one hand, and its heavily Russian-speaking and industrially developed eastern side with strong economic and social ties to Russia on the other. Ukraine is the largest country in Europe by territory, bigger than even France. Its historical connections to Poland play a role in its current position as well. Many Ukrainians are slowly realizing that for better or for worse, they are already part of a greater Europe; however, they are also not exactly free from their mutual history with Russia (see, e.g., Figure 9.2). Ukraine and Russia formally delineated their land borders in 2007, but they dispute the exact location of the border in the Kerch Strait and the status of the Sevastopol naval base. Therefore, the present situation in Ukraine is ambiguous. In general, Russia and Ukraine could be compared to the United States and Canada: One is larger and monolingual; the other is smaller and bilingual. Future relations between the former two are not likely to be as friendly as those between the latter two, however.

Although the Georgian and Russian cultures have been greatly influenced by the Orthodox Church and have much in common, recent political relations between Georgia and Russia have been turbulent. After the fall of Communism, the brief rule of the ultranationalist Zviad Gamsakhurdia in Georgia led to a disastrous war in

FIGURE 9.2. A monument to the famine victims of 1932–1933 in Kiev, casting its shadow on a church wall. The famine happened when Stalinist-forced collectivization deprived millions of their land and livestock. Almost 2 million people died in Ukraine; hundreds of thousands more died in the Volga region of southern Russia. *Photo:* J. Lindsey.

Abkhazia and the rapid secession of Abkhazia from Georgia in 1992. After this loss, the Gamsakhurdia regime promptly collapsed. Russian peacekeepers were positioned in both Abkhazia and South Ossetiya as part of a U.N. peacekeeping force. Separatisms within Georgia are encouraged by Russia, and the escalation of conflict in South Ossetiya in August 2008 brought renewed international attention to the unresolved issue of maintaining peace in the self-proclaimed republics. Despite being tied to Russia by electricity transmission, gas shipments, and much foreign commerce as well, Georgia remains fiercely nationalistic at present. Its Western-educated president, Mikheil Saakashvili, is maneuvering between outright allegiance to the United States and the need to trade with its less accommodating but more immediate neighbor, Russia. While winning elections on an anticorruption ticket, he has done little to fix problems in his past few years in power. Apparently one major improvement has been in the traffic police force: The corrupt staff of the former Soviet police was sacked and replaced with young, better-paid, more mobile units with no ties to the past. Saakashvili is frequently accused by the opposition of not fulfilling his duties in defending the true national interests of the country, however. The United States supports Georgia's need for territorial integrity, but the precedent of Kosovo's recognition has now led to official recognition of Abkhazia and South Ossetiya by Russia, and the situation is far from being permanently resolved.

Kazakhstan is the richest of all the Central Asian states and is craftily treading a middle ground among Russia, the West, and China at the moment—a tricky business indeed. It hopes to attract massive investment in its western Caspian oil fields from U.S., European, Chinese, and Russian companies. It is building oil and gas pipelines into China. It is dependent on Russia for many manufactured goods, as well as for engineering talent and transportation options. It also has a large minority of Russian speakers—mainly in the north and east, where Russians constitute a majority of the population in many industrialized cities (e.g., Ust-Kamenogorsk, Petropavlovsk, and Pavlodar). Russia and Kazakhstan share the longest common border in the FSU (7,200 km). Kazakhstan is a buffer country between Russia and volatile Central Asia. A major negative impact of Kazakh independence from Russia's perspective is the dissection of the historically Russian-settled central Siberian corridor along the south branch of the Trans-Siberian Railroad by two international borders. This is not simply a political issue; it is a major economic inconvenience, because more than half of all freight and electric energy from Europe to Siberia used to flow through the Petropavlovsk corridor during Soviet times. Now passenger and freight trains must stop twice at each of the two international borders to be searched by the customs officers of both countries.

Azerbaijan is almost 100% dependent on petroleum exports for foreign revenue. The completion of the Baku–Tbilisi–Ceyhan pipeline in 2005 now allows direct shipments of its petroleum to Turkey through Georgia, bypassing Russia. A large number of Azerbaijanis live all over Russia and in other FSU republics; their economic specialty is flower and vegetable trade in farmers' markets. Many experience prejudice and outright harassment from the locals. By contrast, relations between Russia and Azerbaijan at the state level remain pragmatic and reasonably friendly. More Azerbaijanis live in Iran than in Azerbaijan, thus necessitating close relations with the southern neighbor as well. Turkey, Iran, and Pakistan supported the acceptance of Azerbaijan into the Middle East economic community. The country remains at a cease-fire in its war with Armenia over the control of the Nagorno-Karabakh region, which Azerbaijan effectively lost in the early 1990s military conflict with Armenia.

Mongolia and China have extensive land borders with Russia (3,005 and 4,300 km, respectively). Mongolia was sometimes dubbed "the 16th Soviet republic" because of the extent of its integration into the Soviet economy. Recently Mongolia has become more interested in developing ties with other countries, including China and the United States. It receives about 95% of its petroleum from Russia, but China is a larger trade partner now than Russia. Mongolia remains a poorly developed, arid, landlocked country with very little political or economic capital. China has a very short common border with Russia in the Altay, and a much longer one along the Amur River. Some portions of this border were disputed in the 1960 and 1970s, but are now firmly fixed. On the grand scale, China thinks of itself as the next superpower, bound to unseat the United States by about 2015 as the world's largest economy (and perhaps by 2030 as the biggest military power as well). Russia is presently viewed by China as a convenient source of military technology (especially missile-, jet-, and space-related) and raw materials (oil, gas, iron ore, metal scrap, timber, etc.). Russia in turn is eager to provide all these products, hoping that any direct political confrontation with its big southern neighbor can be avoided. The demographics are not in Russia's favor; only about 5 million people live in Russia east of Lake Baikal. At the same time, two northeastern provinces of China have over 100 million people living within a day's journey of the Russian border.

As incredible as it may seem, Japan and Russia are still technically at war with each other. At the end of World War II, Russia reclaimed the southern portion of Sakhalin Island (which had been lost to Japan in 1904) and captured all of the Kuril Islands. Japan insists that the four southernmost Kuril Islands—Shikotan, Habomai, Kunashir, and Iturup—must be returned before it will sign a formal peace agreement. Russia does not want to give up either the military advantage that the islands afford (naval bases, early-warning air defense systems) or the fisheries of the northwestern Pacific, which are among the richest in the world. Economically, however, the two countries are on very friendly terms. A quick visit to Siberia reveals that about half of the cars driven on Siberian roads in Russia are used Japanese imports, with the steering wheel on the right side. The Japanese are also eager tourists, and many are attracted to Lake Baikal, Kamchatka, the Trans-Siberian Railroad, and of course Moscow and St. Petersburg. Few Russian tourists go to Japan, because getting Japanese visas is notoriously difficult for outsiders; however, shuttle trading is common along the Pacific Coast.

It may amaze you that the United States is also a country in Tier I. Well, the two countries are merely 20 km apart at the Diomedes Islands in the Bering Strait. In fact, a charter flight on Bering Air from Nome, Alaska, to Uelen, Chukotka, is shorter than the commercial flight from Anchorage to Fairbanks. In contrast, an average commercial flight from New York to Moscow takes about 10 hours across the Atlantic and parts of Europe. The United States and Russia are really very distant on the globe—except where they almost touch in the Bering Strait. The potential for joint exploration of the oil and gas on the Arctic shelf, and even for the construction of a cross-hemisphere railroad tunnel under the strait, exists. Each country, however, is suspicious of the other's intentions. For example, recently the Russian government flatly refused to let foreign companies invest in the development of the massive Shtockmann gas field in the Barents Sea. The Americans have never been keen about letting Russian companies drill in Alaska, either.

Strategically, the United States sees Russia as a convenient counterbalance to China in global affairs and as a partner (among many others) in the war on terrorism. Russia admires many U.S.-made things (ranging from software to bubble gum to Boeing aircraft), but has no problem holding its own line when it comes to the true economic competition: Both countries fiercely compete now in selling military technologies to various regimes around the world. The post-Soviet policy of the United States toward Russia has been pragmatic, but at times too hostile. For example, the very unfair Jackson–Vanik trade amendment of 1974 puts Russia at a huge disadvantage when trading with the United States and has not been repealed by Congress, despite repeated promises from Presidents George W. Bush and Barack Obama. The trade amendment denies Russia most-favored-nation status, because at the time of its ratification, the Soviet Union did not allow free emigration of its Jewish nationals. In addition, the United States unilaterally pulled out of the Anti-Missile Defense Treaty with Russia in 2002 to deploy its missile shield in Alaska, ostensibly against a North Korean missile threat. Also, the recent row in Europe over positioning NATO radiolocation stations in Poland and the Czech Republic, and encouraging Ukraine and Georgia to seek full NATO membership, certainly irritated Moscow. All these things have been done despite the many benefits that the United States has reaped from close cooperation with Russian intelligence in the anti-Taliban war in Afghanistan, or from Russia's support in imposing sanctions against Iran at the U.N. Security Council.

The level of mutual travel and commerce between Russia and the United States is far below what is probably needed. The overall trade balance between the countries in 2008 was $12 billion in Russia's favor: Russia sold almost $27 billion worth of goods to the United States, while the United States sold only $9 billion worth to Russia. This made Russia the 28th most important trade partner of the United States with regard to exports, and 17th in terms of imports. In comparison, the United States bought about $28 billion worth of goods from China per *month* that year. The amount of physical travel is low, too: For every 30 passenger jets leaving U.S. shores for Europe, only 1 flies to Russia. This may change in the future, because Russia and the United States are more similar than many people realize, and the potential for doing business together is very great. At present, Russians and Americans seem happy to cooperate in space research, to visit each other on occasion, and to confer (and often to clash dramatically) at U.N. meetings.

Second-Degree Neighbors (Tier II)

Tier II includes Moldova, Armenia, Turkmenistan, Uzbekistan, Tajikistan, Kyrgyzstan, and Eastern European countries that were former socialist allies (Poland, Slovakia, Hungary, Czech Republic, Romania, Bulgaria, Serbia, Macedonia, and Montenegro). All of these countries retain various degrees of political and cultural ties to Russia, but are no longer as strongly connected as they used to be to the Soviet Union. Tajikistan, Uzbekistan, and Kyrgyzstan are the most closely connected: All use Russian military and economic support, and form part of the Eurasian Economic Community (Evrazes). Armenia is an observer in Evrazes and has friendly relations with Russia. Others are either pragmatic economic partners (Bulgaria) or obstinate political rivals (Poland) of Russia in the new European order. Many are increasingly distant from Russia in terms of politics, but maintain strong economic relations with Russia for pragmatic reasons. Cultural ties among some of these countries are nevertheless deep enough to present many opportunities: Many Poles are fascinated by Russian music and books, for example, while Russians admire Polish fashions and arts.

Distant Nations with Various Strong Ties (Tier III)

Tier III includes the rest of Europe, especially Germany and Cyprus; Cuba, Nicaragua, and Venezuela in Latin America; some African countries with former socialist leanings (Ethiopia, Angola, Mozambique); a few Asian countries (India, Vietnam, Laos, Cambodia); some new trade partners (Turkey, South Korea, Taiwan); and some Middle Eastern states (Israel, Syria, Iran, Iraq, United Arab Emirates, Egypt, Tunisia, Libya). Although this is a very diverse group, all are somehow connected to Russia by past or present political/educational ties, and/or by current economic ties. For example, many professionals in Cuba and Ethiopia were educated at Soviet universities and maintain some connections at their former universities. All of the countries in this category have business ties to Russia at present, which is reflected in favorable political relations. Some of these are underappreciated; for example, few outsiders know just how strong are the economic ties between Russia and Turkey based on tourism and trade, or between Cyprus and Russia based on investment banking. In fact, Russia was Turkey's largest partner in imports, and the sixth largest partner in exports in 2008. Other connections are much discussed—for instance, the Russian military and nuclear ties to Iran. Russia insists that it merely helps Iranians to develop peaceful nuclear power, while NATO suspects that military developments may not be absolutely excluded.

The relations between Israel and Russia are unique. On the one hand, the Soviet Union was one of the chief supporters of the Arab world, and Russia remains a strong supporter of Syria today. On the other hand, about 1 million former Soviet citizens now live in Israel, and these people connect the two countries by countless business and family ties. Israel also has a special significance for Orthodox Christians as the premier worldwide pilgrimage destination, because the holy sites associated with the earthly life of Jesus Christ are located there. In 2007 Russia and Israel mutually abolished visa requirements for their citizens, in a major diplomatic breakthrough aimed at facilitating travel between the two countries.

Other Countries (Tier IV)

Tier IV includes most of Latin America, Canada, most of Africa and the Middle East, the rest of Asia, Australia, and Oceania. Although they are not exactly irrelevant, these countries are only very loosely connected to Russia, as Russia is to them. There are no open conflicts, but also relatively little trade. The main connections are casual tourism and occasional sales of military equipment. There are relatively large Russian and Ukrainian diasporas in Canada, Australia, South Africa, Argentina, and Brazil. Thailand and Indonesia have become popular tropical destinations for Russian tourists. Many foreign students in Russian universities today hail from the poorest countries of Africa, Asia, and Latin America, because of the still relatively low cost of Russian university education and its perceived high quality. In the post-September 11 world, Arab and African students usually have an easier time qualifying for Russian visas than for U.S. visas. Russian military sales to Southeast Asia, the Middle East, and some Latin American countries are growing. On a recent flight from Moscow to Amsterdam, I met two Russian airplane mechanics on their way to Peru to repair a Russian-made fighter jet there.

Is Russia Asian or European?

The perpetual question of Russian foreign policy is where the country fits within Eurasia: Is it a European or an Asian state? This question began to be asked at the time of the Mongol invasions, when Russian princes such as Alexander Nevsky had to choose allegiances between western (Germanic) and eastern (Mongol) realms. Nevsky generally chose the Mongols over the Germans, but he also was an independent-minded ruler who was trying to tread a middle ground. The question again came to the forefront at the time of Peter the Great's Western-style reforms in the early 1700s, and then in a debate between "Westernizers" and "Slavophiles" in post-Napoleon 19th-century Russia. The Westernizers saw Russia as a fundamentally European country, albeit with a backward political system in need of reform. The Slavophiles, in contrast, saw Russia as a Eurasian entity with its own destiny.

In 1915 V. P. Semenov-Tian-Shansky, the most influential Russian geographer of his time, published a monograph on the political geography of Russia. His main thesis was that Russia was more similar to the United States and Canada than to any European or Asian country, in that it represented a "coast-to-coast" system rather than a "Heartland" or a "Rimland." He saw Russia's biggest challenge as developing sufficiently dense settlements in the distant Far East, and he advocated major population shifts toward the empty middle of the country in Siberia as a line of defense against possible invasions from the outside. In the 1930s, the émigré community of exiled Russian philosophers continued debating the question of Russia's "Eurasianness." The geopolitical role of Russia (and of Northern Eurasia generally) in the world has been much debated in the Western political-geographical literature as well, especially in the works of British, German, and U.S. geographers.

Broadly, there are three main viewpoints (I am simplifying them a bit):

1. Russia is part of Western civilization. Its elite is Western-thinking; its society is mostly European in its culture; and its economic patterns of production follow those of Europe, albeit with some variation and usually with a considerable time lag. It is gradually embracing Western democratic ideals and is becoming a more and more fully realized member of the larger European community and the North Atlantic world. This is the view of Westernizers, from Peter the Great to Mikhail Gorbachev.

2. Russia is part of the East (Asia) more than of the West (Europe). It is a politically backward society prone to violence, corruption, political oppression, and heavy top-down control by monarchical, maniacal tyrants. It is not a true democracy and can never become one, because democracy is contrary to its very nature. It will forever be antagonistic to Europe, North America, and the rest of the "free" world. Or, for those who prefer a more positive "spin" on things, Russia is a beacon of moral sanity to the decadent, corrupt West. In one version or the other, this is the view of some

Russian ultranationalists, Soviet-period Communists, American presidential advisors since World War II (e.g., Zbigniew Brzezinski), conservative U.S. talk show hosts, some right-wing politicians in Europe, and conservative economists and political scientists on both sides of the Atlantic (especially in Britain).

3. Russia is neither part of the West nor part of the East, but is its own distinctly "Eurasian" civilization. This is the view of most Russian nationalists, most 19th-century Slavophiles, and a few influential 20th-century Russian thinkers, and it seems to be enjoying the endorsement of the current Putin–Medvedev administration as well. According to this more middle-of-the-road view, Russia has both Western and Eastern traits. More significantly, it has many fused elements and should be recognized as a separate political entity, with a unique identity and interests. Some of these thinkers tend to emphasize the uniqueness of Russian religion, specifically the Orthodox Church, as distinct from both Western Christianity and the Asian religions. To a certain extent, Ukraine and Kazakhstan would also fit this "mixed model." Both are similar to Russia in the fusion of European and Asiatic elements in their cultures, although these elements are not expressed uniformly across the three countries.

This third viewpoint has been particularly popularized in the West by Samuel P. Huntington's book (1996) *The Clash of Civilizations* (Figure 9.3). His main thesis is that

> the fundamental source of conflict in this new world will not be primarily ideological or primarily economic. The great divisions among humankind and the dominating source of conflict will be cultural. Nation states will remain the most powerful actors in world affairs, but the principal conflicts of global politics will occur between nations and groups of different civilizations. The clash of civilizations will dominate global politics. The fault lines between civilizations will be the battle lines of the future. (1996, p. 45)

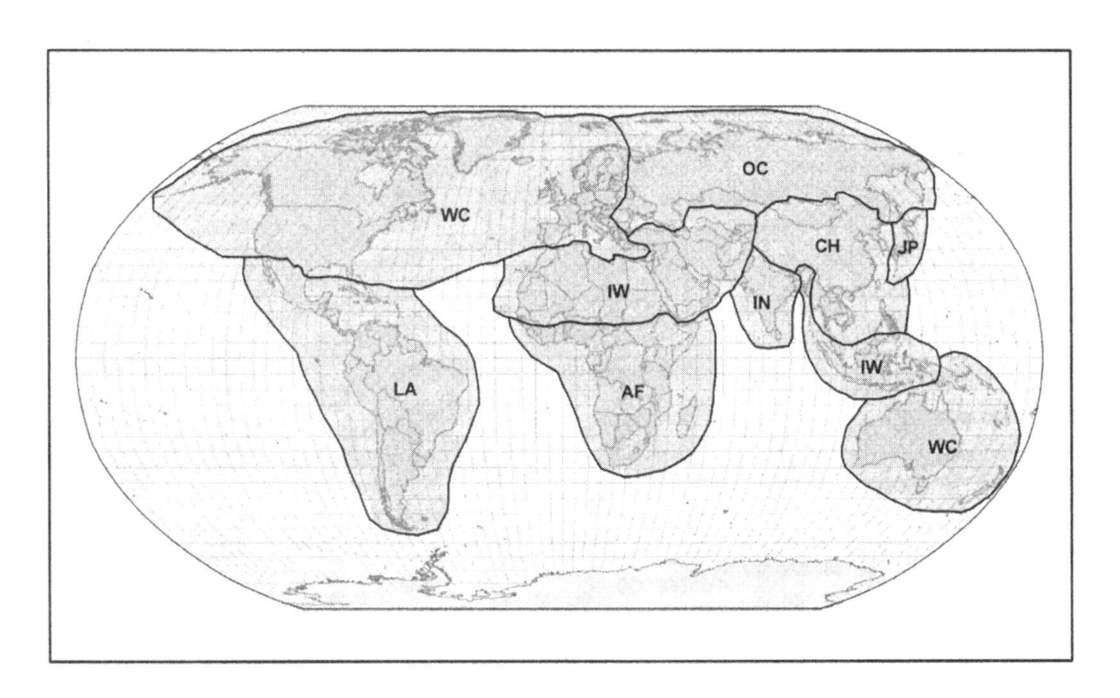

FIGURE 9.3. The eight "civilizations," according to Samuel P. Huntington (with modifications): WC, Western Christendom; OC, Orthodox Christendom; IW, Islamic world; CH, Chinese/Sinic world; LA, Latin America; AF, (Sub-Saharan) Africa; IN, India/Hindu civilization; JP, Japanese civilization. Not labeled are Buddhist Civilization (Burma, Thailand, Bhutan, Mongolia) and two special cases: "Turkey secularism" and "Israeli Zionism" within IW. Huntington's scheme is a controversial and conservative division of the world, but it can be used as a starting point in discussions of the global geopolitical pattern. Map: J. Torguson.

A conservative thinker, Huntington has influenced U.S. foreign policy for the last 15 years. This is significant for Russia for two reasons. First, it vindicates the current posturing of the Kremlin administration vis-à-vis Europeans and Americans in global affairs: "We are equal partners, but not one of you. Even famous Western scientists are saying so." Second, Russia has a "battle line of the future" passing right along its southern border, where the Islamic world meets the Orthodox realm. It is noteworthy that Huntington picked religion as a defining trait of culture. In a largely secularized Western world, this may seem naïve and outdated. However, in all the other civilizations defined by Huntington (see Figure 9.3) except the Chinese/Sinic world, religion continues to play an important role in nation building, national identity, and social cohesion. In the post-Communist societies and Islamic countries of Eurasia, it is actually playing an increasingly important political role (see Chapter 14).

Huntington's model has been much criticized and is, of course, a one-sided and fairly narrow view. Nevertheless, it provides a convenient conceptual map of the world for us to use as we try to understand the present-day political behavior of Russia and other FSU countries. The zone of contact between (Western) Europe and Russia has been contested for centuries and has seen several major wars, including the Napoleonic wars and the two world wars of the 20th century. Cohen (2009) calls this area the "Eurasian Convergence Zone" to indicate its position at the crossroads. He is more optimistic than Huntington that the zone may become the site of a genuine convergence, rather than competition, of world interests. For example, Russia, the United States, NATO, and some Central Asian states are involved as partners in the current efforts to stabilize Afghanistan.

It is interesting to note that while Ukraine and Georgia are Orthodox, they are actually less pro-Russian than nominally Muslim Uzbekistan or Kyrgyzstan, in a contradiction to Huntington's model. The first two countries are geographically on the doorstep of European civilization (the zone of contact between Western and Eastern Christianity); the latter two are in Eurasian hinterlands equally distant from Moscow and Mecca, and clearly in no position even to contemplate a membership in various European alliances. Thus Ukraine and Georgia are justified in their efforts to seek greater rapport with Europe. However, Islamic influences in Central Asia are not particularly strong (because of both 70 years of Soviet atheism and the current rulers' emphatically secular politics), so it could be argued that an alliance with Moscow makes a lot of practical sense for them. Other important zones of contact to watch around Russia are those in the Far East, with the Sinic and Japanese civilizations. Although relationships here are pragmatic and trade-oriented at the moment, these zones of contact are likely to become more contested in the future, as world energy resources become farther depleted.

REVIEW QUESTIONS

1. Which geopolitical changes in post-Soviet Eurasia seem the most significant to you?
2. Discuss the European, Asian, and "Eurasian" viewpoints as defined here. What are the merits of each? Try to find supporting examples in the literature.
3. Discuss the likelihood of three future scenarios: (a) the complete collapse of the Russian Heartland; (b) the emergence of a new strong state that would include most of the FSU; and (c) full integration of Russia within the EU/NATO framework. Which one seems the most plausible to you? Can you think of at least two other alternatives?
4. Some Russian political commentators believe that the country needs to join with the United States in its almost inevitable future conflict with China over dwindling global natural resources (e.g., Middle Eastern petroleum). Others think that Russia should side with China against the United States. Defend both viewpoints.

EXERCISES

1. Stage a classroom role-playing exercise in which Ukraine and Georgia are formally being accepted into NATO over strenuous objections from Russia. Use the following roles: a U.S. representative; a representative of an older NATO member that gets a lot of economic benefits from trade with Russia (e.g., Germany); a representative of a new NATO member

that resents Russia's new influence (e.g., Poland); representatives from Russia, Ukraine, and Georgia; and a NATO secretary whose job is to keep everybody together at the negotiating table.

2. Investigate the borders between Russia and the other FSU republics. Which areas are contentious in any way? Where do you see the greatest potential for future conflict? How can such conflict be resolved?
3. Investigate the actual volume of investments or trade between the following countries, using both online and print sources: Russia and Ukraine (Tier I), Russia and Hungary (Tier II), Russia and Germany (Tier III), and Russia and Canada (Tier IV). To what extent does the four-tier scheme proposed in this chapter holds up when measured in terms of the actual amount of investments or trade between these countries?

Further Reading

Boyd, A., & Comenetz, J. (2007). *An atlas of world affairs* (11th ed.). New York: Routledge.

Cohen, S. B. (2009). *Geopolitics: The geography of international relations* (2nd ed.). Lanham, MD: Rowman & Littlefield.

Huntington, S. P. (1996). *The clash of civilizations and the remaking of world order.* New York: Simon & Schuster.

Kolosov, V. A., & Mironenko, N. S. (2005). *Geopolitika i politicheskaya geografiya.* Moscow: Aspekt Press.

Mackinder, H. J. (1904). The geographical pivot of history. *Geographical Journal, 23*, 421–437.

Ross, C. (Ed.). (2004). *Russian politics under Putin.* Manchester, UK: Manchester University Press.

Spykman, N. (1944). *The geography of the peace.* New York: Harcourt Brace.

Taylor, P. (1993). *Political geography: World-economy, nation-state and locality.* Harlow, UK: Longman.

Treivish, A. (2005). A new Russian heartland: The demographic and economic dimension. *Eurasian Geography and Economics, 46*(2), 123–155.

Websites

www.state.gov/p/eur—Bureau of European and Eurasian Affairs of the U.S. Department of State.

www.gov39.ru—Government of Kaliningrad Oblast. (In Russian only.)

www.globalissues.org/Geopolitics

www.globalsecurity.org

www.hrw.org—Human Rights Watch.

news.bbc.co.uk/2/hi/country_profiles/default.stm—Country briefs on Armenia, Azerbaijan, Georgia, Moldova, and Tajikistan.

PART III

Cultural and Social Geography

CHAPTER 10

Demographics and Population Distribution

"Demography" is the study of populations. The population of a country can be classified by age, gender, occupation, health, and so on. We can look at where people tend to live, and at how and where they move. We also might be interested in the long-term prospects of a given society: Will it have enough resources to sustain its population growth, for example? This chapter deals with the general population distribution over Northern Eurasia/the former Soviet Union (FSU). It also examines changes in population and migration issues. Health and other social characteristics are discussed farther in Chapter 12, and cultural characteristics in Chapter 13.

Population Numbers

Russia is the largest country in the world by area, but it ranked only 9th largest worldwide by population in 2009, with 142 million people—right after Nigeria with 153 million, but ahead of Japan with 128 million. Although 142 million seems like a lot of people, consider that the United States, with less than half the geographic area of Russia, has 305 million people. The European Union (EU) member countries had a total of almost 500 million on a fraction of Russia's land. Bangladesh was even more astonishing: It had over 1,035 people per square kilometer in its land area of 144,000 km^2, while Russia had only 8, the United States 31, and Canada 3. The entire Soviet Union boasted about 290 million people by the early 1990s, of whom only slightly more than half lived in Russia. At that time, about 50 million lived in Ukraine. Uzbekistan, with about 20 million, was then the third largest republic. Today Ukraine and Uzbekistan still have the second and third largest populations in the FSU, while Estonia has the smallest (Figure 10.1 and Table 10.1).

The most important demographic characteristic after the total number is the growth rate. At present, Russia and many other post-Soviet states are actually losing population. In an estimate for 2008, Ukraine had the fastest rate of population decline in the world, at –0.5% per year. Russia, Latvia, and Moldova were at –0.3%; Lithuania at –0.2%; and Moldova at –0.1%. In contrast, the United States was growing by about 0.6% per year through natural increase and by 0.9% with immigration. The Central Asian states are markedly different from Russia or Ukraine, in that they keep growing; for example, Tajikistan was estimated to grow by 2.2% and Kazakhstan by 1.3% in 2008.

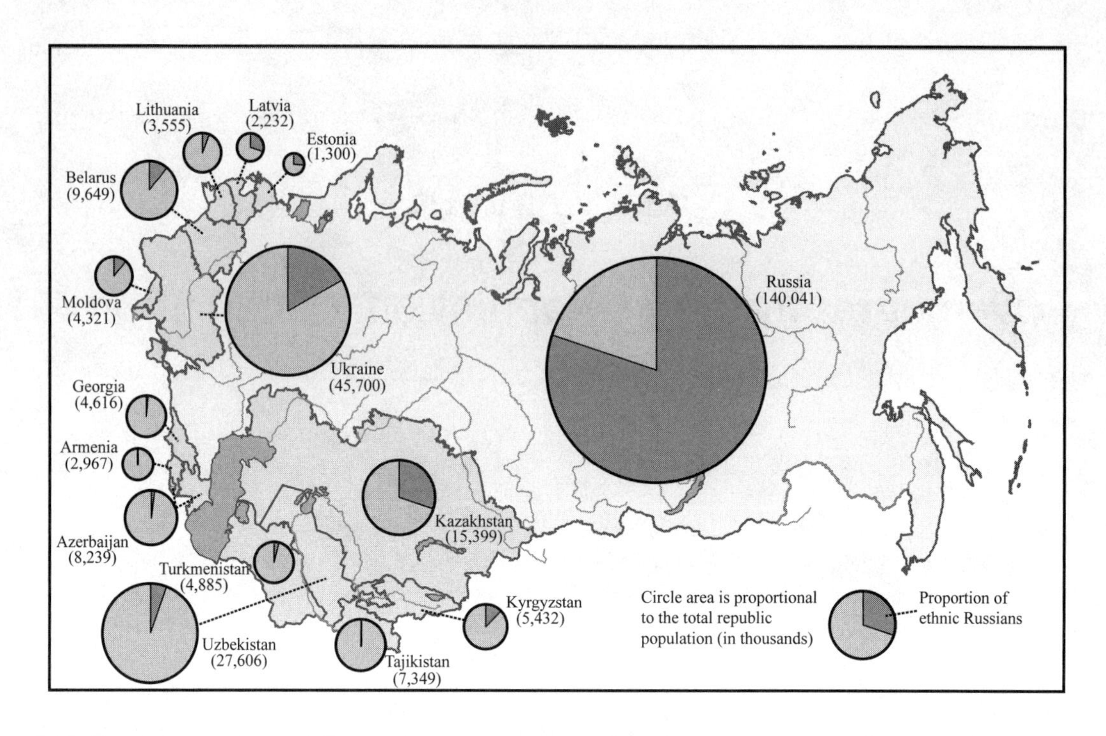

FIGURE 10.1. Population of the FSU republics and percentage of ethnic Russians in each, 2009. Data from CIA World Factbook. Map: J. Torguson.

TABLE 10.1. Comparative Population Statistics for FSU Countries, the United States, and the World (Mid-2009)

Country	Pop. (millions)	Density km^2	Births/ 1,000	Deaths/ 1,000	TFR/ 1,000	IMR/ 1,000	Life expectancy	Urban (%)
Russia	141.8	8	12	15	1.5	9	68	73
Belarus	9.7	47	11	14	1.4	5	70	74
Ukraine	46.0	76	11	16	1.4	10	68	68
Moldova	4.1	122	11	12	1.3	12	69	41
Georgia	4.6	66	11	9	1.4	13	75	53
Armenia	3.1	104	15	10	1.7	25	72	64
Azerbaijan	8.8	101	18	6	2.3	11	72	52
Kazakhstan	15.9	6	23	10	2.7	32	67	53
Kyrgyzstan	5.3	26	24	7	2.8	31	68	35
Uzbekistan	27.6	59	23	5	2.7	58	68	36
Turkmenistan	5.1	11	22	8	2.5	51	65	47
Tajikistan	7.5	50	28	5	3.4	65	67	26
Estonia	1.3	30	12	12	1.7	4.4	73	69
Latvia	2.3	35	10	14	1.4	8.7	72	68
Lithuania	3.3	51	11	13	1.5	4.9	71	67
United States	306.8	32	14	8	2.1	6.6	81	81
World	6,810	50	20	8	2.6	46	69	50

Note. TFR is the total fertility rate (average number of children born to a woman age 15–49; 2.1 is the replacement level. IMR is the infant mortality rate, defined as the average number of babies who are born alive but then die by age 1 per 1,000 newborns (it is primarily a measure of the quality of health services in a country; the lower the better). Data from the Population Reference Bureau 2009 World Population Data Sheet (*www.prb.org*).

Although some European countries have posted declines as well in recent years, the change from positive to negative growth in the FSU was very abrupt. It coincided with the beginning of economic reforms in 1992 and was quite unprecedented, especially considering the lack of a major military conflict. Even with the impacts of two world wars, the civil war of 1917–1922, and the horrors of Stalin's GULAG purges, the population of the U.S.S.R. had never declined between censuses. The period between the 1989 and 2002 censuses was the *only* one in Russia's history when the population actually dropped. Although estimates for war losses are uncertain, it seems likely that at least 14 million lives were lost during the 4 years of World War II. The Soviet estimates of the losses were higher, between 22 and 28 million. However, no census was conducted in 1949, and by 1959 the population had more than rebounded to 117 million from the prewar level of 108 million in 1939. Since 1992, however, Russia has been steadily losing people to the tune of 500,000 or so per year, and this has become a firmly established phenomenon.

Why is population declining in some FSU countries? Demographers assume three general reasons why population can decline: (1) decreased fertility (i.e., fewer babies born per woman), (2) increased mortality, and (3) emigration. Of these three factors, the first two are decreasing population in much of the region today, while the third one varies depending on the country. In Russia, immigration actually exceeds emigration and helps to reduce the losses stemming from the first two. In Moldova and Tajikistan, on the other hand, emigration greatly exceeds immigration. Massive emigration from the entire region was a major concern of Western Europeans at the time of the Soviet Union's collapse; they were expecting a flood of 20 million economic migrants from the former U.S.S.R. into Western Europe by the end of the 20th century. This did not happen, however. Only 1.5 million Russian citizens thus far have left for permanent life abroad since 1992. The vast majority of those who have emigrated left in the first 5 years after the breakup of the Soviet Union, between 1992 and 1997. More than half chose the United States, Canada, Israel, or Australia as their new home rather than Europe. In Europe, Germany received most of the rest. A few hundred thousand Moldovan and Ukrainian residents moved abroad as well.

In their stead, millions of migrants came to settle inside the Russian Federation from other FSU republics—particularly from the economically poor Central Asian states, but also from Moldova, Ukraine, the Caucasus, and the Baltics. Over 25 million ethnic Russians found themselves outside the Russian Federation at the time of the Soviet Union's collapse in 1991 (Heleniak, 2004). People came to Russia to seek jobs, land, health care, and/or education, as well as to move away from conflict zones and from increasingly nationalistic non-Russian governments. This immigration flow now accounts for about 100,000 new arrivals per year, whereas emigration numbers are less than 20,000 per year. Thus emigration is low, immigration is high, and only two other factors remain to account for the decline: a drop in fertility and an increase in mortality.

Decreased fertility is a common phenomenon throughout the world, especially in postindustrial Europe and Japan. "Fertility" is usually defined as the average number of children born to a woman during her lifetime in a given population. The global average 50 years ago was about 5 children; today it is about 2.6, the level in Peru. A fertility level of about 2.1 is called the "replacement level." (Think: Why is it 2.1, if there are two parents? Countries at this level are expected to stop their population growth within a generation, because just enough people are being born to replace their parents' generation in about 20 years.) At present, the United States as a whole is at this level (the level is lower for whites, but slightly higher for Hispanics and blacks). Russia's fertility level is merely 1.5 today, and it is 1.3 in St. Petersburg and some other cities. This means that most mothers have only one child, while some have none at all, and very few have two children or more (Vignette 10.1). The typical American family has two parents and two children, which would be unusual in Russia. Virtually nowhere outside of some religious groups or in the poorest southern republics (Ingushetia, Dagestan) do you see families in Russia with more than two children.

A slight increase in fertility has been noted in Russia in the last 5 years, and this has been attributed to the improved economic conditions.

Vignette 10.1. Portrait of a Typical Russian Family Today

Vladimir and Olga are a typical Russian couple. (Although they are fictitious characters, their story is based on many real ones and represents a common narrative of family life in Russia today.) They are in their late 30s, have been married for 16 years, and have a son who is 15 years old. They live in an industrial city of 500,000 people in the Urals. Their combined income is about 25,000 rubles per month (about $1,000). They live in a formerly state-owned two-room apartment that they fully own, as they were able to privatize it during the 1990s.

Vladimir is a computer programmer. He graduated from a technical university in Yekaterinburg (then Sverdlovsk) when he was 22, and he has two jobs. One is his old job at a state university, which he has had ever since he got out of college; it is a research position supporting computing applications for a department. Another is a job for a local bank, programming its computers and doing some Web design. The first job is full-time and pays about 4,000 rubles per month. The second job is part-time, but pays about 15,000 rubles a month. The reason why he keeps both is that the first provides some security, while the second obviously makes him more money. Because of his first job, he is also able to travel as a scholar to the United States or Germany and to work there periodically on short contracts. His bank job is less secure, because private companies can easily lay off people, and also because his boss is unpredictable and prone to rash decisions.

Olga is a nurse who works at the local hospital. She makes about 6,000 rubles per month. Her job is very secure, but tiring and time-consuming. She has very little time or energy to read books or to see plays at the local theater, which she would like to do. However, she admits that life could be far worse. Because she and Vladimir own their apartment and have only one child, they have enough money left for food, clothes, and limited entertainment. They do not own a car, so they do not need to worry about gas or maintenance. However, recent steep increases in utility and public transportation rates have made her worry. Privately, Olga hopes that Vladimir lands a 3-year contract to work abroad, so that he can make a lot more money. While he is working abroad, she thinks, she can study full time and pass the required tests to get certification as a registered nurse, which may help her find a job in the United States. However, she is not sure that she will be able to leave her native country for a very long time. Their son is in 10th grade. He wants to become a computer scientist, like his dad. However, he is also interested in playing chess and has won some regional tournaments. He does not think he will want to marry until he is 25 or older.

However, this increase is still not enough to change the trend. In this sense, Russia is a typical European country: Fertility rates for Europe range from a high of 2.1 in Iceland to a low of 1.2 in Bosnia. The average is 1.6, the rate of Luxembourg. Children are still wanted in Europe, but having more than one is frequently viewed as an economic liability rather than an asset. In postindustrial societies both parents typically work, and additional children provide no economic benefit to a family, as they do in primarily agrarian societies. With modern contraceptive methods, it is easy for people to minimize the considerable economic sacrifices that additional children impose.

A few years ago, the Russian Duma approved an interesting proposal, which took effect in 2007: The government of Russia will pay parents the equivalent in rubles of about $10,000 for the birth of a second or third child. The hope is that this will increase the birth rate. However, no money will be given away at birth—it will be placed in some savings trust as a "mother's capital" to be cashed in later for a mortgage down payment or a child's education—so the overall impact of this legislation is likely to be insignificant. Ukraine is already making smaller, but immediate, payments to new mothers for every child they bear. This is controversial though, because it raises the possibility that people will have children just to get the money and then abandon the children.

What are notorious about Russia, Ukraine, and Belarus are not their low fertility rates, but rather their high mortality rates. The rates for men in particular approach the levels of the poorest Asian or even African countries. The three Slavic states of the FSU lead the industrial world in high mortality for middle-aged men between the ages of 30 and 60. The average American man is expected to live 75 years, and the average American woman about 80. In contrast, the average Russian man is expected to live only 61 years, and the average Russian woman 74. The reasons for this discrepancy are complex, but the factor most commonly cited is the high rate of alcoholism among Russian men—which increases not only the rates of cardiovascular, liver, and other diseases, but the rates of suicides, accidents, and homicides. Some of these latter are not necessarily due to alcoholism, but also to the overall increase in violence in the post-Soviet period; still, alcohol consumption remains a leading cause.

Much of the alcohol consumed in Russia today is of inferior quality—low-quality locally made vodka and even moonshine liquor. In relative terms, vodka and beer today are more affordable in Russia than they ever were in the Soviet period and are widely available at ever-present neighborhood street kiosks both day and night (Figure 10.2). A 0.5-L bottle of vodka today costs about $4, whereas it was about $20 in the Soviet period if one adjusts for purchase parity. The legal drinking age is 18, but most teens are able to buy alcohol at the kiosks without too much trouble. Particularly worrisome are the very high rates of drinking as well as drug use among early teens, estimated in some communities at 20–30% for drinking and 5% for drug use among children as young as 14.

FIGURE 10.2. Alcohol is readily available in many roadside kiosks throughout FSU cities. *Photo:* J. Kurzeka.

Many health conditions are not directly related to alcohol consumption, however, but are more the result of the crumbling health care network. Certain expensive, but routine, operations that save countless lives of Western men between the ages of 50 and 70 (e.g., cardiac bypass surgery or pacemaker installations) are available only to wealthy clients in private clinics. Indeed, Russia's elite prefers to have these types of medical procedures done in clinics in Switzerland and Germany, just to be safe. Another medical factor is the very slow response rate of ambulances. In many cities in the West, residents are accustomed to seeing someone about 5 minutes after they dial the emergency number. In most Russian cities today, it requires over an hour for an ambulance to appear, if it shows up at all. In rural areas, many people's only recourse is their closest neighbor with a drivable car.

Very high abortion rates constitute another grim factor that depresses fertility rates and increases mortality rates throughout the FSU. Abortion was legal and free in the Soviet Union for most of the post-World War II period, while modern contraception methods were slow to appear. Eventually abortions became the main contraceptive tool, although not necessarily by choice, for the majority of Soviet women. Recent reports cited cases in which women had over 15 abortions in less than 10 years! Although all traditional religions of the U.S.S.R. opposed abortions on moral grounds (with the Russian Orthodox Church and Islam being emphatically pro-life), their role in lowering abortion rates was minimal in the Soviet period because of the state's official atheism. Even today religion has a low impact, due to the low numbers of adherents and the separation of church and state. Russia's abortion rate today remains among the highest in the world (48 per 1,000 for women ages 15–49, as compared to the Bulgarian rate of 30, the U.K. and U.S. rate of 12, and the Belgian rate of 6). The general recent trend has been toward much greater use of modern contraception methods among Russian women: In 2008 47% of women used modern contraception in Russia, as compared to 68% in the United States. Of these Russian women, 8% used the pill, 14% used intrauterine devices, a few percent chose sterilization, and the balance relied on condoms. Nevertheless, Russia's abortion rate is still four times the U.S. rate. About 90% of these are early-term abortions (under 12 weeks) and are not medically necessary. The abortion rates are highest in the rural north and center of the country, where unemployment is high and societal pressure to perform an abortion is great. The rates are lowest in affluent Moscow and in the poor but traditional Muslim republics of the Caucasus.

In summary, the current level of demographic imbalance is such that Russia only replaces 62% of the workforce it needs and is bound to continue to lose population for decades to come. According to some recent projections, the country will have only 110 million people by 2050 (which would be the level of 1939), down from 150 million in 1988.

Can immigration solve the problem? Aside from an apocalyptic scenario (feared by some Russian nationalists) of a massive Chinese stampede into eastern Siberia, much more immigration would be needed to offset the current imbalance. Recall that only about 100,000 legal migrants come to Russia each year, while about 500,000 people are lost per year due to the fertility–mortality imbalance. However, there are indications that hundreds of thousands more are entering the country illegally. In a recent statement, the Russian authorities claimed that Russia has over 10 million undocumented immigrants, the second highest number after the United States in the world. Many of these are ethnic Russians from Ukraine, Kazakhstan, and elsewhere, but quite a few are migrants of other ethnicities from Moldova, Azerbaijan, Uzbekistan, Tajikistan, and other FSU states (Chin & Kaiser, 1996). Others come from Afghanistan, Vietnam, China, and even Africa. Whereas the Soviet Union's border with the outside world was a true "iron curtain" of thousands of miles of barbed wire and gun-toting border guards, today's Russia's border is relatively open to all of its previous satellite countries (except the Baltics, which are now members of the EU and have tight border security). Crossing from Kazakhstan into Russia on foot is not much different from crossing from one U.S. state to the next, and is easier than crossing from the United States into Canada. You do need to present travel documents on trains and planes, of course, but much of the border is not demarcated well, and many options exist for persuading the border guards to look the other way. However, only a portion of the demographic loss will be realistically compensated for migration in the years to come. Below are some other demographic observations that you may find interesting. (Many of the same statistics apply to Ukraine and Belarus as well.)

- Overall, the Russian population is older than that of the United States, but younger than that of Europe or Japan, and there are more women than men (Figure 10.3).
- Women in Russia are more educated than men and live 13 years longer on average.
- About 16% of the Russian population has completed a college education (vs. 28% in the United States).
- The average Russian household has only 2.71 members, and about 22% live alone.
- About 22% of Russian households are single-parent households, while dual-parent households make up 55%. These figures are very similar to the corresponding U.S. statistics.
- Sixty-six percent of Russia's population live in apartments, while only 26% live in single-family homes, and most of those are in rural areas.
- On average, one person has 19 m^2 in which to live (only 200 square feet!).
- Only three-quarters of all households in Russia have running water, while only 71% have flush toilets.
- 82% of urban dwellers have central heat provided by a power plant, while 50% of rural dwellers depend on wood-burning brick ovens or on coal boilers.
- The average age at first marriage (Figure 10.4) continues to rise: It is now 26 years for men and a little over 23 for women. Just a generation ago, in the mid-1980s, these ages were 24 and 22, respectively.
- Many more Russians today stay unmarried longer: Among 30-year-olds, the percentage who have never married is now 30% for men and 20% for women (as compared to 22% and 13%, respectively, just 20 years earlier).

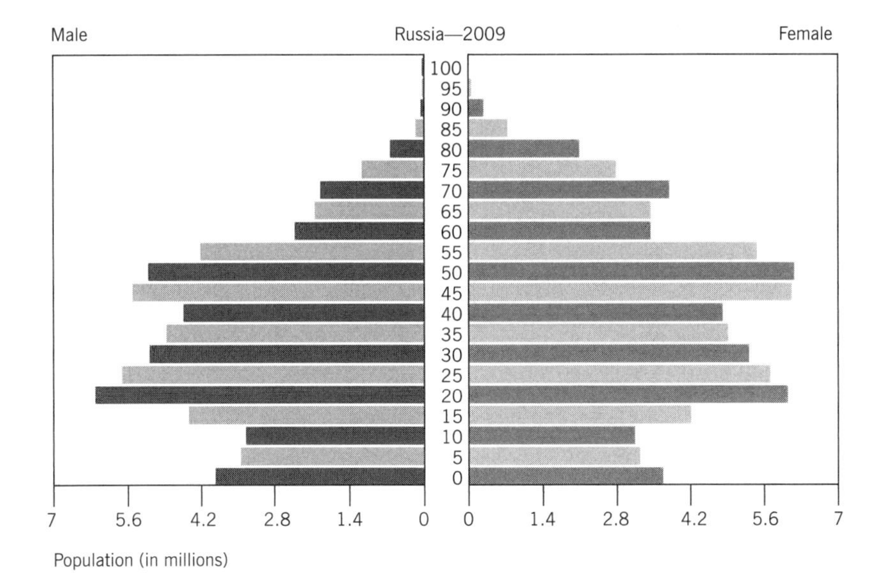

FIGURE 10.3. Population pyramid for Russia in 2009. The impact of World War II is noticeable in the prevalence of women over men in the oldest age groups (although the higher male mortality rate at these ages also plays a role). Note also that there are fewer children (ages 60–65) and grandchildren (ages 35–40) of that generation. Data from U.S. Census Bureau.

- Between the 1989 and 2002 censuses, Russia lost about 1.3 million people to emigration (most went to only seven countries: Germany, the United States, Israel, the United Kingdom, France, Canada, and Australia, in that order). It gained about 6.8 million (almost all from the other FSU republics).

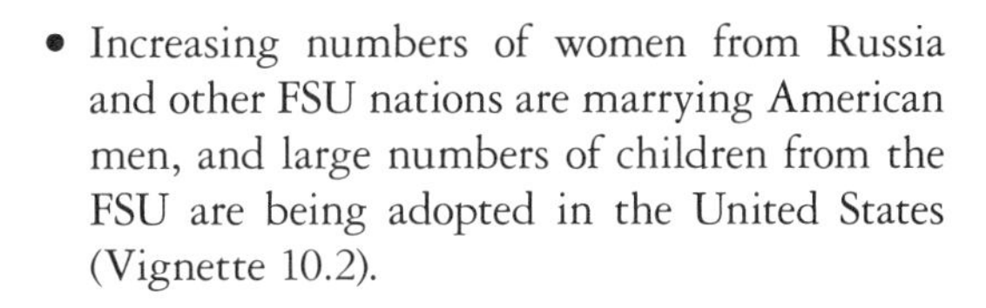

- Increasing numbers of women from Russia and other FSU nations are marrying American men, and large numbers of children from the FSU are being adopted in the United States (Vignette 10.2).

FIGURE 10.4. A young Russian couple on their wedding day in Tomsk. According to the 2002 census, 54.9% of Russians lived as married couples—a decline from 67% only a decade earlier. *Photo:* A. Fristad.

Population Distribution

Where do all these people live? A quick look at a population distribution map reveals an interesting pattern: Three-quarters of all Russians live on one-quarter of the landmass. This populated land is west of the Urals, in the European part of the country (Figure 10.5). The effective national territory of Russia covers about a third of its landmass, stretching from Belarus and Ukraine east across the Urals toward Lake Baikal in a tapering-off triangle (Cohen, 2009). Only a quarter live on the three-quarters of land east of the Urals in Siberia. Moreover, there has been a definite trend recently toward migration from Siberia to the west and south of the country (Heleniak, 2004).

In addition, the population is patchily distributed in Russia. Most people live in big and me-

Vignette 10.2. Brides and Adopted Children from the FSU in the United States

I have met more than a dozen American men who have married Russian wives over the past few years. I also know a few people who have adopted children from one of the FSU. Most of these families are genuinely happy, and I am very glad for them. They reflect the widespread post-Soviet phenomenon of connecting American and Russian/other FSU societies through marriage or adoption. Just how widespread this phenomenon is can be objectively tracked down through statistics released by the U.S. Department of State on visas issued to the brides and adopted children. With respect to adoptions, Russia is one of the three leading suppliers of children to American families in recent years, behind only China and Guatemala (Figure 1): from a low of 746 children adopted in 1993 to a high of 5,865 in 2004. Why has Russia (and the rest of the FSU) become such a popular source of adopted children? First, the FSU nations have a large number of orphans (over 1 million in Russia), who are under the poor care of the state system and in need of families. Second, Russia and other FSU nations allow their children to be adopted, whereas many other countries do not. Third, these children are usually white and of European origin, which may be a preference for U.S. parents from European backgrounds. Most other countries that have children available for adoption are Asian or Latin American countries. Also, there are quite a few American families who have an interest in Russia or other FSU republics because of the family history or religion (Orthodox Christian or Jewish). Most children who are adopted from the FSU by U.S. parents, however, will rapidly lose their native language. A few may retain it if adopted later in life and/or if given plenty of opportunities to practice it.

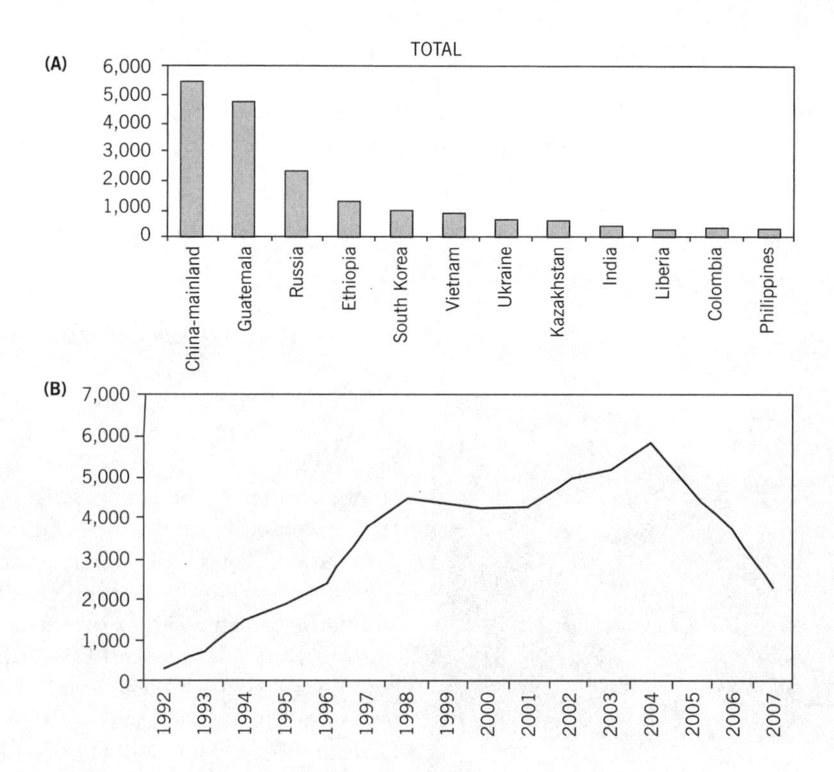

FIGURE 1. (a) Total U.S. adoptions by country of origin in 2007. (b) Total number of children adopted into U.S. families from Russia by year. Data from U.S. Department of State.

(cont.)

A cross-cultural marriage is an even more complex affair than an adoption. Making any marriage work requires a great deal of both partners. More than half of all marriages in the United States end up in divorce, and international marriages are even more complicated and have an even greater probability of failure than intranational marriages, because the spouses do not share a common culture. When an American husband wants to watch baseball with a Ukrainian wife, for example, she has no clue about what goes on in the game, because she does not know the rules; when she makes him a delicious (to her) Ukrainian borscht, he may think it is too meaty or too salty; and so forth. Of course, cultural differences may also be to a couple's advantage and make the marriage strong and long-lasting.

Of particular concern, however, are the cases involving so-called mail-order brides (Osipovich, 2005), especially those that result in immigration fraud (as in arranged fake marriages) or domestic abuse. By definition, mail-order marriages are arranged by a third party, usually a matchmaking agency. They account for a small proportion of all international marriages (perhaps only 4%), but they can create social problems for all involved. For example, some agencies are little more than temporary "bait" Websites that con men into paying money up front and then disappear without a trace. One such long-standing scam involved luring unsuspecting American men with provocative photos of Valeria, a married Russian pop singer, into making up-front payments of a few thousand dollars for her tickets and U.S. visa. Another problem is that even if the woman at the other end is real, her motives or personality may be different from what is advertised. There are books published in America about how to avoid being a victim of such scams, just as there are books written in Russia about how to catch a wealthy American guy to get the coveted U.S. "green card" and then dump the husband, citing marital problems, abuse, or worse.

Conversely, of course, many of these women suffer genuine abuse from their foreign husbands through physical or verbal assault and intimidation. Typical Western men seeking wives abroad are middle-aged, are not physically attractive, and have at least one unsuccessful marriage behind them. Or, they may have some social handicaps that have prevented them from ever having a spouse in the first place. They also may have lower-than-average incomes and low self-esteem, and although they are wealthier than the average FSU citizen, they are not in a position to provide the glamorous lifestyle that some brides may envision (Osipovich, 2005). Given all these factors, it is not surprising that domestic strife and outright abuse often occur.

Documented exploitation of foreign wives has recently led the U.S. Congress to adjust the immigration law to assist women trapped in abusive relationships without jeopardizing their residency status. One stereotype that many Western men, and the Western mass media, have about women from the FSU is that these women are models of old-fashioned femininity—undemanding, quiet, and compliant. This is simply not true and does not help at all. However, a foreign wife may indeed experience greater difficulty than a native-born wife in communicating her needs to her husband, or reporting his abuses to the authorities, because of the language barrier. Osipovich (2005) highlights the fact that many American-born women also experience abusive relationships, but that they may be better able to deal with these situations because they have a better knowledge of English and of U.S. society.

dium-sized cities. The overall urbanization rate is 73%. Of particular note are clusters of population in and near Moscow (about 15 million), St. Petersburg (6 million), Novosibirsk (2 million), and a few 1-million-plus cities in the Volga basin (Nizhny Novgorod, Samara, Saratov, Volgograd, Kazan, Perm) and in the Urals (Yekaterinburg, Chelyabinsk, Ufa). All of these cities are major political, financial, industrial, service, and transportation hubs for their regions. No cities larger than 1 million exist east of Novosibirsk, although both the Krasnoyarsk and Irkutsk agglomerations come pretty close. The current population centroid for Russia is located south of Ufa and north of Orenburg in the southern Urals (Treivish, 2005; see also Figure 10.5).

In Siberia, only 25 million people live east of the Yenisei, and many of them are eager to move to warmer western and southern areas. In contrast, the northeast provinces of China closest to Russia have about 130 million people. The future will tell us whether the increasing demographic

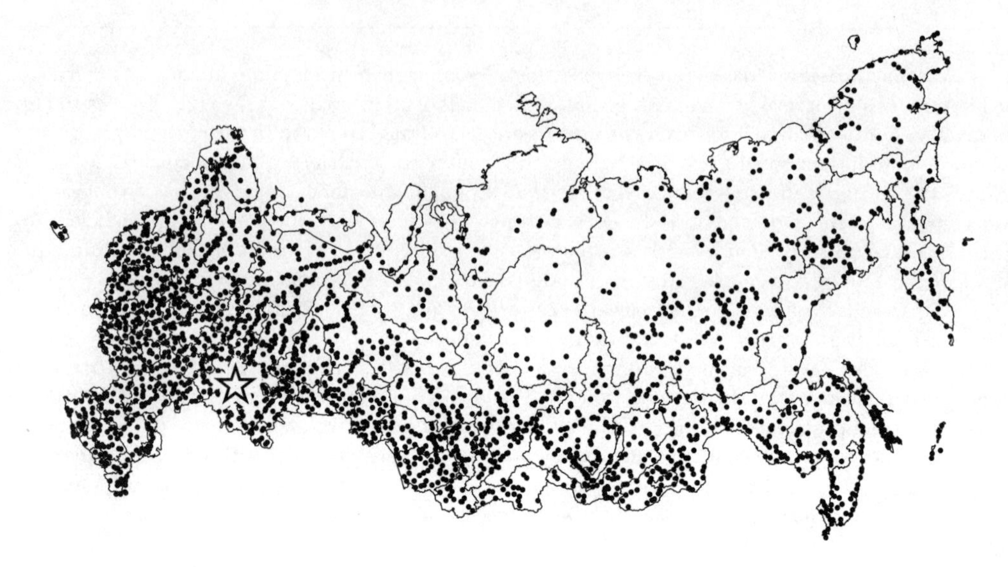

FIGURE 10.5. Population distribution in Russia, as shown by mapping about 5,000 urban settlements. Note the strings of settlements along railroads and rivers in Siberia and northern European Russia. The star marks the center of population distribution (population centroid) at the time of the last census (2002), as presented in Treivish (2005).

imbalance between the two countries in general, and in this area in particular, will lead to any actual confrontation between them.

It is tempting to look at the population distribution of Russia as similar to that of the United States or Canada, with populated coasts and a relatively empty middle. However, even a cursory look at Figure 10.5 reveals that this is not the case. In North America, the majority of the population indeed lives along either the east or the west coast, with relatively few people in the middle of the country. Russia, on the other hand, has very few people on the Pacific side and a great many in the European part. The major Pacific seaport of Vladivostok has only a little over 600,000 people, and there are no other big cities nearby. The biggest city of Russia, Moscow, is not on a coast. The only two significant coastal cities in the European part are St. Petersburg and Kaliningrad on the Baltic Sea, and there is also the northern and remote Murmansk on the Kola Peninsula. Of the two North American countries, Russia resembles Canada much more than it does the United States in terms of its population distribution. Most Canadians live in the southern part of their giant country, within about 100 km of the U.S. border. Russia has a similar pattern of settlement in the Asian part (Siberia), with all the cities there strung along the Trans-Siberian Railroad in the south. Unlike Canada, however, Russia does have substantial cities of a few hundred thousand people in the Far North. The biggest of those is Norilsk with a population of over 300,000, located at 70°N—well above the Arctic Circle! In contrast, Fairbanks, Alaska, has only 80,000 people about 2° south of the Arctic Circle. The biggest city in northern Canada, Whitehorse, Yukon, has 20,000. Both Fairbanks and Whitehorse are located south of Norilsk. The distribution of the rural population is similar to the urban distribution depicted in Figure 10.5.

Hill and Gaddy (2003, p. 227) illustrate the difference between the distribution of the Russian population and those of Canada, Sweden, the United States, and other countries by using an interesting measure of population density called "temperature per capita." Instead of merely looking at the overall distribution, they look at where cities are in relation to the average temperature on the list of 100 coldest cities over 100,000 population:

> [There are] 85 Russian, 10 Canadian, and 5 U.S. cities. The first Canadian city to appear on the list (Winnipeg) would be in 22nd place. The coldest

> U.S. city (Fargo) would rank 58th. Americans are accustomed to thinking of Alaska as the ultimate cold region. But Anchorage, Alaska, would not appear on a list of the coldest Russian and North American cities of over 100,000 until position number 135, outranked by no fewer than 112 Russian cities. The explanation for this result is not that Alaska isn't cold. It is. It's just that Americans don't build large cities there. . . . The United States has only one metro area over half a million (Minneapolis–St. Paul) that has a mean January temperature colder than –8°C. Russia has 30 cities that big and that cold.

In other words, Russians do live under much colder conditions overall than even Canadians do, let alone Americans. The biggest cities are found at convenient locations on rivers, which were historically conducive to defense and shipping. In the Soviet period, many cities were built as factory and mining towns or as sites for GULAG camps.

Outside Russia, the heaviest concentrations of people are found in Ukraine and the fertile Fergana Valley in Uzbekistan, Kyrgyzstan, and Tajikistan. Even in those republics, population density remains low (ranging from 77 people/km^2 in Ukraine to 27 in Kyrgyzstan). For comparison, Portugal's density is 115 and India's is 344. Local densities near cities can, of course, be much higher.

Recent processes that have been discussed with respect to Russia's population include increasing age and spatial migration within the country. First, the population of Russia is beginning to age faster, although at the time of the 2002 census the proportion of retirement-age persons was about the same as in the EU and Japan (20.5%), and only marginally higher than in the United States. (However, this conceals the sad fact that few seniors in Russia are living very long in retirement. Many men die at about the age of entering retirement, currently set at 60. The retirement age for Russia's women remains 55, but there are proposals now to raise this age for both sexes.) Between the 1989 and 2002 censuses, the proportion of people over age 40 has grown from 34.5% to 42.2%. Continued low fertility and the spread of HIV among younger people are bound to increase the average age even farther (Chapter 12).

The spatial pattern of settlement is also beginning to change. The most pronounced trend in Russia is depopulation of the Far East and the north (Heleniak, 2007). Between the two censuses, five units in these areas—Chukotka, Kamchatka, Yakutia, and two autonomous okrugs (now part of Krasnoyarsky Kray)—lost between 15 and 60% of their population. All subjects of federation east of the Yenisei and north of the Arctic Circle lost 10–15% on average. Much of that loss (about 80%) was due to domestic migration to warmer regions of Russia, primarily to the central European part and the Caucasus. About 50,000 people per year are collectively lost to migration from northern regions and the Far East, and the process continues unabated. A particularly alarming aspect of this loss is that the proportion of children in these areas decreased by half. In effect, those moving away are not seniors (like the Americans moving from the Rust Belt to the Sun Belt), but younger families who want a better future for their children. Many people who move are actually relatively wealthy. A personal interview with a successful businessman in Yakutsk revealed the reason: Although his family is economically secure there, the cold, dark nights of winter are sometimes more than his family can handle. Also, with skyrocketing airfares, many families find even temporary vacations to warmer places out of their reach, so the wealthier and healthier segments of the population are moving away for good. This, of course, means that the older, sicker, and poorer segments of the population are more likely to stay. Eventually this trend may dramatically reshape the human fabric of the vast hinterland of Russia. The only exceptions to the trend at the moment are the oil-rich Tyumen and Tomsk Oblasts, which are gaining population.

Nationalities

Russia is a multiethnic country. The whole United States is called one "nation," but we talk about racial, ethnic, or linguistic groups in America as "African Americans," "Asian Americans," "Americans of Norwegian ancestry," and so forth. Except for the Native Americans, these groups are all descendants of immigrants. In Russia one talks about "ethnicities" or "nationalities," which are by and large indigenous. Russians constitute the majority, about 80% of the total. However, members of many other groups call Russia home, hold Russian citizenship, and (for the most part) speak Russian as their first language, but are ethnically distinct from the Russians. According to the 2002 census, there were 182 such "ethnicities." The U.S.S.R. (as the heir of the bigger Russian Empire) was even more diverse, with as many as 200 ethnicities represented, although only 128 were officially recognized.

We have already seen in Chapter 7 that during Soviet rule, *some* of the largest ethnic groups were given individual Soviet Socialist Republics to themselves. For example, Ukraine was created in the areas where Ukrainians primarily lived, Uzbekistan for the Uzbeks, Georgia for the Georgians, and so on. Russians were also present in large numbers in some of these republics (notably in Estonia, Latvia, Belarus, Ukraine, Kazakhstan, and Kyrgyzstan). Many of these people have been politically marginalized in the past 15 years and have chosen to leave for the Russian Federation.

The Russian Federation (or its predecessor, the R.S.F.S.R.) has always been the most complex of all units of the FSU (or the U.S.S.R.). Table 10.2 lists its main nationalities today, their numbers in 2002, and where they live in Russia. Most of these groups (except the Ukrainians, Belarusians, Germans, Kazakhs, Armenians, and Azerbaijanis) have their own ethnic autonomous republics, 21 of which are incorporated into the Russian Federation. Absent from the table are

TABLE 10.2. Ethnicities of Russian Federation in the Most Recent Census (2002)

Ethnicity	Total number (thousands)	Percent of total	Where they live in Russia
Russian	115,889	79.8	Everywhere
Tatar	5,555	3.8	Tatarstan, Bashkortostan, Moscow
Ukrainian	2,943	2.0	Southern European part, Siberia
Bashkir	1,673	1.2	Bashkortostan
Chuvash	1,637	1.1	Chuvash Republic (Volga)
Chechen	1,360	0.9	Chechnya, Moscow
Armenian	1,130	0.8	Moscow and other big cities
Mordva	843	0.6	Mordovia Republic (Volga)
Avar	815	0.6	Dagestan Republic (Caucasus)
Belarusian	808	0.6	Moscow, western European Russia
Kazakh	654	0.45	South central Siberia
Udmurt	637	0.44	Udmurtiya Republic (Volga)
Azerbaijani	622	0.43	Moscow, Dagestan Republic
Mari	604	0.42	Mari-El Republic (Volga)
German	597	0.41	Volga, Siberia
Kabarda	520	0.36	Kabardino-Balkaria Republic (Caucasus)
Ossetian	515	0.35	North Ossetiya Republic (Caucasus)
Dargin	510	0.35	Dagestan Republic (Caucasus)
Buryat	445	0.31	Buryat Republic (eastern Siberia)
Yakut	444	0.31	Sakha Republic (eastern Siberia)
Kumyk	422	0.29	Dagestan Republic (Caucasus)
Ingush	413	0.28	Ingushetia Republic (Caucasus)
Lezgin	412	0.28	Dagestan Republic (Caucasus)
All others	4,257	2.9	

Note. Data from the Russian census of 2002 (*www.perepis2002.ru*).

some small but important groups such as the Jews, the Roma (Gypsies), and various northern peoples (the Chukchi, Nenets, Komi, Karelians, and others). These are not listed in Table 10.2 because their populations are below a threshold of 400,000. Preferential emigration for some of these groups—especially for Jews to Israel, Europe, and North America, and for Germans to Germany since the fall of the Soviet Union—dramatically lowered their numbers within the FSU. Other groups, especially the Azerbaijanis, Armenians, Moldovans, and Tajiks, have greatly increased their presence in Russia in recent times. Most of these are economic migrants to cities in search of work; they typically come as temporary workers and then become permanent residents.

Notice also that despite all this diversity, the ethnic Russians remain by far the dominant ethnic group; every four out of five people in Russia are Russians. Virtually everybody in Russia (99%) speaks fluent Russian as a first or second language, and college education is available only in Russian.

Demographics in Other FSU Republics

Table 10.1 shows that the demographic situation in about half of the other FSU republics virtually mirrors that of Russia, while in the other half the situation is quite different. That is, in the former group the population is rapidly declining as a result of death rates far exceeding birth rates. This is the case even in prosperous Estonia. The situation is most alarming in Ukraine, which was the country with the fastest worldwide decline in 2008. Many of the same factors as in Russia play a role in these republics as well: low fertility among urban women, and high male mortality (especially among middle-aged men) due to alcoholism, depression, accidents, homicides, and suicides. The HIV/AIDS epidemic is becoming a very serious issue in all of the republics with already declining population (Chapter 12).

On the other hand, the Central Asian republics and the Caucasus have a positive demographic balance. Although fertility in these countries is not high in the global sense, it is sufficiently high to offset the mortality. In Tajikistan, for instance, the fertility rate is 3.4 children per woman—about the level of Oman or Gabon, and 25% higher than the world's average.

Some of the FSU republics are currently struggling to keep their citizens. It is estimated that about 1 million ethnic Georgians now live outside Georgia, and more than 1.5 million Tajiks and Moldovans live outside their respective republics as well. Much of this population shift has occurred since 1992. In contrast, Armenia always had a very large international diaspora in the Middle East, Europe, and parts of North America. Russia serves as a magnet for those from Central Asia and the Caucasus, as well as Moldova, while Western Europe does the same for the Baltic states.

REVIEW QUESTIONS

1. Describe the size of Russia relative to other countries in the FSU and the world, using Table 10.1 and additional data from *www.prb.org*.
2. What are the main factors that limit Russian fertility today?
3. What are the main factors that increase Russian mortality?
4. Which countries of the FSU are growing in population? Why?
5. Where do most Russian people live? Why?
6. What are the top five ethnic groups inside Russia? Where are they found and why?

EXERCISES

1. Pick any country from Table 10.1 and compare its demographics to those of the United States and the world. Can you explain the differences?
2. Look at Figure 10.3. Explain what you see. In particular, what accounts for the very low numbers of people in the 60–64 and 35–39 age categories? Why is the top so skewed toward women?
3. Research any of the following cities online by looking up recent news stories: Samara, Yekaterinburg, Nizhny Novgorod, Novosibirsk. Can you find any mention of ethnic tensions? Can you find any indication that these cities are either doing well or struggling economically? For any one of them, research

its main economic strengths and try to propose a counterpart city in the United States that would be similar in size, type of economic activity, location, and/or climate.

4. How can you explain the empty areas in Figure 10.5? How can you explain the long strings of cities found in some areas?
5. Why do you think there are so few American women who want to marry Russian men? Why do you think Russian, Ukrainian, or Moldovan women attract the attention of American men—are there any compelling cultural or social reasons? Watch a recent movie that discusses "mail-order brides" from the FSU (e.g., *Birthday Girl*, 2002, or *Eastern Promises*, 2007). What are some of the stereotypes that they seem to perpetuate? Can you improve their story lines to make them more realistic?

Further Reading

Chin, J., & Kaiser, R. (1996). *Russians as the new minority: Ethnicity and nationalism in the Soviet successor states.* Boulder, CO: Westview Press.

Cohen, S. B. (2009). *Geopolitics: The geography of international relations* (2nd ed.). Lanham, MD: Rowman & Littlefield.

Heleniak, T. (2004). Migration of the Russian diaspora after the breakup of the Soviet Union. *Journal of International Affairs, 57*(2), 99–117.

Heleniak, T. (2007, April). *Migration in the Russian Far North during the 1990s.* Paper presented at the Annual Meeting of the Association of American Geographers, San Francisco.

Hill, F., & Gaddy, C. (2003). The Siberian curse: Does Russia's geography doom its chances for market reform? *Brookings Review, 21*, 23–27.

Osipovich, T. (2005). Russian mail-order brides in U.S. public discourse: Sex, crime, and cultural stereotypes. In A. Stulhofer & T. Sandfort (Eds.), *Sexuality and gender in postcommunist Eastern Europe and Russia* (pp. 231–242). Binghamton, NY: Haworth Press.

Rybakovskii, L. L. (2006). The demographic future of Russia and processes of migration. *Sociological Research, 45*(6), 6–25.

Treivish, A. (2005). A new Russian heartland: The demographic and economic transition. *Eurasian Geography and Economics, 46*(2), 123–155.

Vishnevsky, A. G. (Ed.). (2006). *Naselenie Rossii.* Moscow: Nauka.

Websites

www.census.gov/ipc/www/idb—International Database of the U.S. Census Bureau.

www.perepis2002.ru—Russian Census 2002 Data Portal. (In Russian only.)

www.prb.org—Population Reference Bureau.

CHAPTER 11

Cities and Villages

This chapter examines settlements of Northern Eurasia, with the main focus on Russia as usual. A major distinction must be made between urban (city) and rural (village) settlements. In the United States, urbanized areas generally have over 1,000 people per square mile (400 per square kilometer). An informal way to think about the urban–rural distinction is to look at the services available to residents. You live in a city if you are getting "city services": water, sewer, natural gas, and curbside recycling. You live in a rural area when you have a well, a septic tank, a propane tank, and no recycling.

Soviet geographers recognized seven types of settlements (Table 11.1), two of which were rural and five were urban. Besides numbers of people, the difference between a town and a big village was based on the main economic activity: either nonfarming or farming, fishing, and/or forestry. In Russia today, a city has at least 12,000 people, at least 85% of whom must have nonfarming occupations (the corresponding percentage is 65% in U.S. metropolitan statistical areas). The size thresholds for cities are lower in more agrarian Ukraine (10,000) or Georgia (5,000). Large villages of a few thousand residents are still fairly common in Ukraine, but are now rare in Russia.

The classification in Table 11.1 was derived in part from differences in mass transit needs. The Soviet system of transportation heavily favored mass transit to move masses of workers cheaply. It was presumed that few people would ever

TABLE 11.1. The Soviet Typology of Settlements

Settlement type	Russian name	Population size	% of total population (1994)
Largest city	*Krupnejshij gorod*	1–10 million	17%
Large city	*Krupnyj gorod*	100,000–1 million	30%
Medium city	*Srednij gorod*	50,000–100,000	8%
Small city	*Malyj gorod*	20,000–50,000	8%
Town	*Poselok gorodskogo tipa*	5,000–20,000	9%
Big village	*Selo*	1,000–5,000	20%
Regular village	*Derevnya*	<1,000	7%

own a car. In a village one could walk or bicycle almost anywhere, and motorcycles and tractors were also frequently used on the rutted, unpaved roads. In a small town of 10,000 people, a bus would take workers to the nearby factory or state farm. In a city of over 20,000 but under 50,000, there would be a few different bus routes. In a city of over 100,000, an electric tram or trolley would be available in addition to buses; and in a city approaching 1 million, a subway (*metro*) system could be built. The distinction between a *selo* and a *derevnya* was historical: Before the Soviet period, the largest of about five villages would get a parish church and thus achieve the status of a selo. In many cases, the local landlord's mansion would be located not far from the selo as well, although not directly in it. When churches were closed by the Communists, many were converted into village clubs, thus ensuring continuation of the selo's higher status. The headquarters of the local state farm would later be located there as well.

There is another important difference between U.S. and FSU cities. All cities in the U.S.S.R. were developed under comprehensive plans focused on maximum efficiency in housing and transporting large numbers of workers. By contrast, each U.S. jurisdiction has different zoning rules pertaining to planning, and the development is market-driven. Thus a Soviet-built city of 50,000 will look very different from its American counterpart.

History of Urbanization and City Functional Types in Russia and the U.S.S.R.

As in the rest of Europe, many cities in European Russia, Ukraine, Belarus, and Moldova are old. Although none approach Rome or Marseilles in age (2,600+ years), some are over 1,000 years old, with an unmistakably medieval core (a fortified kremlin) and a more recent periphery. The oldest cities, however, are in Georgia (Tbilisi, Batumi), Armenia (Yerevan), Uzbekistan (Samarkand, Bukhara), and other parts of Central Asia; these cities date back 1,500–3,000 years. The Greeks built fortified colonies along the Black Sea coast at Korsun ("Chersonesos") and Kerch in the Crimea, and at Sukhumi and Batumi in Georgia. These cities are over 2,000 years old, but only a few ruins of the original settlements remain (Figure 11.1). A few of these were consumed by the sea as a result of land subsidence or sea level rise (e.g., parts of the famous archeological site Olvia, east of Odessa). On the other hand, in much of Siberia, the Russian Pacific, and Kazakhstan, cities are recent phenomena. The traditional inhabitants of those lands lived a nomadic lifestyle until the early 20th century and did not create large permanent settlements. The Soviet Union moved millions of people around and created hundreds of new settlements over this eastern frontier.

Some of the oldest Russian cities (Staraya Ladoga, Novgorod, Pskov, Murom, Kiev, and Chernigov) are at least 1,200 years old. They were built before Rus was Christianized under Vladimir the Great in 988 A.D. The second period in which many cities were built was toward the end of the Tatar–Mongol Yoke (1350–1450). Dozens of Russian cities date from that period, including parts of the Moscow Kremlin, which was mainly built under Ivan III in the late 1400s. A few famous monasteries grew in that period, giving rise to new cities around them—Sergiev Posad, Borovsk, and Zvenigorod around Moscow (Figure 11.2). Some cities of the old Asian khanates (Bukhara, Samarkand, Khivy) were renovated during that period as well.

FIGURE 11.1. Ruins of the ancient Greek city of Chersonesos (6th century B.C.) near Sevastopol, the Crimea, Ukraine. *Photo:* O. Voskresensky.

FIGURE 11.2. The Borovsk Monastery of St. Paphnuty in Kaluga Oblast, established in the mid-15th century. *Photo:* Author.

The third peak in city building coincided with the modernizing reforms of Peter I and Catherine II in the 18th century. St. Petersburg (Figure 11.3) and the surrounding cities in the northwest were then built and greatly expanded, as well as some cities of the Urals and Siberia (Yekaterinburg, Chelyabinsk, Tomsk, Irkutsk, Yakutsk). The earliest Siberian cities were established as forts in 1600s, but did not grow much until industrialization began two centuries later.

The fourth period included late-19th-century industrialization (Figure 11.4), when city factories grew rapidly in the developing industrial zones around Moscow and Tula, along the middle Volga, and in the Urals. By 1917, 17% of the country lived in cities. The fifth (Soviet) period brought about massive reconstruction of the old urban cores. Entire neighborhoods with dozens of churches, mansions, cemeteries, and markets were razed to give way to new monuments, plazas, government buildings, and tree-lined avenues suitable for mass transit. Perhaps the most infamous incident involved demolition (in 1931) of the largest church in Russia, Christ the Savior Cathedral in Moscow; the original plan was to replace it with a skyscraper called the Palace of Soviets, which was intended to be taller than the Empire State Building. World War II intervened, however, in the end only a large open swimming pool was built in its foundation. The cathedral

FIGURE 11.3. St. Petersburg is a city of wide streets, canals, big cathedrals, and monuments, most of which were built during the reign of Catherine the Great (1762–1796). *Photo:* S. Blinnikov.

FIGURE 11.4. The Red October chocolate factory, built in the late 19th century south of the Kremlin in Moscow's industrial zone by T. F. von Einem, a German confectioner. *Photo:* Author.

FIGURE 11.5. The reconstructed Christ the Savior Cathedral in Moscow. Originally completed by 1882 to commemorate the war with Napoleon in 1812, the cathedral was razed in 1931. It was rebuilt on its original site but with modern materials by 1997 (Sidorov, 2000). *Photo:* Author.

was eventually reconstructed in the 1990s (Sidorov, 2000; see Figure 11.5). During World War II, entire factories were dismantled and moved away from the European front lines to the Volga region and the Urals, giving birth to new cities there (Figure 11.6). After the war, city construction shifted farther into the Arctic, eastern Siberia, and Central Asia as new deposits of metal ores and fossil fuels had to be exploited there. Table 11.2 provides some examples of various types of cities, based on their historical function and period of construction (arranged from oldest to newest). For each type, an analogous Western city is provided.

FIGURE 11.6. Panorama of Saratov, a typical large city on the Volga, which was greatly expanded during World War II by the building of factories to accommodate military needs. *Photo:* S. Blinnikov.

TABLE 11.2. Functional Types of Russian/Other FSU Cities

City type	Examples	Foreign analogues
Ancient walled city	Novgorod, Pskov, Moscow	Paris, France
Medieval city built around a monastery	Sergiev Posad, Murom	Carcassonne, France
Old administrative centers	Penza, Tambov, Saratov	Philadelphia, PA
Early industrial centers	Tula, Nizhniy Tagil, Ivanovo	Pittsburgh, PA
Transportation hubs	Novosibirsk, Nizhniy Novgorod	Chicago, IL
Seaports	Murmansk, Novorossiysk	New Orleans, LA
Soviet GULAG centers (mining)	Magadan, Karaganda, Vorkuta	Fairbanks, AK; Hibbing, MN
Soviet new industrial centers	Norilsk, Magnitogorsk	Gary, IN
Science towns and "secret cities"	Obninsk, Arzamas-16	Los Alamos, NM; Argonne, IL
Resorts	Sochi, Yalta	Palm Beach, FL
New capital	Astana	Canberra, Australia

The Soviet Union planned urban development not only at the level of individual cities, but for the entire country. If the economy demanded, new cities could be created in the middle of nowhere. At the same time, population flows into the largest, most desirable cities could be controlled through a system of mandatory residence permits (*propiska*). This system had certain advantages over a market-driven, locally controlled model of urban development, because the resources had to be mobilized quickly and to achieve certain uniformity with respect to living standards. At the same time, the system was insensitive to the local variations in cultures and led to increasingly homogenous urban designs, with the same basic apartment buildings mass-produced for the whole country. For example, sanitary norms set in 1922 dictated the size of the minimal livable space at 9 m^2 (about 100 ft^2) per person. This remained unchanged over the entire Soviet period and without respect to local needs (e.g., in regions with more severe climates). As illustrated in Bater (1996), the actual space available toward the end of the U.S.S.R. ranged from 13 m^2 in Estonia to 7 m^2 in Turkmenistan, with 10 m^2 being the national average. In practice, not only the central planners or the local governments, but primarily the various Soviet ministries determined the actual city layouts, apartment configurations, and materials used in construction. Some of the best-designed cities were the ones built by the wealthier industries (e.g., mining, oil/gas, and nuclear energy).

Urban Demographics

The FSU/Northern Eurasia is a fairly urbanized region (Chapter 10, Table 10.1). The average level of urbanization in the FSU (64%) is above the world's average (50%), but is considerably below the European (74%) or North American (79%) levels. Russia and Belarus are the two most urbanized countries in the region, while Tajikistan is the least urbanized. In some republics there is only one major city, and the majority of the population outside this city is distinctly rural. The level of urbanization rose through the 20th century: In 1900 almost 80% of the Russian Empire consisted of peasants; in 1950 the U.S.S.R. had an urbanization level of 52%; in 1970 it was 62%; and since 1990 Russia's level has been 74%. Within Russia today, the highest urbanization levels are observed in Slavic-settled, economically developed regions (e.g., Moscow Oblast, with 79%) and in the Urals (e.g., Khanty-Mansy Autonomous Okrug, where over 90% of the population is urban). The lowest urbanization levels are observed in the ethnic republics of the northern Caucasus (43–45% are common); in the Tyva (51%) and Altay (26%) Republics in Siberia; and in some northern autonomous districts.

In the most recent population census of 2002, there were a total of 2,938 "urban centers" in Russia. Of these, 13 had over 1 million people, while another 20 had over 500,000 people. Most, however, lost population between the 1989 and 2002 censuses—some as much as 10%. The biggest cities of Russia are primarily concentrated in the European part; Siberia has only one city, Novosibirsk, with over 1 million people. Moscow is similar in population size to Paris, London, Los Angeles, or Chicago; St. Petersburg to Toronto; and Novosibirsk and Nizhniy Novgorod to the Milwaukee, Wisconsin, or Memphis, Tennessee, metropolitan areas.

Keep in mind that almost all Russian cities are unicentric and compact, while the majority of American metropolitan areas are polycentric and sprawling. Because of the Soviet emphasis on high-rise apartments and centralized services, Moscow, with 11 million residents, covers about as much area as Minneapolis–St. Paul, with only 3 million; the city of Barnaul, with 600,000 people, about the same footprint as St. Cloud, Minnesota, with 60,000! Of the cities listed in Table 11.3, only four occur in polycentric urban agglomerations: Samara, Togliatti, Novokuznetsk, and Izhevsk. Among the top 30 American metropolitan statistical areas, most are polycentric (e.g., New York–Newark–Bridgeport, Washington–Baltimore, San Francisco–San Jose–Oakland, Dallas–Fort Worth, and Minneapolis–St. Paul). The monocentric areas are in a distinct minority, perhaps five or six in all (e.g., Chicago, Houston, Atlanta). In Russia, the monocentric Moscow agglomeration includes over 70 cities and 13 million residents, almost 10% of the national total. Moscow is therefore the primate city of Russia, capturing more population than the second and third biggest cities combined. This, however, is a much lower share than those of greater Paris and London, which include almost 20% of the population of France and the United Kingdom, respectively. It is also noteworthy that the same monocentricity is expressed strongly at the local level: In each *rayon* (the equivalent of the U.S. county), the main city is always the largest. In fact, most rayons have only one city; the rest are towns and villages.

Outside Russia, the biggest cities of the FSU are invariably national capitals. The Soviet system of government greatly favored the concentration of political power, economic institutions, higher education, health services, and the arts in one place in each region. Thus Tallinn, Kiev, Tbilisi, Baku, Tashkent, and so forth are all indisputable primate cities in their respective republics. The only exception is the new capital of Kazakhstan, Astana; with only 600,000 inhabitants, it is half the size of the former capital, Almaty, with 1,200,000.

TABLE 11.3. Biggest Cities in Russia in 2002 and 2008

	Population (thousands)	
City	2002	Early 2008
Moscow	10,101.5	10,470
St. Petersburg	4,669.4	4,568
Novosibirsk	1,425.6	1,391
Nizhniy Novgorod	1,311.2	1,275
Yekaterinburg	1,293.0	1,323
Samara	1,158.1	1,135
Omsk	1,133.9	1,131
Kazan	1,105.3	1,120
Chelyabinsk	1,078.3	1,093
Rostov-on-Don	1,070.2	1,049
Ufa	1,042.4	1,022
Volgograd	1,012.8	984
Perm	1000.1	987
Krasnoyarsk	911.7	936
Saratov	873.5	836
Voronezh	848.7	840
Togliatti	701.9	706
Krasnodar	644.8	710
Ulyanovsk	635.6	628
Izhevsk	632.1	613
Yaroslavl	613.2	605
Barnaul	603.5	597
Irkutsk	593.4	576
Vladivostok	591.8	579
Khabarovsk	582.7	577
Novokuznetsk	550.1	562
Orenburg	548.8	526
Ryazan	521.7	511
Penza	518.2	508
Tyumen	510.7	560
Naberezhnye Chelny	510.0	506
Astrakhan	506.4	503
Lipetsk	506.0	503

Note. Data from the Russian census of 2002 (*www.perepis2002.ru*) and the *Moj Gorod* online encyclopedia (*www.mojgorod.ru/cities/pop2008_1.html*).

Urban Structure

Historically, Russian cities were centered around a kremlin—a fortified settlement high on a river bank, frequently on an easily defensible hill at a confluence of two rivers. For example, the Kremlin in Moscow is located high on Borovitsky Hill between the Moscow and Neglinnaya Rivers, and the kremlin in Nizhniy Novgorod is situated between the Oka and the Volga. Such locations made sense, because rivers served as transportation arteries, while the hill between two river valleys was easy to defend.

Inside the kremlin, the local prince's palace would be on a big square with churches, along with the armory, warehouses, some noblemen's houses, and soldiers' quarters. Outside the kremlin, a large square (e.g., Red Square in Moscow) would form the main market area. The lands beyond the square would be settled by artisans, merchants, ambassadors, and other professionals and skilled workers in the part of town called the *posad*. Sometimes the posad would get an additional fortified wall later on (e.g., Kitaygorod in Moscow). The peasants would live still farther away, but would regularly come to the city for market and in times of troubles. When enemies attacked, the entire local population would find shelter behind the kremlin walls. A few dozen cities in Russia have a kremlin, or at least a central square with some remaining walls adjacent to it. Pskov, Novgorod, Vladimir, and Yaroslavl all boast impressive kremlins worth a visit. Some monastery-based towns, like Sergiev Posad and Murom, have monasteries in the middle instead.

No kremlins were built after the 16th century. The cities built after that period would have a more expansive modern design, with broader streets and no walls. A lot of old Siberian cities started as small forts, but these were quickly outgrown and a large, grid-like network of streets was laid out, not unlike that of many cities in the American Midwest. Some Russian cities were developed in this period along rivers in a linear fashion. For instance, Volgograd stretches along the Volga for over 60 km but is very narrow, being constrained by the Privolzhsky Hills from the west and by the floodplain from the east. St. Petersburg was built in the early 1700s on a flat marsh at the mouth of the Neva with a distinct diagonal pattern of tree-lined avenues—a pattern similar to that of Paris or Washington, D.C. In fact, architects from France and Italy contributed heavily to the construction of both the American and the Russian capitals in the 18th century. Because of its unique history, St. Petersburg retains a wide-open plan, unlike Moscow with its curving and congested streets.

Soviet-era cities were frequently built from scratch around a factory, mine, or GULAG camp. Some were built as scientific cities to house important laboratories and institutes, frequently ones associated with the Russian Academy of Sciences and the Ministry of Defense. Such cities, such as Novosibirsk, would utterly lack an old core (Figure 11.7).

Within most Russian cities in the Soviet era, old or new, a few typical districts could be distinguished: the historical city center (the core or downtown area); the old periphery (in the cities built before the Revolution); the industrial belt of the Soviet period; and sleeping quarters for the workers, connected to the industrial belt and the center by bus lines and by a subway in the biggest cities (Bater, 2006). Beyond the sleeping quarters there is usually a sharp city growth boundary in the form of a beltway, and beyond that is countryside, with scattered villages, summer dacha cabins on tiny plots, collective farms, and forests. Until very recently, the model was practically uniform (Figure 11.8). The main difference was in the size of the apartment houses: In the biggest cities these would have 9, 12, or even 24 floors, while in the smaller cities they would have only 3–5. A lot of cities also included village-like wooden houses built over 100 years earlier, and poorly built temporary barracks for construction workers that became permanent dwellings. Beginning in the 1990s, because of land privatization and the new possibility of owning a private home, many newly rich residents began to flee the city for suburbia in the familiar pattern of suburban sprawl. This phenomenon is well documented not only in Moscow or Novosibirsk, but also in Tallinn, Almaty, and Kiev.

City Center

The city center in the old cities almost always housed the kremlin or a big cathedral with

FIGURE 11.7. The center of Novosibirsk is less than 100 years old, including a church built just a few years ago to mark the "midpoint" of Russia along the Trans-Siberian Railroad. *Photo:* P. Safonov.

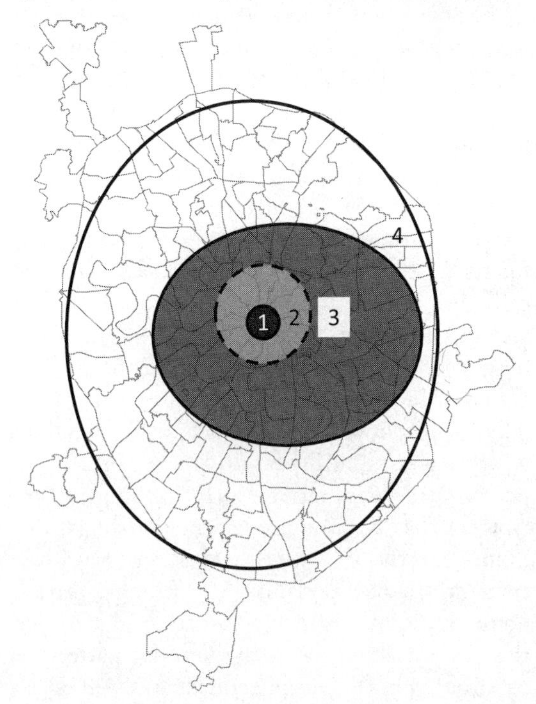

FIGURE 11.8. Moscow's four functional zones: (1) the historical core (pre-1800); (2) the old periphery (19th century); (3) the Soviet industrial belt (1920–1960); (4) sleeping quarters and parks (post-1960). Note the slight asymmetry caused by prevailing winds from the west; more factories were located east than west of downtown.

a large square next to it. In the Soviet period, many churches were destroyed and replaced with large government buildings (with an obligatory statue of Lenin in front). Some prerevolutionary homes of the center would house museums or government buildings; others would have communal flats, with as many as five or more families each having one room and sharing a common kitchen and bathroom. In modern Russia, virtually all such flats have been converted into the company offices, and some new office buildings have been constructed in the historic city core. Almost all but a handful of the most elite residents (or, conversely, the homeless) now live outside this area. Today the city center houses government buildings, banks, offices, the most expensive boutiques, the oldest theaters, some urban universities and colleges, and some quiet pedestrian areas.

Old Periphery

The old periphery area, with homes built at least 100 years ago, would be immediately outside the old city center limits. In contrast to many North American cities, where there is usually a "zone of discard" between downtown and the residential areas, Russian cities would have this zone of reasonably well-maintained large residential homes, train stations, markets, and shops. Usually this would be the most desirable place to live. Today much of this area is undergoing rapid construction and gentrification, with new condos, shops, and office towers quickly moving in.

Industrial Belt

Mainly developed in the 1930s, the industrial belts of Soviet cities would accommodate the factories. In Moscow the belt literally surrounds the center, with only a slight asymmetry; in other cities it could be located off to one side of the city, usually downwind from downtown to minimize air pollution. Many of the old industries are now in decline, and some cities are now removing the old factories and replacing them with new residential districts and commercial centers.

Sleeping Quarters

Sleeping quarters (*microrayony*) were built to accommodate the people who would work in the industrial belt. The later microrayonys of the 1970s came close to embodying the Soviet planners' ideal of self-contained residential units, with everything but work available locally (Figures 11.9 and 11.10). A typical microrayon would be a city area of about 35 ha in size, surrounded by streets with mass transit (buses, trolleys, sometimes trams). It would include about 10–12 large apartment buildings; 6–8 stores; a school; a clinic; and perhaps a library or a small stadium surrounded by playgrounds, tree-covered areas, and flowerbeds. Workers who lived here would still need to get to their work by mass transit, but much of their lives (and almost all their children's lives; see Vignette 11.1) could be lived inside the microrayon. There was enough distance allowed between buildings to let air and sunlight

FIGURE 11.9. Yasenevo, a typical late Soviet microrayon, built in the 1980s on the periphery of Moscow. It has multistory apartment buildings, playgrounds, day care centers, schools, clinics, and shops along the periphery. The retail kiosks date from the 1990s. *Photo:* I. Blinnikova.

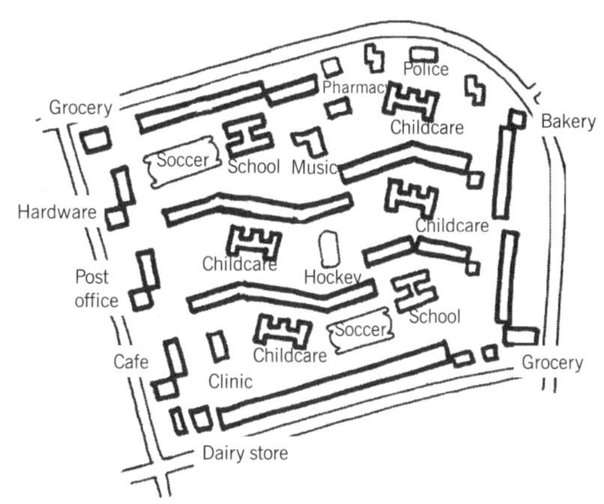

FIGURE 11.10. Plan of a Soviet microrayon of 1975, about 800 by 600 m in size. Apartment buildings range from 9 to 22 stories. The small squares are stores, a post office, a café, and so on. There are four child care facilities, two schools (grades K–10), and one health care clinic. Trees, playgrounds, and garages occupy the spaces between the buildings. Such a microrayon would house 15,000 to 20,000 people. *Drawing:* I. Blinnikova.

Vignette 11.1. Typical Daytime Travels of a Sixth-Grade Student in a Large Soviet City in the 1970s

Alexander gets up at 7:15 A.M. He lives with his mother and father in a two-bedroom apartment on the second floor of a typical nine-story building in the microrayon Zvezda located 15 km away from the city center. His mom is getting his breakfast ready. He leaves home at 8:05 A.M. and walks to his school, across the courtyard from the apartment building. His school houses grades K–8; the school day starts at 8:30 A.M. Alexander spends 6 hours at school, including lunch break. His physical education class requires him to run outside for 15 minutes, which everybody does at the school's soccer field, adjacent to the main school building.

At 2:00 P.M. Alexander goes home to an empty apartment. His mother left for her job at a government office (4 km away by city bus) at 8:20 A.M. Her work day is from 9 A.M. to 6 P.M., with 1 hour allowed for lunch. His father left even earlier, at 7:30 A.M.; he needed to take the same bus route and then transfer to another one to reach his factory (10 km away). He is a leading engineer, and he frequently works late. Alexander has three chores to do today, besides his homework: water the houseplants, buy bread, and mail a postcard. He can water the plants quickly while snacking on some leftover food right after school. He then walks over to the bakery shop on the microrayon corner, which is only 10 minutes away. He needs to cross the street to get to the post office, where he buys some envelopes and drops off the postcard. He is back home at 4:30 P.M. in time to watch some TV, exercise a little, and go outside to play with his friends until his mother comes home at 6:30 P.M. and calls him home for dinner. He walks a total distance during the day of about 1.5 km, almost all of which is within his own microrayon. How does this routine differ from what you experienced in sixth grade? How far did you have to travel during the day?

in, but very little space allocated for parking. This made sense, because the car ownership rate was under 10%. Although Westerners often referred to the "drab appearance" of the apartment complexes, most were in fact painted in pretty shades of white, pink, light blue, green, or yellow, and some were covered in colorful glazed ceramic tiles.

Post-Soviet Changes

The carefully planned Soviet cities have been undergoing rapid transformation as the new post-Soviet economic realities have set in. The literature in the **Further Reading** list at the end of this chapter provides more details about specific patterns and processes. Here I am only briefly going to mention several tendencies that are discussed in the current research on the topic.

Soviet-style planning has not completely disappeared. Old traditions die hard, and many of the same people who planned the Soviet cities are still around. In fact, the Soviet planning of the urban areas was exemplary in its attention to public needs, green spaces, mass transit, health, and other pertinent topics. The slums and squatter settlements or ghettos so common in less developed countries were nonexistent. Nevertheless, the old system underpinning the planning process is gone. Some of the municipal layouts of the 1980s and 1990s are still being reproduced around the country, but many changes based on economics and local politics are also being made. For example, instead of complete subdivisions built according to a few basic designs, smaller, more expensive, individualized projects appeared in many cities in the 1990s. Frequently these were funded by a private developer and underwritten by a large company, such as Gazprom or the city government.

Many of the older core areas are undergoing rapid transformation. Old residential areas are being replaced with renovated office buildings, elite boutique shops, new high-rises for the truly rich (with condominiums costing over $1,000,000 in some parts of Moscow), and new corporate buildings. This process is nearly complete in Moscow and St. Petersburg, but is still in progress in more peripheral cities.

The industrial belt in many cities is undergoing renovation; for example, the ZIL truck plant is moving out of Moscow. However, many of the old factories remain, especially in the cities where they are the single main employers. Perhaps the largest renovation of recent years in an industrial belt is the high-rise business center built west of downtown Moscow on the banks of the Moscow River, with over 2 million m^2 of finished office and an equal amount of elite retail space (Blinnikov & Dixon, 2010).

Many Soviet cities' sleeping quarters are likewise being redone. Much better retail services are becoming available. Some individual homes and office towers are being built, filling existing gaps in the construction of these but frequently encroaching on public spaces, parks, and squares, which leads to vocal protests from the local residents.

The suburbs of virtually all post-Soviet big cities are being rapidly privatized, as modern, detached, single-family suburban homes for the rich are being built (Figure 11.11). Much of this development is illegal or poorly regulated, and is occurring in floodplains or in forested recreation areas, which is against the law. Most of these new developments are gated communities, with 24-hour surveillance and private security guards to exclude "undesirables" (Blinnikov et al., 2006).

FIGURE 11.11. Suburban housing for the rich is proliferating around Moscow, St. Petersburg, Novosibirsk, and some of the other biggest cities. An average house in this development west of Moscow has about 300 m^2 of finished space and was worth between $500,000 and $1,000,000 in 2005. *Photo:* Author.

In many cases, recent industrial development on the cities' periphery has caused increasing pollution of water, air, and soil. The old sewer, heating, and electricity systems designed for different production patterns are not adequate for the increased load and frequently break down. At the same time, many small and medium-size cities in the European north, Siberia, or the Far East are rapidly depopulating. People either move out to better climates or die trying. Therefore, in many cities the top priority is not confining or channeling growth, but preserving the existing infrastructure. The booming cities, on the other hand, are found in the European part of the country (Moscow, Samara) and in the Urals federal district (Surgut, Tyumen).

Formerly cheap city services are rapidly increasing in price. For example, electricity and garbage disposal rates are increasing at a rate much higher than inflation (between 20 and 30% per year in some municipalities). Subsidies for these are theoretically available for some categories of residents, but they are difficult to obtain in practice, because one has to wait in long lines at a municipal office to present appropriate paperwork (and/or a bribe). There is also an increase in xenophobia in most Russian cities, and in many cities in the other FSU republics, against recent arrivals from other parts of the FSU—typically migrant workers or refugees. For example, many landscaping services in Moscow are provided by migrant Tajiks, while construction is done by large contingents of Moldovan, Belarusian, or Turkish workers. In Khabarovsk and Vladivostok, Chinese and Vietnamese workers are more common. There are as yet no ethnic ghettos or slums comparable to those in Latin American or Asian cities, but this may be changing in the near future.

Rural Settlements: The Woes of the Russian Village

Russian village life has always been hard. For centuries, peasants formed the majority of the country's population. The old village life focused on the extended family, with husband, wife, many children, grandparents, and frequently also younger siblings living under the same roof. The land was owned communally, with specific parcels allocated each year to households, depending on the size of families. A local census every 10 years or so ensured that each family had enough land to meet its needs. During the 17th century, however, serfdom became the mechanism that tied peasants to specific landlords. This was done to ensure that migration to the newly opened frontier lands along the lower Volga and in Siberia would not depopulate the central parts of the country. As a result, the serfs were not able to move, to own property, or even to decide whom to marry. Eventually their condition became little better than that of the slaves in North America.

For two centuries, between 1650 and 1861, the Russian serfs were treated as the property of their landlords. A handful of free (actually, state-owned) peasants lived in remote villages in Siberia and along the lower Volga. When serfdom was abolished by Alexander II in 1861, very little land was available to the newly freed peasants in central Russia; the majority continued to work, now for a fee, at their former masters' estates. Villages remained essentially peasant communes, cultivating common fields, with little private property of any sort available. The climate was harsh, the technology was primitive, and the harvests were correspondingly meager. Many Western commentators have explained the plight of the Russian village as a consequence of three things: the harsh environment, the archaic feudal production system, and ubiquitous drinking. One can also add the great distances that separated villages from each other or from more "civilized" urban life. Despite Stolypin's short-lived pre-World War I reforms, which attempted to redistribute land away from many poor to a few wealthier farmers, only a handful of areas had seen a rise in such independent family farms. Thus, unlike in much of 19th-century Europe and North America, virtually no independent private family farms existed in Russia on the eve of the Communist revolution. Many peasants were eager to support the Bolsheviks, who promised free land to all—something that would never come about in reality.

As explained in Chapter 7, the Soviet collectivization disposed of private farms altogether, and millions of the best farmers were sent to languish in Siberia as part of Stalin's plan to strip them of their property. The remaining poorest villagers, many of whom drank heavily or were lazy, were herded into the new system of large collective farms (*kolkhozy*). After the great famine of 1930–1932, in which about 2 million lives were lost, the Soviet system of collective agriculture was born (again, see Chapter 7). Each state farm (*kolkhoz*) included about 65 km^2 (6,500 ha or 14,000 acres) of farmland in the vicinity of a few villages (Figure 11.12). The central part of the farm would have a relatively modern tractor repair station, a club, a medical facility, and a school. The production was decidedly large-scale: Hundreds of cows would be kept in massive barns, and fields were cultivated with large combines. It is interesting that very large farms also emerged in North America at about this time, but under a completely different political system. In 1941 there were almost 250,000 state farms in the U.S.S.R., but after the war many were combined to create even larger units, so by the late 1970s only about 26,000 remained in existence. Despite the upbeat Soviet propaganda, farm productivity improved little; morale was therefore low, especially in the late Soviet period with its chronic shortages of food, feed, seed, fertilizer, equipment, and other necessities.

Yeltsin's reforms made the situation on the farms dramatically worse. On the one hand, virtually all of the Soviet subsidies were abruptly terminated. On the other, no incentives or credits were put forth to provide for the creation of independent private farms. Few people volunteered to become private farmers in the absence of clear laws or government-backed loans. Those few who did experienced tremendous physical and economic hardship, as well as derision and even outright hostility from envious neighbors. Even by 2005, only 7% of the total agricultural output in Russia was produced on private farms. The kolkhozy were restructured into joint-stock cooperative ventures, but their management practices remained essentially unchanged. Although the workers collectively own each enterprise now, the head manager typically has the controlling vote, and the enterprise continues to be inefficient. In 2005, the output of the Russian agricultural sector was 40% *less* than in 1990; the sown acreage had decreased at least 30%; and the number of cattle had decreased by 46%. Russia today imports a little less than half of the food

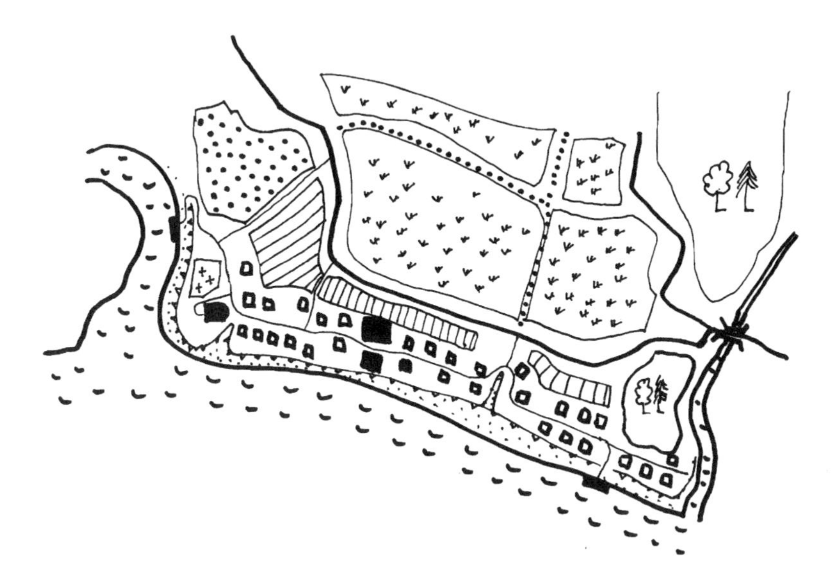

FIGURE 11.12. A typical small collective farm of the late Soviet period, with two villages stretched along the river, a few fields to the north, a forest, an orchard (large dots), a central office, and a tractor station (shaded blocks in the middle). The old village church (on the left near the cemetery) was remodeled into a club. Homes have small garden plots next to them. Between the homes and the fields are communal potato plots.

it needs to feed its own population—one of the highest rates of foreign-food dependency in the world—although it now has some surplus grain to export (Chapter 20).

Today Russia has about 150,000 villages, compared to only about 2,500 urban areas. The most striking fact about the villages is that most of them are rapidly dwindling or even disappearing. A similar process of loss in the U.S. farm belt started later but is somewhat similar. About half of all villages in Russia are now very small, defined as having fewer than 50 people; they include a mere 3% of the rural population. Many such dying or even ghost villages are scattered in the forested areas of European Russia, especially north of Moscow in Pskov, Novgorod, and Kostroma Oblasts, where farming has never been particularly strong. However, the regions with better agricultural potential in the forest–steppe belt south of the capital also have large depressed areas, referred to as "agricultural black holes" (the western Bryansk region, the eastern Ryazan and Tambov region, etc.; Ioffe et al., 2004). In contrast, 48% of the rural population of Russia live in the largest villages (each with at least 1,000 people), which constitute only 5% of the total number of villages. The services are of course better in these villages, and typically each is at the center of a collective farm. In addition to the agricultural villages, about 10% of small rural settlements in Russia house workers engaged in forestry, small-scale mining, or transportation/retail services.

Russian villages do not have many of the services that all American small towns do (e.g., natural gas, sewer, or water), but electricity is typically available. The main street is unpaved and is little more than a barely drivable rutted dirt road along which log houses are located (Figure 11.13). There are about 100–500 people living close to each other in small individual homes with two or three rooms each (Figure 11.14). Every house has a bit of land for a garden. There may be a central house in the village for the local administration, and a library or village club across from the primary school. In a bigger village there may also be a church, frequently in ruins now. The village is surrounded by agricultural fields or forests. Villagers are not individual farmers or city workers who like to live in the country. The majority are working in the same agricultural enterprise (a former kolkhoz). Both in times of serfdom and

FIGURE 11.13. A typical Siberian village house (Novosibirsk Oblast). The street in front is not paved. The house has electricity and sometimes natural gas, but no running water or flush toilet. *Photo:* Author.

under the Soviet system of collective agriculture, this arrangement made sense.

A separate type of rural settlement is a dacha development (these developments are called *dachny poselok*). These are enclaves of summer cabins or more permanent homes used primarily for recreation. Hundreds of these exist along scenic waterways, lakeshores, and suburban forest edges near the biggest cities, especially Moscow and St. Petersburg. The Soviet-period dachas were little more than plywood cabins on about 0.06 ha of land each, just about enough to grow a few rows of cabbage and tomatoes. The more recent developments of true year-round suburban housing (see "Post-Soviet Changes," above) created, for the first time in Russian history, suburban residential gated communities virtually indistinguishable from their counterparts in California or Virginia (Blinnikov et al., 2006). Less discussed, but also noticeable, is the out-migration of long-time city residents who want to try living in the country for personal, spiritual, or economic reasons. For example, dozens of villages in Central Russia have recently been taken over by urban residents who have created communes, with themes ranging from strict Russian Orthodox family life to organic agriculture.

Cities and Villages in Other Countries of the FSU

The urbanization from levels of the other FSU republics range 73% in Belarus to 26% in Tajikistan; all these are below the levels of either Russia or any Western country. The republican capitals are large: Kiev has over 2.5 million people; Minsk, Tashkent, and Baku have about 2 million each; and Tbilisi, Yerevan, and Bishkek have about 1 million each. The capitals of the Baltic states, Moldova, and Tajikistan have about 500,000 people apiece. Almaty is no longer the capital of Kazakhstan, but is still its largest city

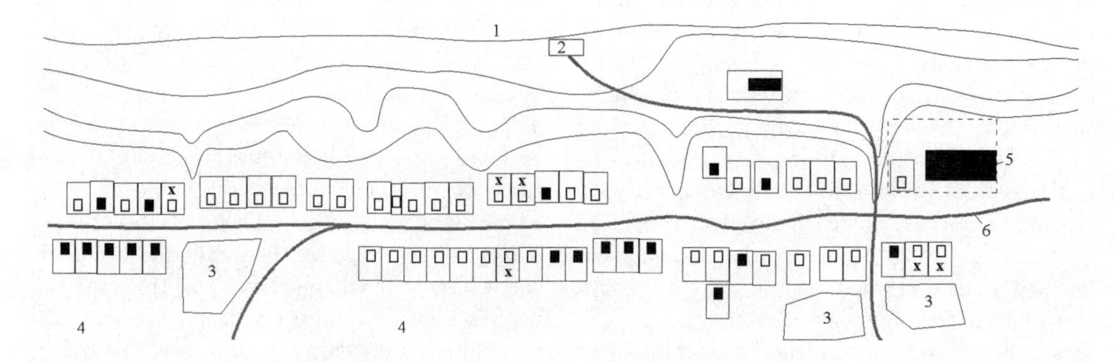

FIGURE 11.14. A typical village in Central Russia (Ivanovo Oblast) lined up along a main street, which in this case runs parallel to the Volga River. (1) The Volga River; (2) ferry dock; (3) vegetable plots; (4) abandoned agricultural fields overgrown with shrubs and birch trees; (5) the former school and library. Houses with open squares are occupied by local people; houses with shaded squares are used as summer cabins by city residents; houses that were abandoned and burned down are marked with an x.

at over 1 million. By contrast, the provincial and district centers in all the republics rarely have more than 100,000 people, with the notable exceptions being large industrial cities in Ukraine (Kharkov, Odessa, Dnepropetrovsk, Donetsk).

Some urban centers in these other FSU republics are very old (e.g., Tbilisi is about 1,600 years old; Kiev is at least 1,200 years old). Such cities have a distinct old core with either a kremlin or a cathedral square. Some cities in Central Asia date back over 1,000 years, whereas others are much more recent (Almaty was established as a Russian frontier fort in 1854, Ashgabat in 1818, etc.).

As far as rural settlements are concerned, types similar to those in Russia exist in Ukraine, Belarus, and Moldova. In much of Ukraine, the villages tend to be much larger than those in Russia. Many have a few thousand inhabitants and stretch for kilometers along river valleys. The Baltic states spent less time under Communism and had a prior history of small family farms; thus the kolkhoz period made less of an impact on them. Indeed, stronger ties to the land and a good work ethic made the Baltic farms exceptionally productive during Soviet times.

In Central Asia and the Caucasus, village life is especially important. Traditions run deep. People settle close to each other, in extended families. Most families now have at least one member who lives in a city, but the life around the old village houses is always lively. A strikingly unusual situation exists in Tbilisi, where an essentially rural population with close ties to the land lives in the middle of the city (Van Assche et al., 2009). A peculiarity of Georgia's and Armenia's urban architecture is the presence of extensive self-designed structures (e.g., balconies and verandas) that extend the living space outward, but are not formally approved by the local government. They existed even in the Soviet period, but are more common now. They reflect the creative and informal spirit of the tenants, as well as a real need for more family space.

In Central Asia, especially in Kyrgyzstan, Kazakhstan, and parts of the northern Caucasus, many people led a nomadic lifestyle until the 20th century (Figure 11.15). Some continue limited seasonal nomadism even today. People in these cultures would not normally live in urban settings. The majority were forced to settle in villages and cities during the Soviet period, because migratory nomads could not be easily tracked by the state. Although distinct patterns vary from country to country, most Central Asian cities have at least one main square in front of the administrative building, a nearby market, and a bus terminal. Many now also have prominently placed mosques. The traditional forms of architecture (e.g., Kazakh yurts) were largely replaced during Soviet times with generic log cabins in the Russian style, or with concrete apartment blocks. There is now a resurgence of interest in traditional architectural models; many new administrative buildings, banking centers, train depots, and so on follow such models but are built with modern materials. Some of the best examples are found in Ashgabat, Turkmenistan, and in the ambitious projects in the new capital of Kazakhstan, Astana (Chapter 31).

As in Russia, suburbanization is now common in all other FSU republics. Some well-documented examples include communities near Tallinn, Tbilisi, Almaty, and Kiev. As in the West, gated communities and other exclusive subdivisions are much talked about. However, the main form of suburbanization in these republics is the proliferation of cheap lodging options for the urban poor. The workers typically employed in construction or services, most of whom are migrants from rural areas, would find

FIGURE 11.15. Nomadic Kazakhs continued to live in yurts until the mid-20th century. *Photo:* Author.

it impossible to afford to live in the inner city and would endure long commutes to get to and from the inner city now.

REVIEW QUESTIONS

1. Summarize the main differences in urban structure between Western and Russian cities.
2. Use the Soviet concept microrayon to propose a new development in your city. Where would you locate it? How big is it going to be? What services will be placed inside and outside the microrayon?
3. What are the essential differences between urban and rural lifestyles in your country? How do you think this compares to the situation in Russia or the other FSU republics?
4. What are some of the common trends in the post-Soviet development of cities and villages mentioned in this chapter?

EXERCISES

1. Use online research and the categories in Table 11.2 to identify the city types to which the following Russian cities belong: Cherepovets, Suzdal, Nakhodka, Kirillov, Gelendzhik.
2. Use a map of any large Soviet city from an atlas in the library to identify the main functional zones.

Further Reading

Axenov, K., Brade, I., & Bondarchuk, E. (2006). *The transformation of urban space in post-Soviet Russia.* London: Routledge.

Bater, J. H. (1996). *Russia and the post-Soviet scene.* London: Arnold.

Bater, J. H. (2006). Central St. Petersburg: Continuity and change in privilege and place. *Eurasian Geography and Economics, 47*(1), 4–27.

Bater, J. H., Amelin, V. N., & Degtyarev, A. A. (1998). Market reform and the central city: Moscow revisited. *Post-Soviet Geography and Economics, 39*(1), 247–265.

Blinnikov, M., & Dixon, M. (2010). Mega-engineering projects in Russia: Examples from Moscow and St. Petersburg. In S. Brunn (Ed.), *Engineering earth: The impacts of mega-engineering projects.* Dordrecht, The Netherlands: Kluwer.

Blinnikov, M., Shanin, A., Sobolev, N., & Volkova, L. (2006). Gated communities of the Moscow green belt: Newly segregated landscapes and the suburban Russian environment. *GeoJournal, 66,* 65–81.

Boentje, J., & Blinnikov, M. (2007). Post-Soviet forest fragmentation and loss in the green belt around Moscow, Russia (1991–2001): A remote sensing perspective. *Landscape and Urban Planning, 82,* 208–221.

Demko, G. J. (1987). The Soviet settlement system: Current issues and future prospects. *Soviet Geography, 28*(10), 707–717.

Ioffe, G., & Nefedova, T. (1998). Environs of Russian cities: A case study of Moscow. *Europe–Asia Studies, 50*(8), 1325–1356.

Ioffe, G., Nefedova, T., & Zaslavsky, I. (2004). From spatial continuity to fragmentation: The case of Russian farming. *Annals of the Association of American Geographers, 94*(4), 913–943.

Iyer, S. D. (2003). Increasing unevenness in the distribution of city sizes in post-Soviet Russia. *Eurasian Geography and Economics, 44*(5), 348–367.

Nefedova, T., & Treivish, A. (2003). Differential urbanisation in Russia. *Tijdschrift voor Economische en Sociale Geografie, 94*(1), 75–88.

O'Loughlin, J., & Kolossov, V. (2002). Moscow: Post-Soviet developments and challenges. *Eurasian Geography and Economics, 43*(3), 161–169.

Saushkin, Y. G., & Glushkova, V. G. (1983). *Moskva sredi gorodov mira.* Moscow: Mysl.

Sidorov, D. (2000). National monumentalization and the politics of scale: The resurrections of the Cathedral of Christ the Savior in Moscow. *Annals of the Association of American Geographers, 90*(3), 548–572.

Van Assche, K., Salukvadze, J., Duineveld, M., & Verschraegen, G. (2009). Would planners be as sweet by any other name? Roles in transitional planning system: Tbilisi, Georgia. In K. Van Assche, J. Salukvadze, & N. Shavishvili (Eds.), *City culture and city planning in Tbilisi* (pp. 243–317). Lewiston, NY: Edwin Mellen Press.

Websites

www.kreml.ru—The Kremlin Website (not the president's Website, which is *www.kremlin.ru*, but the site for the Kremlin Museum).

engl.mosmetro.ru—Official site of the Moscow–metro (subway) system.

www.mosmuseum.ru/eng—Moscow City Museum.

www.saint-petersburg.com—St. Petersburg for the English-speaking crowd.

CHAPTER 12

Social Issues

HEALTH, WEALTH, POVERTY, AND CRIME

"Social geography" looks at many aspects of people's daily lives as expressed in their engagement with and movements through space. It encompasses both traditional customs and modern developments. At the outset, it is important to note that social issues can be understood both objectively (as, for example, when one looks at statistics on health or crime) and subjectively (when one perceives things in a certain way, based on his or her cultural upbringing, information received from peers or the mass media, and personal biases). Take the media impact, for example. The former Soviet Union (FSU) is frequently portrayed in the Western media as a rough place plagued by crime, drug use, violence, and corruption. To an American or Western European these days, the very word "Russia" conjures up images of roaming street gangs, ubiquitous disease, corrupt autocratic leaders, and hopeless human misery. Consider this, however: If your only image of New York City was formed by TV reports of gang wars in Queens, would you consider spending any time there? Reality is usually multifaceted, and this truth is nowhere as obvious as in the geography of social issues—be it health, disease, wealth, poverty, crime, or any other issue.

This chapter focuses primarily on the objective patterns of three main social issues in Northern Eurasia today. For an understanding of how the FSU got to this point in regard to these issues, see the chapter on Yeltsin's and Putin's reforms (Chapter 8), as well as Chapters 13–21. If you travel through the region, or when you read the travel accounts of others, you will have a chance to form a more personal view of the social situation there. There are also many research articles available written by scholars who have lived and observed social issues in the FSU, some of which are listed at the end of this chapter. Although many different social patterns in the countries of the FSU could be discussed, this chapter looks only (because of space limitations) at three major ones: health, income distribution (and associated unemployment and gender issues), and crime.

The Soviet Health System

The Soviet Union had what was arguably one of the best health care systems in the world. Surprised? If you have seen Michael Moore's film *Sicko*, you may not be: Moore depicts Cuba as an example of a socialist state with a free, universal health care system that has produced impressive results. This is something many Americans and even some Europeans have a hard time imagining. First, if all this is free, then who is paying the bill? Also, if all of this is universal, how are priorities set? Who gets treated first or most, for example? Is the quality of care adequate? Do people need to wait in a long line to see a doctor? Do the doctors make a decent living? Are the nurses caring and well trained? Do the patients have a choice of doctors or clinics? How is all of this possible?

During the Soviet period, the socialist government owned and ran everything, including the entire health care system. The right to free health care was listed as one of the fundamental human rights in the constitution. The state paid for it, because it made political, economic, and social sense to do so. Sick workers do not work well; sick teachers do not teach well; sick soldiers do not fight well. Instead of forcing people to choose among clinics or doctors based on their income, insurance policies, or personal taste, the system simply provided all with basic care through either their place of residence or their employment. By and large, the care was decent. A Soviet worker who came down with flu, for example, just needed to dial the local clinic's phone in the morning and stay in bed; the physician on call would come and visit the worker *at home*, usually later that same day. Physicians were accustomed to spending about half of their workday making house calls. Typically, with a common illness, one could receive a doctor-approved excuse from work, while keeping 100% of pay for 7 days. If something more serious was detected, the doctor could prescribe home rest for 2–3 weeks, or send the person to the hospital.

As far as the choice of clinics was concerned, one could go only to the local polyclinic with multiple doctors of various specialties right in one's neighborhood (rather like a health maintenance organization [HMO] in the United States today), or get treated at the factory or institute clinic. Some highly specialized treatments (e.g., laser eye surgery, pioneered by the famous Feodorov Clinic in Moscow in the 1970s) had long waiting lists, but were available on a referral basis for free. Life-threatening diseases would be treated right away, however. The Soviet doctors received free education (Chapter 15), so they had no student loans to pay back, but they were expected to work long hours for relatively low pay at the clinics to which they were assigned. Transfers and promotions were rare. An average doctor's salary was comparable to that of many qualified factory workers, about 200–250 rubles per month for a good specialist; nurses received about half that. Many world-class surgeons in the best national hospitals in Moscow or Kiev would work for a small fraction of the possible pay in the West, but their jobs were guaranteed and there were no threats of litigation. The quality of their work was very high, although medical equipment and drugs (with few exceptions) were less advanced than in the West. In reference to the poor quality of after-surgery care in many hospitals, the common late Soviet joke was that the doctors would save your life, but the nurses would kill you. Relatives of patients undergoing major surgery would typically bring a small gift to the surgeon (a bottle of good brandy or a box of chocolates was common). Of course, the party elite had their own clinics, sanatoria, doctors, and the very best equipment, purchased for hard currency in the West.

By the end of the Soviet period, the U.S.S.R. had the highest ratio of doctors to patients in the world, about 1 physician per 233 people (the United States has about 1 per 435 today). Eighty-six percent of the medical staff were female: The Soviet system encouraged women to consider medical careers early on, and the prevailing culture favored that idea too, because of the stereotype that women are more compassionate and better suited for caregiving than men (Hughes, 2005). The predominance of women was also related in part to the relatively low wages Soviet medical specialists received in exchange for a lot of very hard work. Surgeons were usually male, but family physicians and nurses were overwhelmingly female.

The Soviet Union also had one of the longest average hospital stays in the world, because home care was viewed as inherently inferior, while hospital beds were free. A typical hospitalization would last for 2–3 weeks, and frequently over a month. Another common feature was an emphasis on prophylaxis: Vaccination rates were among the highest in the world, and every child and adult was expected to have a physical checkup and a dental exam at least once per year. All of this did not make the U.S.S.R. the healthiest place on the planet; environmental pollution, stress, poor working conditions, and high alcoholism rates all took their toll. The average Soviet lifespan in 1990 was 69.5 years—well below the U.S. rate of 75 years in 1990, but respectably in the upper third of the world, and well above the expectancies in most African, Asian, or Latin American countries.

Post-Soviet Declines in Health Care and Health

Since the fall of the Soviet Union, there has been a major decline in health care availability. The results in Russia have included a huge slump in life expectancy (Chapter 10); an increase in most diseases; the reemergence of previously suppressed diseases such as tuberculosis (TB), polio, and diphtheria, due to a decline in vaccinations; a surge in HIV/AIDS; and many other indicators suggesting a full-blown crisis (Figure 12.1).

The main reason for this was pure economics. The health care system went through a major restructuring on short notice, with support from the state abruptly declining to a fraction of its former amount due to rising inflation rates and to unwillingness or inability to pay more. Other factors included emigration of some of the best doctors to Western countries, restructuring of the Soviet pharmaceutical and medical industries, and disruptions in the production of medical drugs and equipment. Thus, although the post-Soviet states remained committed in theory to free, universal health care, in reality there were increasingly fewer doctors, fewer supplies, less equipment, and fewer opportunities to provide the level of care needed. With inflation at over 20% per year for much of the 1990s, and without comparable pay raises, state-paid doctors' salaries dropped from being in the upper third of all sal-

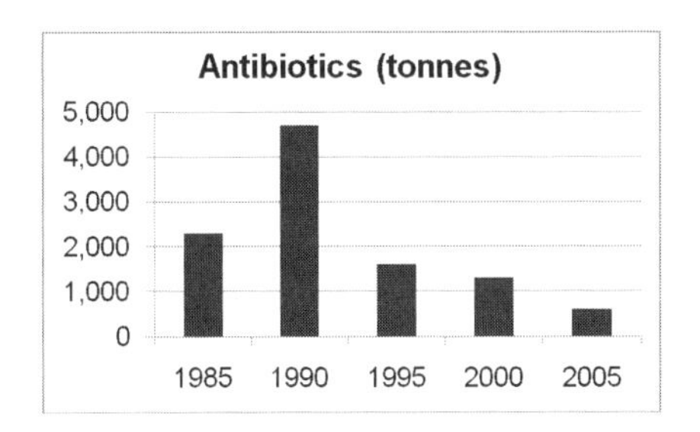

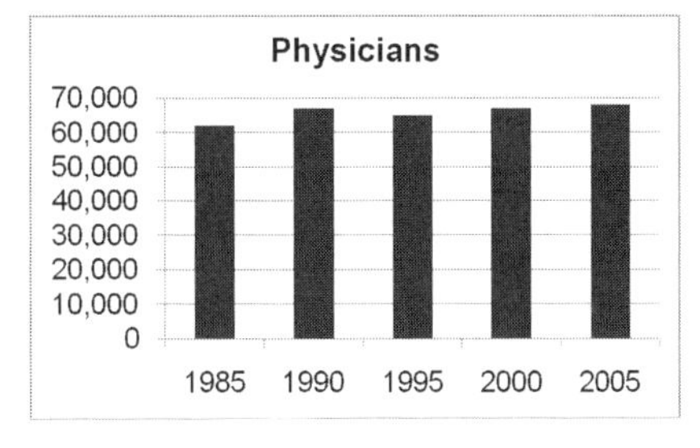

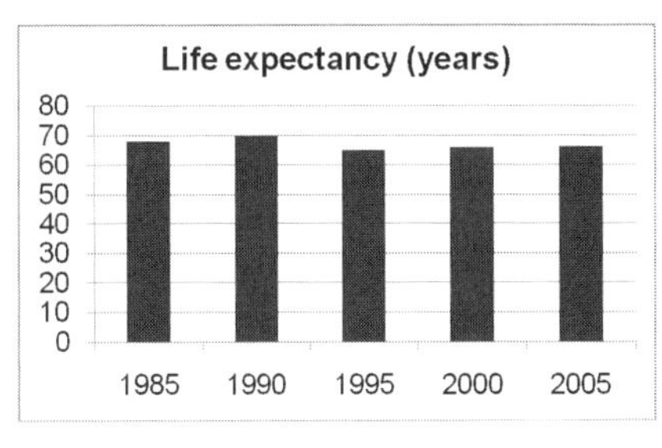

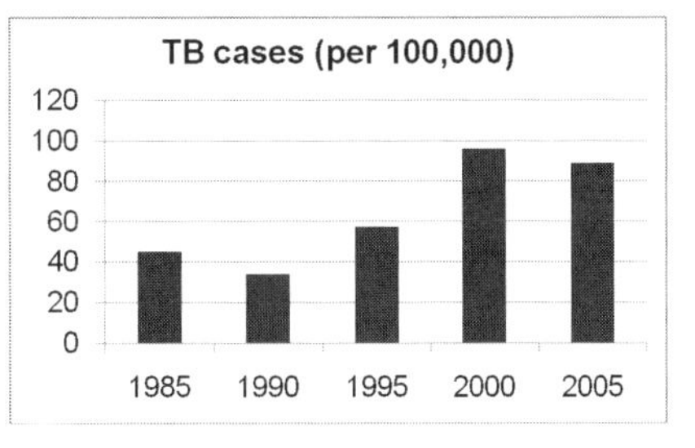

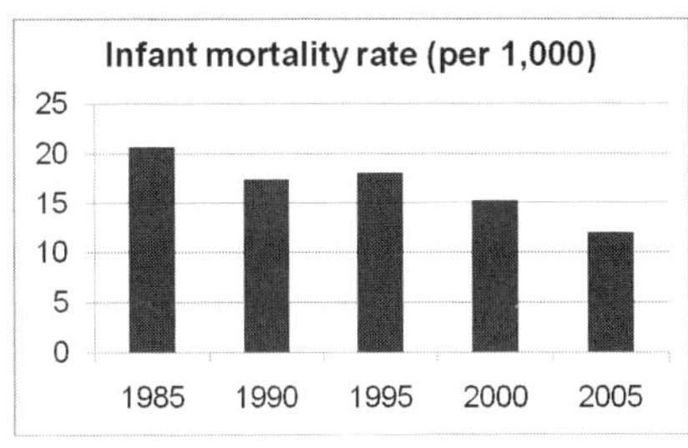

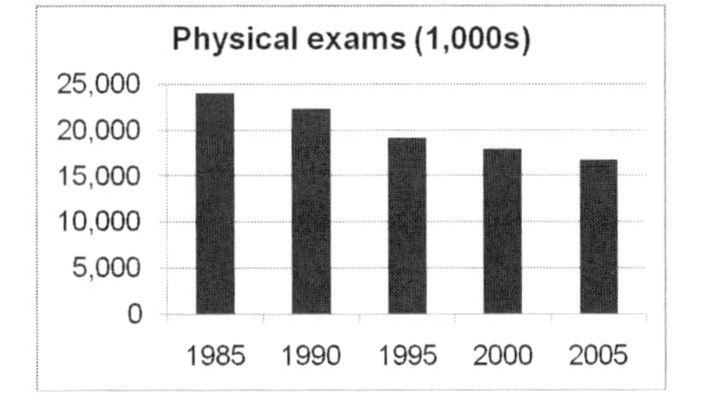

FIGURE 12.1. Some trends related to health in Russia between 1985 and 2005. Data from the Federal Service of State Statistics, Russian Federation, the Population Reference Bureau, and the World Health Organization. Data for physical exams for 1985 are estimates by author.

aries in the country to the bottom 10%. A janitor at a local McDonald's was making more money in the late 1990s than a doctor at a local health clinic. Private clinics emerged to compensate in part for the decline in state care; however, they were only able to provide care to the wealthiest 15–20% of the population.

Although infant mortality has been steadily improving in most recent years in Russia, adult mortality has not. In fact, after reaching an all-time high rate of about 15.7/1,000 per year in 1994 (up from only 11.3 in 1985), it decreased only modestly to 13.6 by 1998 before rising again after the default of 1998. The current adult mortality level in Russia is 15/1,000, which is comparable to that in such countries as Mali, Tanzania, or South Africa. By comparison, the U.S. mortality rate is 8/1,000, the U.K. rate is 9, and the Croatian rate is 12. Haiti has 10, the highest rate in the Americas. Only 12 countries in the world (most in sub-Saharan Africa, but also Ukraine) had higher mortality rates in 2008 than Russia. Within Russia, the highest mortality rates are found north of Moscow (e.g., Ivanovo, Pskov, and Tver Oblasts), mostly due to the older populations there. The lowest mortality is found in the demographically young republics of the northern Caucasus, where large families are more common. Mortality rates in the countryside are about 8% higher than in the urban areas, thus placing Russian rural mortality on a par with the rate in Sierra Leone, the world's worst.

In a recent study (Vishnevsky, 2006), 12 factors were identified out of a possible 175 that were primarily responsible for increased mortality rates between the Soviet period (1965–1984) and years 2000–2003 in Russia. The top ones were heart attacks and strokes; alcohol-related liver poisoning and stomach cancer; and lung cancer, TB, and pneumonia. Some additional causes were accidents, high blood pressure, and neurological diseases. Clearly, a lot of these are directly related to a decline in health care and/or environmental quality.

The increase in TB is particularly alarming, because TB is a highly preventable disease. Bacterial in origin, it occurs largely in individuals who live in chronically poor conditions, lacking vitamins, adequate nutrition, water, or exercise. Although it can be highly contagious, many individuals who come into contact with a TB-infected individual do not get infected right away, if they practice proper hygiene. TB was common in 19th-century Europe among the urban poor who lived in damp basements, worked under dreadful conditions in large factories, and were chronically malnourished. In the Soviet Union, widespread vaccination against TB, better hygiene, and preventive screenings had all but eradicated it by 1960, with a significant exception being the massive prison populations. Unfortunately, the release of thousands of infected inmates under Gorbachev, coupled with a sharp drop in vaccinations, rapidly led to an increase of TB throughout Russia (Figure 12.1). Worst of all, a new, highly drug-resistant form of TB emerged that is now accounting for over 20% of all new cases. Russia currently ranks 12th among 22 countries with high TB burden worldwide; about 166,000 new cases are reported every year. The TB infection rate is double that of the late Soviet period. Improved surveillance and detection made possible by aid programs from international agencies in the 1990s, especially in the prison system, have made some headway toward reducing the spread of new infections. Nevertheless, TB continues to spread in many places, such as hospitals and day care centers; it can even be contracted by sharing a compartment with an infected person on a long-distance train.

Another infection that scares a lot of people is, of course, HIV. Virtually absent from the Soviet Union, it spread in the countries of the FSU via various channels in the late 1980s. The first HIV cases were found among foreign students from Africa; the first case in the Russian population was detected in 1987. Hundreds of people became infected in regional hospitals via blood transfusions tainted with HIV-positive blood. Soviet hospitals at the time were not equipped with disposable syringes; instead, they relied on autoclave sterilization, which was not sometimes done according to proper standards and allowed some transmission of HIV via dirty needles. Also, no rapid tests were available at the time to check all of the incoming donated blood. By the mid-1990s, disposable supplies and modern, more accurate tests ensured a much higher level of safety in blood transfusions.

Despite these measures, HIV infection in Russia has increased explosively via direct person-to-person transmission—from fewer than 20,000 total cases in 1996 to 448,000 in 2008, according to the Russian Ministry of Health. Unofficial estimates by Western specialists suggest a much higher level, approaching 1.5 million infected persons in 2008. Ukraine is thought to have an even higher rate of infection (about 1.6%, relative to Russia's 1.1%). The main channels of infection are now heterosexual (25%) and homosexual (4%) contacts; transmission from pregnant mother to baby (6%); and, by far the biggest one, the sharing of needles among intravenous drug users (65% of all new cases in 2008). The incidence rate in Russia is still low compared to that of South Africa or Botswana, but it is expected to grow rapidly (Baker & Glasser, 2005). It is already about 10 times the rate of an average European country. Most other FSU states have infection rates ranging between 0.1 and 0.3%, with the exceptions of Estonia (1.3%) and Ukraine (see above). According to the Population Reference Bureau (see **Websites** at the end of this chapter), the world's average HIV infection rate in 2008 was 0.8%, with the U.S. rate at 0.6%, Senegal's at 1.0%, Haiti's at 2.2%, Kenya's at 7.4%, and rates in some southern African nations approaching 25%.

The highest numbers of HIV-infected individuals in Russia are observed in a few big cities: Moscow, St. Petersburg, Yekaterinburg, Samara, Irkutsk, Chelyabinsk, Orenburg, Tyumen, Kemerovo, and Saratov. Such cities tend to have high rates of drug use and prostitution. HIV is not uniformly spread among these cities however. Irkutsk, for example, has an unusually high rate that has to do with the early pattern of spread there among intravenous drug users; some other comparably sized cities have much lower infection rates. The most alarming recent trend is the rapid increase of infection via heterosexual contacts among persons who do not use drugs. More relaxed attitudes toward casual sex among young Russian adults play a big role in the spread of all sexually transmitted diseases, including AIDS (Figure 12.2; Haavio-Manilla et al., 2005). As might be expected, HIV's impact is greatest among the young population, with 80% of Russia's infected persons being between the ages of 15 and 30. More than 40% of new reported HIV infections in 2005 were among women, and the majority of those are thought to have acquired the virus through unprotected sex with an infected male partner, not through unsterilized drug injections (Joint United Nations Programme on AIDS, 2006). Since 2005, approximately 95% of

FIGURE 12.2. Young Russian adults on a city street in Mosocw. *Photo:* Author.

all infected persons in Russia have been receiving antiviral drugs. Nevertheless, the number of new cases grew by 20% in 2008, suggesting that prophylaxis is lacking. Russian scientists are participating in efforts to develop an HIV vaccine.

The national and especially local governments in Russia have made some attempts to improve financing of the health care system in recent years. The right to health care is guaranteed by Russia's new constitution. Article 41.1 says:

> Everyone shall have the right to health protection and medical aid. Medical aid in state and municipal health establishments shall be rendered to individuals gratis, at the expense of the corresponding budget, insurance contributions, and other proceeds.

Local and regional governments still run free clinics that anyone can use (although a state insurance card is now required). They also subsidize medical drug expenses for seniors and the poor. The exact quality and level of care, of course, depend on geography. The wealthiest regions, including Moscow and Tyumen, will have considerably better care, more modern equipment, more diverse clinic choices, and higher subsidies. Some of the poorest and/or most remote regions have a very low quality of care indeed (Tyva, Altay, most of the northern Caucasus, and many underperforming regions in European Russia).

Another big change since 1991 is the appearance of private clinics. Some now function similarly to the U.S. HMOs, trying to do everything in house except major surgeries, while requiring annual payments in advance. Others provide on-the-spot care for cash. The cost of visiting a private clinic varies dramatically from region to region. At the time of this writing, the average clinic visit to a physician in Moscow costs $25–$30. This does not include any lab work or drug costs, which may be considerable. Patients will be charged a smaller fee for return visits, but a treatment that requires antibiotics and a few blood or urine tests may cost a total of $100–$200. This may sound like a bargain to U.S. residents, but bear in mind that the average salary in Moscow is still below $1,000 per month. One unquestionable improvement over the Soviet system is the elimination of long queues at the clinics. Most are now open from 9 A.M. to 9 P.M., 7 days a week. The same is true for the dental clinics, among which there is a considerable amount of competition. Whereas a U.S. patient has to wait sometimes more than 2 weeks to get into a dentist's chair, one can call a local clinic in Russia and get in within 2–3 hours, including on weekends! The cost of private dental care in Russia is currently about 20% of the U.S. level, with essentially the same level of care, Western-made fillings and prosthetics, and adequate pain control. The main difference is in the cost of dentists' labor, which is considerably lower in Russia. A recent increase in medical and dental malpractice suits may change this for the worse in the near future, however.

A final interesting aspect of health care in the FSU is the wide availability of alternative care, including acupuncture, herbal medicine, and homeopathy. All are very popular, as they also are in the West. The Altay Mountains of central Siberia produce a large share of medical supplements and herbs. Traditional Chinese and Tibetan practitioners can be easily found in major cities. Also, there are many unregulated herbalists and shadowy psychics, who advertise their services on TV, in press, and online. As in the West, many are little more than charlatans, so buyers must beware.

Income and Wealth Distribution

In 2007, the *Forbes* magazine list of dollar billionaires (see **Websites** at the end of this chapter) listed 53 billionaires from Russia, out of a total of 946 worldwide. Only the United States and Germany had more. However, Russia's mostly young, self-made tycoons are catching up to Germany's often aging heirs and heiresses. Russia was two people shy of Germany's total, but the Russian oligarchs were worth a collective $282 billion in 2007, $37 billion *more* than Germany's richest. It is worth remembering that *Forbes* does not include government officials or royals on its list, to avoid political trouble; if one were to include wealth controlled by people from Putin's circle or by some regional governors, a few more billionaires would undoubtedly be added. The majority of Russian billionaires live in Russia, with a few important exceptions. The richest in 2007, Roman Abramovich (worth $18.7 billion), widely believed to be the personal banker of the

Yeltsin family during the late 1990s, now lives in England after having cashed in most of his Russia-held assets in 2000. Boris Berezovsky, worth a paltry $1.1 billion, likewise makes his home in the United Kingdom, where he is enjoying the British government's protection against the arrest warrants repeatedly issued by the Russian chief prosecutor's office. (Berezovsky is wanted on corruption charges back home; see Chapter 8.)

The next nine Russian "sharks" on the *Forbes* list made their fortunes in the 1990s, mostly in steel and nonferrous metals, petroleum, telecommunications, and banking. Some participated in the infamous auctions that allowed quick privatization of the most lucrative state assets for a fraction of the real price (again, see Chapter 8). Some of the smaller and more recent "fish" on the list made their fortunes in real estate, construction, information technologies, and retail. A few of these are more modest, and even religious people, known for their philanthropic work. All in all, Russia's position in the top three countries with billionaires is remarkable, given that it was only the 9th largest world economy in 2007 after adjustment for purchasing power parity (PPP). France, in 7th place, had an economy about 10% larger than Russia's, but only 15 billionaires. This, of course, indicates a highly uneven post-Soviet distribution of wealth. Ukraine had 7 billionaires on the *Forbes* list, and Kazakhstan had five. The main sources of wealth for those people were steel and coal, as well as oil and banking, and (most importantly) personal connections to the ruling elites of those republics.

Economically speaking, none of the FSU countries are yet giants in terms of personal wealth. According to the CIA World Factbook (see **Websites**) Russia's gross domestic product (GDP) per capita (adjusted for PPP) for 2007 was merely $14,600—about the same as that of Botswana ($14,700) or Malaysia ($14,400). It did move up from 2006 to 75th place from 82nd among 229 countries, a modest accomplishment. All the FSU countries, except the Baltics, were considerably below this level. The poorest country in Western Europe, Portugal, on the other hand, had had a GDP PPP of $21,800 (in 55th place); the EU average was $32,900; and the United States was at $46,000. With the world's average GDP PPP per capita at $10,000 that year, Russia was barely above the middle mark. Thus, generally speaking, it fits into the category of countries with a slightly higher-than-average income, but not a truly high one. According to the CIA data, about 60% of Russia's labor force in 2007 was occupied in services, 29% in industry, and 11% in agriculture and forestry. In 1940, over 50% of the workforce was still in agriculture, with one-quarter in industry, and only 24% in services. For comparison, over 80% of the U.K. workforce in 2007 was in services, 19% in industry, and less than 2% in agriculture. The official unemployment rate in Russia is quite low (only about 6%), but in reality there is a lot of underemployment and underreporting, especially among undocumented immigrants.

Russia has a very uneven distribution of wealth (Figure 12.3)—similar to that in the United States, but quite unlike those of its European neighbors. Its "Gini index," which measures inequality of family income, is 41.5 (0 = perfect equality, 100 = perfect inequality). For comparison, one of the most equitable countries in the world, Denmark, has a Gini index of 24; the U.S. index is 45; and Brazil's is 56.7. To describe this situation in another way, the income distribution in both the United States and Russia is very "top-heavy," with over 30% of all wealth concentrated in the hands of about 6% of the households. In Finland or Denmark, on the other hand, the same one-third of wealth will be distributed over 12% of all households. Of course, being in the top 10% in the United States vs. Russia means different things.

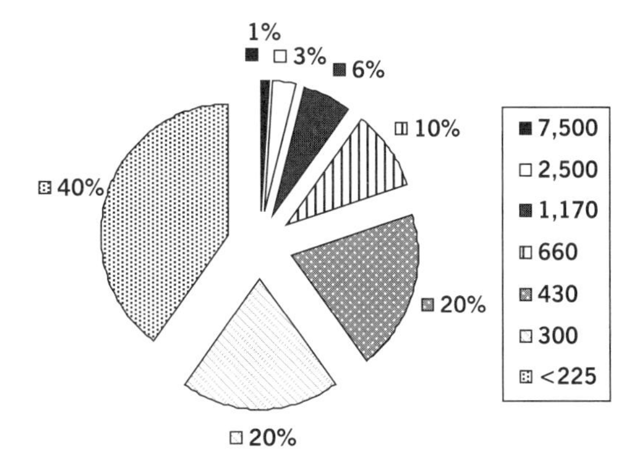

FIGURE 12.3. Average household income distribution (in U.S. dollars) per month in Russia in 2002, based on adjusted official Goskomstat data and additional social research. The average household size in Russia is 2.7 people. Data from Berezin (2002).

The top 5% of all U.S. households in 2006 earned over $160,000 per year. For Russia, these earnings were only $20,000—but the income inequality in Russia is still striking, because the poorest 10% in Russia made under $1,000 per year per household, or only $3 per day.

Before we discuss money farther, we need to bear in mind that in Russia it is customary to express earnings not as hourly or annual rates, but as monthly rates. Soviet salaries were paid out in cash twice per month, and the paid vacation period for white-collar workers was also about 1 month (24 work days; a shorter period for blue-collar workers). Hence a common statement in job advertisements was "Possibility of earning over $2,000 per month," not "$24,000 per year." Because of the high inflation over the past 15 years, it has also been common to express salary amounts in dollars or euros, although today they must be paid in rubles (usually to a bank account accessed via a debit bank card). In the 1990s it became fairly common for cash salaries to be paid in U.S. dollars to avoid taxes. In 2001, about 35% of all salaries (amounting to about $60 billion) were not properly declared (so-called black-cash). More transparent tax regulations put forth in the first year of Putin's administration, plus a crackdown on illegal tendering of dollars, reduced the black cash somewhat.

The inequality in income distribution increased greatly after the fall of the Soviet Union. Money did not mean much in the U.S.S.R., because many goods were not available for money at all. Social capital was needed to obtain those scarce, often imported goods (e.g., nice shoes, modern kitchen equipment, caviar) through government distribution channels. Even then, no one, including party officials, was allowed to own planes, yachts, or palaces. Although the state property was managed by the nomenklatura, they could not bequeath state assets to their heirs (except their housing) or publicly flaunt their wealth. All that changed with the advent of the Russian *kapitalism*. Private wealth appeared suddenly and with vigor. Visits to night clubs and casinos, exotic cars, and regular weekend trips to St. Tropez or Davos suddenly became both attainable and visible for the lucky few. In modern Russia, two former friends who graduated from the same high school in the 1980s may be as different in their levels of income or social positions today as a Hollywood star is different from an undocumented Mexican janitor in Los Angeles—a situation inconceivable even a generation ago.

In fact, the members of the generation who entered perestroika as 20- to 30-somethings ("Generation X" would be the U.S. equivalent) were precisely those who attained the highest or the lowest levels of income in subsequent years. Their parents, roughly comparable to the U.S. "baby boomers," by and large entered the new system in middle age or close to retirement age. Their lifestyles or occupations did not change much, and their incomes remained relatively unchanged (i.e., low). Likewise, the youngest workers in Russia today started working after the major economic shift occurred, so again their incomes are more comparable to each other, although usually much higher than those of the older workers. Middle-aged Russians today, on the other hand, either sank to the very bottom or floated close to the top in the mid-1990s. Many of my high school classmates (class of 1987) have very comfortable lifestyles today, but others unfortunately do not. In fact, a few are no longer even around, because they fell victims to violent crime or drugs (at least 2 from a class of 30; also, at least 1 of my 10 or so high school teachers died in an armed robbery).

Who have the new wealth in Russia today? Besides the oligarchs and other rich business owners and executives, they also unquestionably include government officials at all levels (e.g., members of the parliament, governors, mayors, the highest-ranking police officers, and customs/border patrol officials), mainly because of widespread corruption. Also among the wealthy are many professionals who work for foreign or the best Russian companies (e.g., BP-TNK, Gazprom, Alfa Bank). Although official statistics systematically underestimate the actual level of Russia's personal income, some recent research paints the following, more realistic picture (Berezin, 2002; see also Figure 12.3).

In 2002, about 1% of Russia's households had incomes approaching or exceeding that of the average American middle-class family ($7,500 per month; all figures in this passage are given in U.S. dollars). The average income in this group was $90,000 per year. About half of these households lived in and around Moscow, and made up about 10% of all Moscow households. Some of these people had considerably higher incomes, in the millions of dollars. The presence of these people makes Moscow one of the most expensive cities in the world. Another 3%, the so-called Russian upper middle class, had household incomes between $20,000 and $45,000 per year.

Next in line are the "true middle class" (about 6%) with incomes between $10,000 and $20,000 per year, averaging $1,170 per month. Many of these people, unlike their Western counterparts, cannot afford a private home, but most own an apartment with all the modern essentials and one car per family. (A typical modest two-bedroom Soviet apartment today costs over $100,000 in most major cities, and more than triple that in Moscow.) A typical car is either a 5-year-old Russian Lada or an 8-year-old used German or Japanese sedan. A typical family from this class can afford a nice, but short, annual vacation abroad in Turkey or Egypt; it eats out a few times per month; and it can save a little extra cash for music lessons for the children and an occasional theater visit or rock concert for the adults. Another 10% are the "lower middle class," with incomes of $6,000 to $10,000 per year (averaging $660 per month). These people and those in the "true middle class" buy most of the durable goods in Russia (e.g., refrigerators, TVs, and other appliances). They have some extra money to spend, but little to invest. The mass consumption boom in major cities is largely driven by them (Figure 12.4).

FIGURE 12.4. The renovated shopping mall at the most famous store in Moscow, the venerable GUM near Red Square, attracts shoppers from the consumption-oriented middle class. *Photo:* A. Fristad.

A large portion of Russia's society (about 40%) belongs to the "moderately-low-income" and "low-income" categories with about $3,600 to $4,800 available per household per year (or $300–400 per month). The average Russian household consists of 2.7 persons, so these incomes are not too bad, as long as people do not have to rent living space or own a car. The vast majority of residents in the provincial towns and cities, and the wealthier villagers, fit into these two categories. The remaining 40% are "truly poor" people. About one-quarter of these, or 10% of all households, live in abject poverty on $75 per month per household. This is the global poverty level currently defined as $2 per day by the United Nations. Many of the truly poor are recent migrants or refugees from the Asian and Caucasian FSU republics, Moldova, or Ukraine; they include some ethnic minorities, but also many people of Russian descent from the same regions. This category also includes native Russian citizens who are disabled, unemployed, alcoholic/addicted, war veterans, single women, and/or pensioners (the average Russian pension for seniors is about $100 per month).

The wealthiest regions in Russia are the same as the most economically productive or active: The top five by average per capita income in 2007 were the city of Moscow and the Nenets, Yamalo-Nenets, Chukotsky, and Khanty-Mansi Autonomous Okrugs. The lowest were some of the republics of the Northern Caucasus. The income in Moscow was eight and a half times greater than that in the poorest subject of federation, the Ingushetiya Republic.

Poverty and Welfare

We have just learned that about 20% of Russian residents live well or very well, but that the other 80% do not. Poverty in the new Russia takes diverse forms: Many people struggling with chronic illness, unemployment, single parenthood, recent migration, or drug addiction also struggle to make ends meet. A striking image of the early 1990s was that of hungry grandmas who appeared near subway stations or newly opened Western restaurants, gnawing on scraps of food from the trash. Typically pensioners on fixed incomes had lost their life savings in a series of ruble devaluations during the late 1980s. One old relative of mine had 10,000 rubles saved from all the tips she earned while checking coats for over 40 years at the famous Academy Theater in Moscow. In 1988, the amount was sufficient to buy a new car, or even an apartment. Then, without a warning, the value of the ruble fell sharply, while the savings accounts nationwide were frozen on orders from the government. Virtually all of her savings turned worthless in a few months because of inflation, never to be recovered.

A particularly infamous problem of the Yeltsin period also caused many people to see a decline in income: delays in payment of state pensions and salaries to the workers. Newly privatized banks, many of which were run by people close to the Kremlin, were deliberately put in charge of the state payroll accounts. They would deliberately hold on to the money for a few weeks to accrue some interest. With inflation raging in double digits over much of the 1990s, even a few days' delay would considerably lessen workers' purchasing power. The people in charge of the payrolls made millions of dollars from late payments in this manner; some of them are still prominent in Russia today.

The unraveling social net also contributed to the rise of the new poor. In the Soviet Union, there was no unemployment (at least not officially). People were used to guaranteed, lifelong jobs with benefits. There were no temporary job offices, no welfare, and no soup kitchens. People were not used to sleeping in the streets. In fact, if they tried, they'd be promptly picked up by the cops and sent either to a mental clinic for evaluation or to jail, and then to their original homes, if they had any. People lacked the professional skills required to find a new job in a dynamic, Western-style job market—skills such as resumé writing, networking, or business etiquette. Lack of a job or pension in the early reform period meant that many people swelled the ranks of the poor or very poor and spilled into the streets (see Chapter 10).

There were also some benefits of the new situation. The greatest benefit that the state bestowed upon its ordinary citizens was to allow free privatization of the formerly state-owned apartments. This was an underappreciated aspect of Yeltsin's period of reforms (Åslund, 2007). The families that did have apartments on the eve of the reforms were lucky, because the market rate for these quickly skyrocketed. Effectively, a new class of apartment owners, a majority of the population, was created overnight. Those people who did not have apartments or were waiting for apartments from the state were out of luck; they had to pay the new market rate to rent. In addition, some elderly and/or chronically ill people lost their apartments in elaborate con schemes. A signature on the dotted line sometimes meant that instead of obtaining in-home care until death, an elderly pensioner was signing over his or her only piece of real estate to a shadowy company that would come and harass him or her later to give up the apartment altogether. In some cases, seniors simply disappeared without a trace after signing what was in effect their death sentence. Eldar Ryazanov's dark comedy *Old Hags* (2000) provides an insider's look at this type of situation.

A society can be judged the best by how it treats its most vulnerable members: the elderly, the sick, and the children. The earliest versions of this statement are ascribed to Confucius; more recently, it has been attributed to Dostoevsky or Mahatma Gandhi. Like the sick and the elderly, children fared poorly in the new Russia, with orphans hit the hardest (see the discussion on adoptions in Chapter 10, Vignette 10.2). The Soviet orphanages were never very good, but at least they provided a level of stability and shelter. Russia today has some of the worst-run orphanages in the world, as well as one of the highest ratios of child abandonment. Many children also live with a single parent, a grandparent, or even alone. Whereas 63% of all Soviet children lived with both of their parents in 1970, only 54% in Russia do now. The rate of single-parent households has correspondingly risen from 16 to 21.6%. A common scenario is that of an alcoholic parent dumping the child(ren) on the doorstep of a state orphanage or having parental rights revoked after a particularly egregious case of repeated child abuse. Many real-life stories are too painful to be presented here, but a quick search online are cause for true concern.

The fate of children cannot be separated from that of their parents, especially their mothers. Much has been written about gender relations and issues in post-Soviet societies (e.g., Engel, 2004; Hughes, 2005). The women in these societies have experienced a disproportionate rise in inequality based on ageism, sexism, and related prejudices. For example, it is very common now to see female workers recruited for boutique jobs on the basis of their height or age; typically, only taller women under age 30 are encouraged to apply. In the Soviet system, the vast majority of women worked full-time. With perestroika, some were forced to scale back their hours by their employers, whereas others were simply terminated. Yet others were attracted to wealthier, older men who promised them carefree lives as housewives, only to be abandoned later. Today women in Russia receive fewer maternity benefits than under Soviet rule and are accustomed to lower-level office positions, typically with sexist male bosses. A small number of women have nevertheless succeeded as business leaders, but these remain a small minority in the generally male-dominated world of commerce. Fewer than 10% of Russia's legislature members are female. Among top state or business managers, men outnumber women by a 3:2 ratio. At the same time, 79% of women are employed in the FSU overall—a higher percentage than in Europe (72%) or Latin America (65%).

Women remain the primary providers of services at home, essentially working two jobs. Russian men play with children less than American or Western European fathers do (fewer than 30 minutes per day on average). They also cook or wash dishes less commonly than their Western counterparts do. At the same time, they expect their spouses to remain physically attractive, fit, slim, and so on without necessarily living up to that ideal themselves, as is amply demonstrated by observations of Russian women and men in public places. Of course, many men in the region do not fit this uncaring stereotype, but unfortunately many others do. The number of single mothers in Russia today, therefore, does not tell the whole story. The existing, but uninvolved, partners are partly to blame for the heavy burdens imposed on women.

Which regions are the hardest hit with social ills in Russia? One way to find this out is to look at the available unemployment statistics, bearing in mind that underreporting is commonplace. In 1998, as described in Chapter 8, the post-Soviet economy hit bottom. The highest estimated unemployment rates for that year were found in some of the national republics of the south (e.g., Kalmykia, 31%; Dagestan, 30%; Karachaevo–Cherkessiya, 25%). There were no numbers available for Chechnya, but neighboring Ingushetia had 50% unemployment. Some Siberian republics had slightly lower rates (Buryatia, 21%; Tyva, 20%; Altay, 18.5%). Most Russian regions had rates close to 13%; the lowest unemployment was in Moscow (4.8%) and the Central district (8–10%). The general situation has since improved somewhat. In 2007 the official national unemployment rate was 5.6%, although independent social research suggests that the actual unemployment was two or three times higher than that.

One important statistic with respect to post-Soviet unemployment is that men are about one-third more likely to be unemployed than women, although women are typically laid off first. Why may this be the case? One factor is massive layoffs among male-dominated professions, such as factory workers or military officers. A lot of men in these professions were abruptly laid off in 1993–1994, as the economic reforms got fully under way, and have not been rehired since. Another factor is that women have remained in the jobs that men have left in their search for other opportunities. Also, many women were and are underemployed, but they do not enter official unemployment counts because they remain in the part-time workforce or do not officially declare unemployment. Men typically work full-time and, if they are laid off, search only for full-time jobs. An important coping mechanism for households is simultaneous participation in multiple economies, both formal and informal (Pavlovskaya, 2004). The proportion of families with diverse sources of income has been rising, at least in the cities, in the post-Soviet period. Many families are able to survive by working multiple part-time jobs, or by receiving the direct benefits of informal services from relatives and friends.

Education is no guarantee of employment in the new Russia: About one-third of Russia's unemployed are people with 2-year technical college degrees, and about 10% have university diplomas. Generally, however, educated people have fared better in banking, finance, information technology, and management, while less educated ones have failed to make the required adjustments. Their two most obvious deficiencies are in foreign-language skills and computer skills. Very few of those who were over 40 in 1991 had managed to master either skill set, but those who did fared considerably better than their peers. For example, a sociological study from 2003 (Gorshkov & Tikhonova, 2004) found that 31% of the Russian upper class had professional computer skills, versus only 4% among those with lower incomes. Likewise, 7.5% of the rich had a foreign-language skill (even this is a low figure by European standards), versus only 3% among the poor. On the other hand, some of the most gainfully employed people or new businessmen came into the field with only minimal education—some after being released from Soviet prisons, using their underworld backgrounds as an advantage. Nevertheless, the richest in modern Russia are also very well educated, typically with PhDs (or the equivalent) or advanced engineering degrees. For the young generation, having a university diploma is a must, although the few top programs are the most competitive and the hardest to get into.

Crime and Punishment

Sharp differences in the levels of income in a society can be dangerous. If the most basic needs of the poor are not met, a revolt may happen; this is, of course, the basic tenet of Leninist ideology. Up to a point, some inequality is inevitable, and it can even be stimulating when there are opportunities available to improve one's life. However, excessive inequality and a high degree of perceived injustice will bring about social unrest. Gorshkov and Tikhonova (2004) report that the new poor have a high degree of awareness about the new rich. The most often cited items that the latter have but the former don't include better living conditions (such as a large modern apartment or a new suburban home), opportunities to travel abroad, the ability to buy new furniture and major home appliances, better health care, and better education for children. In a society where inequality is prominent, it is small wonder that the poorer people respond with jealousy and sometimes with criminal behavior. At the same time, the rich are more sheltered from everyday violence than the poor, and so with the increasing crime rates, the most typical victims are the poor themselves. Generally, crime rates in Russia increased sharply soon after the start of the economic reforms, but have abated a bit since Putin came into power. Some of the factors that led to the increase were as follows:

- The rapid release of many thousands of criminals under Gorbachev to save money.
- A sharp drop in government spending on law enforcement.
- Criminalization and corruption of the police.
- Gang wars erupting over privatization of the most lucrative state assets, such as metal smelters and oil refineries.
- The appearance of private security contractors who would replace state law enforcement and sometimes clash violently with the latter.
- The initial appearance of rich people without spatial segregation from the poor.
- The appearance of expensive shops and restaurants in prominent locations.
- The emergence of casinos, brothels, escort services, drug-dealing networks, and other trappings of Western decadence that were previously either nonexistent or only secretly run.
- The appearance of large numbers of clueless Western tourists and business travelers, eager to spend money and easily conned.
- Yeltsin's revisions to the Soviet penal code and suspension of the death penalty, so that Russia could join the Council of Europe.
- The federal government's general inability to control the situation in the regions, and its noninterference in the doings of the regional elites.

Just how bad the crime situation in Russia is today is a matter of much debate. In the late 1990s, Russia had the 5th highest rate of murders per capita among 62 countries tracked by the U.N. Survey of Crime Trends—behind countries such as Colombia, South Africa, Jamaica, or Venezuela, but ahead of Mexico, Zimbabwe, and all other FSU republics. (Among the latter, interestingly, the prosperous and democratic Baltic states led the pack—not Ukraine or Kyrgyzstan, for example. The latter nations may have underreported serious crimes, however.) The U.S. rate of 0.04/1,000 was only 20% of Russia's rate of 0.2/1,000 per year. This data set, however, did not include any countries in central Africa or in much of Asia and South America. In robberies, on the other hand, Russia's rate was only 66% of the U.K., U.S., or Mexican levels, and was more similar to that of Germany or Canada. The United States also topped the charts with a total of 23 million recorded crimes per year; Russia was only in 6th place worldwide, with a total just short of 6 million. This, by the way, does not automatically mean that the situation in the United States is that much worse: The difference can be explained in part by better U.S. law enforcement statistics (and of course by the much bigger U.S. population).

Russia had over 1 million prisoners in 1995, and about 872,000 10 years later. Seven percent of the inmates in 2005 were women, and about 17% were repeat offenders. Based on crime rates alone, your chance of being killed or mugged in Moscow is about as high as in New York, but higher than in most European capitals. There are lots of perfectly safe neighborhoods; however, there are also late-night train rides and walks through dark areas of the city periphery where safety is questionable. One positive development recently has been a drop in the most violent crimes (this is also true in the West). For example, there were 31,800 murders and attempted murders in Russia in 2000, versus only 22,200 in 2007. The majority of contract killings were perpetrated by the mob against prominent businessmen and journalists in the mid-1990s (Volkov, 1999); such attacks are now rare. Most domestic homicides happen between spouses and involve alcohol. Random drive-by shootings, bomb explosions, and so on are mercifully very rare, although widely publicized. Another common target is anyone who is perceived as different, especially migrants from the Caucasus and some categories of people of color. However, the vast majority of Russians are friendly people, and outside a few big cities you are unlikely to experience much trouble. This is not to say that you should not remain alert at all times, of course.

Besides personal crime, there is economic crime: tax fraud, embezzlements, bribery, and government corruption of all kinds. The Transparency International organization's global Corruption Perception Index for 2007 ranked Russia very much near the bottom, in 143rd place out of 179 countries—right above Togo, but below Indonesia. For comparison, the worst three were Somalia, Myanmar, and Iraq, while the best three were Denmark, Finland, and New Zealand; the United States ranked only 20th, and the United Kingdom was in 12th place. There are many reasons for widespread corruption in Russia, including but not limited to these:

- A historical tradition of pervasive government corruption; as a result, people find such corruption acceptable or even inevitable (see Gogol's or Saltykov-Shchedrin's satirical descriptions of 19th-century Russian bureaucrats).
- The low pay scale for some categories of government workers relative to businessmen, which encourages these workers to demand bribes.
- The regulations requiring numerous government permits to do almost anything.
- Civil servants' enjoyment of their power to make or break deals and to make money.
- Deliberately confusing and frequently changing laws.
- A lack of independent, objective courts or arbitration and a lack of transparency in general.
- The sheer scale of the country: Distant places are less open to the scrutiny of the central government, while a scarcity of local resources encourages profiting on the side.
- The abundance of resources, natural or economic, which actually encourages corruption.
- Inability and/or unwillingness to enforce existing anticorruption laws.

Some forms of economic crimes can be arbitrarily prosecuted, as a means of reprisal against politically inconvenient businesses. For example, the Yukos affair of 2003–2005 (clearly perpetrated by the Russian courts on behalf of Putin's government) bankrupted the richest and one of the most transparent companies to benefit a small circle of Kremlin insiders, and imprisoned the top executives who had become too independent-minded (Baker & Glasser, 2005; Åslund, 2007).

What part of Russia has the highest crime rates (Figure 12.5)? There is no apparent pattern. Some regions with the highest rates are in the distant coastal Pacific (Magadan, Jewish Autonomous Okrug, Sakhalin), as well as in the Urals (Kurgan, Perm) and Siberia (Buryatia, Tomsk). Many penal colonies and long-term reformato-

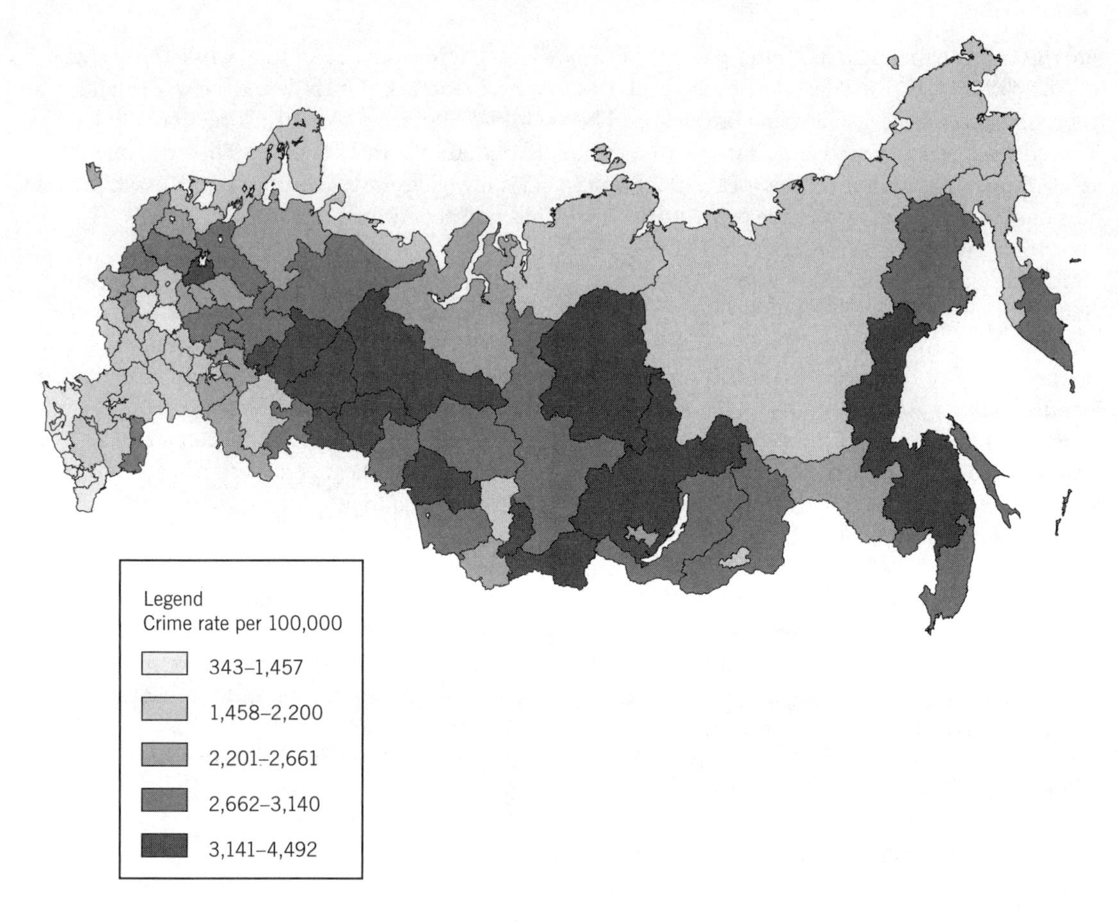

FIGURE 12.5. Crime rates in Russia, 2005. Data from the Federal Service of State Statistics, Russian Federation.

ries are located in remote corners of Russia—for example, the Krasnokamensk facility on the border with Mongolia, where Yukos chairman Mikhail Khodorkovsky was sent. Northern European Russia and the Urals have many old prisons, which are still functioning with few changes since Soviet times (Figure 12.6).

Health, Income Distribution, and Crime in Other FSU Republics

Health, wealth, and crime trends in the other FSU republics have been comparable to Russia's in many ways, with some regional differentiation (Table 12.1). The richest economies of the Baltics tend to have lower poverty, higher wealth, and more equitable income distribution; at the same time, the Baltics have relatively high crime rates—higher total rates than Russia's, in fact. The Central Asian states, of course, have high poverty, low wealth, and less equitable income distribution. Only 2 of the 15 FSU republics besides Russia (Ukraine and Kazakhstan) have resident billionaires; however, many very wealthy Armenians, Georgians, and Azerbaijanis undoubtedly exist, with many living comfortably abroad in Russia, Europe, or the United States. In fact, among the crop of billionaires on the 2007 *Forbes* list were one Belarusian, two Georgians, three Armenians, four Azerbaijanis, and six Ukrainians—all residing in Russia. The poorest countries of the FSU, Moldova and Tajikistan, have surprisingly low crime rates; perhaps this is due to underreporting, but it also may reflect the possibility that since people's means are truly

FIGURE 12.6. A still-active prison camp in northern European Russia, not far from a major railroad. *Photo:* Author.

TABLE 12.1. Selected Social Statistics for the FSU Countries and Some Comparison Countries

Country	CPI	Billionaires	Crime per capita	Physicians/1,000
Russia	143	50	21	4.25
Belarus	150	0	13	4.55
Ukraine	118	7	12	2.95
Moldova	111	0	9	2.64
Georgia	79	0	3	4.09
Armenia	99	0	4	3.59
Azerbaijan	150	0	2	3.55
Kazakhstan	150	7	NA	3.54
Kyrgyzstan	150	0	8	2.51
Uzbekistan	175	0	NA	2.74
Turkmenistan	162	0	NA	4.18
Tajikistan	150	0	NA	2.03
Estonia	28	0	43	4.48
Latvia	51	0	22	3.01
Lithuania	51	0	23	3.97
United States	20	410	80	2.56
United Kingdom	12	41	86	2.30
Japan	17	25	19	1.98
China	72	19	NA	1.06
India	72	33	2	0.60
South Africa	43	2	77	0.77
Colombia	68	2	5	1.35
Mexico	72	10	13	1.98

Note. CPI, Corruption Perception Index (rank worldwide; the higher the number, the worse the corruption in the country; Transparency International data). Data on billionaires from *Forbes* list by country of residence (2007). Data for China do not include Hong Kong. Crime data from *nationmaster.com*; data on physicians from the World Health Organization Statistical Information System (2007). NA, data not available.

modest, there is less incentive for crime. Some of the most corrupt countries in the world are Uzbekistan (175th) on the Transparency International list, and Turkmenistan (162nd). The least corrupt nation in the FSU is Estonia (28th), but it is still below most Western European countries or the United States.

REVIEW QUESTIONS

1. What health issues plague Russia and other FSU republics?
2. What is the general distribution of income in the new Russia? How is it different from that in your country?
3. What were the main sources of wealth for the Russian billionaires on the 2007 *Forbes* magazine list?
4. What generation in Russia today has seen the greatest polarization of incomes? Is this a problem? Why or why not?
5. What Russian crime statistics surprise you? Why?

EXERCISES

1. Study the current *Forbes* magazine list of the richest Russians. What industries (sources of wealth) are most frequently represented on the list?
2. Have a class discussion about the merits of the recent proposal to compensate people who lost savings in the 1980s ruble devaluation by paying them compensation out of the state stabilization fund, which accumulates petroleum and gas taxes in foreign accounts.
3. Create a tourist brochure outlining the health and crime risks involved in visiting Moscow or another large city in the FSU. Do additional research to make sure you have some reliable statistics to back up your claims.
4. What would your parents say if you decided to go to Russia on a class trip? Ask them.

Further Reading

Åslund, A. (2007). *Russia's capitalist revolution: Why market reform succeeded and democracy failed.* Washington, DC: Peterson Institute for International Economics.

Baker, P., & Glasser, S. (2005). *The Kremlin rising: Vladimir Putin's Russia and the end of revolution.* New York: Scribner.

Berezin, I. S. (2002). *Distribution of incomes in the population of Russia in 2001* [in Russian]. Retrieved October 1, 2008, from *www.marketing,spb.ru/mr/social/income-01.htm*

Engel, B. A. (2004). *Women in Russia, 1700–2000.* New York: Cambridge University Press.

Gorshkov, M. K., & Tikhonova, N. Y. (2004). *The new social reality of Russia: The rich, the poor, the middle class* [in Russian]. Moscow: Nauka.

Haavio-Manilla, E., Rotkirch, A., & Kontula, O. (2005). Contradictory trends in sexual life in St. Petersburg, Estonia, and Finland. In A. Štulhofer & T. Sandfort (Eds.), *Sexuality and gender in postcommunist Eastern Europe and Russia* (pp. 317–363). Binghamton, NY: Haworth Press.

Hughes, D. M. (2005). Supplying women for the sex industry: Trafficking from the Russian Federation. In A. Štulhofer & T. Sandfort (Eds.), *Sexuality and gender in postcommunist Eastern Europe and Russia* (pp. 209–230). Binghamton, NY: Haworth Press.

Joint United Nations Programme on AIDS. (2006). Global AIDS report 2006. Retrieved March 12, 2008, from *www.unaids.ru/en/HIV_data*

Pavlovskaya, M. (2004). Other transitions: Multiple economies of Moscow households in the 1990s. *Annals of the Association of American Geographers, 94*(2), 329–351.

Vishnevsky, A. (Ed.). (2006). *Naselenie Rossii.* Moscow: Nauka.

Volkov, V. (1999). Violent entrepreneurship in post-Communist Russia. *Europe–Asia Studies, 51*, 741–754.

Websites

www.prb.org—Population Reference Bureau; publishes the annual World Population Data Sheet.

www.who.int—World Health Organization; has a variety of health statistics by country.

www.forbes.com—*Forbes* magazine; keeps track of the world's billionaires.

www.unaids.ru—The Joint United Nations Programme on AIDS, dealing with the AIDS epidemic worldwide.

www.transparency.org—Transparency International; tracks corruption worldwide.

www.cia.gov/library/publications/the-world-Factbook—The CIA World Factbook.

CHAPTER 13

Cultures and Languages

"Culture" is an elusive concept that is hard to define. A good working definition is "a shared set of meanings that are lived through the material and symbolic practices of everyday life" (Knox & Marston, 2007, p. 29). Culture is learned primarily in early childhood, but also throughout one's life. It includes nonmaterial items, such as language and beliefs; material objects called "artifacts," such as clothing and housing; and everyday practices, such as shopping and commuting. Because culture is learned early in life, it is resistant to change. Cultural geographers study cultures from a spatial perspective: core areas of cultures; cultural realms and their boundaries; diffusion of major cultural traits; the production and transformation of cultural landscapes; and cultural adaptations to the environment.

Many regions in the world are defined on the basis of their dominant culture(s). For example, north Africa and the Middle East are defined by the prevalence of Islam; Latin America is defined by Spanish and Portuguese colonial influences; and so on. Is there a unifying cultural trait that would apply to our region of study, Northern Eurasia/the former Soviet Union (FSU)? Is it the Orthodox religion? Communist ideology? Adaptations to the cold climate? The defining trait that works best today, in my view, is the presence of the Russian language throughout this region. Although it is home to over 200 ethnic groups, all of these groups had to communicate with each other in Russian during Soviet times. Many ethnic Russians continue to live throughout the region, providing a common cultural milieu. Thus this chapter focuses primarily on languages. Chapter 14 then discusses some other important cultural elements: religion, diet, and dress. In addition, examples of cultural landscapes from the region are provided throughout this book.

Languages of Northern Eurasia/ the FSU

East meets West in Northern Eurasia, and consequently the languages spoken and understood there are either Western (Slavic or Baltic, which are branches of the Indo-European family) or Eastern (branches of the Altaic and Uralic families). Some languages are spoken by millions; others by just a handful of speakers. The most common by far is of course the Russian language, spoken and understood today by at least 255 million people worldwide. Russian is the 8th most common native and 6th most common overall language in the world (Table 13.1). In the United States at any given moment, there are about 25,000 learners of Russian in more than 400 universities and

TABLE 13.1. The Top 10 Languages in the World

Language	Spoken as native language (millions)	Spoken total (millions)	Language family
Mandarin Chinese	873	1,051	Sino-Tibetan
Spanish	322	500	Indo-European
English	309	1,100	Indo-European
Arabic	206	323	Semitic
Hindi	181	948	Indo-European
Portuguese	178	223	Indo-European
Bengali	171	196	Indo-European
Russian	145	255	Indo-European
Japanese	122	130	Japonic (Altaic)[a]
German	95	170	Indo-European

Note. Data from Gordon (2005).

[a]Some linguists consider Japanese a language isolate in its own (Japonic) family; others place it within a broadly defined Altaic family.

colleges, and about 3 million people who can speak and understand it. Although the popularity of the Russian language in the United States has declined since the end of the Cold War, it remains one of the top 10 most studied foreign languages in America.

Most of the languages of Northern Eurasia belong to three language families: Indo-European, Altaic, and Uralic (Figure 13.1). The **Indo-European** languages are spoken by the majority of people in Latvia, Lithuania, Belarus, Ukraine, Moldova, Armenia, Tajikistan, and Russia. (Again, because Russian was the official language of the Soviet Union, it is widely spoken in all FSU republics.) The **Altaic** languages, particularly the Turkic group, are common in Russia along the Volga and in Siberia, in Azerbaijan, and in four of the five Central Asian states. The **Uralic** languages are spoken in the northern and eastern parts of European Russia and in western Siberia. Some unique languages in the Caucasus and in northeastern Siberia belong to the **Caucasian** and **Paleoasiatic** families, respectively.

Indo-European Languages

Because you are reading this textbook in English, you already know at least one Indo-European language. Of the top 10 languages in the world (Table 13.1), 7 are Indo-European. Their origin can be traced to a single mother tongue spoken somewhere in the middle of Eurasia more than 4,000 years ago. Some scholars believe that it originated among the people of the steppes northeast of the Black Sea, along the Don River and the lower Volga. These were people of the mysterious culture that left us distinct burial mounds, called "kurgans." Other scholars (e.g., most American researchers) suggest Asia Minor as another possible center of origin. It is certain that the languages we call Indo-European today spread as far west as northern Europe and the British Isles, and as far east and south as central and eastern India, by 3,000 years ago. The Indo-Europeans were efficient colonizers. Evidently they were farmers and animal breeders with superior technology, which allowed them to wipe out or absorb whatever local populations they encountered very quickly. For example, in all of Europe today, the only native language that has survived from the times before the Indo-European conquest is Basque. Hungarian, Finnish, and Estonian belong to the Uralic family; however, these languages are more recent arrivals in Europe. Among the most significant technologies of the Indo-Europeans is probably the domestication of the earliest type of horse around 4000 B.C., somewhere in what is Ukraine today. Worship of the sun was another distinctive cultural characteristic of these people (Figure 13.2).

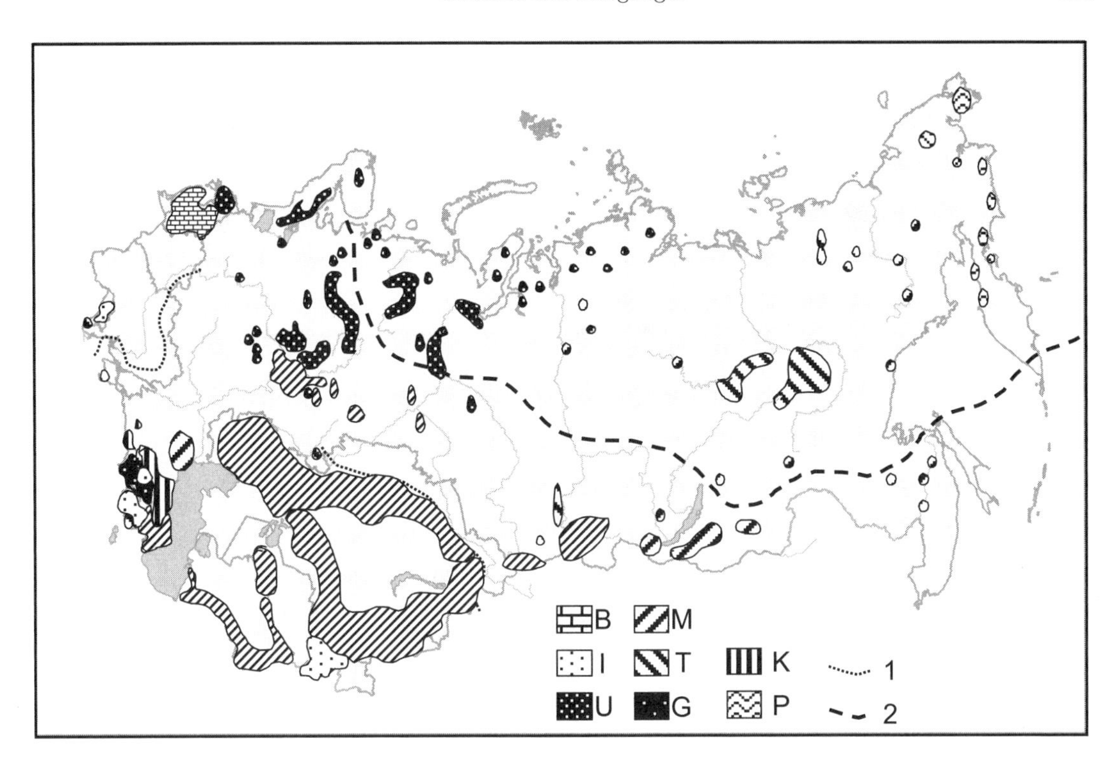

FIGURE 13.1. Main language groups of Northern Eurasia. B, Baltic languages; I, other Indo-European (Moldavian, Ossetian, Armenian, Tajik); U, Uralic (Estonian, Karelian, Komi, Mordvinian, Mari); M, Mongolian-Tunguss (Buryat, Kalmyk, Even, Evenk, Nanay, Udege); T, Turkic (Kazakh, Uzbek, Turkmen, Azeri, Tatar, Bashkir, Chuvash); G, Georgian; K, other Caucasian (Circassian, Chechen, Ingush, and Dagestani); P, Paleoasiatic (Chukchi and Koryak). Russian, Ukrainian, and Belarusian are spoken within the borders of their respective countries. Line 1 shows the western limit of a majority of the population speaking Russian in Ukraine, and the southern limit in Kazakhstan. Line 2 shows the northern limit of contiguous settlements speaking Russian.

All Indo-European languages share word roots and grammatical features that are thought to have been retained from the ancestral tongue—so-called shared retentions. For example, just to use words starting with the letter "B," the words for "bean," "bee," "brown," "brother," "bottom," and "birch" sound very similar in most of the Indo-European languages. The words that are common tend to describe basic concepts—divinities, family members, numbers, heavenly bodies, materials, minerals, colors, plants, and animals—thus attesting to the protolanguage's antiquity. As an example, the English word for "snow" is *sneha* in Sanskrit (a now-extinct language of the Indian group), *snaēža* in Avestan (Iranian), *sniegas* in Lithuanian (Baltic), *sneg* in Russian (Slavic), *nipha* in Greek (the initial *s* disappeared), *snjór* in Old Norse (Germanic), and *nyf* in Welsh (Celtic).

The sacred language of the Hindu Vedas scriptures, Sanskrit, is thought to be one of the earliest languages and perhaps the closest to the hypothetical mother tongue of the Indo-Europeans that we know. Lithuanians claim that theirs is the most similar to that language in Europe today. In terms of grammar, Indo-European languages tend to have gendered nouns (male, female, or neutral), with English being a major exception. They also tend to have cases for nouns (flexes); that is, the ending of the noun changes, depending on its position and function in a sentence. In English it only survives in some pronouns—as, for example, in "*He* came to them," but "They came to *him.*" In most Indo-European languages,

FIGURE 13.2. Sun symbols from many world cultures are on display at the Museum of the Sun in Novosibirsk. In Eurasia, worship of the sun was particularly important for the pre-Christian Eastern Slavs, the pre-Muslim Kazakhs, the Mongols, and the Altay people. *Photo:* Author.

including Russian, flexing is very complex: Endings can change along three or more patterns, each grouped into six cases! A foreign learner is thus forced to memorize up to 18 possible combinations of endings for every noun. Verb systems also tend to be complex, with suffixes or supplemental verbs indicating changes in tense, gender, and number. Adjectives tend to agree with nouns in gender and number, and this means that they also change their endings.

The majority of the existing 70 Indo-European languages use the Latin alphabet. Some Slavic languages use the Cyrillic alphabet (Russian, Ukrainian, Belarusian, Bulgarian, Serbian), while others use their own alphabets (Greek, Armenian, Hindi) or Arabic script (Farsi, Tajik). In addition, Cyrillic continues to be used in Kazakh, Kyrgyz, Tatar, Bashkir, Chuvash, and all Uralic languages within Russia. It was also used during the Soviet period in the Azeri, Uzbek, Turkmen, and Moldovan languages, which have since been Latinized. The Tajik language went from Cyrillic back to Arabic script. While a different alphabet may present an initial difficulty for the foreign-language learner, commonality in grammar and words makes studying an Indo-European language a much simpler task for an English speaker than mastering some non-Indo-European languages of Africa, Asia, or the Americas.

In Northern Eurasia, the following nine languages belong to the Indo-European family, with the approximate number (in millions) of native speakers given in parentheses:

- Slavic: Russian (145), Ukrainian (40), and Belarusian (9).
- Baltic: Latvian (2.5) and Lithuanian (4).
- Romance: Moldovan (a dialect of Romanian; 3).
- Armenian (6).
- Iranian: Tajik (5) and Ossetian (<1).

The Russian Language

The Russian language belongs to the Eastern Slavic group of the Slavic branch, along with Ukrainian and Belarusian. All three are very similar, mutually intelligible languages sharing over 80% of their basic vocabulary and the Cyrillic alphabet. Before the time of the Tatar–Mongol Yoke (the 13th–15th centuries; see Chapter 6), the three languages were practically one. Until the late 18th century, they could be thought of as dialects of the same broad language, with regionalisms that were borrowings from surrounding nations. For example, Russian has many words borrowed from the Tatar language; Ukrainian has words borrowed from, or shared with, Polish. The spoken languages were formalized as written literature evolved in the 19th century. The works of Alexander Pushkin (1799–1837) are frequently cited as the first literature written in a truly contemporary Russian, not in the pompous Latinized prose used before him (e.g., by the poets Derzhavin and Zhukovsky). In modern Belarus and Ukraine, virtually everyone still understands Russian with little trouble—a legacy of the Soviet period. Moreover, over 40% of Ukraine's population and over 60% of people in Belarus still speak Russian as their first language, even though they may consider themselves Ukrainians or Belarusians, respectively.

The Russian alphabet has 33 characters (Figure 13.3), many of which were borrowed from Greek. It is a good idea to learn how to read Russian if you intend to travel in or do research on the FSU. In fact, once the characters are mastered, reading Russian or Ukrainian does not present as much

Аа	"Ah" as in "cAr"	Ии	"I" as in "sEE"	Сс	"S" as in "See"	Ъъ	Hard stop
Бб	"Be" as in "Bill"	Йй	"Y" as in "boY"	Тт	"T" as in "Tip"	Ыы	"y" as in "Phyl"
Вв	"Ve" as in "Vine"	Кк	"K" as in "Kitten"	Уу	"Oo" as in "bOOt"	Ьь	Soft stop
Гг	"Ge" as in "Get"	Лл	"L" as in "Lamp"	Фф	"F" as in "Face"	Ээ	"E" as in "mEt"
Дд	"De" as in "Do"	Мм	"M" as in "Me"	Хх	"Kh" as in "loCH"	Юю	"Yu" as in "dUke"
Ее	"Ye" as in "Yet"	Нн	"N" as in "Nose"	Цц	"Ts" as in "piZZa"	Яя	"Ya" as in "Yard"
Ёё	"Yo" as in "Yoke"	Оо	"O" as in "pOt"	Чч	"Ch" as in "CHip"		
Жж	"Zh" as in "pleaSure"	Пп	"P" as in "Pot"	Шш	"Sh" as in "SHop"		
Зз	"Z" as in "Zebra"	Рр	"R" trilled (like Scottish "R")	Щщ	"Shch"— "sh" followed by "ch," merged in one sound		

FIGURE 13.3. The Russian alphabet. Only approximate pronunciations are given. Hard and soft stops are not pronounced, but modify the letters that precede or follow them.

difficulty for an English speaker as, for example, French does; there are few silent letters and few exceptions to the reading rules in Slavic languages. In this sense, Russian is more phonetic, like Spanish, with almost a one-to-one correspondence between sounds and letters. Some letters were specifically invented for the language; therefore, Russian uses few diacritical marks or combinations of letters to represent sounds. Instead of a "ts," there is a separate letter, ц; instead of "sh," ш; and so on. Russian vowels are not divided into long or short. Russian consonants, however, may be either hard- or soft-palatalized. In transliterated Russian place names, the palatalized form is frequently rendered as an apostrophe—"the Ob' River." The "b' " at the end is not a "B," but rather a "Bee" sound with an extremely short "ee." I have chosen not to use the apostrophe in this book in such cases.

The difficulty of learning Russian lies not in its alphabet, but in its complex grammar. Although its verb *system* is simpler than that of English or French (with fewer tenses), the language has many verb *forms*, both perfect and imperfect, in three tenses. Even worse, there are three declensions with six cases for nouns (Latin has five declensions and may be an even harder language to learn), and nouns belong to one of three genders. Thus one needs to memorize lots of noun endings to be able to speak correctly. For a native speaker of English, this is very cumbersome, because no exact equivalents of these endings exist in English. On the other hand, Russian learners of English struggle with using the articles ("a," "an," and "the"), which simply do not exist in Russian. Also, Russian is phonetically close to some European languages (Italian, Spanish), but is very different from English: Any Hollywood movie depicting Russian mobsters makes this "pRetty cleaR" (with a rolling "R"). Italians who come to Russia have a very slight accent in their Russian, but all Russians who come to North America are instantly recognizable by their strong accents in English.

The Russian language has produced one of the greatest literatures in the world, with many excellent writers and poets. This literature continues to generate interest in the Russian language worldwide (Vignette 13.1). In addition, the Russian language emerged as one of the languages of international communication in space exploration, chess, several scientific fields, and politics in the second half of the 20th century, because of the Soviet achievements in those fields. Russian is one of only six official working languages at the United Nations, along with English, French, Spanish, Chinese, and Arabic. Although Russia's political clout has greatly weakened since the end of the Cold War, it remains a strategic language for U.S. intelligence, along with Chinese, Persian, Arabic, and Hindi. George Weber (1997) has listed Russian as the fourth most influential language in the world after English, French, and Spanish, and just ahead of Arabic, Chinese, and German. His system rates languages according to the number of native speakers, secondary speakers, number of countries where the language is used, number of areas of human activity in which the language is used, the main country's economic power, and social/literary prestige.

As in English, there are dialects in the Russian language. Most of the country now speaks a fairly uniform language, as can be heard on Russian TV. The greatest linguistic diversity within Russian is found in the rural areas in the old European part. The southern dialects have a soft "Gh" sound, commonly heard in the 1990s from Mikhail Gorbachev, who grew up in the Stavropol region. The northerners say the hard "G" instead. The Ivanovo and Vologda regions of Russia are notorious for their unstressed "O," pronounced as "O," not "Ah." The Ryazan region turns "Ahs" into "Yahs." Moscovites tend to overemphasize "A" at the expense of "O" in unstressed positions—something that people from "B(ah)ston" do in the United States.

Before the advent of TV and jet travel, the vocabulary varied tremendously from region to region. Vladimir Dal composed a famous dictionary of the Russian language in the late 1800s, with 250,000 words in four volumes. Much of the dictionary consisted of regionalisms—words that were only used in a few places and are now lost. Only 20,000–30,000 of those words are probably used today. For instance, *bulka* in Moscow means "sweet roll," but in St. Petersburg it denotes "white bread" as opposed to "rye bread."

Vignette 13.1. Twenty-Five Russian Authors You May Wish to Read

All of the titles listed below are available in English. Some exist in numerous translations, although some translations are better than others. Each of the authors listed here wrote more than just the titles listed, but I have picked my particular favorites. The authors are listed in chronological order. I have only picked some of the most famous writers; there are many other great authors to consider.

1. Alexander Pushkin: *Belkin Tales*, lots of poems.
2. Mikhail Lermontov: *A Hero of Our Time*, lots of poems.
3. Nikolai Gogol: *Taras Bulba, Dead Souls, Evenings in Dikanka.*
4. Ivan Turgenev: *Fathers and Sons* and *Nobleman's Nest.*
5. Nikolai Leskov: Many short stories.
6. Fyodor Dostoevsky: *The Brothers Karamazov* and *The Possessed* (or *Demons*).
7. Leo Tolstoy: *Anna Karenina.*
8. Anton Chekhov: Many short stories and plays.
9. Marina Tsvetaeva: *Collected Poems* and some great short prose pieces.
10. Evgeny Zamyatin: *We.*
11. Alexander Grin: *Crimson Sails* and short stories.
12. Anna Akhmatova: *Collected Poems.*
13. Alexander Solzhenitsyn: *One Day in the Life of Ivan Denisovich* and *Cancer Ward.*
14. Mikhail Bulgakov: *Master and Margarita.*
15. Andrei Platonov: *The Foundation Pit* and *Epiphansky Locks.*
16. Viktor Nekrasov: *Front-Line Stalingrad.*
17. Boris Pasternak: *Doctor Zhivago* and many poems.
18. Anatoly Rybakov: *Arbat's Children.*
19. Vasily Aksenov: *Island Crimea* and *Burn.*
20. Sasha Sokolov: *A School for Fools.*
21. Sergei Dovlatov: Short stories.
22. Victor Pelevin: *The Life of Insects, Omon Ra.*
23. Lyudmila Ulitskaya: Any novel.
24. Tatiana Tolstaya: *Kys* and many short stories.
25. Viktoria Tokareva: Any novel.

Slang and Modern Influences in Russian

Russian street slang is extensive. The use of expletives is unfortunately very common; they can be heard everywhere—in the streets, on TV, and even in the Duma's parliamentary proceedings! Much of the offensive vocabulary comes from the prison slang perpetrated in the Soviet GULAG. Not only the patently taboo words (such as *mat*, derived from the word for "mother"), but even common, slightly off-color phrases like *mochit v sortire* ("drown [terrorists] in a toilet," popularized by President Putin during the second Chechen campaign), come straight from the street language. About 10 words in Russian are customarily "bleeped out" on TV; 4 of them are considered particularly bad. Their origin is obscure. A popular legend suggests that the words were borrowed from the Tatar invaders in the 13th–15th centuries, though no serious scholar has ever supported such a claim. A more realistic idea is that such originally meaningful Russian words describing human parts and relations were used in sacred spells during the spring seeding season, and later became the swear words in low usage.

Thousands of Russian words are obsolete. Alexander Solzhenitsyn made a heroic effort to bring some of these back from oblivion in his *Dictionary of Extended Russian Language*, with over 5,000 entries. His own prose was greatly enriched by these, but in the modern age of rapid online communications the trend is toward fewer and simpler words. Of course, there are hundreds of new words in Russian that have been borrowed straight from English, especially in the computer and business worlds. Some of these words (*scanner, Internet, flashka*) did not exist before, and their borrowed use is justified. Some perfectly reasonable old Russian terms, however, have been jettisoned in favor of the trendy new ones; for example, no one in the business community in Russia today would want to work in a *kontora* (itself a foreign word based on "counter"), but rather in an *office*.

The Russian Language Abroad

Russian is a global language. In the Russian Federation, 20% of the population belong to 1 of 182 non-Russian ethnicities, but all of these people are semifluent to fluent in Russian as well. In addition, Russian is spoken by most people in the other FSU republics who grew up during the Soviet period, and by Russians who live abroad. In the FSU today, the lowest proportions of non-native Russian speakers are observed in the Baltics, where the young generation has been lured to study English since independence was achieved. However, even there more than half of non-Russian adults are able to communicate in Russian, although not everyone would admit it. It is easier to use Russian in Latvia and Lithuania than in Estonia. About 30% of Estonia's population is Russian, but most ethnic Estonians would much prefer to use English to communicate with foreigners. Paradoxically, the young people in Estonia now are actually *more* likely to study Russian than their peers in the other republics. Nevertheless, older people will definitely have a higher degree of proficiency in it than the generation born after 1991.

In Ukraine, Belarus, and Moldova, virtually everyone can still understand Russian, but there are an estimated 8 million Ukrainians who are not able to speak Russian. In Armenia, Azerbaijan, and Central Asia, most people over the age of 20 can speak some Russian; in Kazakhstan, some urban Kazakhs will even speak Russian better than the Kazakh language. In Georgia, however, proficiency in Russian is rapidly dropping. A recent study by the Russian Academy of Sciences (Berkutova et al., 2007) estimated that about 1 million Georgians today (20%) have no proficiency in Russian. Most of these grew up in the post-Soviet period.

Russian language and culture extend around the globe. Four waves of Russian emigration have scattered millions of Russian speakers abroad. The first wave began before World War I and lasted until after the Revolution of 1917; the second took place after World War II; the third consisted of Jewish and other religious groups' emigration from the Soviet Union in the 1970s; and the fourth has consisted of professional emigration since the fall of the Soviet Union. The first was the most extensive wave, numbering in the millions. In the first two waves, hundreds of thousands went to Europe, although many also ended up in Canada, the United States, China, Australia, Brazil, Argentina, South Africa, and New Zealand. Not all of those Russian speakers were ethnic Russians; Ukrainians and Jewish emigrants were usually bilingual and were sometimes classified as "Russians" abroad. In the third wave, the majority went to Israel, but many also went to the United States, Canada, Germany, and France (de Tinguy, 2004).

The fourth wave has produced about 1.4 million emigrants, the majority of whom have ended up in just three countries: Israel, Germany, and the United States. The United States has received about 650,000 to date, including 16,000 athletes; 4,000 actors; and thousands of computer programmers, mathematicians, engineers, and scientists. Russian mathematicians and physicists now account for about 3–4% of the total numbers in their respective professions in the United States. Unlike the other three waves, the last wave contained primarily professional emigrants with a good knowledge of foreign languages and excellent employment prospects. Some were ethnic German and Jewish migrants, but quite a few were Russian and Ukrainian. They greatly enriched the countries they migrated to, but also caused concerns about "brain drain" back home. Effectively, the university education given to these people in the Soviet Union for free has been imported by the advanced Western economies. However, unlike those in the previous waves, the most recent emigrants remain connected to home; most travel back at least once per year and collaborate with their peers left behind. The United States received about 13,000 new permanent residents from Russia in 2006, and an even higher number from Ukraine, 17,000 (a much higher proportion than Russia's, considering its smaller population). If Ukraine and Russia were combined, they would represent the eighth largest source of immigrants into the United States in 2006, after Mexico, China, the Philippines, India, Cuba, Colombia, and the Dominican Republic.

In the United States, large communities of Russian speakers (including, importantly, Soviet-period Jews, some Ukrainians, Germans of Russian descent, and others for whom Russian is the first language, but who are not ethnically Russian) exist in New York City, Los Angeles, San Francisco, Chicago, Boston, Seattle, Portland, Miami, Houston, Minneapolis, and other large metropolitan areas (Hardwick, 2007). Many of these communities have thriving Russian grocery stores, pharmacies, restaurants, bookstores, nursing homes, real estate offices, car dealerships, theaters, schools, newspapers, and radio/TV stations. The total number of persons age 5 and older speaking Russian at home, according to the 2000 U.S. census, was a little over 706,000. In 1990 there were only 241,000 such speakers; the difference highlights the surge of Russian immigration after the end of the Cold War. It allows us to estimate the total influx in these 10 years at about half a million. Today Russian has become the 10th most common language in the United States, after English, Spanish, Chinese, French, German, Tagalog, Vietnamese, Italian, and Korean.

Outside the United States, particularly large Russian communities exist in Israel (750,000), Germany, the United Kingdom, France, Canada, Australia, Brazil, and Argentina, roughly in this order. Whereas France attracted a lot of refugees after the two world wars, and Israel and Germany drew a million Soviet Jews in the 1980s and 1990s, today the United Kingdom has become the biggest magnet for rich and/or professional Russians in Europe. One reason is simple: It is the only major European Union (EU) country that uses English, the main foreign language studied by the Russians. Moreover, U.K. admission policies encourage the immigration of rich and educated people. The traditional magnets, Canada and the United States, also lose out because of the geographic distance: It takes only 3½ hours to get from London to Moscow by plane, but over 10 from New York or Toronto, which is an important consideration for the jet-dependent business elite. The new immigrants make regular trips back home for business or family reasons. Finally, after September 11, 2001, new visa and immigration policies have discouraged access to the United States; the job market there has also been lackluster, resulting in a steady drop in the immigrant flow over the past 5 years.

Other Indo-European Languages: Latvian, Lithuanian, Moldovan, Armenian, and Tajik

Latvian and Lithuanian are two existing Baltic languages. The Old Prussian language was spoken 500 years ago in what is today northeastern Poland, but is now extinct. The Baltic languages are ancient and complex. Their vocabulary and grammar places them somewhere between Germanic and Slavic languages. As noted earlier, Lithuanians claim that their language has much in common with Sanskrit, the old Indian language that appears to be the closest to the presumed Indo-European mother tongue. The Baltic languages are written in the Latin alphabet. Although both Latvia and Lithuania are anxious to preserve their languages, the geographic reality is such that most young people are learning at least English as the language of international communication, and perhaps German or French to open up farther opportunities within the EU. Many Russian speakers also live in these two countries, interspersed among the locals (7% in Lithuania, about 30% in Latvia); in addition, much business and trade are now done with Russia and Poland, and to a lesser extent with Ukraine and Belarus, thus necessitating at least some knowledge of Russian and/or Polish for business purposes.

Moldova speaks the Moldovan language, which is a northern dialect of Romanian, an Indo-European language of the Romance branch. Until recently, the Moldovan language used Cyrillic script, but it is now Latinized to conform to Romanian. With about 10% Slavic and 90% Latin roots, Moldovan has a vocabulary that is easy for other Romance-speaking peoples of Europe (French, Italians, etc.) to learn. However, its grammar is rather complex, owing in part to borrowings from the surrounding Slavic languages. If you have ever wondered why people in this part of the world would speak a Latin-derived language, look at the historical maps showing the eastern Roman Empire in about 400 A.D., and you will understand. The Romanians are descendants of the indigenous Wallachians, who became culturally Romanized.

The Armenian language is in its own separate group and has a unique alphabet (Figure 13.4). The culture and language of Armenia are very old, dating back more than a thousand years. According to local legends, Armenia was supposedly the place where the original Garden of Eden was located, as well as the site where Noah's Ark landed (Mt. Ararat in modern-day Turkey). There are some parallels between the Armenian and the Greek languages. Due to its proximity to Iran, the Armenian language was influenced by Persian (also known as Farsi). Although Armenian is a distinct language, it still has some words that speakers of Indo-European languages would recognize. For example, "cow" is *kav* (*korova* in Russian), while "daughter" is *dostr* (*doch* in Russian). Besides the Jews and the Roma, the Armenians are easily the most widespread people internationally today, with twice as many Armenians living abroad as in Armenia proper. California alone had 78,000 Armenians in 2000. There are about 300,000 Armenians in the United States (look up local listings in your city; their last names typically end with "-ian"). More than 3 million live worldwide in places as diverse and distant as Israel, Argentina, Brazil, France, Syria, Lebanon, Iran, and Australia.

FIGURE 13.4. Armenian characters depicted on the side of a church. *Photo:* K. Van Assche.

The Tajik language is also Indo-European, belonging to the Iranian branch and thus related to Persian/Farsi. It is the only main language of Indo-European origin in the five "-stans" of the former Soviet Central Asia. To the south of Tajikistan, however, the Pashtu and Urdu people of Afghanistan and Pakistan speak related languages, with Farsi-speaking Iran being also not too far away. In the northern Caucasus, Ossetian is a related Iranian language spoken by a few hundred thousand Ossetians, who claim to be descendants of the ancient Scythians.

Altaic Languages

The second largest language family by numbers of speakers in the FSU is the Altaic family. With about 66 languages, it is almost as large and as diverse as the Indo-European family. Its origins are believed to be in the Altay Mountains of Siberia (Figure 13.5). Today this region still has

FIGURE 13.5. Artifacts of the prehistoric Altay culture on display at an ethnographic museum in Biysk. The Altay Mountains are believed to be the source of the Altaic language family, which includes the Mongolian and Turkic languages. *Photo:* Author.

the ethnic Altaytsy, who resemble Mongolians in appearance and speak an Altaic language. The largest branch within the family is Turkic. Languages in this group spoken inside Russia include Tatar, Bashkir, Karachay-Balkar, Chuvash, Yakut, Tuvin, Altay, and a few others. In the broader FSU, there are also Azeri, Kazakh, Kyrgyz, Karakalpak, Uzbek, Gagauz, and a few others. Other important languages from this family are those of the Tungusic branch, spoken by the Evens and the Evenks of northern Siberia, and by the Nanay and Udege of the Russian Pacific. The Mongolian branch is represented in Russia by the Buryat language spoken near Lake Baikal, and by the Kalmyk language spoken north of the Caspian Sea. Outside the FSU, the Korean and Japanese languages are sometimes also included in this family, although this inclusion is debatable (Starostin et al., 2003).

Like the Indo-Europeans, the Altaic people in all probability share a history. They are likely to have originated somewhere near the geographic center of Eurasia in the Mesolithic Age. They probably then spread away from the Altay and other Central Asian mountains to the north and west in waves of successful migrations about 8000 to 6000 B.C., bringing in important technologies (e.g., the bow and arrow) and domesticating hunting dogs. Some Altaic groups today are more Asian in their physical appearance, while others are more European. Certainly they represent a mixture of Asian and European groups. Although the Altaic people may have originated in the mountains, today their languages are more common in distinctively steppe-based cultures formed around a nomadic lifestyle and traditionally dependent on horses and sheep.

The Altaic languages share some common words, including roots for some numerals and most common things. Within the Turkic branch many words may be virtually identical. For example, "mountain" is *tau* or *tay*; "lake" is *kol* or *kul*; "water" is *su*; and so on. In the Middle Ages, many of the Turkic cultures came into contact with Islam, and the Arabic language and script were introduced via Persian and Tajik scholars. Therefore, the most common greeting in these languages today is not a Turkic phrase, but some variant of the Arabic *Salam aleykum*, meaning "Peace be unto you." (Several versions of this greeting, along with "Hello" in several other languages of the FSU, are presented in Vignette 13.2.)

The Turkic languages are the largest branch of the Altaic family, numbering about 30. They played an important role historically because they were used by the Tatar–Mongol invaders during the Middle Ages. Today they are mainly shared by civilizations shaped by farming (Uzbekistan, Turkey), but were originally languages of the nomads. Many are mutually intelligible, at least with some basic study (e.g., Kyrgyz and Kazakh, or Azeri and Turkic). The Turkic languages can be subdivided into the southwestern Oghuz group, which includes Turkic, Azeri, and Turkmen; the northwestern group (Kipchak), which includes Kazakh, Kyrgyz, Tatar, and Bashkir; the northeastern group of Yakut-related Siberian languages; and the southeastern group, with Uzbek being most prominent.

Many Altaic languages had no writing systems until the late 19th or even the early 20th century, and a few remain spoken-only languages today (e.g., Gagauz in Moldova). During the Soviet period, most written Turkic languages were converted from Arabic to Latin script on Lenin's orders, and later to Cyrillic under Stalin. When users of Tatar, Uzbek, Azeri, Kazakh, Kyrgyz, the languages of the northern Caucasian peoples, Buryat, and Kalmyk were all required to make Cyrillic the basis for their alphabets, new letters had to be added for sounds that do not occur in Russian. Why was this conversion ordered? First, it was important to the Soviet authorities for people to be unable to read documents written in the pre-Soviet era. A way to do this was to change their alphabets, thereby making old documents unreadable by young people. It also facilitated the cultural Sovietization of these cultures (see Chapter 7), because new Marxist and scientific terminology borrowed from Russian could now be introduced. At the same time, some Persian and Arabic words were replaced with Russian equivalents. Today Kazakh, Kyrgyz, and the Turkic languages in Russia continue to use Cyrillic, whereas the Azeri and Turkmen languages have been converted to the Latin alphabet to facilitate economic and political integration with Latinized Turkey and the broader world. This, of course, may now lead to the cultural exclu-

Vignette 13.2. How to Say "Hello" in 15 Languages of the FSU

Indo-European family

Slavic branch	
Russian	*Zdrastvujte!* (formal), *Preevet!* (informal)
Ukrainian	*Dobri den!* (formal), *Pryvit!* (informal)
Belarusian	*Pryvitáni! Zdarow!*
Romance branch	
Moldovan	*Salut!*
Baltic branch	
Lithuanian	*Labbas!*
Latvian	*Labdien!*
Armenian branch	*Barev!*
Iranian branch	
Tajik	*Assalom u aleykum!* (from Arabic, a Semitic language)

Uralic family

Estonian	*Tervist!*

Caucasian family

Georgian	*Gamardjobah!*

Altaic family

Turkic branch	
Azeri	*Salam æleykǘm!* (from Arabic)
Turkmen	*Salam aleykum!* (from Arabic)
Uzbek	*Salaam aleikhem!* (from Arabic)
Kazakh	*Asalamu alaykim!* (from Arabic)
Kyrgyz	*Salam aleykum!* (from Arabic) or *Kandisiz!*

sion of older people who can no longer read the newspapers.

The most amazing, and perhaps extreme, example of a writing system's transformation involved the Uzbek language. Before 1928, written Uzbek used Arabic script borrowed from Muslim Arab scholars during the Middle Ages. The new Uzbek language was written, taught, and enforced in Latin script from 1928 until the enforced switch to Cyrillic in 1940. Between 1940 and 1992, Uzbek was written primarily in Cyrillic, but the newly independent Uzbekistan officially reintroduced Latin script in 1992. Nevertheless, old traditions die hard, and Cyrillic still continues to be widely used. Currency, street signs, educational programs, and governmental communications are being gradually switched to Latin script, however. Although it may seem practical in our increasingly globalized world to use the most widespread alphabet, it also clearly reflects the political orientation of the Uzbek leadership in recent years away from Russia and toward Europe, Turkey, and the United States.

Who Are the Tatars?

Anyone who likes the history of Russia is fascinated by the Tatars. The name stood for different groups of people at different periods in history, actually. The Tatars were known to the Western Europeans during the Middle Ages as "Tartars," based on the belief that they came straight from the underworld (*Tartarus*, in Greek mythology). The various groups of Altaic people of the Turkic branch came to be known as Tatars over centuries, beginning as early as 500 A.D. It must be stressed that since their participation in the Tatar–Mongol occupation of Rus in the 13th–15th centuries, the Tatars had undergone big changes with respect to their lifestyle, language, and customs. The Tatars adopted Islam and mixed in with many tribes that they encountered farther west. In Russia today the Tatars are the second largest ethnic group after the Russians, numbering over 6 million people. They live primarily in three areas: Tatarstan in the middle Volga, Astrakhan near the Caspian Sea, and parts of western Siberia. The Bashkirs are closely related to the Tatars. Other Turkic speakers living nearby are the Chuvash (Figure 13.6), formerly known as the Bolgars. They are distantly related to the Bulgarians in Moldova and Bulgaria; both groups have mixed Slavic and Turkic ancestry, with some Uralic influences as well. Also significant are the Crimean Tatars in Ukraine. The majority of Tatars today profess Islam, but the Chuvash are Orthodox Christians.

The Mongolian Connection

A few peoples in Russia speak languages that are related to the Mongolian or Manchu languages of northeastern China. These are the Buryats, who live near the Mongolian border around Lake Baikal, and also the Kalmyks of the north Caspian steppe. Like true Mongolians, they historically depended on horses, lived in movable yurts, and led a nomadic pastoral lifestyle. Most lead a settled life today. The Buryats in particular have adapted very well to the cold Siberian conditions by learning agriculture and cattle ranching from the Russian settlers, while continuing with fishing and hunting to supplement the ranching.

The Evens and the Evenks are closely related groups living throughout eastern Siberia. They are taiga hunters and fishermen, who travel hundreds of kilometers on sleighs pulled by reindeer or dogs. They survive in the least hospitable climates on earth, including the upper Yana basin, where winter temperatures routinely plunge below –50°C. If an Even(k) child is born in winter, he or she receives a first bath in . . . snow. The

FIGURE 13.6. The capital of Chuvash Republic, Cheboksary, has its name given in Russian (left) and English (right) on top of the river ferry terminal, and in the Chuvash language in the Cyrillic alphabet on the lawn in front of the building. *Photo:* S. Blinnikov.

Evenks tend to live along the Yenisei and in taiga north of the Amur River, while the Evens live in small communities farther north and east in the Lena and Indigirka basins.

Uralic Languages

As the name suggests, the Uralic languages originated somewhere near the Ural Mountains. Today these languages are spoken as far south and west as Hungary, and as far north and east as the Lena River delta by the Yukaghir people, but also in Finland and Estonia; in the Karelia, Mordovia, Udmurtiya, Mari El, and Komi Republics of Russia; and in a few places in western Siberia and along the Arctic Ocean's shores. There are about 20 million Uralic speakers worldwide, with perhaps 4 million in Russia. In contrast with the steppe- and mountain-based Altaic speakers discussed above, the Uralic speakers are peoples of the forests, river banks, and seacoasts. Note that in many older textbooks and atlases the Uralic languages are still placed in the same family as the Altaic in a joint Altaic–Uralic family; however, linguists have not considered this correct for over 50 years now.

The Uralic languages have a very complicated grammar, with many cases for nouns in particular. In some dialects of Komi, there may be 27 cases; in Estonian, 14; in Mordvinian, 13; and so on. In comparison, Russian has only 6 and German 4, while French and English have none. The Uralic languages, however, do not have gendered nouns—or, curiously, the verb "to have." They also commonly have negative verbs (i.e., a verb form that combines "no" as a suffix with the verb stem, as in the English "don't"). J. R. R. Tolkien was so taken in with the beauty and the unusual grammar of the Finnish language that he based his invented Quenya tongue in *The Lord of the Rings* on it.

Because Uralic languages are so difficult for outsiders to learn, and because their native speakers are few in number and are scattered across vast northern forests, the future of many of these languages is currently in question. These people were assimilated much earlier and more thoroughly than the Altaic people, and thus they tend to speak Russian as their first language now. Of the approximately 25 such languages spoken in Russia, 13 are endangered. For example, the Saami of the Kola Peninsula, also called the Lapps in northern Norway, number fewer than 2,000. About 70,000 live in Finland and Norway, so the overall group is unlikely to go extinct soon, but the Kola dialects are dying out. Incidentally, these are the people who gave the world the word "tundra" and (along with others) domesticated reindeer. It would be a great tragedy if their language and culture completely disappeared. There are a few other Uralic languages in Russia with only a few hundred speakers.

The Komi people have lived with the Russians the longest—about 600 years, in the Pechora River basin. Novgorod merchants traded with them and provided needed technology in the early stages of the settlement of the Russian north. Their region of northeastern European Russia was the first place where nonferrous metals began to be mined in the Russian Empire. Coal mines and the military provide much employment to them now.

The Karelians live along the Finnish border. Their language is only spoken, not written, and does closely resemble Finnish. The Estonians likewise live close to Finland (across the Baltic Sea) and have a similar language. Other large groups of the related Finno-Ugric branch of the Uralic people live in the basin of the Volga and have their own republics within Russia—Udmurtiya, Mari El, and Mordovia (two ethnic groups). Most are heavily Russianized. Unlike most of the Turkic people considered above, the Uralic people were converted to Orthodox Christianity in the 15th and 16th centuries, which facilitated their acculturation.

The shores of the Arctic Ocean in the European part of Russia are settled by the Nenets people. You may have heard of the Samoyed dog. *Samoyed* literally means "self-eating" and was a derogatory name that the early Russian settlers used to designate the Nenets population, because they incorrectly believed that the Nenets were cannibalistic savages. Like the closely related Saami of the Kola Peninsula, these people domesticated reindeer and practiced subsistence hunting, fishing, and berry gathering in the tundra.

The lower Ob basin in western Siberia is settled by the Khanty and the Mansy, two closely related peoples of the Uralic group. These are true forest dwellers dependent on forest game hunting and fishing. As is typical of the Uralic peoples, they are small in stature and have dark hair, but light-colored eyes. There is an opinion that these people are the last remaining representatives of the formerly mighty people Sybir, who gave Siberia its name. They may be also related to the Huns, who helped destroy the Roman Empire and contributed to the Magyars of Hungary.

On the frozen shores of the East Siberian Sea, a few hundred Yukaghir ("ice people") survive as a remnant of a more widespread, apparently indigenous Uralic tribe that went farthest east of its original home in the Urals. However, some sources suggest that they are more closely related to the Chukchi people, who speak a Paleoasiatic language, and are not Uralic at all.

Estonian is a Uralic language closely related to Finnish. The economic openness of Estonia and its desire to attract foreign investment have made it the most English-speaking of any FSU republic—much more so than neighboring Latvia and Lithuania. German is also widely studied. However, the influx of Russian tourists and businessmen and the continued presence of Russian speakers ensure that Russian will remain understood. Over 58% of Estonian children studied Russian in school in 2006, although only 30% spoke Russian at home.

Other Languages

The Caucasus is a melting pot of languages and cultures. Some languages spoken in the region today are from the Indo-European family (Russian, Ossetian, Armenian) or the Altaic family (Karachay-Balkar, Kalmyk). Most, however, belong to the distinct and indigenous Caucasian family. The Georgian (Kartli), Vainakh (Chechen and Ingush), and Circassian languages are worth mentioning here.

Georgians (their name for themselves is Kartli) are ancient inhabitants of the Caucasus Mountains' south slope; they number only 4 million, with a deep traditional culture. Their language has a distinct alphabet. The Mingrelian, Laz, and Svan people have related languages inside Georgia. The Georgian alphabet, called Mkhedruli, is over 1,000 years old and has 38 letters. The letters follow the order of those in the Greek alphabet, but they are highly unique in style. Curiously, only uppercase letters are used. The Ossetian and Abkhaz people sometimes used it in the past for their languages. The reading is straightforward, with both consonants and vowels spelled out, and is done from left to right. Unlike the nearby Armenians, few Georgians live abroad. One of the distinctions of the Georgian language is that many consonants are frequently grouped together, which makes it difficult for foreign learners to pronounce.

Other indigenous inhabitants of the Caucasus are the Circassians (Kabardins, true Circassians, Adygs, Abkhaz) and the Vainakhs (Chechens and Ingush). They live primarily on the northern slopes of the mountains. The Circassian culture is originally steppe-based and in many ways resembles that of the Altaic people; however, their language is not Altaic. Their main occupation has been sheep and horse ranching, along with some agriculture. The Abkhaz people of the disputed separatist republic are related to the Circassians on the north slope of the Caucasus in Russia and are unrelated to Georgians. They use the Cyrillic alphabet. According to one hypothesis, Circassians may be the closest living relatives of the Basque people in Spain. The Vainakhs are typically mountain dwellers, living high up river valleys near the snow-capped peaks. Historically, they were hunters and warriors. The Dagestan Republic of Russia has over 30 somewhat related languages, and is the most linguistically diverse part of Russia today. The main groups there are the Avars, the Lezghins, the Dargins, and the Lakhs. Many Lezghins live farther south inside Azerbaijan as well. The languages of Dagestan are hard to classify into a specific family or language group, but are apparently indigenous to the Caucasus.

In the extreme Far East of Russia, in Kamchatka and Chukotka, small groups of Paleoasian people speak ancient and complex languages that are related to the languages of Native Americans. These groups are the Chukchi, the Koryaks, the Itelmens, the Inuit (Eskimo), and the Aleuts.

REVIEW QUESTIONS

1. Name the main three language families of Northern Eurasia. Which of these are represented in your country and your community?
2. What are the three or four most common languages spoken in Russia today?
3. What alphabets were used in the territory of the U.S.S.R.?
4. What are the advantages and disadvantages of moving from one to another alphabet (consider Arabic -> Cyrillic, Cyrillic -> Latin moves)?
5. What Indo-European languages are spoken in the countries of Central Asia? In Europe? In the Middle East?
6. Describe the differences between the Altaic and the Uralic languages and the geography of their distribution. Which group has historically been closer to the Russians?

EXERCISES

1. Compare one Altaic and one Uralic language by using online sources (a good place to start may be *www.omniglot.com*). Compare and contrast their alphabets, grammar, and syntax. Which one do you think would be an easier one for you to learn?
2. Investigate any endangered language of the FSU by using the web portal for the Red Book of the Peoples of the Russian Empire (*www.eki.ee/books/redbook*). Where do these people live? What are the numerical trends? Why the decline?
3. Find out whether there are any community resources (schools, after-school programs, universities, libraries, clubs, etc.) available for people who want to study the following languages where you live: Russian, Ukrainian, Georgian, Armenian, Kazakh, Tatar (or any other Turkic language), Estonian, Lithuanian, Latvian, Romanian (Moldovan), and Tajik.
4. A famous Russian geographer, V. P. Semenov-Tien-Shansky, wrote a book called *Earth Colors* in the early 20th century, in which he argued that languages of all cultures are heavily influenced by the natural environment in which they develop. For example, the Russian language has two words for blue (*goluboy*, "azure blue," and *siniy*, "dark blue"), but it makes poor distinctions among the shades of red, orange, and yellow, which are uncommon colors in the wintry northern landscape. Some northern peoples (e.g., the Chukchi) have a dozen names for various shades of white. Investigate any language that you know well (starting with English) to see what colors are described the best (i.e., have the highest number of synonyms). Why do you think these particular colors are common in this particular language?

Further Reading

Berkutova, A., Kosobokova, T., & Tsaregorodtseva, I. (2007). Russkij sdaet pozitsii. RBC Daily. Dec. 20, 2007. Retrieved December 20, 2007 from *www.rbcdaily.ru/2007/12/20/focus/310285*.

de Tinguy, A. (2004). *La grande migration: La Russie et les Russes depuis l'ouverture du Rideau de Fer.* Paris: Plon.

Gordon, R. G., Jr. (2005). *Ethnologue: Languages of the world* (15th ed.). Dallas, TX: SIL International. (Online version: *www.ethnologue.com*)

Hardwick, S. W. (2007). *Far from home: Slavic refugees and the changing face of Oregon.* Portland, OR: Oregon Council for the Humanities.

Knox, P. L., & Marston, S. A. (2007). *Places and regions in the global context* (4th ed.). Upper Saddle River, NJ: Pearson/Prentice-Hall.

Starostin, S., Dybo, A. V., & Mudrak, O. A. (2003). *An etymological dictionary of the Altaic languages.* Leiden, The Netherlands: Brill Academic.

Weber, G. (1997). The world's ten most influential languages. *Language Today*, 2.

Wixman, R. (1984). *The peoples of the USSR: An ethnographic handbook.* Armonk, NY: Sharpe.

Websites

www.eki.ee/books/redbook—The Red Book of the Peoples of the Russian Empire.

www.omniglot.com—World languages with notes about their spelling rules, alphabets, and grammar.

www.creeca.wisc.edu—The Russian and East European Studies Center at the University of Wisconsin–Madison.

www.russianamerica.com—A Russian expatriates' Web portal.

CHAPTER 14

Religion, Diet, and Dress

This chapter focuses on other cultural elements, besides languages, that are important in the geography of the former Soviet Union (FSU). Cultural geographers are frequently interested in learning about the influences of religious beliefs on the organization of space in human societies (Park, 1994). The major religions of the world have left an indelible mark on many cultural landscapes and facets of human life. How people think and what they do are determined, among other things, by their beliefs. The geography of religion is not the study of theology, but the study of how beliefs shape and transform cultural landscapes, politics, economics, and social relations. Think of your hometown: What are some of the marks of the predominant religion on the local landscape—houses of worship, cemeteries, and so forth? If you live in North America or Europe, chances are that the biggest impact you see is that of Christianity (Catholicism and/or Protestantism, depending on the region). In parts of New York City, it may be Judaism. In a few communities on either coast, Islam, Baha'i, or another faith may be the most visible. Overall, though, most Western countries today are religiously pluralistic societies. Also, because of the separation of church and state in the West, no governmental endorsement is given to any faith; the cultural landscapes thus reflect more the popular, not official religion in these countries.

Some common influences of religion on geography include the following:

- Architecture, especially places of worship (churches, synagogues, temples).
- City and village layouts.
- The imagery and language on street signs.
- Religious art (or prohibition of religious imagery in art).
- The local calendar (e.g., weekly closures on Sunday, seasonal festivals).
- Cemeteries' location, configuration, and appearance.
- Dress, especially the gender, class, and age differences expressed in it.
- Diet (in the sense of food that can or cannot be eaten on certain days, or ever, and the rituals associated with consumption of food).
- Pilgrimage sites and associated economic activities (e.g., sales of religious cards or other artifacts).
- Festivals (e.g., a Christmas parade or harvest pageant).
- Political restrictions on, or contested space among certain religious groups.
- Direct or indirect influences on patterns of

production and trade (e.g., pork cannot be produced or sold in Saudi Arabia, or beef in many parts of India).

This chapter discusses Orthodox Christianity, the most historically influential religion in Russia and several other FSU states, in the greatest detail. Somewhat briefer discussions are provided of Islam and other important religions of the region. The topics of diet and dress are considered separately at the end of the chapter.

Main Religions of Russia and Other FSU Countries

There are two main patterns in our region of study—one that is visible on maps, and another that is not. The first one is the predominance of Orthodox Christianity in much of Russia, Ukraine, Belarus, Moldova, and Georgia, with Sunni Islam being common in all of the Central Asian states and in parts of the Volga and northern Caucasus regions of Russia populated by the Turkic cultures (Figure 14.1). Lithuania is predominantly Roman Catholic; Estonia and Latvia are mostly Lutheran; Armenia has its own Christian Apostolic church (related to both the Orthodox and Roman Catholic Churches); and Shia Islam predominates in Azerbaijan. In addition to these major groups, Roman Catholic, Lutheran, and Jewish communities are found in the biggest cities, and Buddhism is practiced in the Kalmykia and Buryatia Republics of Russia. Some new Protestant communities and alternative religious movements (e.g., Hare Krishna, Aum Shinrikyo, Scientology, etc.) can be found in most urban areas as well. Parts of Siberia and the Russian north have had small communities of Orthodox Old Believers (*staroobryadtsy*) since the 17th-century church schism over liturgical reforms under Patriarch Nikon.

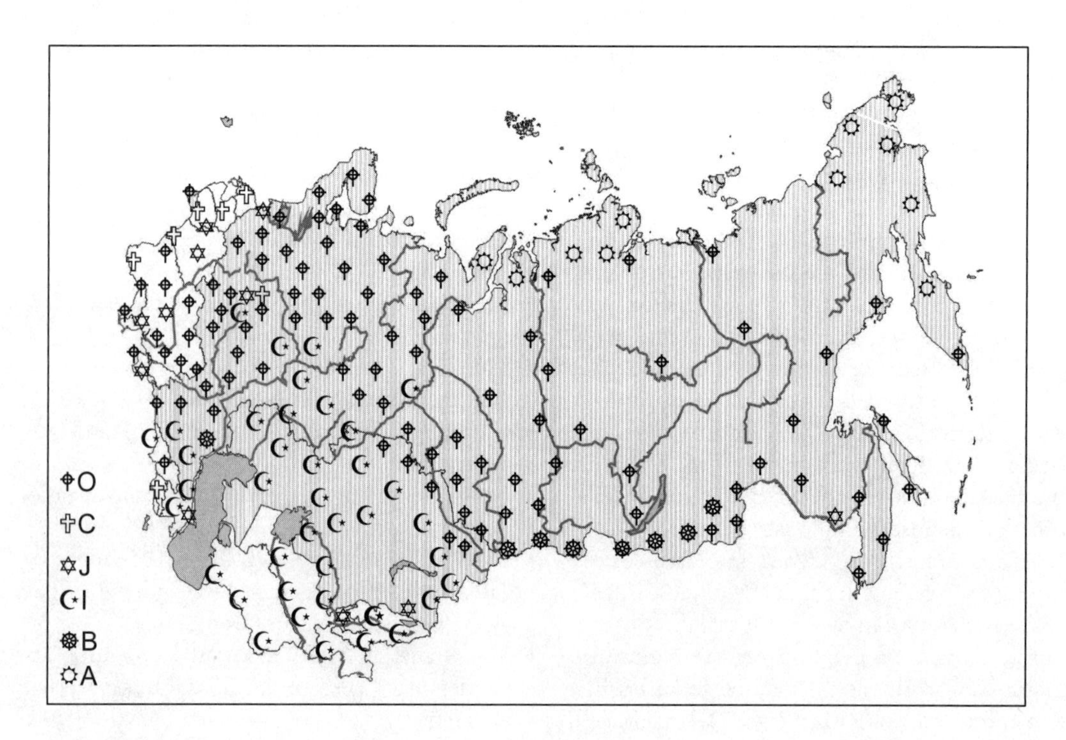

FIGURE 14.1. The main religions of Northern Eurasia: O, Orthodox Christianity (Russia, Belarus, Ukraine, Moldova, Georgia, some parts of Kazakhstan, and the Baltics); C, other Christian churches (Roman Catholic in Lithuania, Byzantine Catholic in western Ukraine and Belarus, Lutheran in Estonia and Latvia, Armenian Apostolic in Armenia); J, Judaism; I, Islam (Shia in Azerbaijan, Sunni in other areas); B, Buddhism (Buryatiya and Kalmykiya) and Burkhanism (the Altay); A, animism/shamanism. Map: J. Torguson.

The second pattern, which is not easily depicted on the map, is the secularism that is common to the entire region. "Secularism" is the absence of religion in people's lives. It is not the same as atheism; it simply means that formal religious observances play no role in people's daily routines. The presence of secularism is well documented in most developed parts of the world, especially Europe and Japan. Polls in the United States indicate that the fastest-growing group there in terms of religion is people without any formal religious affiliation. In the FSU, the impact of Communism during the Soviet period (1917–1991) and the general modernization of life resulted in high numbers of nonreligious people. Although the majority of people in the FSU call themselves religious, only a small minority actually practice a religion. For example, in Russia about 80% of people have been baptized in the Orthodox faith, but only 44% profess belief in a God, and merely 12% attend church on a monthly basis. Fewer still participate in the sacraments (e.g., communion or confession) that are required according to the church's teaching. Practicing Muslims make up less than 4% of the Russian Federation's population, although nominally about 16% are Muslims. Fewer than 1% each are Jewish or Buddhist. About 7% of Russians believe in supernatural forces other than a God, while a whopping 22% are agnostics who are not sure whether there is a God, and about 22% call themselves atheists.

By comparison, in the United States about 75% of people consider themselves Christians, and about 40% attend a religious ceremony at least once a month. Only 14% do not have any religious affiliation at all, although this is the fastest-growing group now, as noted above. The Russian pattern of religious adherence is thus closer to that in most European countries, which overall tend to have a higher proportion of nonreligious people than the United States. On the other hand, Japan and the United Kingdom are even more secularized than Russia. In recent estimates from Britain, there are fewer than 2 million practicing Anglicans now, out of about 60 million people.

A common history of religious persecution under the Soviet regime is shared by all faiths of the FSU. Particularly affected are the generations who were born and raised before 1991, which are overwhelmingly nonreligious. Among the younger people, there is actually a higher interest in practicing their new-found faith.

The Orthodox Church: Origins and Beliefs

The main religion of Russia and its allied Slavic states, as well as of Georgia and Moldova, is Eastern Orthodox Christianity. Georgia and Russia have their own national Orthodox churches headed by patriarchs; Ukraine, Estonia, and Moldova have some communities under the Moscow Patriarchate, and others under their own national church leaders. Other former republics have mainly parishes under Moscow's leadership. Georgia has been Orthodox since the 4th century; Rus became Orthodox in 988 A.D., when Prince Vladimir of Kiev converted to Orthodox Christianity and married the Byzantine emperor's sister. Vladimir's choice was partly based on politics: By choosing Orthodoxy, he aligned himself with the powerful state of Byzantium. He also considered Islam, Judaism as practiced by the Khazars, and Roman Catholicism, but he reportedly chose the Orthodox religion because of the beauty of the Orthodox liturgy.

The Orthodox Church, like the Roman Catholic Church, has had an uninterrupted succession of bishops since the time of the Apostles. The Roman Catholic and the Orthodox faiths separated in 1054 A.D.; each claims to be the "true church," not merely a part of the church, while seeing the other as in error on a number of theological points. Orthodoxy comprises a worldwide communion of national churches, all of which share theology and sacraments, but which have different sets of governing bishops. There is no Pope for all. The important decisions are made by councils, not by individual hierarchs. The Orthodox Church uses the same Bible and Creed as the Roman Church (with a few small exceptions). Numerous books are now available in English for those who wish to learn more about the practices and traditions of Orthodox Christianity.

The Orthodox Church stresses belief in the Holy Trinity (Father, Son, and Holy Spirit); asserts the true bodily Resurrection of Jesus Christ after death; venerates sacred images (icons); prays to God, but also to the Virgin Mary and the saints; has very elaborate, long services; observes a complicated calendar of feasts and fasts; has a strong monastic tradition; and differs in many practices from either contemporary Catholicism or Protestantism. The main service of the day is called Divine Liturgy and is analogous to the Mass or Eucharist of Western Christians. People stand through this entire service, singing *a capella* responses, crossing and bowing, surrounded by icons, candles, and fragrant incense smoke (Figure 14.2). One can become a member by baptism in the name of the Holy Trinity, with full triple immersion. Children are given communion after baptism, which usually happens at 40 days of age. The Eucharist is believed to be literally the body and blood of Christ, not a symbol. The Orthodox Church has an all-male clergy in three ranks: bishops, priests, and deacons. In contrast with the Catholics, married men become Orthodox priests. Bishops, however, must be celibate and are chosen from the ranks of the monastic clergy.

Orthodox Religious Landscapes

The most notable features of Orthodox religious landscapes are, of course, the churches themselves. The Orthodox call churches *khramy* (temples), to stress that Divine Eucharist is actually offered there as a form of bloodless sacrifice. Each church is laid out according to a standard plan with a theological meaning (Figure 14.3). People always enter from the west under a bell tower (Figure 14.4) into the *narthex* (vestibule), and from there into the elongated *nave*. There are no pews, since people are expected to stand (and sometimes prostrate themselves on the floor, in a fashion somewhat similar to that of the Muslims). At the east end of the building is the raised sanctuary with an altar table hidden behind the curtain in the *iconostasis* (icon screen). The screen has three sets of doors in it, which are closed between services. Only priests and male altar servers are allowed inside the sanctuary. The altar itself is a square table covered with richly embroidered cloth; a candelabrum, the Gospel Book, the Tabernacle, the Cross, and various other holy objects are placed on it.

The biggest churches are called "cathedrals," with bishops serving in those. Small chapels can be found at most cemeteries and in other locations. Chapels typically do not have altars and cannot be used for celebration of the Eucharist, but are suitable for saying prayers for the dead or for reading daily services. Every big village in Russia used to have a church; such a village was

FIGURE 14.2. The Orthodox Divine Liturgy is an elaborate and ancient service, with most parts unchanged since the 4th century. Note the icons on the walls, the icon screen, the candles, and the vigil lamps. *Photo:* Author.

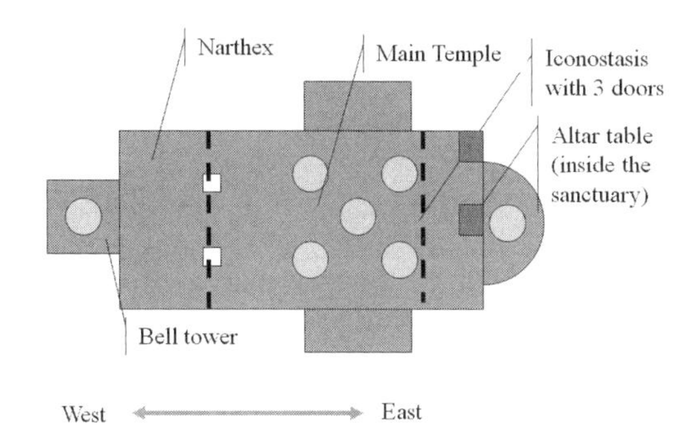

FIGURE 14.3. A typical plan of an Orthodox church (temple). People enter under the bell tower into the narthex and proceed to the main temple (nave). The sanctuary with the altar is hidden behind the iconostasis screen. The five circles indicate domes on the roof.

called a *selo*, as opposed to a village without a church (*derevnya*). Over 50,000 churches existed in the Russian Empire before the Revolution. Today there are over 15,000 parishes operating in Russia, and about half as many in Ukraine; much of Siberia and the Far East have relatively few churches, however.

In an old Russian city the biggest cathedral would typically be found inside the walls of the local kremlin. The main cathedral in Moscow is

FIGURE 14.4. The 17th-century Church of the Annunciation in Murom is a fine example of stone architecture of the pre-Baroque period. Note the location of the porch under the bell tower on the left (west) and the entrance into the narthex on the right (east). *Photo:* Author.

the one dedicated to the Dormition of the Theotokos (Virgin Mary) inside the Kremlin, not St. Basil's. Built by the Italian architect Fioravanti in the late 1400s, it is a soaring white-stone edifice of incomparable beauty, both inside and outside (see Chapter 6, Figure 6.5). In a village, the church typically anchors one end of the main village street, with the cemetery located immediately behind the church's altar wall. In some cases, prominent church and state leaders were buried inside a church or cathedral itself, under its floor or in niches along the walls. For example, Archangel Michael's Cathedral in the Moscow Kremlin has dozens of graves of the Rurikid dynasty, ending with the sons of Ivan the Terrible.

On the outside, the most striking features of Orthodox churches are their golden or blue onion-shaped domes. Usually there are 5 of these, but sometimes there are 7, 9, or even 13 of them (Figure 14.5). Each number, always odd, has some significance—5 domes symbolize Christ and the 4 evangelists, while 13 represent Christ with the 12 apostles. The bell towers are relatively late additions, borrowed from the Catholics. Prior to the 14th century, Russians used flat metal *bila* for ringing.

Monasteries can be very large and prominent, usually fortress-like, built many centuries ago to protect the monks from physical attacks by invaders. Inside are numerous churches, monastic cells, refectories, warehouses, and other buildings. Russia had about 1,000 monasteries a century ago; today a few hundred are open. The most famous monasteries (and one famous convent) are these:

- Kiev Caves Lavra in Kiev, Ukraine (among the oldest; not in Russia any more, but still part of the Moscow-based Russian Church).
- St. Sergius Trinity Lavra in Sergiev Posad, about 1 hour by bus or train northwest of Moscow (Figure 14.6).
- Valaam, on islands at the northern end of Lake Ladoga.
- Pskov Caves Monastery near the Estonian border (the only monastery on Russian territory that did not close during Soviet times, because it was under Estonian rule before World War II).
- Optina Hermitage near Kozelsk, Kaluga

FIGURE 14.5. Transfiguration church in Kizhi cemetery on Lake Onega. This World Heritage Site is a celebrated example of the wooden architecture of the Russian North. *Photo:* S. Blinnikov.

FIGURE 14.6. Holy Trinity Monastery, founded by St. Sergius of Radonezh (d. 1392), is the most prominent monastery in Russia today. It houses a few museums, a library, and the Moscow Theological Seminary and Academy. *Photo:* Author.

Oblast, about 4 hours by car southwest of Moscow.
- Solovki Monastery on the Solovetsky Islands in the distant White Sea (also a museum of the infamous nearby GULAG camp, where thousands of priests and bishops were executed in the early 1930s).
- Diveevo Convent near Arzamas in Nizhniy Novgorod Oblast (the most famous and largest convent in Russia, associated with the great 19th-century mystic St. Seraphim of Sarov).

These are all major pilgrimage centers receiving hundreds of pilgrims on an average day. Of the seven, six are for men, and the convent (Diveevo) is for women. There are actually many more convents in Russia today than men's monasteries, but few of the convents are well known.

As in the West, the medieval monasteries were major centers of learning and arts. However, relatively less emphasis was placed in the East on manuscript copying and more on icon making. Also, no monasteries became university centers, except in a specialized sense as theological academies. Their biggest impact on the economy today is serving as focal points for religious pilgrimages. They also publish books, make icons and other items for worship, and house beautiful museums. Some are involved in charitable work in the surrounding communities (e.g., supporting orphanages).

Orthodox faith is also visible in the cemeteries. Russian cemeteries look and feel very different from most of those in Western Europe or North America. They occupy high points in the landscape, both to avoid flooding and for spiritual reasons. The biggest difference from typical Western cemetery landscapes is the presence of lots of shady trees and wild, uncut grass. From a distance, Russian cemeteries look like dense forests. Graves were formerly adorned with large eight-pointed Orthodox crosses, not with tombstones. In the Soviet period, however, the crosses were joined by granite or marble headstones with five-pointed stars for Communists and unbelievers. The grass would be allowed to grow tall and free (Figure 14.7). Graves would be fenced off to create a sense of privacy. (In a way, a grave site was the only truly private space that a person could count on.) Cremation is generally prohibited by the Orthodox Church, but in the Soviet period, with space being very limited near big cities, it became a common practice. The most famous cemetery of Russia is that of Novodevichy Convent in Moscow, where hundreds of Soviet-era dignitaries are buried (the ashes of many are interred inside the walls).

FIGURE 14.7. A cemetery near Moscow resembles an overgrown forest more than a lawn. Cemeteries in Russia combine Soviet and Orthodox symbols, reflecting changes in attitudes about the afterlife. *Photo:* Author.

Many other signs of Orthodoxy are visible in the countryside: roadside shrines to saints; chapels over holy springs; frescoed icons on cliffs; sacred caves and groves; and other sites. In recent times there has also been a proliferation of churches and chapels as monuments or memorials: a cathedral in southeastern Moscow dedicated to the Millennium of Christianity, chapels commemorating heroes fallen in wars, and a chapel in Novosibirsk that marks the "midpoint" of Russia (see Chapter 11, Figure 11.7).

Many old churches have been restored. The most famous example of such restoration is Christ the Savior Cathedral in Moscow (see Chapter 11, Figure 11.5). Described in detail by Sidorov (2000), this is a premier example of "national monumentalization"—a process in which, consciously or subconsciously, buildings are manipulated for the state's political aims. Other famous buildings recently restored in Moscow include Our Lady of Kazan Cathedral on Red Square and the Iveron Chapel nearby, both housing important religious artifacts. Many of the restored churches actually had to be rebuilt from scratch by using historical photographs and blueprints.

The Impact of Orthodoxy (and Other Religions) on Culture in the FSU

The Orthodox Church shaped Eastern Slavic culture for about 10 centuries (and even longer in the case of Georgia), and its impact is thus profound. Virtually all Russian classical music masterpieces, and a great deal of classical Russian literature until the end of the 19th century, were informed by and infused with Orthodox values. For example, Glinka, Tchaikovsky, Bortnyansky, Balakirev, Rachmaninov, and Rimsky-Korsakov produced stunning choral, piano, and orchestral masterpieces as parts of actual church services (e.g., Rachmaninov's famous *Vespers*). Many Western readers have first encountered Orthodoxy through the writings of Feodor Dostoevsky, Nikolai Leskov, or Nikolai Gogol.

Orthodoxy has had its strongest impact on the visual arts, because icons and other forms of religious art are ubiquitous in Orthodox worship. In addition, the Russian language itself bears many imprints of the Orthodox worldview. For example, "Thank you" in Russian is *Spasibo*, which literally means "May God save you!" Curiously, the early Byzantine missionaries Cyril and Methodius preached the gospel in the Slavic lands in the vernacular—the Old Bulgarian language widely understood at the time—but the Russian church today uses an archaic Church Slavonic language (still based on that Old Bulgarian) in its worship. Although Church Slavonic is very precise and beautiful, it is not the language commonly spoken by the people.

The Orthodox Church exalts obedience and humility and frowns upon common vices, such as greed, lust, gluttony, malice, and pride. It is doctrinally one of the most conservative of the Christian churches, with beliefs changing little from one century to the next. Most of the contemporary social topics debated by Western Christians (female priesthood, homosexuals in the church, abortion, euthanasia, etc.) rarely appear in the Orthodox discourse. The opinion of the church is formed by the church councils, not by individual Popes or the believers themselves. Some historians believe that downplaying the earthly elements and elevating the eternal questions in church have placed Orthodox lands at a disadvantage in making the transition to a modern market economy, relative to its Protestant and even Catholic counterparts. For example, the work ethic of Western Europeans was greatly influenced by the Protestant concept of individual freedom, including the freedom to become rich and the need to take responsibility for one's own actions. Orthodox believers are more oriented toward the public good; the idea of owning a business strictly to enrich oneself is seen as a vice. Orthodox believers are also more used to a hierarchy in both the state and the church, and are thus less likely to take up individual political initiatives.

The fasting rules of the Orthodox Church are rigorous. A "fast day" means eating vegan food (not complete abstinence from food): No animal products can be consumed, although seafood is sometimes allowed. Vegetables, fruits, and grains may be eaten in moderation. About *half* of the Orthodox calendar falls on fast days—including the periods of Great Lent, Advent, and two additional fasts in summer, and almost every Wednesday and Friday. Imagine the impact of fasting on the patterns of agricultural production and retail in a country in which the vast majority of people were Orthodox. In Great Lent, people did not eat (and restaurants did not serve) meat or dairy at all, so the producers of those foods would have to be flexible in timing their production. The demand for meat and dairy would skyrocket after Pascha (Orthodox Easter), which usually happens in April. The church calendar of fasts and feasts dictated when and what crops would be planted and harvested, when marriages could be performed, when people would get a break from work, and so on. At the same time—unlike in some other religions, where certain foods are entirely forbidden—there are no "unclean" animals or plants on non-fast days. Slavic cultures are fond of pork, for example, whereas the diet for observant Muslims or Jews in the FSU would obviously exclude pork.

Dress has been likewise influenced by the Orthodox, Muslim, and Jewish cultures throughout the FSU. It is hard to notice this now, after 70 years of Communist rule and 20 years of post-

Soviet regimes, since Western dress is the common contemporary choice. However, the Orthodox religious rules require women to wear skirts or long dresses in church, and to cover their heads with scarves. Men are supposed to remove their hats when stepping into a church, and to be likewise modestly dressed (in long pants, long-sleeved shirts, etc.). In the Muslim community, women's traditional coverings in public were forbidden during the Soviet period, and so even today an Uzbek or a Tajik woman is much less likely to wear a *hijab* or *burqa* in public than a woman in much of the Middle East or North Africa. The faithful are still expected to observe correct dress code in mosques, however: Shoes must be removed and ritual ablutions performed. The Communist influence on dress in both Muslim and Orthodox cultures was thus one of modernization.

There are some other subtle Orthodox and Muslim influences on culture in the FSU. For example, the bright smiles so common among Americans and some Western Europeans are rare in Russia, because there is a cultural taboo against "showing oneself off." Although a big, open smile is a friendly sign in the West, it is not as common in the East. The tone of voice likewise is supposed to be subdued in public. When people meet, they may hug each other and exchange light kisses on the cheeks three times, the same way worshipers commonly do in church; handshakes are much more common now, however. Both Orthodox Christianity and Islam call on people to be more communal and less individualistic. This results in a preference for large family gatherings, for public forums, and for special treatment of household guests. In fact, the hospitality of most Eurasian cultures is legendary. The cult of the collective, in the opinions of some conservative researchers, also influenced the political life of the region: A single autocratic ruler presiding over a community of citizens is seen as an extension of the divine rule of God on earth, and as the normative political structure for Russia and the Central Asian states. The community is seen as united in submission to this ruler, just as believers are supposed to be in submission to God. A differing viewpoint suggests that while Eurasians have been accustomed to autocratic rulers, each generation chooses to reproduce this accommodation without necessarily thinking about divinity at all, and that many people would much rather embrace more individualistic behavior if they were given a choice.

Islam in Russia and Other FSU States

The second most common faith worldwide, Islam, is also the second most common religion of the FSU. The majority of the Turkic people in the region have been Muslim since the 12th–13th centuries. Persian Tajiks adopted Islam from Arab missionaries from the Middle East in the 14th–15th centuries, or the Chechens in the Caucasus in the 18th century. During the 16th–17th centuries, the powerful Tatar khanates of Kazan and Astrakhan became Muslim (Figure 14.8). Khivy, Bukhara, and Samarkand arose as Islamic states in what is today Uzbekistan. Like Christianity, Islam is a "universalizing" religion; this means that anyone can potentially become a Muslim by

FIGURE 14.8. A brand-new, impressive mosque in Kazan attracts thousands of Muslim worshippers. *Photo:* S. Blinnikov.

conversion, and that the goal is to convert all humankind to the one true faith. Therefore, Islam has the potential for rapid expansion.

Today about 15% of Russia's population are nominally Muslim (although fewer than 4% of the people actually practice Islam), with about 47% of the population being Muslim in Kazakhstan, 75% in Kyrgyzstan, 88% in Uzbekistan, 89% in Turkmenistan, and 93% in Azerbaijan. Like the Christians in the region, most Muslims do not practice their religion daily, but have only a nominal affiliation. As an example, a young police officer from Kazakhstan explained that although he observes the Islamic teachings in principle, he likes to drink beer and does not like to spend his Fridays going to the mosque, so his religion is "not up to code." He still considers himself a Muslim, but not by the traditional standards. Nevertheless, the influence of Islam on culture in Central Asia and in the Muslim parts of Russia has undoubtedly grown in the past 20 years.

The heaviest concentrations of Muslims in Russia are observed in Tatarstan, Bashkortostan, and the republics of the northern Caucasus. According to the Law on Religions passed by the Duma, Islam is recognized along with Catholic, Lutheran, Jewish, and Buddhist religions as a traditional faith of the Russian Federation, and thus does not require special permits or scrutiny from the authorities (unlike various "nontraditional faiths," such as Mormonism, Baha'i, or Pentecostal Christianity). Sizable Muslim minorities also live in Ukraine (especially Tatars in the Crimea) and Georgia (in the separatist Abkhaz Republic).

It is significant that the resurgence of interest in Islam is highest not among poor people in villages, but among the more educated, younger urban people. Some Arab nations, particularly Saudi Arabia, have made major investments in the building of mosques and the printing of Qurans and other religious literature for the Central Asian states. In Turkic-speaking Azerbaijan, Turkey is heavily involved in promoting its own agenda, which may include elements of Islam. Iran has an even greater influence there, because millions of Azerbaijanis who live within its limits; it also shares the Shia version of Islam with Azerbaijan, unlike Turkey, which is both more secular and Sunni.

Most cities in Central Asia, the northern Caucasus, and the Volga region of Russia now have at least one mosque. Some recently built ones rival the biggest Russian cathedrals in size, and are quite beautiful, durable, and modern structures (Figure 14.8). Islamic religious schools and culture centers are likewise now common. Compared with much of the Middle East, however, the post-Soviet Muslim states remain relatively secular. There are no openly Islamic governments in any, and in fact radical Islam is viewed with tremendous suspicion by the leaders of all. Of the six Muslim states in the FSU, only Uzbekistan and Tajikistan have a recent history of radical Islamist movements' causing trouble. In Russia, the Chechnya, Ingushetiya, and Kabardino-Balkariya Republics have known Wahhabi cells.

The most common cultural imprint of Islam on the landscape is undoubtedly the presence of mosques surrounded by slender minaret towers. The prohibition against imagery in Islam may be noticeable in street advertisements, which will use heavy ornamentation, but less revealing or conspicuous imagery. Also common are cemeteries with tombs or mausoleums designed according to Muslim principles.

Other Faiths in the FSU

Roman Catholicism is traditional in Lithuania as well as in western Ukraine and Belarus, where one can visit splendidly decorated Gothic churches. The early Lithuanian kings vacillated between Catholicism and Orthodoxy, choosing the former by the late 14th century, primarily because of the political situation at the time. For a few centuries there was a strong Polish–Lithuanian kingdom that rivaled Russia and Germany in strength. Since the Vatican II council, Catholic prayers have been said in the vernacular (Lithuanian), with a few parishes remaining faithful to the traditional Latin Mass. Compared to U.S. or French Catholics, Lithuanians are more traditional in worship, dress, and political opinions. There are few who question the Pope's authority in such matters as contraception, women's roles in the church, or contemporary worship styles. In fact, some of the splinter Old Catholic and pre-Vatican II Catholics in the United States have Lithuanian backgrounds. In this sense, Lithuania resembles neighboring Poland, one of the two most traditional Catholic countries in Europe (along with Ireland). Although the state in Lithuania is secular, the religion is recognized as important, and there is a lot of popular respect and support for the church (about 50% of the people consider themselves Christian, which is a higher proportion than in most FSU countries).

Occupying a position somewhere between Roman Catholicism and Orthodoxy in the matters of doctrine, the Armenian Apostolic Church is the traditional religion of the Armenians (Figure 14.9). It is believed to have been established by two apostles, Thaddeus and Bartholomew. The Armenians separated from the Orthodox Church after one of the early church councils that discussed the presence of two natures in Christ (the Armenians, along with the Ethiopian and Coptic churches of Egypt, subscribe to the view that there is only one divine nature in Christ—a position known as "monophysitism"). Armenians who live worldwide have their spiritual leader in Lebanon, while the post-Soviet Armenians have theirs in Armenia proper. Interestingly, the externals of the Armenian Church have a lot in common with those of the Catholic Church as the latter looked at the time the two churches separated (about 600 A.D.). For example, Armenian bishops wear mitres very similar to those of Catholic, but not Orthodox, bishops. The Armenian Christians do not have a full icon screen in the churches, but rather a curtain. Their liturgical music is a distinct Armenian chant.

FIGURE 14.9. Armenian churches have a distinct visual style. The religious complex at Etchmiadzin is the worldwide spiritual center of the Armenian Church, where the Chief Hierarch (*catholicos*) resides. *Photo:* K. Van Assche.

Lutheranism became widespread in Estonia and Latvia as the Germans and the Swedes extended their reach over the Baltic region in the 16th–17th centuries. Some sizable pockets of Lutheranism also exist along the Volga River and in parts of Central Asia, where Germans began to settle in the 18th century. However, many of those settlers were actually members of religious minorities who were persecuted by the mainstream Lutherans in Germany. Therefore, German communities in Kazakhstan or Kyrgyzstan today may have distinct Pentecostal, Baptist, or other non-Lutheran Protestant affiliations.

Anglicans have had a presence in Russia for several centuries, as England always needed someone to meet the religious needs of its political and trade representatives in Russia. A beautiful Anglican church located in downtown Moscow looks indistinguishable from some in England itself, but most of its parishioners are visiting British citizens.

All other major Protestant churches are represented in Russia, Ukraine, and some other countries of the FSU. There are also some "homegrown" groups, such as the Russian Evangelical Baptists and the Moscow Church of Christ, but also more recent foreign imports, such as Seventh-Day Adventists, various Evangelical and Pentecostal groups, the Latter-Day Saints (Mormons), and Jehovah's Witnesses.

Buddhists traditionally lived in Buryatia and Kalmykia. Both these republics within Russia are areas of Mongolian settlement as a result of the Tatar–Mongol conquest and later migrations from Central Asia. The specific version of Buddhism primarily practiced in Russia is Lamaism. Buddhism was first officially recognized as a traditional religion in Russia by a decree of Empress Elizabeth in 1741. The largest Buddhist complex in Russia, Ivolginsky *daitsan*, is located near Ulan-Ude in Buryatia (Figure 14.10).

The traditional religion of Siberian indigenous peoples is "animism," also known as "shamanism." Siberian shamanism is broadly similar to the religion of Native Americans, with many

FIGURE 14.10. Buddhism is widespread in Buryatiya Republic, on the border with Mongolia. The *daitsan* shown here has been recently constructed near Ulan-Ude. *Photo:* P. Safonov.

FIGURE 14.11. Burkhanism in the Altay combines shamanistic and lamaistic elements. Prayer flags are common near holy springs and waterfalls and signify offers to the local spirits. *Photo:* Author.

of the same elements of spirit worship through dance, trance, and sacrifice. The same powerful animals and plants are worshiped on both sides of the Pacific (wolf, eagle, bear, whale, walrus, pine, oak). A handful of people in the distant corners of Chukotka Peninsula and in Yakutia may still be found who actually practice it. Generally, the fate of this religion's adherents was conversion to Christianity first, and then to Soviet atheism later. Shamanistic beliefs also survive in the southern mountains of Siberia, especially in the Altay and the Sayans, where they are combined with Buddhist and Christian elements—as, for example, in Burkhanism, practiced by the Altay people (Figure 14.11). Recently there has been a resurgence in shamanism among the young urban people in Siberia, frequently mixed with nationalism.

Judaism

In the western urban centers of the present-day FSU, especially in Ukraine, Belarus, Moldova, and western Russia, Judaism played a traditionally important role from the Middle Ages onward. Jewish settlements existed primarily in the western part of the Russian Empire, because Judaism diffused into the region primarily through Western and Central Europe, where it existed uneasily amidst the predominantly Christian population. (An earlier kingdom of the nonethnically Jewish Khazars, who practiced Judaism in the 7th–8th centuries, existed in and around the Crimea.) The Pale of Settlement law of the tsarist period allowed permanent Jewish settlements only in the western part of the country, pretty much confining them to Lithuania, Belarus, Ukraine, Moldova, and Poland. The law was first created by Catherine the Great in 1791, a German, who was afraid of the rising influence of the educated Jewish middle class. Jews could abandon their religion and become Christians, in which case all the benefits of Russian citizenship would be conferred on them, and they would then be able to leave the Pale. Some took full advantage of the opportunity, but many did not. Historically, there were large Jewish communities in the big cities of Central Asia and the Caucasus as well, and some small pockets of the Jewish faithful remain there even today.

In the early 20th century, hundreds of thousands of mostly urban poor Jews left the increasingly anti-Semitic Russian Empire to avoid pogroms. Some went to Western Europe, but the majority ended up in North America, particularly New York City. The remaining communities (*shtetls*) were decimated by the civil war of 1917–1922, collectivization, and finally the Holocaust of World War II. It is estimated that over 1 million of the 6 million or so of the Holocaust's victims came from the Soviet Union, mainly from Belarus, Lithuania, and western Ukraine.

The Soviet Union abolished all inequalities based on religion in theory, but not in reality. The Jewish Autonomous Oblast in the Russian Far East was created by Stalin with an idea of relocating the Jews from Central Russia to a new "homeland" along the Amur River. Today its population is only 1.2% Jewish, but it does house some important Jewish cultural elements (including a theater, a university, and a museum) in Birobidzhan. In the early Soviet period, many Soviet leaders were actually of Jewish ethnicity (e.g., Trotsky, Zinovyev, and Kamenev). However, after the purges of 1937–1940 the party leadership was decidedly not Jewish any more, and there was much personal antagonism between the Russians and the Jews at the local level as well. Because few of the ethnic Jews of the Soviet period were religious, the anti-Jewish prejudice was really more against the distinct ethnicity than against Judaism as a religion.

In the 1970s and 1980s, on the other hand, many people from Jewish backgrounds had a chance to emigrate to Israel and other countries because they were sponsored by the Jewish communities there, whereas it was not possible for ethnic Russians to leave the country. Fewer than 1 million Jews remain in today's FSU, and most do not practice their religion. Over 1 million emigrated to Israel, and a few hundred thousand to Germany and the United States (most of the latter in the 10-year period between 1988 and 1998). Nevertheless, large synagogues exist in Moscow, Nizhniy Novgorod, Minsk, Kiev, Odessa, Kishinev, Tashkent, and other major cities. Russian Judaism is united in the All-Russia Jewish Council, with a chief rabbi in Moscow. Unlike in the United States, the majority of synagogues in Russia are centers of Orthodox, not Reform Judaism.

Anti-Semitism, though illegal, is still common in Russia today. In fact, several prominent members of the Duma and regional governors have made openly anti-Semitic remarks on numerous occasions. Even more ominous is the rise in openly xenophobic hate groups, including real "skinheads," in the new Russia. At the same time, the vast majority of people in the region remain tolerant, and more inclusive environments are being created at workplaces and in schools.

Nonreligious People and the Politics of Religion in Russia Today

In recent Russian history, there has been some controversy over the role religion should play in the politics of the state. On the one hand, the Russian state today is explicitly secular, with full separation between church and state since 1917. On the other hand, some religions are defined as "traditional" for the peoples of Russia, and others are not. As noted above, the traditional religions include Orthodox Christianity, Roman Catholicism, Lutheranism, Islam, Judaism, and Buddhism. Although Orthodoxy is not a state religion, Russia's recent leadership has been frequently seen at various church functions and ceremonies, and many members of the Putin–Medvedev government claim to practice their religion regularly. There is also no doubt that a lot of public funding, however defined, has gone into restoring churches and monasteries around Russia. In other FSU republics, the construction of mosques and other structures may likewise be partially funded by central or local governments. This is justified in part by the argument that the atheistic state destroyed many religious landmarks over the course of Soviet history and is now expected to make reparations. At the same time, many people question the exact nature and extent of the state's meddling in religious affairs.

In a society as corrupt as Russia's today, with most of the leaders representing only one religion, serious religious bias may result. In fact, when the Law on Religions was initially passed during Yeltsin's presidency in the mid-1990s, many Western observers were led to believe that very shortly thereafter there would be a widespread crackdown against all forms of religions not explicitly sanctioned by the state. This has not happened. Some particularly notorious sects, including the suicidal Japanese cult of Aum Shinrikyo, were in fact shut down, and some Western-sponsored groups indeed experienced increased difficulties with their official registration. However, no major crackdown on religious freedoms has occurred, as far as any observers can tell. In fact, when visiting any big city in Russia today, you are likely to be greeted by religious tract pushers of one sort or another at the entrance to any subway station.

Representatives of the Russian Orthodox Church claim that it receives very little support (financial or otherwise) from government officials. Early in the Yeltsin period, the church received the privilege of importing some Western goods duty-free, as a way to sponsor its rebuilding activities at home. Although this was not a bad idea in itself, most of the money was made through importation of cigarettes, which arguably was not the healthiest arrangement. Also, other nonprofit groups complained that the church received an unfair privilege, shared by only some sports' and veterans' groups.

There has been much discussion of how much religious instruction can or should be allowed in Russian public schools. Religious ideas could be conceivably taught in Russia in the context of a "religious culture" class, whereby it is recognized as a cultural tradition and permitted by the constitution. There is much public support for including some religious ideas, whether Christian, Muslim, or Jewish, in a course focusing on ethics. However, questions arise as to what the exact content of the class will be, who will be qualified to teach it (clergy or regular teachers), and what to do about students who may wish not to be included in such a course. There is an ongoing debate on what would be best for the nation as a whole at the moment, but generally the idea of religious instruction at schools meets with considerable public opposition.

Indeed, the majority of the population in Russia today leads a distinctly nonreligious lifestyle. Although the number of self-professed nonbelievers (22%) is low, it is higher than the number of those actively practicing Orthodoxy (8–12%). Many of the least religious people grew up in the Soviet period. Atheists in Russia have gained publicity in recent years, as when the Nobel Prize laureate academician V. L. Ginzburg went public with his denunciation of the religious worldview in general as counterproductive medieval gibberish.

Also, many people in Russia today embrace dual religious identities—practicing astrology and Christianity together, for example. About 25% embrace a vague syncretic worldview that recognizes the existence of spirits, karma, and reincarnation, and affirms divination, talismans, tarot, and yoga as legitimate practices, while simultaneously professing adherence to the Russian Orthodox Church (which vehemently condemns all of these things). Even among "real" believers, the adherence can be pretty minimal. Some people show up in midservice just to light a candle, without staying for more than 5 minutes out of the 2-hour long liturgy.

Explicitly religious conflicts in Russia, or anywhere else in the FSU, are thankfully rare. Members of the clergy are sometimes targeted as victims of hate crimes (e.g., the murder of a prominent missionary priest in Moscow in the fall of 2009 received much attention, because the alleged reason for the killing was the priest's work with Muslim converts to Christianity). Although the continuing conflict in Chechnya is frequently cast in the light of Christian–Islamic antagonism, it is clearly a political struggle primarily focused on control over the land and minds of Chechnya's inhabitants. The major Chechen warlords did receive support from many international Islamic sources (some as notorious as Al–Qaeda), but their main goal, at least in the early stages, was political independence rather than creation of an Islamic state of Ichkeriya per se. However, once the conflict began, it was very hard to avoid references to the identifying religion on both sides, as frequently happens in many wars around the world.

My grandmother comes from the city of Kasimov in Ryazan Oblast, Central Russia, where for centuries Muslim Tatars lived alongside Orthodox Russians in peaceful coexistence. There were churches and mosques in town, and while Christians prayed on Sundays and Muslims on Fridays, members of both groups met each other at the city market on Saturdays. This model worked for centuries, and in fact it is much more normative in the region than the occasional conflicts that plague newly established frontiers, despite the international news coverage of only the latter.

In addition to Muslim–Orthodox and Orthodox–Protestant relations, the two lines of religious antagonism typical of the FSU are residual anti-Semitism (see "Judaism," above) and

Orthodox–Catholic relations. The relations between the two largest Christian bodies, Orthodoxy and Roman Catholicism, have never been particularly warm since the Great Schism of 1054 A.D.—and especially not since the Catholic sack of the Byzantine capital, Constantinople, in the Fourth Crusade in 1204 A.D. Attempts at unity were made repeatedly in the Middle Ages, primarily upon the initiative of the Popes, but all of these were rebuffed by Orthodox leaders on the grounds that the Popes wanted unity primarily for political rather than theological reasons. Besides some real theological disagreements—for instance, belief in the Holy Spirit as proceeding from the Father *and* the Son in the West (only the Father in the East), the new Catholic dogmas of papal infallibility and the immaculate conception of the Virgin Mary, and the questions of indulgences and purgatory (none of which the Eastern Churches recognize)—there were some very real geopolitical motives at play as well.

FIGURE 14.12. Orthodox or Byzantine Catholic? In Belarus and western Ukraine, many church buildings have repeatedly passed back and forth from Orthodox to Byzantine Catholic control. The Byzantine Catholics, or Uniates, retain the Orthodox liturgy but recognize the Pope of Rome as their spiritual head. *Photo:* P. Miltenoff.

Some Orthodox Christians came into full union with Rome in 1596 A.D. in the Act of Union at Brest-Litovsk. Known now as the Uniates, or Byzantine Catholics, these Christians—primarily living in western Ukraine, Slovakia, and parts of Moldova and Belarus—were accepted into full communion with Rome, but were allowed to keep their Orthodox liturgy, icons, and married priesthood. However, they were denounced by the Orthodox bishops in Russia and Greece as schismatics and were marginalized in the Russian Empire. Since the Uniates lived between primarily Catholic Poland and Austro-Hungary and primarily Orthodox Russia, their fate was either good or bad, depending on who was in charge of their land at a particular moment in time. Both the Uniates and the Orthodox Church were persecuted during the Soviet period, but after World War II, when many Eastern European lands were absorbed into the Soviet Union, the Soviets actually encouraged the Orthodox communities there to seize some of the Uniate churches (Figure 14.12). The fall of Communism provided a hope that all sides would be finally able to practice their religions alongside each other in the newly independent nations. This did not happen, because both the Russian and Roman churches would openly support their respective sides, trying to win the local authorities to their cause. Moreover, several Orthodox churches in western Ukraine were seized by the Uniates in the 1990s, with the full complicity of the local authorities. In Lvov, for example, many Orthodox parishes lost their buildings without any compensation. At the same time, the Uniates are suspect in Russia proper. The Uniate issue remains one of the main reasons why the Pope of Rome and the Patriarch of Moscow have yet to meet in person.

At the same time, the Roman Catholic Church has been trying to extend its reach across Russia. It is establishing new parishes and dioceses, while arguing that for decades it was deprived of the opportunity to serve existing Catholics, especially in Siberia and the Far East. Although it is not barred from active religious practice by law in Russia and is even recognized as a "traditional" faith, the Catholic Church is viewed with the utmost suspicion by the Orthodox Church, because it is perceived as a powerful political organization influenced by secularized Western ideas of what the church should be like. Of course, the Catholic Church is also perceived as a strong competitor for the souls, the minds, and the purses of the faithful. Orthodox leaders are concerned about potential defections of their own members to the Western faith.

In practice, struggling Orthodox or Catholic parishes in, for example, remote Siberia have much in common: Both are poor and short of priests, with the faithful scattered over a huge, inhospitable terrain. In some cities (e.g., Vladivostok), local Orthodox and Catholic parishes actually join forces for noble humanitarian causes, such as providing food and shelter to the homeless or helping orphaned children. The mistrust, however, runs very deep and is farther reinforced by the anti-Western rhetoric of many regional and federal politicians in Russia.

Diet

As explained above, religion clearly influences many choices in people's lives, including things they eat and wear. With respect to both, however, climate plays an even greater role. Many plants and animals cannot survive cold winters, thus limiting food choices. At the same time, the cold weather has made warm winter clothing, primarily made of wool and furs, a necessity for the Russians and other inhabitants of the region. This section focuses primarily on the Russian diet; Turkic/Central Asian variants are briefly mentioned at the end.

The Russian cuisine is of legendary quality. In fact, in a recent international poll it was rated among the top three tastiest worldwide, along with Italian and Japanese. Its main ingredients are wheat, beef, and dairy, so it is not greatly different from the mainstream European or American diets; all are direct descendants of the diet of the Middle East/Asia Minor, the region where both wheat and cattle were domesticated. The staple grains are rye, barley, and oats in the north, and wheat, buckwheat, and corn in the south. Soy is becoming more commonly used too, but is not a component of any traditional meal.

The Russian diet is generally heavy on carbohydrates and fats, both important for providing energy during the cold winter months. For example, the classic Russian open-faced sandwich (*buterbrod*) consists of white bread, a thin layer of pure unsalted butter, and a slice of either cheese or sausage on top. Many hearty soups are beef-based, such as *borscht* (which also includes cabbage and beets) and *schi* (which includes cabbage only). Wild game (e.g., deer, boar, bear, rabbit, goose, duck, snipes, partridge, grouse, and quail) would traditionally complement the meats obtained from cows, pigs, and sheep. The choice of vegetables and fruit is very limited, because few of these can be grown in Russia. The staple vegetables are cabbage, beets, green peas, carrots, squash, and turnips. Turnips were the main starchy food before the potato was introduced during the reign of Peter the Great in the early 18th century. Tomatoes and cucumbers are very common in salads. The essential two herbs are parsley and dill.

The main fruits are wild berries (raspberries, strawberries, lingonberries, blueberries, cranberries), as well as apples, pears, plums, and (in the south) apricots, peaches, and cherries. Berries from the forest are processed into sweet *varenye* (boiled fruit in very heavy syrup, but no pectin), which is added to tea. Russians are very fond of fish and other seafood. Over 50 kinds of fish (both freshwater and saltwater) were commonly eaten before the Revolution, as evidenced by the stories of Shmelev, Leskov, Turgenev, and others. Some of this bounty, especially eel and sturgeon, is now threatened with extinction. Another important food item is mushrooms, which are collected wild in the forest. A few dozen species are eaten fried, boiled, or pickled.

The traditional drinks include *kvass* (a mildly alcoholic fermented rye malt beverage), vodka (the best is made from rye and wheat filtered through birch charcoal), and hot black tea. Juice was not commonly available in winter, so the Russians invented the *compote* (a drink consisting of boiled dried fruit) and added fruit to tea as described above. Some more exotic drinks from the old times include *sbiten*, made from honey and spices, and *kisel*, made from cranberries. Although Russians have now developed quite a taste for beer, wine, coffee, and soda, consumption of those beverages was very limited even 20 years ago.

The Russian culture gives high importance to food. Traditionally, three meals a day are eaten, somewhat later than is customary in North America or northern Europe (e.g., breakfast at 8:00 A.M., dinner at 1:00 P.M., and supper at 7:00–8:00 P.M.). The midday meal is the biggest, consisting of a salad, soup, a main course with meat, and compote or varenye. People would formerly spend about an hour at the midday dinner, with leisurely conversations over food. This is no longer as common now, because Western-style office schedules reduce available time. Until recently, very little processed food was used in cooking; this required more time for food preparation at home, but resulted in a much healthier diet and more satisfactory taste. Frozen TV dinners are still viewed with suspicion by many Russians as "fake food," but are now commonly available in stores.

Most food in the Russian diet is grown domestically. In the past few years, an increasing proportion of staples have had to be imported (e.g., dill from Europe or pickles from India), reflecting the poor state of domestic agriculture. Some tropical items, most notably black tea, are imported from India or Sri Lanka. Russia has limited tea plantations near Sochi along the Black Sea coast. Sugar comes either from domestic sugar beets or from tropical sugar cane. Of course, all tropical fruits must be imported. Russian food is well balanced with respect to spices; it is "just enough" salty, sweet, or spicy for most people. However, the southern regions of the FSU, especially Georgia, have notoriously spicy food that rivals some South Asian foods in hotness.

The Ukrainian diet is generally very similar to the Russian, with some specialties shared by both cultures (e.g., borscht and the ravioli-like *pelmeni*). One famous Ukrainian food is *salo*, which is basically salted pig fat consumed raw as a snack, sometimes accompanied by shots of *horylka* (Ukrainian vodka). The Ukrainian diet has more dairy and fresh produce items than the Russian.

Central Asia and the Caucasus have their own unique diets, which emphasize lamb, goat, local spices, olive oil, flatbread (*lavash*), vegetables, and fruit. In many respects, the Georgian and Armenian cuisines are simply versions of the famous "Mediterranean diet." Red wine is a Georgian specialty, made from unique grape varieties grown only in this country—especially the legendary *saperavi* grapes, with a semisweet, exotic taste, and the darkest color of any grape. Also common are fermented milk beverages and foods (e.g., kefir and cheeses). In the Muslim regions of Central Asia and the Caucasus, pork, of course, cannot be eaten by the observant Muslims or Jews, so lamb (mutton) and goat are the most common meats. Beef may be eaten too, but is usually too expensive to produce on the dry rangelands. Another notable type of Central Asian meat is horsemeat, generally eaten either boiled or dried among the Kazakh and Mongolian cultures. Fermented mares' milk, called *kumys*, is both traditional and popular. In Turkmenistan, camels' milk is consumed too. Also, members of Central Asian cultures drink a lot of hot green tea with milk and butter, to stave off thirst. You might expect iced tea to work better, but for centuries Central Asians have used the old recipe, and it always works.

Dress

Both religion and climate have historically shaped what people wear. In Russia, the main dress today is essentially European, with little noticeable difference between Moscow and Paris (Figure 14.13). In the provinces, however, people may still wear workers' clothing left over from Soviet times (e.g., oversized cotton-stuffed jackets in winter, striped sailors' shirts in summer, and huge rubber or felt boots—a necessity, given the absence of pavement). In winter, men wear fur hats with ear flaps, called *ushanka*. These

FIGURE 14.13. Russians wear modern, European-style dress, whether casual or formal. The Moscow dress code is a bit stricter than in an average U.S. city, but is generally not very formal. However, great variety exists among different groups of people in the provinces. *Photo:* Author.

hats are made of rabbit, dog, fox, or wolf hides, and (for much higher prices) of beaver, mink, or even sable. Women would traditionally cover their heads with woolen or silk scarves or shawls; today they wear anything that looks nice and is in fashion (Figure 14.14). Many prefer to let their hair show and wear no head covering at all, despite the cold.

Warm overcoats are a necessity in winter. The traditional ones (*tulup*) were made of sheepskins and were very warm, but heavy. The nobility could afford beaver, mink, weasel, or even sable fur coats. Even today, you are much more likely to see a Russian than a Western European dressed in real fur, both as a fashion statement and also as a necessity, given the climate.

The pants worn in Russia are usually long. Shorts are not commonly worn even in the warmest months, and frankly it never gets warm enough in much of the country to require them. Women would traditionally wear dresses and skirts, but since Soviet times they increasingly wear much the same clothing as men—including long pants or trousers, as dictated by the needs of the working class or by an overt attempt to create gender neutrality. Skirts are still required in Orthodox and many other churches. In the old Russia, each region would have its own dress embroidery style. These survive today primarily only in ethnographic museums, although you may have luck finding some people still wearing traditionally embroidered clothing in remote villages in Ukraine or Belarus (Figure 14.15).

In the Muslim cultures of the FSU, the traditional costume would be likewise long, with ample head and other coverings for women (Figure 14.16). The decades of Soviet rule changed this rather radically, with very few people wearing any ethnic clothing outside of some cultural events. However, there is a growing trend toward wearing national dress for fun and for religious observances among the new wealthy elites in Kazakhstan, Uzbekistan, and some other republics, as well as among ardent new followers of Islam. Men in the Caucasus wear long coats with belts (to which daggers are strapped), and long, tailored pants underneath. In Central Asia, given its warmer climate, long yellow or white robes are more common. The head cover is either a tall sheepskin hat of a distinctive type (*papakha*) in the Caucasus, or a round or square thin black skullcap (*tyubeteika*) in Central Asia, especially in Uzbekistan and Tajikistan. Traditional Jewish

FIGURE 14.14. Russians have to dress warmly in winter; long goose-down jackets and fur or wool hats are a must. This picture was taken in Yekaterinburg in early March. *Photo:* I. Tarabrina.

FIGURE 14.15. Traditional Belarusian long dress with embroidery. *Photo:* P. Miltenoff.

FIGURE 14.16. Kazakh traditional dress on display in an ethnographic museum. *Photo:* Author.

dress has long ago disappeared. What we now think of as "Jewish" attire for Orthodox Jewish men is in fact a costume based on the dress of 18th-century Polish urban dwellers (black hats, jackets, etc.). Few Russian Jews wear religious clothing even to the synagogues.

REVIEW QUESTIONS

1. Name the main religions of the FSU. Where are they found?
2. Describe the elements of an Orthodox cultural landscape.
3. Which ethnic groups in Russia are "polyconfessional" (i.e., may belong to more than one religion), and which religious groups are "polyethnic" (i.e., embrace members of more than one ethnicity)?
4. Speculate on your future as a restaurant owner in any republic of the FSU. Make sure to investigate the republic's religious makeup before proposing menus tailored to the predominant population.
5. How is your diet similar to or different from the typical Russian diet described in this chapter?
6. Explain why horse meat is generally an uncommon food choice in the United States. How would you feel about someone offering you a piece of dried horse over dinner in a Central Asian country? What would you do?
7. What is the stereotypical dress of the Soviet period, according to Hollywood? Do you think that this is an accurate representation? If it is, how do you think dress has changed in Russia since the fall of Communism?

EXERCISES

1. Research the history of a particular monastery (you can use one on the list in the "Orthodox Religious Landscapes" section). Try to determine the geographic factors that led to its establishment at its site.
2. Schedule a visit to an Orthodox church in the city where you live. Look in the Yellow Pages under "Orthodox–Eastern" churches to find one. You can also use an online locator (*www.orthodoxyinamerica.org*).
3. Do additional research and a classroom presentation on some other religion of the FSU: Sunni or Shia Islam, the Armenian Apostolic Church, Lutheranism, Roman Catholicism, Judaism, or Buddhism. What impact has this religion had on the cultural landscape of the region(s) where it is found?

Further Reading

Billington, J. H. (2004). *Russia in search of itself.* Washington, DC: Woodrow Wilson Center Press.

Luxmoore, J., & Babiuch, J. (1999). *The Vatican and the red flag: The struggle for the soul of Eastern Europe.* New York: Chapman.

Magosci, P. R. (2007). *Ukraine: An illustrated history.* Seattle: University of Washington Press.

Park, C. C. (1994). *Sacred worlds: An introduction to geography and religion.* London: Routledge.

Pospielovsky, D. (1984). *The Orthodox Church in the history of Russia.* New York: St. Vladimir Seminary Press.

Robson, R. R. (2007). *Old believers in modern Russia.* DeKalb: Northern Illinois University Press.

Ro'i, Y. (2000). *Islam in the Soviet Union: From the second World War to Gorbachev.* New York: Columbia University Press.

Rothenberg, J. (1971). *The Jewish religion in the Soviet Union.* New York: Ktav.

Shirley, E. B., Jr., & Rowe, M. (Eds.). (1989). *Candle in the wind: Religion in the Soviet Union.* Washington, DC: Ethics and Public Policy Center.

Sidorov, D. (2000). National monumentalization and the politics of scale: The resurrections of the Cathedral of Christ the Savior in Moscow. *Annals of the Association of American Geographers, 90*(3), 548–572.

Sidorov, D. (2001). *Orthodoxy and difference: Essays on the geography of the Russian Orthodox Church(es) in the 20th century*. San Jose, CA: Pickwick.

Snelling, J. (1993). *Buddhism in Russia: The story of Agvan Dorzhiev, Lhasa's emissary to the Tsar.* Rockport, MA: Element Books.

Ware, T. (1993). *The Orthodox Church.* London: Penguin.

Websites

www.adherents.com—Published statistics on the number of followers of any religion.

www.mospat.ru/en—Moscow Patriarchate of Russia.

www.orthodoxinfo.com—General information about all aspects of the Orthodox Church for English-language speakers.

www.russianfoods.com—An online Russian food store; you can check out traditional food selections, as well as some cultural items.

CHAPTER 15

Education, Arts, Sciences, and Sports

Education, together with its outcomes in arts, sciences, and sports, is an important subject of geographic research. Each country and region has its own distinct style of education and its own educational system. Comparisons among countries, and among regions within each country, must be made if we are to understand the particular nature of each place. It is impossible for us to comprehend what is happening in politics or economics, for example, unless we also know the educational background of the society in question.

The Soviet Union was proud to be one of the most educated societies on earth, achieving virtually 100% literacy by the early 1970s. Soviet education was universal, public, comprehensive, and free; what still astonishes many Americans is that it was free all the way through college. The U.S.S.R. also had a world-class scientific research program and was famous for its accomplishments in arts and sports, although these were not uniformly distributed. However, much has changed since the breakup of the Soviet Union. Today one can still get a free education in any republic, but there are new hidden or indirect costs that used to be either nonexistent (textbooks or tuition) or very low (paper and other school supplies) in the Soviet period. This chapter first considers the Soviet educational system and the changes made to it in the post-Soviet era. It then considers achievements in the areas of arts, sciences, and sports. As usual, these are discussed primarily from the perspective of Russia today, with some examples drawn from other former Soviet Union (FSU) republics.

Education

The Soviet system of education was based on the old, tsarist-period model, which was good but incomplete. In 19th-century Russia, only the privileged classes had a chance of receiving an education through college. The education of the nobility during this period was of excellent quality. Youth from noble families were educated by private tutors at home in early childhood. It was common for aristocrats' children to grow up speaking fluent French, some German, a little English, and only occasionally Russian. Boys would then enroll in a "gymnasium" or "lyceum" at the high school level. After this, some would join the army's cadet corps to become career officers. Others would enroll at a university, the first one in the country having been established in Moscow in 1755 by a decree of Empress Elizabeth. Estonia has the oldest university in the FSU at Tartu, which was established by King

Gustavus Adolphus of Sweden in 1632. Girls had few higher education options until the late 19th century, when college-level classes became available to them. The clergy, who constituted their own class of society, prepared their sons to become clergymen through parochial schools and seminaries. In contrast, the working class, and especially peasants, received very little formal schooling—at best, 4 years at a local parochial school. After the liberal reforms of 1861, it became fashionable for landlords to establish secular local schools run by the *zemstvo* (the local council), as well described in Tolstoy's *Anna Karenina*. Despite all this, less than half of the total population was literate by the time of the Bolshevik Revolution.

The Soviet System of Primary and Secondary Education

The Soviet government had very progressive ideas about universal schooling for all, to ensure both a qualified workforce and compliant Marxist citizens. Universal, compulsory 8-year education became the norm by the 1930s, and 10-year education by the 1950s. The normal school week lasted 6 days (including Saturdays), but school days were shorter than in the United States, with classes out by 1:30 in the afternoon. After-school programs were also available. There were many specialized schools (with emphases on math, physics, arts, languages, etc.) and, in remote areas, boarding schools with a 5-day week. In the early Soviet period, the schools were coed; they were then replaced with separate classrooms for boys and girls in 1943, but then went coed again by the mid-1950s. Experimentation with the curriculum was continual. Anton Makarenko, one of such experimenters in the 1920s, emphasized collaborative learning environments.

Of course, the main emphasis of the Soviet school system was on raising loyal citizens of the socialist state. To that effect, classes on the Soviet version of world history, the Marxist theory of economics, and Marxist philosophy, as well as antireligion classes, were offered. In high school, basic military training was also provided to both men and women. The rest of the curriculum emphasized mathematics, Russian language and literature, natural sciences, history, geography, and foreign languages. Social sciences (sociology, psychology, economics) were taught very little, because they were thought to be too subjective, reactionary, and contradictory to Marxist precepts. Foreign-language instruction generally started in the fifth grade. About 80% of students learned English, with substantial minorities learning German or French. Other world languages (Arabic, Japanese, Hindi, or Spanish) were available at a few specialized language schools, which one could enter on a competitive basis.

One great advantage of the Soviet curriculum was its uniformity. This ensured that all the material was learned everywhere in the country in the same grade, so that students who moved from one school district to another would still be literally "on the same page." In addition, all students wore school uniforms, patterned after those of the prerevolutionary gymnasia. Boys wore dark blue pants, white shirts, and dark blue jackets; girls wore brown dresses with aprons (black on regular school days, and white for major state holidays). Uniforms reduced the anxiety associated with deciding on what to wear and instilled respect for authority. The choice of regular children's clothes in stores was notoriously limited anyway, so having uniforms was helpful to parents.

The difficulty of the curriculum was increased gradually. Schooling in demanding subjects (math, physics, and biology) started early, usually in fifth grade. For example, in biology classes, botany would be taught in fifth grade, zoology in sixth, and human anatomy in seventh. Moreover, in contrast to some U.S. curricula in which the natural sciences are taught all together even in high school, the U.S.S.R. system would add subjects while continuing to teach the earlier ones. For example, physics would start in the sixth grade; when chemistry was added in the seventh, more physics would also be taught.

After 8 years of schooling and after passing exit exams, students would graduate from middle school. Depending on their academic aptitude and aspirations, they would then either enroll in high school (9th–10th grades; 11th grade was added in the late 1980s) or enter a professional–technical school (known by its Russian acronym, PTU). A PTU gave its students exposure to the high school material but in a less demanding way, while additionally providing the necessary skills for a blue-collar profession. Typical PTUs would train factory workers, carpenters, bus drivers, auto mechanics, and the like. Some programs were in high demand, such as the ones training jewelers or restaurant chefs—both lucrative professions with possibilities of making private money on the black market. Students who stayed at the regular high school would typically attempt to enter university upon graduation. There were exit exams in a few subjects at the end of the 10th grade that had to be successfully completed. A typical set of questions for 10th-grade graduation was comparable in difficulty to a moderately difficult exam for a freshman-level U.S. college course. There was also a third option: a technical college (*technikum*). Students would enter them after eight grades, as in the case of a PTU, but continue studies a year longer to learn a more advanced profession (e.g., accountant or electrician). Graduates of many technical colleges were allowed to transfer to universities.

Besides the official school program, there was a wealth of after-school opportunities, ranging from music programs to sports camps to young-naturalist clubs. Many of these were conducted by enthusiastic teachers at the regular schools after hours. Others were conducted at independently run youth clubs. Every large city had at least one of those, commonly known as the City Young Pioneer Palace (or, nowadays, the Palace of Youth). Many smaller municipal districts had one as well. Even in the countryside there were similar opportunities provided by local municipal units or state farms. In fact, some of the best schooling and after-school opportunities were offered not in the biggest cities, but frequently in medium-size provincial towns (e.g., Penza or Murmansk), where there were more demanding teachers and more incentive to try harder to make it to a big university someplace else.

An important aspect of the Soviet education was the Young Pioneer movement. The Young Pioneers were a Communist version of Scouts. Primary school children were automatically enrolled in *Oktyabryata* (Young October Youth) in the first grade by being given a five-pointed red star badge with a picture of the young Lenin on it to wear, and told to love the Motherland and Lenin. In third or fourth grade, virtually all children would then be enrolled as Young Pioneers. The Young Pioneers wore bright red neckties (Figure 15.1), and were supposed to swear an oath of loyalty to the Soviet state. This presented a problem for a handful of religious youth, who would sometimes object to the oath on religious grounds. The repercussions of doing so could be severe, all the way to expulsion from school; parents could also be sanctioned by their employers. Not surprisingly, then, over 95% of all schoolchildren of the Soviet period were Young Pioneers. When students turned fourteen, they could join *Komsomol* (the Young Communist League). This required passing a test on the basic history of the movement and swearing another oath. Eventually some Komsomol members would end up as full members of the Communist Party. The main incentives to join Komsomol in the late Soviet period were career advancement and easier access to the best university programs.

It is worth noting that although many Young Pioneer and Komsomol projects involved indoctrination in Communist ideology, most emphasized developing a collective spirit while engaged in useful and even fun activities. Many worthy social initiatives were carried out under the Komsomol banner. For example, there was a tradition of collecting scrap metal and newspapers for recycling once or twice a year. Schools would compete, winners would receive prizes, junk would be cleared out of local neighborhoods, and

FIGURE 15.1. A Soviet photo (ca. 1988) showing a middle school class with its teacher in Biysk, Altaysky Kray. Notice the Young Pioneers' neckties. *Photo:* I. Tarabrina.

FIGURE 15.2. A World War II memorial in Sergiev Posad. In the Soviet period, local schools would typically take pride in maintaining such monuments at no expense to the government. Patriotic education is again emerging as a priority. *Photo:* Author.

of course the environment would benefit. Other worthy projects included after-school poetry and art classes, agricultural experiments in the school garden, sports events, summer camps, concerts and plays, and charitable work to help war veterans or needy families (Figure 15.2).

The Soviet System of University Education

A Soviet college education was offered free of charge to all qualified students who could pass the entrance exams. There were universities offering 5-year degrees in all of the humanities and sciences, engineering, law, medicine, and so forth; there were also technical/engineering institutes (many are now known as technical universities). The difference between a university and an institute was in the breadth of the programs offered. Moscow State University (MSU), for example, had 29 schools, called "faculties," offering degrees in every imaginable subject (Figure 15.3). Moscow Physical Technical Institute, on the other hand, offered mainly programs in physics, chemistry, or engineering, but not necessarily programs in history or foreign languages. The total enrollment at MSU (among the largest in the country) was about 35,000—smaller than some of the biggest U.S. universities, but still large. Regional universities would typically enroll between 3,000 and 10,000 students.

In order to take advantage of the free college education, one needed to prove oneself. This was done in a few ways. High school seniors graduating with honors (a near-perfect grade point average) received a gold medal, were exempt from college entrance exams, and could apply to any university they chose. The vast majority had to take between three and four entrance exams. There was no national test similar to the SAT in the United States, and it was not possible to apply to more than one university at a time, so the choice had to be made very carefully. Each university had its own system of tests designed and administered by the faculty. For example, the School of Biology at MSU would test applicants in math (a written test with five very difficult problems to be worked out in about 4 hours), the Russian language (a critical written composition based on a choice of three topics pertaining to Russian literature), chemistry (an oral exam based on three broad questions in organic and inorganic chemistry), and biology (an

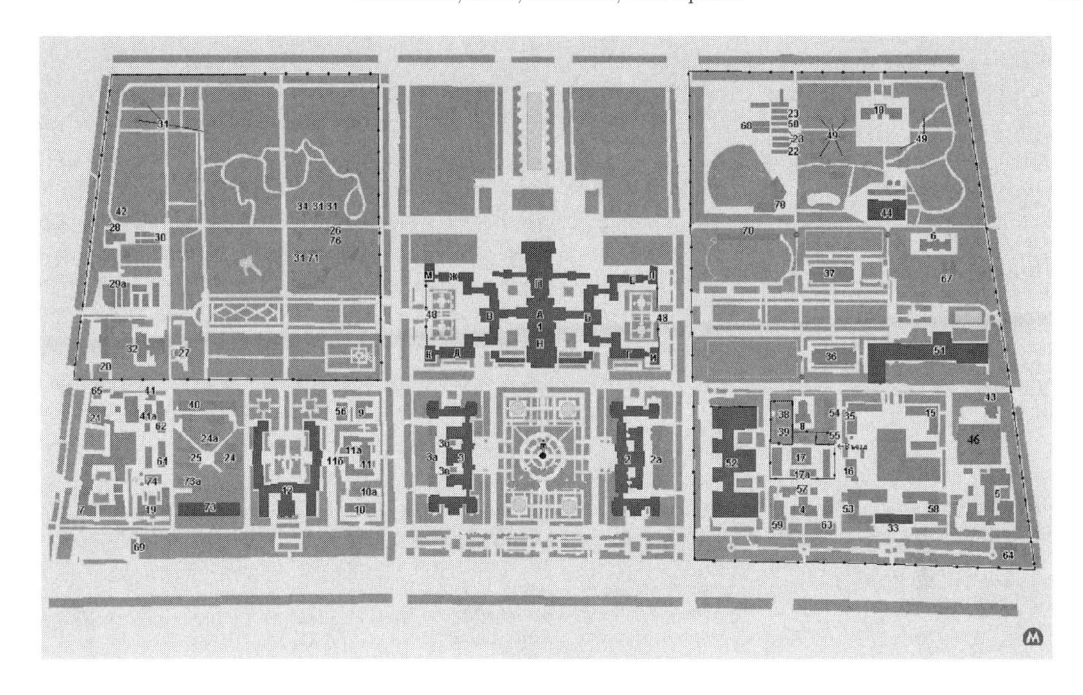

FIGURE 15.3. The Moscow State University (MSU) campus (built 1948–1953) covers 200 ha. The main building, which at 240 m is the tallest university building in the world according to *The Guinness Book of World Records*, houses the math, geology, and geography schools (see Chapter 7, Vignette 7.1). Additional buildings house the schools of physics, chemistry, biology, economics, history, philology, computer science, and others, for a current total of 26 schools. There are also affiliated research institutes, dormitories, sports facilities, and a botanical garden on the premises.

oral exam similarly based on three broad questions in biology). Each year the School of Biology at MSU would accept about 200 new students from a number of applicants ranging between 1,500 and 2,000 in a typical year. About one-third would flunk the math test; another third would be eliminated by the written composition test; and the remainder would struggle with the chemistry and biology oral exams. Eventual winners were those who not only did not fail any of the tests, but generally received a minimum of two A's and two B's. The students who failed could try again a year later, or apply to other, less competitive schools with lower standards of admission. Because MSU was so competitive, one needed to know considerably more than was taught in a regular school curriculum, so hiring a private tutor in high school was almost a necessity. At least in my own experience, the critical test was the math exam, for which I had to prepare for about 2 years by spending between 3 and 4 hours per week solving sample problems.

The U.S.S.R. was one of the top five providers of college education worldwide, with over 126,000 foreign students enrolled in 1990. Only the United States, the United Kingdom, Germany, and France had more international students than the U.S.S.R. that year. The majority of foreign students came from the socialist countries of Europe (Poland, Bulgaria, East Germany), Communist Asia (China, Vietnam), socialist Africa (Angola, Mozambique, Ethiopia, Rwanda), and Communist or socialist-leaning Latin America (Cuba, Nicaragua, Peru, Brazil). There were admission quotas in place for the foreign students, but their education was free, provided that they could pass the entrance exams in Russian. Although students from the capitalist countries were not explicitly excluded, it was harder for them to apply, due to the logistical difficulties of getting Soviet visas; the fear of living in a hostile country under an oppressive government also deterred many Americans and Western Europeans. MSU, the top school, had students from over 60 nations in the late Soviet period. Other popular schools for foreign students were medical and dental academies, language colleges, and the technical/engineering institutes.

Once accepted at a university, students would typically study for 5 years to earn a diploma. Students chose a broad field of study (e.g., biology or physics) when they applied to the university. Some specialties were considered priorities for the state (e.g., physics and civil engineering); these would have very attractive stipends, in addition to the free tuition. The Soviet universities had very few general education courses and virtually no electives. This may seem strange, but given the rigorous high school curriculum, a good general education had already been acquired by the age of 17. Electives were not available because the experienced faculty in each field of study had already figured out all the necessary coursework. This was an undemocratic but efficient approach, ensuring that free resources would not be wasted on teaching subjects that the students might never need later on. The results were young specialists with narrow, but deep, technical knowledge in their subject areas. Additional cultural breadth could be acquired by reading good books or going to concerts, museums, and theaters, which Soviet youth commonly did.

Beside full-time university programs, there were many evening college programs for working adults, as well as correspondence courses for those living in remote locations. Unlike in the United States, one could not enter a university much later in life; only evening and correspondence programs were available for students over 30.

Textbooks were obtained from the university library for free (they were loaned out for 1 year). They were usually not new, but adequate. The rooms at the university-run dormitories were free, and an allowance provided for some food (about enough for one meal per day); the rest was a student's responsibility. If you consider that health care was likewise free, and that bus transit cost almost nothing (the equivalent of 50 U.S. cents in big cities and less in the provinces), you can imagine that being a university student in the U.S.S.R. was not a bad thing at all.

It is important to stress that because of the rigorous testing and limited state resources, far fewer students were enrolled in the Soviet university system than is common in most Western countries. In the heyday of the Soviet Union, only about 20% of all young adults (ages 18–25) were enrolled. In the United States, the initial college enrollment rate today is about 50%; however, there is also a correspondingly high dropout rate of about 30% in the first 2 years. Therefore, less than 35% of the total U.S. population actually graduates from college in a given cohort, and just over 25% of those in the general population have college degrees. In the Soviet system, family culture definitely reinforced the need to be in college (as is common today in many Asian countries), and dropouts were rare.

Young men had an additional incentive for staying in school: They were required to serve at least 2 years in the military, unless enrolled at a university full-time. In a few dozen of the best universities, male students could go through military training while enrolled in their academic programs, and would graduate with a specialty and rank without ever being required to do active duty. The exact specialty depended on the university and the program: Physicists were trained as artillery or radio communications specialists, biologists in germ warfare, and linguists in the foreign languages most needed for military purposes.

A university education resulted in better employment opportunities, although the wage differential in the Soviet Union was lower than that commonly found in Western countries. For example, wages for a qualified worker in some occupations (metallurgy, mining) were essentially the same as (or even higher than) those of an assistant professor with a PhD, or a physician. However, education had many other benefits, including better working conditions, more interesting jobs, social connections, and usually longer vacations. To ensure that the graduates stayed in their profession, the state had a placement program that guaranteed employment to the young specialists for 3 years upon graduation. However, frequently people were placed in less than desirable companies, and sometimes in cities other than where they had been born or attended school. Muscovites and Leningraders were especially affected by the transfers to different places, because life in their two cities was so much better than in other parts of the country. On the other hand, attending a university or getting a job placement in a different city afforded one of the few sure ways of changing one's place of residence in the U.S.S.R. Graduates were also sometimes placed in jobs where they had studied, which allowed many to receive residence permits to stay in Moscow, Leningrad, Sverdlovsk, Novosibirsk, and other desirable big cities.

Schooling of Ethnic Minorities under the Soviet System

The Soviet system of primary, secondary, and university education was remarkably uniform. The other republics of the U.S.S.R. and ethnic/national units within Russia had additional language instruction in the local language, especially in primary and middle school. Most of the high school classes were taught in Russian, to ensure cultural Sovietization and to facilitate career opportunities in adult life. This was not done (as often erroneously assumed by Western scholars) to promote Russian culture or language per se, or to oppress the minorities. It simply made practical sense to the state to use one common language of communication, just as is done in the United States or United Kingdom with English. Minorities had the options of studying in their own republics/regions all the way through college or going to another one of their choice. The best universities in the country had a small number of seats reserved for talented minority group members who were recruited through their republican/regional boards of education, as a form of affirmative action. In many cases, the system was rigged in favor of the local party bosses' children, but genuinely talented ethnically non-Russian students could usually make it through. Bribery was not uncommon; particularly notorious in this regard were Azerbaijan and Uzbekistan, where (rumor had it) entire university diplomas could be sometimes purchased, and certainly admission into the most prestigious schools could.

In the distant villages of Siberia and the north, it was not possible to provide adequate schooling to very small, scattered populations. Therefore, members of many ethnic minorities would send their children to Russian-language boarding schools for weeks on end. It was beneficial with respect to education, but it also severed the critical ties between the older and younger generations, and precluded the passing down of oral traditions. The overall impact of the Soviet period on these cultures was not much better or worse than that of mainstream U.S. culture on the Alaskan natives.

Changes after the Fall of Socialism

The basic system of education described above is still in place, in Russia as well as in other FSU republics. Some pertinent comparisons with other countries worldwide are provided in Table 15.1. Although there is no single rating of the best universities, one such rating is provided in Table 15.2. Many traditionally well-known but nontechnical universities are not included, because the rating was made to reflect the probability that recent graduates will be employed

TABLE 15.1. Russian Educational Achievements Compared to Those of Other Countries

	Russia	United States	France	China	Brazil	Nigeria
Literacy rate (%)[a]	99.4	99.0	99.0	90.9	88.6	68.0
Spending (% GDP)[a]	3.8	5.3	5.7	1.9	4.0	0.9
Number of world-class universities[b]	2	168	21	8	4	0
Scientific articles published[c]	14,000	211,000	32,000	29,000	8,700	400
Primary teachers/1,000 students[c]	2.226	5.885	3.353	4.434	4.836	4.554
Nobel Prize winners in science[d]	15	190	30	4	1	0

[a]Data from CIA World Factbook (2007, 2009).
[b]Data from *nationmaster.com*.
[c]Data from World Development Indicators Database (2003).
[d]Data from Nobel Prize Committee (by country of origin).

TABLE 15.2. Top Universities of Russia, Based on Employment Prospects of Recent Graduates (2007)

Top tier—Moscow
State University—Higher School of Economics
Moscow State Construction University
Bauman Moscow Technical University
Lomonosov Moscow State University
Gubkin Russian State University of Oil and Gas
Financial Academy of the Government of the Russian Federation
Top tier—Regions
Shukhov Belgorod State Technological University
Voronezh State University
Irkutsk State Technical University
Kuzbass State Technical University
St. Petersburg State Architecture and Construction University
St. Petersburg State University
Tula State University
Tyumen State Oil and Gas University
Ufa State Oil Chemistry University
South Urals State University
Second tier—Moscow
Moscow Aviation Institute
Moscow Automotive and Road Construction Institute
Moscow Institute of Steel and Alloys
Moscow University of Food Production
Moscow Energy Institute
Second tier—Regions
Kazan State Architecture and Construction University
Tupolev Kazan State Technical University
Kuban State Technical University
Novosibirsk State Technical University
Omsk State Technical University
Perm State Technical University
Peter the Great St. Petersburg Polytechnic University
Saratov State Technical University
North Caucasus State Technical University
Siberian State Automotive and Road Academy
Tver State Technical University
Ural State Technical University

Note. Data from RosBusinessConsulting (*www.rbc.ru*), 2007.

in today's Russia (based on expert opinions), not on the quality of the education per se. Besides MSU and St. Petersburg State University, other very good general schools include Tomsk, Kazan, Yekaterinburg, Nizhniy Novgorod, and Novosibirsk State Universities; the People's Friendship University in Moscow; the Foreign Languages University; the University of International Relations; the First and Second Medical Universities in Moscow; Moscow State Pedagogical University; and a few others. Not surprisingly, many universities are located in Moscow and St. Petersburg: In 2000, 171 (19%) were found in Moscow and 77 (8%) in St. Petersburg, with a total of 914 colleges and universities, public and private, in the entire country. In 2004–2005, 3.4 million students attended universities in Russia, or a little over 20% of the college-age group. (Figure 15.4 gives the 2008–2009 enrollment figures for students at all levels.). The overall enrollments in universities have more than doubled, from only

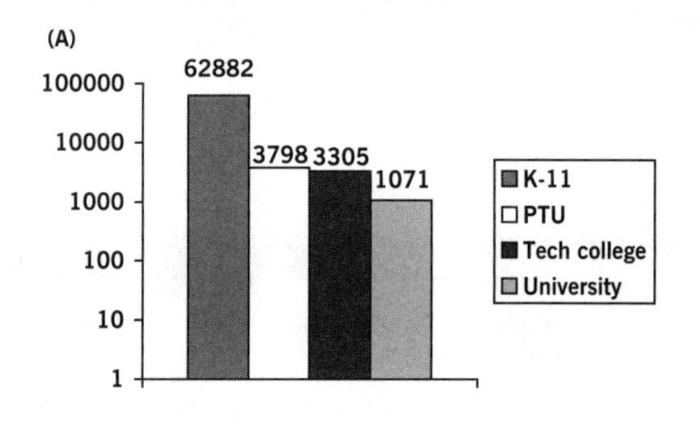

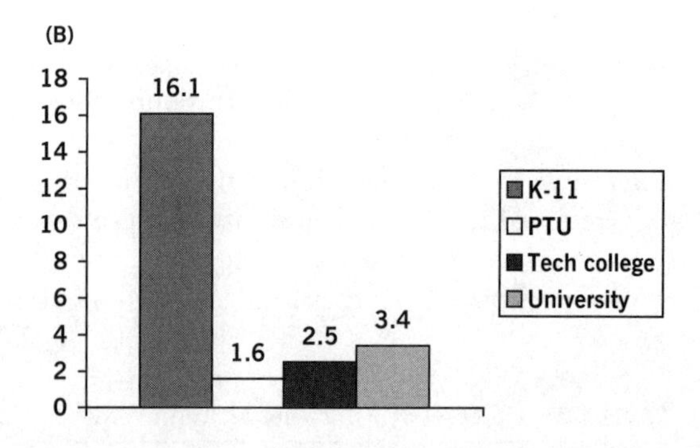

FIGURE 15.4. Some Russian educational statistics for the 2008–2009 school year: (a) Number of schools; (b) number of students (in millions). In grades K–11, an overwhelming majority of students (over 99%) attend public schools. Only 70,000 students attend private schools, most for the very elite. A PTU is a technical high school that trains skilled workers. A tech college (*technikum*) is typically a 2-year program, while a university is a 5- or 6-year program. Only regular daytime university students are included; an additional 2.9 million were taking evening or correspondence courses. Data from *stat.edu.ru*.

1.6 million in 1995; this reflects improved economic conditions, better state subsidies, and the increasing importance of having a university diploma to land a good job.

In 2002, 19.1% of adults in Russia had some college-level education (it was about 27% in the U.S. in the same year, although the latter nation tracks 4-year bachelor's-degree programs, not 5-year diplomas). In the same year, however, it was revealed that for the first time in 70 years, 1.6% of school-age children did not attend *any* primary or secondary school—a scandalous and sad admission in a country that had formerly prided itself on its 100% literacy rate. It is interesting to note that women in Russia are now better educated than men. For example, 16.6% of women but only 15.8% of men had completed a college degree, according to the 2002 census. Universities are concentrated in the largest cities, in distinct contrast to the United States, but similar to Canada, Australia, and some European countries. In 2009, 12 universities besides MSU and St. Petersburg State were proclaimed "federal" universities, with corresponding increases in financing levels, as a strategic move to prevent declines in the quality of university education. Of these, about half are in the distant regions.

There have been many changes, both good and bad, in the educational systems of Russia and its neighbors since the fall of the Soviet Union. Some specific changes include the following:

- There is greater emphasis on the local languages in the newly independent republics. In some (e.g., Ukraine), the university language of instruction is now Ukrainian, while many local schools continue to teach in Russian in grades K–11—a reversal of the Soviet policy, when university instruction was given primarily in Russian. In other countries (e.g., Kyrgyzstan and Kazakhstan), quite a bit of Russian instruction is still allowed at the university level. The Baltic states are now part of the European Union (EU), and have adopted many European policies and standards with respect to education.
- All FSU republics have literacy rates over 90%, and some have rates over 99%. However, armed conflicts in Tajikistan, Georgia, Armenia, Azerbaijan, and Moldova have led to a decrease in schooling in those republics. Undoubtedly there are now more illiterate adults in these countries than before the breakup of the U.S.S.R.
- An increased interest in studying foreign languages, typically English, is observed throughout the FSU. Although teaching English has always been common in the past, many programs now approach Western levels of quality, in particular because better texts and native-English-speaking tutors are now available. The Soviet intelligentsia could read in English, but few could speak it well, due to a lack of practice. Now, with opportunities for foreign travel readily available, many people have taken up studying English, German, French, Italian, Spanish, Turkic, Arabic, Thai, Hindi, and other useful languages for traveling abroad. A working knowledge of English (and also German and French) helps people to get better-paying jobs with Western firms or Russian companies doing business internationally, as well as to get promotions. About 25% of youth in the big cities in Russia, Ukraine, and Kazakhstan now have at least some command of spoken English, and the percentage is higher in the Baltics. (Even in the Baltics, however, the rate is much lower than in the Scandinavian countries.)
- In Russia, a national standardized test known as the EGE (modeled after the SAT) has recently been introduced, supposedly to root out corruption and bribes among the college examination boards. MSU and a few other top schools in Moscow and St. Petersburg refuse to accept it, considering it a short-lived fad and an unnecessary concession to Western standards of assessment based on multiple-choice tests.
- There is greater conformity to international standards of awarding degrees, mainly based on the U.S. model. Specifically, in 2007 universities in Russia switched to the familiar system of 4-year bachelor's degrees and 2-year master's degrees. The traditional 5-year diploma system currently coexists with the bachelor's/master's system, but it makes it difficult for Russian graduates to compete for admission into schools abroad. Nevertheless, the professors in Russia have opposed the move on the grounds that the old system worked just fine; they believe that requiring another year over 5 may strain their budgets, and that teaching for only 4 years is not enough.

• An alarming recent trend is the simplification of the K–11 school curricula. The old Soviet system emphasized natural sciences, but deemphasized social sciences (e.g., cultural geography was not taught at all, and was considered part of economic geography). In what many consider a misguided quest to "do as the Americans do," the Russian Ministry of Education has been systematically cutting the numbers of hours allocated to traditional disciplines since the late 1990s. This dilutes what is perceived as "too difficult coursework" in math, physics, biology, history, and literature with courses in such trendy subjects as human relations, home economics, psychology, family studies, and applied art. Although they are of course useful, the latter subjects cannot replace the classics and the "hard" sciences, and it is becoming clear that present-day Russian schoolchildren already know less than their predecessors did merely a decade ago about the traditional subjects.

• Privatization of education is ongoing and takes many forms. Private schools and colleges (most of dubious quality, but with high tuition rates) exist now in Russia along with the public ones. Even in the latter, some things now cost a lot of money: The costs of textbooks, school supplies, and new equipment have risen astronomically. Many upper-class parents simply choose to send their children to elite British, Swiss, or French schools now, to avoid the hassle of dealing with the quality issues at home. To get into any good university still requires a lot of tutoring and cramming, usually accomplished by paying hard cash to tutors (in the best case) and to admission committees as bribery (in the worst). The majority of state universities continue to offer tuition-free education to about half of all students, based on merit. At MSU, tuition is already approaching the levels charged by the cheapest state universities in the United States ($5,000–$6,000 per year), in a country with only one-quarter of the average U.S. income per capita.

• For obvious reasons, Communist youth organizations have declined. There are no longer Young Pioneers in red ties. However, some new alternatives now exist: Major political parties (e.g., the Communist Party and the pro-government United Russia) offer experiences for youth through their respective nationwide programs.

• The decline in international student attendance is unfortunate. A recent report suggested that there are now only 96,000 foreign students studying in Russia, compared to 126,000 in 1990. The country has dropped to sixth place worldwide in the number of foreign students—behind not only the United States, the United Kingdom, Germany, and France, but also Australia, and just barely ahead of Japan. Most of these students also now come from the poorest countries in Africa, Asia, and Latin America, rather than from the former socialist countries of Europe.

• Ominously, the number of hate crimes involving foreign students, especially black students, has greatly increased. Although individual cases of verbal threats or physical assaults may not always be racially motivated, visual appearance, ethnicity, and foreign status clearly play a role in many attackers' decision to pursue these students. The worst and most frequent cases of physical attacks, some fatal, have been reported in Voronezh, St. Petersburg, Tver, Vladimir, and Rostov-on-Don—all predominantly working-class, Russian-populated, almost 100% white areas, with high rates of unemployment and many struggling households. Moscow sees some attacks as well, but Muscovites are also much more used to seeing people of color and seem to accept foreigners more readily. Russian citizens of the northern Caucasus republics are, on the other hand, favorite targets everywhere in Russia, partially as a backlash from the Chechen conflicts (only a few of these students are in fact ethnic Chechens, but Russian youth may not be able to tell the difference). Overt racism was rare in the Soviet Union, where students were always taught that workers of all lands are brothers. Given the new realities, why do the international students still keep coming to Russia? Because a Russian university education remains among the best in the world and is still very affordable. Many provincial universities have plenty of room available for foreign students and have low living costs and tuition relative to universities in Europe or North America.

Let's talk now about the fruits of education: arts, sciences, and sports.

Arts

The importance of Russian literature has already been discussed in the context of cultures and languages (see Chapter 13). Other Russian arts worthy of note are visual arts (painting and sculpture), performing arts (opera, ballet, drama), and cinema. There are hundreds of art galleries, museums, and theaters in Russia, and thousands of movie theaters. Their geographic distribution is discussed here, along with some specific highlights of the Soviet and post-Soviet periods.

The arts were well supported during the Soviet period. Especially promoted were the art types associated with socialist realism (e.g., monumental paintings of workers and peasants, socialist films), as well as the classics. Artists, directors, writers, and musicians were supported by state salaries and benefits distributed through the professional unions. To become a member of one of these unions required considerable talent, personal connections, and a bit of luck. Some of the best talents were to be found in the informal sectors (e.g., local artist clubs or youth organizations). Because the socialist state had free education and a low cost of living, some gifted artists would work in dead-end official jobs and create their pieces in their spare time. State support for the arts has receded with the post-Soviet economic reforms, although it has improved somewhat since 2000. There are now also many philanthropic private foundations and corporate sponsors supporting the arts, as is common in the West.

Visual Arts

The main collections of Russian visual arts are concentrated in Moscow and St. Petersburg. Moscow's Tretyakov Gallery and St. Petersburg's Russian Museum house the premier collections of Russian art, including ancient icons, classical paintings of the 18th and 19th centuries, and modern art. Icons can also be viewed at the Rublev Museum in Moscow, and in many old churches, especially inside the kremlins. After the icons, the best Russian paintings are either those by the late-19th-century realist artists (Ge, Kramskoy, Kuindzhi, Levitan, Perov, Repin, Savrasov, Shishkin, Surikov, Polenov, Vasnetsov) or those by avant-garde artists of the early 20th century (Kandinsky, Malevich, Chagall), depending on your taste.

World art is on display at Pushkin Museum in Moscow and the State Hermitage Museum in St. Petersburg (Vignette 15.1). Other great art museums of the FSU include the State Art Museum of Belarus in Minsk, the National Art Museum of Ukraine in Kiev, the Research Museum of the Russian Academy of Art in St. Petersburg, and art galleries in other major cities. In addition, some former country estates where painters used to live have been turned into museums: The Polenov Museum in Tula region and Abramtsevo near Moscow attract thousands of visitors annually (Figure 15.5; see also Chapter 4, Vignette 4.1).

Nationwide, in 2005 St. Petersburg had the highest number of museum visitations per 1,000 people per year (3,658), with Yaroslavl in second place, Volgograd in third, Vladimir in fourth, and Moscow in fifth. In Siberia, the highest attendance was noted in Khakassia Republic and the Krasnoyarsk region. Not all of the museums attended are art museums, but these data do give an idea about the distribution of the greatest interest in and opportunities for museum attendance. Even in St. Petersburg, the attendance level is only about 60% of the 1990 level, which may be indicative of the social changes going on. On the one hand, museums may now be too expensive for some people. On the other, more well-off citizens now have many more entertainment options, including eating out, visiting amusement parks, boating, golfing, or driving around for pleasure.

Many museums of local studies (*kraevedcheskie*) contain wonderful collections of local folk art, including wood carvings, dolls, toys, ceramics, porcelain, lacquer boxes, embroideries, mosaics, metal designs, jewelry, samovars, and more (Figure 15.6). Small towns in the European part of Russia (Palekh, Zhostovo, Gzhel, Pavlovsky Posad, Dymkovo) continue making traditional wooden and porcelain souvenirs, as well as tableware, shawls, and toys, for sale (Figure 15.7); many of these are exported, and some are counterfeited. Artisans in the Urals specialize in stone cutting, producing goblets, eggs, and fine jewelry from malachite, jasper, opal, chalcedony, and other semiprecious stones. The Caucasus has many local specialties: Dagestani metal engravings,

Vignette 15.1. The Russian Ark: The State Hermitage Museum

According to *The Guinness Book of World Records*, the world's largest art gallery is the State Hermitage Museum in St. Petersburg, Russia. Visitors would have to walk 15 miles to see all 322 galleries, which house nearly 3 million works of art. At any given time, only 5% of the collection is on display. The Hermitage occupies six magnificent buildings situated along the embankment of the Neva River, right in the heart of St. Petersburg. The main and most famous building is of course the Winter Palace, used for over 150 years by the Romanovs as their main winter residence. It was designed by the famed Italian architect F. B. Rastrelli, and constructed in 1754–1762 during the time of Empress Elizabeth as the fourth attempt at building something magnificent on this site (the three other palaces did not completely satisfy the rulers). The Winter Palace alone has over 1,000 rooms and 117 staircases. Over 3 million people visit the museum annually.

The Hermitage's collections reflect the development of world culture and art from the Stone Age to the 20th century. The main contributors to the collections were Catherine the Great (an avid collector of European art who frequently outbid German and English royals) and the Soviet government (which added the entire collections of the Stroganovs, Sheremetyevs, Shuvalovs, Yusupovs, and other aristocratic millionaires, as well as some art stolen from the German museums seized in World War II). The Hermitage today boasts two *Madonnas* by Leonardo da Vinci, a few works by Titian and Raphael, many more by Van Dyck and Rubens, 26 Rembrandts, and a lot of Impressionist art, mainly from the famed Shchukin collection (e.g., 35 canvases by Matisse). Although the Hermitage is primarily a museum of Western art, there are fine ancient Egyptian, Greco-Roman, and Scythian collections; exhibits from Russian prehistory and early history; and art from other corners of the world.

If you are not able to visit in person, a nice vicarious look at the Hermitage is provided by a visually stunning, record-setting film by A. Sokurov, *The Russian Ark* (2002). It was the first feature film ever to be shot in a single take; it was filmed using a single 90-minute Steadicam tracking shot (a feat that required four attempts). The film displays 33 rooms of the museum, which are filled with a cast of over 2,000 actors. Many events and characters from Russian history are represented. The plot may be a bit too abstract for the average American audience to appreciate, but the message is clear: The museum is the ark preserving the riches of the Russian past.

FIGURE 15.5. Abramtsevo Museum in Moscow Oblast showcases carved wooden architecture created by the famous 19th-century artist V. Vasnetsov and others. *Photo:* Author.

FIGURE 15.6. The V. Bianki Museum of Local Studies in Biysk shows a collection of samovars, including some made in the region. *Photo:* Author.

FIGURE 15.7. These *matryoshka* dolls are for sale at souvenir shops on Arbat Street in Moscow. Traditional crafts are frequently counterfeited. *Photo:* A. Fristad.

Ossetian wood carvings, Adygei embroidery, and so on. Kazakh and Kyrgyz specialties include highly decorated dresses, woolen rugs, pillows, and cushions. Uzbek, Turkmen, and Tajik crafts include robes of cotton or silk, Persian-style rugs, and pottery.

Theaters

Theaters in Russia are mainly concentrated in Moscow and St. Petersburg. The Bolshoi Theater is world-famous, specializing in classical opera (*Eugene Onegin, The Queen of Spades*) and ballet (*The Nutcracker, Giselle, Swan Lake*). Its counterpart in St. Petersburg is the Kirov, or Mariinsky, Theater, whose school of ballet is considered to be one of the world's finest. The Maly Theater is right across the street from the Bolshoi in Moscow, specializing in dramatic productions. Other excellent choices in Moscow include the Chekhov Art Theater MKhAT, the Theater of Nemirovich-Danchenko, and the Taganka Drama Theater. In the past 20 years, many small drama studios have sprung up, producing plays to suit the tastes of an increasingly discriminating public. Some theaters in Moscow are so popular that tickets are sold out months in advance. St. Petersburg, true to its nickname as "the cultural capital of Russia," leads the country in the number of theater visits per year (576 per 1,000 people in 2005), trailed by the Moscow, Omsk, Udmurtiya, and Tomsk regions. The Opera and Ballet Theater in Novosibirsk occupies the biggest theater building in Russia (Figure 15.8). In each FSU republic, the capital city will typically have at least one main drama theater and frequently an opera/ballet house.

Cinema

Lenin famously proclaimed that for the Bolshevik state, the main form of art would be cinema. He correctly recognized the propaganda potential of the new art form. The Soviet films were lavish productions heavily promoted by, and serving the interests of, the state. Perhaps the best known are the epics *The Battleship Potemkin* (1925) and *October* (1928) by Sergei Eisenstein, although the musical comedies *Vesyolye Rebyata* (1934) and *Volga-Volga* (1938) by Grigory Aleksandrov are also great Soviet classics. The Internet Movie Database (IMDB) lists over 6,700 entries for Soviet films (i.e., those made prior to 1991). Many great movies were made about World War II (e.g., *Soldier's Ballad* and *The Cranes Are Flying*). Only a handful of Soviet or Russian films have ever been nominated for American Academy Awards; this is not surprising, given the specific expectations of American film critics about movie making, and given the political situation that existed during the Cold War. The most recent two to receive the Best Foreign Film Award were *Moscow Does Not Believe in Tears* (1980) by Vladimir Menshov and *Burnt by the Sun* (1994) by Nikita Mikhalkov. More deserving movies by Andrei Tarkovsky, Eldar Ryazanov, or Georgy Danelia were never nominated for Oscars (Vignette 15.2). Soviet and Russian movies have fared considerably better at the main European film festivals (Cannes, Berlin, Vienna).

In the late 1970s, over 150 full-length movies were made in the U.S.S.R. per year. Russian film production practically ceased in 1992–1996 due to lack of funding, with merely 20–30 produced per year; it began again in the mid-1990s with Hollywood-wannabe gangster flicks sponsored by shady businessmen. By comparison,

FIGURE 15.8. The Opera and Ballet Theater in Novosibirsk (opened in 1945) is the largest building of this type in Russia, with almost 12,000 m^2 of area and a 60-m dome. Its grand hall can seat almost 2,000 spectators. *Photo:* A. Fristad.

Vignette 15.2. A List of Must-See Russian Films

The Battleship Potemkin (1925) and *October* (1928) by Sergei Eisenstein
Vesyolye Rebyata (1934) and *Volga-Volga* (1938) by Grigory Aleksandrov
Carnival Night (1956) and *Promised Heaven* (1991) by Eldar Ryazanov
The Cranes Are Flying (1957) by Mikhail Kalatozov
Ivan's Childhood (1962) and *Andrei Rublev* (1966) by Andrei Tarkovsky
Common Fascism (1965) by Mikhail Romm (documentary)
War and Peace (1968) by Sergei Bondarchuk
The White Sun of the Desert (1969) by Vladimir Motyl
Caucasus Prisoner (1971) and *Ivan Vasilievich Changes His Occupation* (1973) by Leonid Gaidai
Sherlock Holmes and Dr. Watson series (1979–1980) by Igor Maslennikov
Love's Formula (1984) by Mark Zakharov
Kin-Dza-Dza (1986) by Georgy Danelia
Repentance (1987) by Tengiz Abuladze
Little Vera (1988) by Vasily Pichul
American Daughter (1995) by Karen Shakhnazarov
Idiot (2003) and *Master and Margarita* (2006) by Vladimir Bortko
The Island (2006) by Pavel Lungin
Nu Pogodi (1969/1993) and *Cheburashka* (1971/1974) (animated films for children)
A Long Goodbye (1971) by K. Muratova
Pushkin's Duel (2006) by N. Bondarchuk
Prince Vladimir (2005) by Yuri Kulakov (animation)
Any animations by Yuri Norstein and his pupil Andrei Petrov

Hollywood produced over 400 movies in 1996. However, since the late 1990s there has been a resurgence of genuinely good films in Russia, due to an increase in state funding, better corporate sponsorship, and growth in DVD sales. About 120 new movies come out every year in Russia now, according to the IMDB; this puts Russia in a tie with Germany, but behind India, the United States, Japan, China, France, Spain, or the United Kingdom. Some movies are also now being produced by post-Soviet filmmakers abroad, especially in France. Also famous are films of the highest artistic quality produced by Georgian and Armenian filmmakers (Vignette 15.2). The number of modern multiplex cinemas in Russia went up from 8 in 1995 to 185 in 2001, and DVDs are available everywhere in street kiosks, although few are licensed copies.

Television, Radio, and Newspapers

A few comments can be made about TV and radio as well. In the Soviet period, both were very popular, and indeed indispensable—for the state to control the masses on the one hand, and for the masses to gain access to information and culture on the other. The most conspicuous feature for a Westerner at this period would have been the lack of commercials, because all channels were state owned. The information was carefully censored, of course, but the news coverage was very thorough. The main prime-time news program *Vremya* (*Time*), on TV Channel One, lasted 45 minutes, and approximately 20 of these minutes were spent on covering a range of international topics from many countries in the world—not just one or two main stories of the day, as on U.S. television. When the Soviet Union or its allies were depicted, only positive achievements were highlighted. Life in the West was typically shown as consisting of unemployment lines, urban pollution, drug-related violence, and war or other conflicts, with only occasional glimpses of nature in some famous national parks. The interiors of Western stores were cleverly never portrayed until the late Gorbachev era, when these became popular. However, many American and European movies were available in cinemas, thus giving Soviet citizens a glimpse into many aspects of contemporary Western lifestyles anyway.

The puritanical Soviet attitude was reflected in the TV programming: There was absolutely no nudity or profanity, and very little violence. Many feature movies were shown, including wartime dramas, contemporary comedies, and even international classics (although the latter were edited for mature content). Only four or five TV channels were commonly available via air broadcast, however, and there was no cable TV until the late 1980s. Radio was ubiquitous in city parks, at work, and at home. Much classical drama, poetry, and music could be heard.

Soviet citizens also read a lot of newspapers, many of which were posted on billboards in city parks. In short, the mass media worked toward making Soviet citizens a very literate population. Today TV channels are much more numerous than in the Soviet period, but they remain heavily controlled by the authorities after a brief period of less control during the Yeltsin period. About 20 channels are commonly available on local cable, and hundreds of international ones via satellite dish. Much TV and radio production is also now heavily commercialized, with as many commercials as in the West.

Sciences

Major Accomplishments of Russian and Soviet Scientists

Russian scientists made famous discoveries in all major scientific fields—from physics to biology, from anthropology to history, from chemistry to geography—both before and after the Communist Revolution. One of the earliest was a self-taught peasant, Mikhail Lomonosov (1711–1765), who would become a secretary of state, a cofounder of MSU, a president of St. Petersburg University, and a codiscoverer of oxygen. He made contributions in physics, chemistry, geography, astronomy, linguistics, and history, along with some major accomplishments in poetry and art.

Some major scientific names from the late 19th and early 20th centuries include Nikolai Lobachevsky, Pafnuty Chebyshev, and Alexander Lyapunov in mathematics; Dimitry Mendeleev, Alexander Borodin, and Alexander Butlerov in chemistry; Vasily Dokuchaev in soil science (see also Chapter 4, Vignette 4.2); Pyotr Semenov-Tyan-Shansky in geography; Ivan Pavlov, Kliment Timiryazev, and Ilya Mechnikov in biology; and many others. There were also dozens of explorers on both sea and land serving under the Russian crown, including Vitus Bering, who discovered Alaska in 1741; I. Kruzenstern, who circumnavigated the globe and made important oceanographic studies; Y. Lisyansky, who studied the Pacific islands; and so on.

As described earlier in this chapter, the Soviet Union put a heavy emphasis on scientific education, especially in the natural sciences. Math and physics were two areas in which the Soviets traditionally excelled. Both were critical in the creation of better weapons during World War II, as well as in the country's becoming the second nuclear power in the world (1948), the first to put a human-made object in space (1957), and the first to send a man into space (1961). The names of Igor Kurchatov and Sergei Korolev are forever connected with the development of the Soviet nuclear and space programs. Andrei Sakharov and Yakov Zeldovich helped to develop the Soviet thermonuclear weapons, but they also designed the peaceful Tokamak, a bagel-shaped prototype plasma reactor to produce controlled thermonuclear fusion. The Nobel Prize for discovering and developing the first lasers went to Soviet physicists Nikolai Basov and Alexander Prokhorov. Lev Landau and Vitaly Ginzburg won another Nobel Prize for contributions to the field of superconductivity; Landau also coauthored one of the best textbooks in theoretical physics ever written. Chemistry, both organic and inorganic, has been another strong point of Soviet science. For example, Alexander Nesmeyanov and associates developed a new technology in organic chemistry synthesis that allowed the combination of metal atoms with organic compounds, and K. Adrianov was the first in the world to synthesize complex silica–organic structures.

The Soviet Union led the world in genetics research until 1937 with such famous names as Nikolai Koltsov, Nikolai Vavilov, and Nikolai Timofeev-Resovsky, but faltered later as an attack against genetics was launched under Stalin—curiously, in the name of misunderstood Darwinism. The campaign's champion was Trofim Lysenko, a barely literate protégé of Stalin, who claimed that with proper socialist methods of cultivation a pine tree could be forced to produce oak branches, and a cuckoo could beget a hawk. Lysenko wanted to rid "Marxist" biology of Western superstitions, as he understood them. To do this, he started a massive witch hunt that decimated the ranks of Soviet genetics researchers and left the country 20 years behind the rest of the developed world in biology research by 1950. Despite the major setback caused by Lysenko, research in molecular biology and biochemistry was reaching new heights in the U.S.S.R. by the mid-1960s. For example, the Soviet molecular biologists A. Belozersky and A. Spirin predicted the existence of matrix RNA in 1957.

Geography was considered primarily a natural science, and many developments occurred in climatology, geomorphology, glaciology, oceanography, and biogeography. An important contribution of Soviet geographers, especially Lev Berg, to the whole discipline was the development of the landscape science approach (Shaw & Oldfield, 2007). Cultural and human geographies were all treated as expressions of economic geography, which had to be explicitly socialist and Marxist. Among famous Soviet geographers, Vladimir Vernadsky proposed the concept of the "biosphere" as a unified global self-regulating system and the "noosphere" as a new sphere governed by human reason; Berg developed a complex method of researching geographic landscapes; and Y. Gekkel managed to produce the first map of the Arctic Ocean floor. Gekkel's students were at the forefront of oceanographic research in the 1950s and 1960s, and the Soviet Union was the first country to produce a complete atlas of the world's oceans, using extensive submarine research. V. Sukachev studied the biogeography and ecology of "biogeocenoses" (local ecosystems within a given landscape). Boris Polynov made substantial contributions to the understanding of soil evolution and development of physical landscapes. Nikolai Baransky and Nikolai Kolosovsky worked in the areas of socioeconomic development, regionalization of human landscapes, and economic complexes of production. Yuri Saushkin was a prominent urban geographer who made contributions to our understanding how cities evolve in time and space. This list could be greatly expanded, but it gives an idea of the breadth and depth of Soviet geographers' interests.

The social sciences did not fare as well as the natural sciences. Some fields (e.g., sociology) were considered "bourgeois" and thus suspect. In a country with 3% of the people controlling virtually all aspects of the economy, it was dangerous to pry into the class structure of the supposedly "classless" society. Psychology was likewise suspect, given the prominence of Western thinkers (e.g., Freud and Jung) in developing subjective theories of the human mind and supposedly reactionary views on the nature of humanity itself, all of which were disapproved of by the Soviet Marxists. Excellent research was nevertheless carried out in history, archeology, and anthropology, although it had to be conducted under the politically correct Marxist umbrella. The anthropological research included the discoveries of several Paleolithic cultures of Eurasia, thorough anthropological studies of the early Slavs, studies of early Central Asian civilizations, and profound insights into the development of Eurasian cultures. There was also much groundbreaking research in linguistics, helped by the tremendous diversity of native languages in Northern Eurasia (see Chapter 13). Regional studies were done not only inside the U.S.S.R., but in other socialist countries—especially in Africa, Southeast Asia, and the Middle East, where the Soviet Union had many allies.

The Structural Organization of Scientific Research

There were two main forms of research in the Soviet Union: fundamental and applied. The former was carried out by hundreds of universities, and especially by the research institutes and centers of the Soviet Academy of Sciences (today the Russian Academy of Sciences, or RAS). The RAS remains a formidable organization of fundamental science, with over 2,000 institutes under its wing and three regional branches. The Soviet applied research was carried out in hundreds of institutes and construction bureaus run by dozens of government ministries. For example, the Ministry of Oil and Gas had a few institutes devoted solely to natural gas and petroleum research. Quite a bit of research also occurred at the factories themselves. Many world-class discoveries and patents were produced in the applied labs of the Soviet period. For example, Soviet applied scientists invented many of the most modern methods in metallurgy (production of alloys from rare earths, continuous steel pouring), developed artificial diamonds, and made major discoveries in the organic synthesis of plastics.

In 1993, at the beginning of Yeltsin's reforms, there were still over 2,000 institutes, 865 bureaus, 495 research and development (R&D) firms, 29 research factories, 440 universities, and over 340 factories involved in R&D in the Soviet Union. The number of scientists was impressive. The mid-1990s International Science Foundation project sponsored by George Soros counted over 5,000 specialists just in the field of biodiversity in the FSU. Some fields (e.g., physics and chemistry) were represented by 10 times as many people. A total of 1 million scientists were working in Russia in 1990, representing about 18% of the world's total. Hundreds of scientific journals were published in Russian, and thousands of conferences were held each year.

Despite so much apparent activity, the U.S.S.R. produced relatively few Nobel Prize winners (15), and only a fraction of the number of publications per capita produced in the United States at the time. One of the reasons for this was that there was no need to account immediately for the results of state-funded research. In the West, science is driven by competition for limited grants. In the Soviet Union, scientific employment was guaranteed for life; there were no monetary incentives for publishing more; and no one had to compete for research grants in an open, peer-reviewed process. The number of scientific journals in each field was relatively limited, and much of what was published required personal knowledge of the specific academics who were in charge. Due to political constraints, it was hard to publish in foreign journals. Moreover, few Soviet scientists were allowed to travel abroad; their mail was routinely intercepted by the authorities; and no one had the hard currency to pay the page charges. In recent years, Russian scientists have published fewer articles than their French or Chinese counterparts, but many are now published in English (including translations of the top Russian academic journals), so Russia remains a major scientific power.

Geographically, Russian science was of course centered on Moscow, because Moscow had the biggest and best universities, the best-prepared students, and lots of industrial enterprises in need of serious research. In the 1960s, however, "science towns" (*academgorodki*) sprung up away from big cities to allow for the more relaxed lifestyle of the scientific elite, as well as for better control over what type of research was going on (to preclude foreign spying). Such cities are especially common near Moscow, and most were built from scratch (Chernogolovka, Dubna, Troitsk, Obninsk, Pushchino, Protvino, Zelenograd, Zhukovsky). Secondary clusters of scientific research existed near St. Petersburg, Penza, Gorky, Kazan, Sverdlovsk, Ufa, Novosibirsk Akademgorodok, Tomsk, Krasnoyarsk, and Irkutsk in Russia, as well as near Kiev, Kharkov, Odessa, and Dnepropetrovsk in Ukraine, and in most republics' capitals.

Moscow continues to lead Russia in the number of scientific researchers, with 40% of the total; it also has 80 academic institutes and at least as many universities. St. Petersburg is a distant second, with 15% of researchers and about 30 research institutes. Yekaterinburg, Tomsk, and Novosibirsk are three leading research centers farther east. Ten medium-size cities were recognized by the Russian government in 2007 as "science cities" (*naukograds*), and are receiving more budgetary support to revitalize their aging research infrastructure.

Unfortunately, after the fall of the Soviet Union, scientific salaries plummeted and many laboratories literally fell apart (Figure 15.9). This situation was caused in part by the near-sightedness of the Yeltsin period: The government was strapped for cash, and many officials were too busy with personally benefiting from chaotic privatization. Adjustment to the new free-market economic realities was also partly to blame, as were the decreasing societal benefits of being associated with the "knowledge class." In real terms (after adjustment for inflation), the salary of a PhD-level senior researcher decreased by a factor of 10 between 1989 and 1999, whereas many other professions supported by state budgets did not see a comparable decline. Thus, if in the late Soviet period a Moscow city bus driver had a salary slightly lower than that of a physics professor, by the end of the Yeltsin period the bus driver was making five to seven times more than the professor. The result, predictably, was a drastic reduction in the number of scientists. Many older specialists retired or passed away and were not replaced. Middle-aged scientists had three

FIGURE 15.9. Concrete ruins of an abandoned building in the "science town" of Puschino, 2007: A sad example of the lack of state investment in scientific research since the fall of the U.S.S.R. *Photo:* Author.

choices: remaining in science and depending on another family member working in the private sector; leaving for graduate school or a research position abroad; or quitting science altogether and entering the murky waters of business.

The official statistics suggest that the number of researchers was roughly halved between 1992 and 2002—reduced to about 500,000 from over 1 million. These figures included not only scientists themselves, but also lab technicians, assistants, and other staff. The "brain drain" hit all FSU republics hard, particularly the ones with ongoing military conflicts (Armenia, Azerbaijan, Georgia, Moldova, Tajikistan) or strong anti-Russian sentiment (the Baltics, Uzbekistan, Turkmenistan). While Russian scientists left Uzbekistan or Azerbaijan for Russia, their Russia-based colleagues left Russia for opportunities abroad—first and foremost in the United States, but also in Australia, Canada, Germany, France, the United Kingdom, the Netherlands, Sweden, Denmark, Switzerland, and Israel. Although the brain drain was not as dramatic as originally feared, it undoubtedly left the country in a precarious position, because the best and the brightest were typically the ones to leave first. Indirectly, by forcing its best talents abroad, Russia subsidized all the recipient countries to the tune of a few billion dollars.

The overall level of state support for scientific research in Russia remains shamefully low (Figure 15.10). Despite recent proclamations by the Putin/Medvedev government on the need to move from extensive to intensive scientific development, the Russian state spends less than 1.5% of the gross domestic product (GDP) on all scientific research (including military R&D) at this writing. The comparable figure for the United States is about 4%, and of course the U.S. GDP is about seven times greater than Russia's when adjusted for purchase parity. Thus an average university or national lab in the United States today has a budget comparable to the *entire* budget of a major scientific branch in Russia that includes dozens of institutes. For example, in 2005 the RAS operated almost 400 of its institutes on a budget of merely $500 million. An average research university in the United States will have a comparable budget. In 2010, the Russian government announced the creation of a few national research universities and technoparks to boost research and development.

FIGURE 15.10. Over 500,000 researchers left academic science between 1990 and 2000 to pursue careers in business or to work abroad. Those who remain must cope with low salaries, a lack of technology, and the plummeting quality of Russian K–11 education. The best professors manage to get scarce research grants. Many college students are now forced to pay tuition. *Photo:* Author.

The number of scientists with Russian or Soviet roots in the United States today is not very large, but includes some well-known figures: A. Abrikosov received the Nobel Prize in physics in 2003 for research that he began in the Soviet Union. V. Voevodsky was a recent Field's Medal recipient in math. S. Brin, the cofounder of Google, is of Soviet extraction and received his early education in the Soviet Union. There are entire departments at many U.S. or U.K. universities where Russian/ex-Soviet physicists, mathematicians, geophysicists, or molecular biologists constitute 25–50% of all faculty members. Russians make up about 6% of the total number of those holding H1B professional worker visas who enter the United States each year, although many of these are information technology specialists rather than fundamental scientists. Even more remarkable is the situation in Israel, where the number of engineers and researchers per capita exceeded that of the United States by 50% (145 vs. 85 per 10,000 population) by 2005—almost entirely due to the influx of post-Soviet immigrants with advanced university degrees.

Sports

The Soviet Union was very supportive of sports (see Chapter 7). The Soviet schools had 3–5 hours of physical education per week. Schoolchildren were also encouraged to join sports clubs, to compete in district and city tournaments, and to earn GTO badges. GTO was an acronym for "Ready for Labor and Defense" in Russian, and the program included a rigorous series of exercises involving track and field, other athletics, sharpshooting, orientation, first-aid skills, and the like. Earning badges in these areas could help young people gain admission to prestigious sports schools. Moreover, it was expected that virtually all young men would have to serve in the military upon turning 18, so physical fitness was expected. In present-day Russia, there is a renewed interest in youth fitness. Sports programs around the country are enjoying something of a renaissance, stimulated by a fresh infusion of federal and corporate cash.

Although the Soviet Union had all kinds of sports programs, this section focuses only on soccer (called "football" in Russia) and ice hockey as two popular and widespread team games that had, and still have, a strong geographic affiliation with large cities. Ice hockey seems like a natural sport for Russians to play, given Russia's climate. Invented in Canada, it quickly diffused to Europe in the early 20th century, and became especially popular in Czechoslovakia, Finland, Sweden, Estonia, and Russia. The Soviet Union had about a dozen teams in the "super league" and a great international team. The U.S.S.R. ice hockey team won gold medals seven times in the Winter Olympics, from 1956 (Cortina d'Ampezzo) to 1988 (Calgary), and the Commonwealth of Independent States (CIS) team won in 1992. The top league, now called the RHSL, remains very strong in Russia today, despite having lost many of its best players to the U.S.-based National Hockey League (NHL) and some other foreign teams. In 2003 there were at least 57 Russian hockey players (with the total payroll of the best 20 topping $62,000,000) in the NHL, including the famous goalie Khabibullin (Tampa Bay), the Bure brothers (Florida), Gonchar (Washington), Malakhov (New York Rangers), Larionov and Fedorov (Detroit), Yashin and Kvasha (New York Islanders), and others.

Inside Russia, the best RHSL teams are found mainly in the largest industrial centers, where they were traditionally supported by big industry or government ministries. In Soviet times, the perpetual champion was the Moscow-based Red Army team—which naturally would get the best players, as soon as those had been drafted at 18 years of age to serve in the military. Two other Moscow-based teams were Dynamo and Spartak. Russian hockey teams are less dependent than their NHL counterparts on ticket sales, concessions sales, and apparel sales to fund their hockey. They are mainly supported by large corporate sponsors now, though some may continue to rely on state support (Figure 15.11).

Each team's sponsor usually runs a youth sports school as well, to provide fresh talent as the players grow older. Today the best teams come from such industrial cities as Togliatti (where the team is sponsored by the VAZ car factory), Novokuznetsk (sponsored by a steel combine), Magnitogorsk (sponsored by another steel combine), Cherepovets (sponsored by the steel giant of the same name), and Voskresensk (sponsored by a chemical plant).

Soccer has always been popular in Europe, including the U.S.S.R. Most major cities have at least one professional club and a major stadium (Figure 15.12), and the overall sponsorship pattern is similar to that in ice hockey. However, the performance of the international U.S.S.R./Russia soccer team has been much less spectacular than that of the hockey team. The Soviet soccer team won Olympic gold in Melbourne/Stockholm in 1956, in Munich in 1972, in Moscow in 1984, and in Seoul in 1988—a feat that no CIS or Russian team has yet repeated. And no Soviet or Russian team has ever won soccer's World Cup. This is not to say that there have never been enough good, or even great, players; it is just that Brazil, Argentina, Italy, Germany, France, and the United Kingdom have generally managed to play even better. Also, after the fall of the Soviet Union, many of the best players left for abroad. Recently Russian clubs have begun recruiting more players internationally, including some of the best from Latin America. It is curious to note that the richest man in Russia in 2007 according to *Forbes*, Roman Abramovich, chose to invest internationally in U.K. soccer by

FIGURE 15.11. A game of ice hockey in Biysk, Russia. *Photo:* A. Fristad.

buying the famed Chelsea Football Club in 2003. He has since poured almost half a billion British pounds into it—a questionable investment perhaps, but a serious cultural statement. Inside Russia, soccer fields as well as hockey rinks can be found in many city backyards, but few of the boys and girls playing in them will ever rise to be international stars. When you travel in Russia, bear in mind that neither American football nor baseball is particularly well known, although some baseball teams have recently been formed in a few largest cities (usually at universities).

Although hockey and soccer are predominantly male sports, Russian female athletes have excelled in a range of sports—from volleyball to gymnastics to figure skating to swimming. The most recent phenomenon is the rise of excellent professional tennis players. In November 2009, 4 of the top 10 players in the global Women's Tennis Association ranking were from Russia (the most of any country), including Dinara Safina in 2nd place. Maria Sharapova, of considerable tabloid fame, was in 14th place. One player in the top 10 was from Belarus. The Soviet female volleyball team was the perpetual winner of the world championships in the 1950s and 1960s. It remained fairly strong in the 1970s and 1980s, and the Russian team remains one of the best in the world today, including winning the 2006 world championship in a match against Brazil.

FIGURE 15.12. The largest stadium in Russia, Luzhniki, was built in the 1950s, but then greatly updated for the 1980 Moscow Olympics. It can seat over 84,000 people. *Photo:* Author.

REVIEW QUESTIONS

1. What were the biggest differences between Soviet and U.S. education at the primary, secondary, and college levels?
2. What have been some of the changes in Russian education in the past few years?
3. What are some of the most advantageous places in Russia for receiving a good education?
4. Comment on the spatial and structural organization of Soviet scientific research.

5. To what extent was the geographic distribution of Soviet sports teams similar to that in the United States? What were the differences and why?

EXERCISES

1. Using the Internet, research the options available to you for study abroad in Russia. How many of the programs offered are U.S.-based? How many are Russia-based? In which cities are they located? How expensive are the programs? What subjects are being advertised? Are those programs a good choice for you? Why or why not?
2. Investigate the 20th-century history of any major branch of fundamental physical science (physics, chemistry, geology, biology, etc.). How many Russian (Soviet) names do you see mentioned? Where were the biggest contributions?
3. Pick any major team sport that you like, and find out whether any Russian/ex-Soviet athletes play for your country's teams. Find out more about how they were chosen, what their strengths are, and why they are now playing abroad.
4. In class, discuss the pros and cons of the Soviet/Russian model of college education, in which there are very few general courses, lots of subject-specific advanced courses, and virtually no electives.
5. Rent any of the recommended movies from Vignette 15.2, and watch it with friends. Then have a discussion of the film. Did you see any cultural or physical geography of Russia (or any other FSU republic) in the movie? How did this movie depict the country represented in it? How is it different from the typical Hollywood fare? Why do you think this may be the case?

Further Reading

Balzer, H. D. (1989). *Soviet science on the edge of reform.* Boulder, CO: Westview Press.

Graham, L. R. (1993). *Science in Russia and the Soviet Union.* New York: Cambridge University Press.

Medvedev, Z. (1978). *Soviet science.* New York: Norton.

Schott, T. (1992). Soviet science in the scientific world system. *Science Communication, 13*(4), 410–439.

Shaw, D. J. B., & Oldfield, J. D. (2007). Landscape science: A Russian geographical tradition. *Annals of the Association of American Geographers, 97*(1), 111–126.

Back issues of *Soviet Education* (renamed *Russian Education and Society* after 1991), *Soviet Science, Soviet Sports Review.*

Websites

www.artlib.ru—Library of Russian Art online. (In Russian only.)

www.hermitagemuseum.org/html_En—Official site of the State Hermitage Museum.

www.ras.ru—Russian Academy of Sciences. (In Russian only.)

www.msu.ru/en—Official site of Moscow State University.

www.museum.ru—Listing of most museums that exist in Russia. (In Russian only.)

www.rossport.ru—Federal Agency of Sports and Physical Culture of Russia.

CHAPTER 16

Tourism

This chapter discusses the tourism and heritage preservation issues of Northern Eurasia/the former Soviet Union (FSU). As Harvard biologist E. O. Wilson famously said, "Each nation has three kinds of wealth: material, cultural, and biological." Cultural and biological features attract visitors, both domestic and foreign. Unquestionably, the FSU has a treasure trove of both natural and cultural landmarks; however, the tourism potential of this vast landmass is greatly underused. A combination of physical and cultural geographic factors makes the region one of the least visited by international tourists today, and even domestic tourism remains underdeveloped. For much of the 20th century, the U.S.S.R. was a forbidden terrain behind the Iron Curtain. It allowed few foreigners in, and those were tightly chaperoned by Intourist agents and allowed to visit only a dozen or so destinations. Domestic tourism did exist, but with the decline of the Soviet state, much of the infrastructure for it deteriorated rapidly. After independence, every republic went a separate way, none (with the notable exception of Estonia) making development of tourism a high priority. Although some FSU nations now have vibrant domestic tourism (Russia and Kazakhstan), others have internal conflicts that make tourism highly problematic (Georgia, Armenia, Tajikistan, Azerbaijan), and still others simply do not have the resources or political goodwill to invest more in tourism. Whatever the reasons, few republics attract the foreign tourists proportionate to their potential.

Two big factors besides lack of governmental involvement are size and location. The FSU is very large and very remote. For example, if visiting famous Lake Baikal is an objective, an American needs to spend 10 hours on a plane just to get from New York to Moscow (about 14 from Los Angeles), and then an additional 5 on another plane to get to Irkutsk near Lake Baikal. The North American Great Lakes are considerably closer and offer a broadly similar experience. Kamchatka's volcanoes are awesome, but so are the ones in Alaska, and those can be visited at a much lower cost even by Europeans because of better-developed infrastructure and more competition among U.S. tour operators. The tsars' treasures in Moscow are phenomenal, but so are those of the Chinese emperors, and for many developed nations China offers faster transportation and easier access to visas. Without major investments in hospitality infrastructure, and much more spending on advertisements, Northern Eurasia will continue to lag far behind most of its worldwide competition. Russia spends only a few million dollars per year on promoting its tourism—considerably less than an average U.S. state.

The top attractions of Northern Eurasia are cultural. Most tourists come to see the Kremlin and the Tretyakov Gallery in Moscow; the State Hermitage Museum and the Peter and Paul Fortress in St. Petersburg; the Kiev Caves Lavra in Kiev; the medieval Islamic complex in Samarkand, Uzbekistan; and so on. Some come to see natural wonders (e.g., Lake Baikal, Kamchatka Peninsula, the Caucasus, the Pamirs, and other wild places). Others come because of various unique experiences the region can offer: the longest railway ride in the world, a trip to the North Pole on a nuclear icebreaker, or the thrill of flying at 2.0 Mach in an Su-27 fighter jet. I begin by considering the main recreational areas developed during the Soviet period, and then consider major types of tourism and places that are being currently developed for both domestic and international tourists, as well as pertinent social and environmental issues related to tourism.

The Main Recreational Areas of the U.S.S.R.

Tourism is a form of leisure service. The earliest forms of tourism in the U.S.S.R., and in tsarist Russia before it, were visits to the warm sea and spas in a classic form of health tourism. Russia won its access to the Black Sea in a series of bloody wars (mainly with Turkey and the Crimean Tatars) over two centuries, starting in the 1600s and culminating at the time of the U.S. Civil War. The tiny Crimea Peninsula, with an area of 26,200 km², was apparently a crown jewel worth shedding blood for. It has a unique, Italy-like climate with little frost in winter, a warm seacoast sheltered by mountains, and picturesque forests and steppe beyond the mountains. Swimming is possible for about 4 months each year. The Russian nobility built palaces in the Crimea from the early 19th century onward. In the early 20th century, Tsar Nicholas II, Prince Felix Yusupov, and other aristocrats had lavish palaces in Livadia, Foros, and Alupka in the Crimea, forming the so-called Russian Riviera. Asthma and TB sufferers, the writer Anton Chekhov being the most famous among them, would get respite from their diseases in the pine groves near Yalta. The Bolshevik government promoted healthier lifestyles for workers, and for this purpose redeveloped the imperial resorts lining the Black Sea coast and built new sanatoria there (Figure 16.1).

After Georgia and Armenia were incorporated into the Russian Empire (in 1800 and 1813, respectively), the Russian nobility gained access to additional warm sea beaches near Sukhumi and Batumi. In addition, the Russian Cossacks' push into the northern Caucasus in the 19th century opened up the mineral spa areas of Pyatigorsk (beautifully described in *A Hero of Our Time* by Mikhail Lermontov) and the entire coastal stretch from Novorossiysk to Sochi. The Soviet elite continued to develop the Black Sea coast, with numerous sanatoria for the Communist elite and summer camps for youth. The Artek camp at Gurzuf, with the bear-shaped Ayu Dag Mountain as a stunning backdrop, was established shortly before World War II as the first international Communist camp. The old settlement of Eupatoria attracted ailing children to its healing muds and saline inland lakes with

FIGURE 16.1. Black Sea coast in Georgia. *Photo:* K. Van Assche.

unique chemical properties. The Crimea had a resident population of just over 1 million, but this more than doubled in summer (Nikolaenko, 2003). Many visitors came on official sanatoria tickets, but many others came as independent tourists to rent a bed for a few nights from the locals or to camp out on the beach. The Crimea was so popular and crowded that a common joke "curse" for the locals became "May your relatives visit you in summer!"

Other areas with early tourism development in the Soviet period included resorts in Central Russia, skiing and water recreation in the lake country of the Valdai Hills, fishing and swimming along the Volga River, and hiking in the Caucasus and the Altay. The vast majority of tourists were citizens of the Soviet Union, primarily party bosses and privileged workers. Foreigners were allowed in, but only on prepackaged tours, and they were kept separate from the local population at all times. Independent travel by foreigners was not allowed. The early Soviet sanatoria were mainly focused on health; visiting spas, enjoying the forest air, sunbathing, and swimming were the main activities. Some resorts had or developed additional cultural resources—for example, along the famed "Golden Ring" of medieval Russian cities east of Moscow. Other favorites were various forms of active tourism: downhill and cross-country skiing, sailing, mountain trekking, and horseback riding, for example. Also always popular were summer camps for children; 2- to 4-week packages were available for free, largely through parents' places of employment, through schools, or by lottery to the most prestigious destinations like Artek.

At the same time, with the Iron Curtain in place, it was next to impossible for Soviet citizens to travel abroad. Some Communist apparatchiks could occasionally go to Bulgaria or Cuba, but even they required a special clearance from the KGB, which was not easy to obtain. It was almost impossible to visit North America, Western Europe, Asia, or the Pacific Islands as a tourist.

As noted above, the leading domestic tourist destination was the Crimea in Ukraine, with the Black Sea coast of Russia and Georgia trailing closely behind. Also popular were the Baltic capitals of Tallinn and Riga, and the small seaside towns of Yurmala and Palanga farther south. Limited Baltic Sea coast development also existed west and north of Leningrad and in Kaliningrad Oblast in Russia. Azerbaijan had Lenkoran, a resort on the Caspian Sea. The Central Asian republics had areas of mountain tourism (mainly horseback riding and mountaineering), especially near Lake Issyk-Kul, and health resorts in the warm Fergana Valley.

Tourism to and from Russia Today

By the end of the Soviet period, about 30 million people per year took advantage of resorts and sanatoria in the Russian Federation alone, not counting the other republics. Most were domestic tourists. The number of organized tourists in Russia abruptly plunged to a mere 8 million per year following the economic collapse of 1991, however. At the same time, the number of foreign tourists seeking to experience new and exciting opportunities in a previously unseen land rose substantially, but not nearly enough to compensate for the drop in the level of domestic tourism (Figure 16.2). Over 20 million foreigners visit Russia annually, as compared to 46 million visiting the United States, 52 million going to Spain, and 75 million going to France. The majority of foreign visitors come to Russia from Ukraine and Kazakhstan, and many of these are business or family visitors. The number of "true" foreign tourists visiting Russia per year from non-FSU countries is much lower (see below). If all visitors are included, Russia was the 10th most popular world destination in 2004—just ahead of tiny Austria, but behind Germany. Nikolaenko (2003) notes that such statistics are frequently misleading, however, because small countries with porous borders in Europe obviously see many more border crossings than, for example, large and isolated Canada or Russia.

As the economy stabilized after 2001, many Russian and other ex-Soviet citizens realized that they could now travel abroad. Visitors from the FSU became increasingly common in many European capitals, Alpine ski resorts, Mediterranean beaches, and some tropical countries. In 2008 36.5 million Russians crossed the nation's borders; 11 million of these crossings were for tourist trips, and 2 million business trips. The

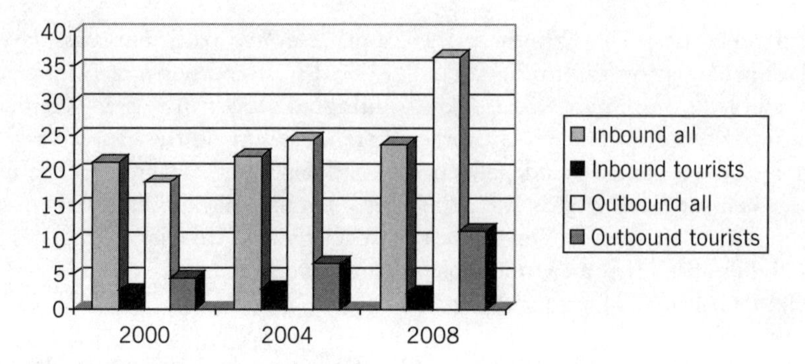

FIGURE 16.2. Number of inbound and outbound travelers (all travelers and tourists only) to and from Russia, in millions. Data from the Federal Tourism Agency of the Russian Federation.

rest were either family visits or regular commutes between FSU republics, primarily crossings to–from Ukraine and Kazakhstan. Russia is in the top 10 nations in terms of both sending tourists abroad and tourists' spending. For example, in 2005 Russian tourists spent $15 billion on foreign trips—more than Belgian or Hong Kong residents, but quite a bit behind the top spenders, Germans ($71 billion) and Americans ($66 billion). The top destinations for Russian tourists going outside the FSU in 2008 were, in descending order, Turkey, China, Egypt, Finland, Italy, Spain, Greece, Germany, Thailand, and France, with Cyprus, Tunisia, the Czech Republic, Bulgaria, and the United Arab Emirates (UAE) commonly in the top 10 in other years. Turkey was the top destination, with over 2.2 million visits (most to the "everything-included" resorts on the Mediterranean). Visits to China have been growing spectacularly—up 25% since 2007, with over 2 million visits, not counting business trips. The European destinations for Russian tourists are traditional ones, except Finland, which is attractive to the Russians because of its proximity, its high level of services, and the relative ease of securing a visa. Russians are also willing to travel to Egypt, Tunisia, Jordan, and other parts of the Middle East where many Westerners would not go, as long as the price is right. Few express concerns about safety. Visits to Israel more than doubled after the introduction of visa waivers there in 2008.

The numbers of tourists coming to Russia from non-FSU nations are drastically lower—only 2.3 million, and an additional 4 million business trips, out of a total of 23.6 million foreign visits in 2008. The rest are family trips by Russians who live abroad. So Russia sends five times as many tourists abroad as it receives. Among the non-FSU countries, the most tourists in 2008 came to Russia from Germany, the United States, Italy, Finland, the United Kingdom, China, Spain, France, Turkey, Canada, and Japan, in descending order. The top sender, Germany, sent merely 333,000 tourists; the United States sent 186,000. Such low numbers reflect a variety of factors: lack of hotel rooms, poor infrastructure, low promotional spending, perceived health and security risks, and difficulty in obtaining entry visas. The numbers for 2009 and 2010 are likely to be even lower because of the global recession.

The Main Forms of Tourism in Russia and Other FSU Countries

Classifying forms of tourism in the FSU is necessarily a subjective endeavor. I follow the classification used by the Russian Federal Tourism Agency, as well as by Kosolapov (2009). The following types of tourism are available for both foreign and domestic tourists: mass tourism in established resorts (at the seaside or near lakes, mineral springs, or mountains) or on cruises; cultural heritage tourism; religious pilgrimages; active and adventure tourism; nature tourism and ecotourism (plus hunting and fishing); and other tourism. Tourism accounts for only 0.5% of Russia's gross domestic products (GDP), as opposed to 1% in the United States and 6% or more in the European Union (EU). Kosolapov (2009) estimates that Russia's environmental potential for tourism is being utilized at a rate of only about 20%. Statistics on levels of domestic tourism are hard to find. The official statistics service only collects data on organized tourists, who account for fewer than half of all tourists countrywide. In established resort areas, perhaps 70% are organized tourists, but in the majority of the regions, only 30–40% are. Experts assess the total number of organized domestic tourists in Russia as smaller than the number of organized tourists traveling abroad. Only 10% of all registered travel agencies in Russia deal with domestic tourism; the rest are involved in the more lucrative outbound foreign tourism.

The capacity of Russian resorts is ridiculously low. The largest resort city in Russia, Sochi, has a total capacity of only 600,000 beds in peak season (although this is likely to change as preparations for the 2014 Winter Olympics are made). If an average vacationer stays there for 1 month, the total visits to Sochi may be estimated at 3 million per year. This is part of the total for the entire Krasnodarsky region of about 5 million per year (including Anapa, Gelendzhik, and Tuapse). An additional 600,000 can be accommodated in the Kavkazsky Mineral Waters area in the Stavropol region. Because of many resorts' low capacity, their facilities are often overwhelmed, resulting in environmental problems of overcrowding; for example, poorly treated sewage frequently closes beaches both in the Crimea and near Sochi in peak season.

In Russia, over 60% of all vacations take place close to home on dachas (private summer cabins), and not as organized tourism in far places. With the deepening recession in 2009, even more people were expected to opt for the cheapest local options.

Mass Tourism at Established Resorts or on Cruises

The traditional packages sold by travel agents in the FSU involve visits to seaside sanatoria, spas, and other resorts, or cruises. These usually take place in the warmer parts of the region, and/or in places regarded as having curative properties of air, water, and scenery. Over half of all organized tourism activity in Russia belongs to this category, as well as a large majority of foreign tourist activity. The top destinations are Krasnodarsky Kray, the Moscow region, St. Petersburg, Stavropolsky Kray, Chelyabinsk, Tatarstan, Kaliningrad Oblast, and the Altay. Also popular are cruises along the Oka and Volga. The Crimea in Ukraine continues to attract visitors and is the top destination in the FSU outside Russia. Each of these regions receives from about 500,000 to a few million tourists a year. Most visits are prepaid, all-inclusive vacations, but there are also considerable numbers of independent tourists. In seaside locations, tanning and swimming are the most popular activities. Inland, the main focus is on water sports in the rivers and lakes, as well as recreational fishing and limited hiking. About 70% of all tourists in Russia have relaxation as their main goal; an additional 6% are taking those trips primarily for health reasons.

A typical sanatorium from the late Soviet period is located on the Black Sea coast, belongs to a government ministry or a large private company, has about 200–300 beds available at any given time, and provides 2- to 3-week packages. The health regimen is taken very seriously. Upon check-in, each guest is evaluated for a range of physical conditions that may be amenable to treatment. Typical procedures include bathing in a sulfuric spa, drinking mineral water, walking outdoors, natural tanning, swimming, gymnastics, yoga, and other activities. Additional cultural events are provided. There are three meals a day. Today many of the old sanatoria are still functioning, although some had deteriorated to the point of no return by the 1990s and were permanently closed. New, smaller facilities are appearing every year in many old locations. Particularly booming at the moment seems to be the Black Sea coast south of Sochi (the future site of the 2014 Winter Olympics, as noted earlier). By contrast, the Crimea Peninsula is experiencing a downturn in activities as a result of fragmented ownership and lack of investment there (Nikolaenko, 2003).

Cultural Heritage Tourism

All FSU countries have multiple prehistoric sites. Some caves in the Altay and Georgia show evidence of human occupation going back to the late Pleistocene, over 100,000 years ago; some of these sites are among the earliest known anywhere outside of Africa to contain anatomically modern human bones and artifacts (e.g., Denisova Cave in the Altay). Archeologists in Ukraine and Russia have found entire villages with buildings constructed of thousands of mammoth bones. Neolithic and Bronze Age sites are common in the Central Asian states, located along the traditional Silk Route from the Middle East to China. The World Heritage Program of the United Nations Educational, Scientific, and Cultural Organization (UNESCO) lists numerous historical sites in the FSU of global importance (Table 16.1). For comparison, some sites with the same recognition in the United States include the Statue of Liberty in New York, Independence Hall in Philadelphia, and the native dwellings of Mesa Verde, Colorado. The listed sites in England include Stonehenge, Canterbury Cathedral, and the Tower of London.

Armenia has three medieval religious complexes dating back over 1,000 years on the World Heritage list. Azerbaijan has the walled city of Baku and the Gobustan rock art cultural landscape. Georgia has three sites, including Bagrati Cathedral, which is about 1,000 years old, in the center of Kutaisi. Estonia has the old part of Tallinn listed; this is a medieval fortified city built by the German Teutonic knights. Parts of the capitals of Latvia and Lithuania have likewise merited World Heritage designations. In Ukraine, there are the Kiev Caves Lavra and St. Sophia Cathedral in Kiev, as well as the historic center of Lvov. In Central Asia, 15th- and 16th-century Islamic complexes in Samarkand and Bukhara, Uzbekistan, and the Mausoleum of Khoja Ahmed Yasawi in Turkestan, Kazakhstan, are recognized as examples of medieval Islamic culture along the Silk Route.

The European countries of the FSU also have a unique geographic artifact, the Struve Geodetic Arc, included in the World Heritage Program. This is a chain of survey triangulation points stretching from Norway to the Black Sea, through 10 countries and over 2,820 km. The survey, carried out between 1816 and 1855 by astronomer F. G. W. Struve, represented the first accurate measuring of a long segment of a meridian on our planet. This helped to establish the exact size and shape of the earth's ellipsoid, and marked an important step in the development of earth sciences and topographic mapping. It is also a rare example of early scientific collaboration among different countries and cultures in Europe.

Russia has 14 historical sites on the World Heritage list, including the Moscow Kremlin, St. Petersburg's city center (its status is now being threatened by the Gazprom office tower development), the wooden churches of Kizhi, and a few stone-walled monasteries and old cities of Central Russia. Although no cultural sites are yet listed in Siberia, there are of course many such attractions there too—simply not on a par with the oldest ones in European Russia. In particular, historical Tobolsk and Tomsk may be worth a visit, and the remains of GULAG camps.

The majority of packaged historical tours of Russia include a few days in Moscow and St. Petersburg, with additional days spent in visiting the cities of the Golden Ring east of Moscow (Suzdal, Vladimir, Rostov, Yaroslavl) or the Valaam and Kizhi islands in Lakes Ladoga and Onega, respectively, east of St. Petersburg. St. Petersburg is the most attractive destination in Russia, in the opinion of Russian travel agents (Kosolapov, 2009).

Another common option is to take a river cruise all the way from Moscow down the Volga, with stops at all major towns on the way. There are few other options as far as historical tours go, unless tourists speak Russian and are adventurous enough to strike out on their own, or so wealthy as to arrange for a personal tour guide with a driver and a car (as in the movie *Everything Is Illuminated*). In Ukraine, Kiev and Lvov attract a lot of historical tourism from abroad, with additional visitors coming to Chernigov, Poltava, Odessa, and a few other of the oldest cities. Some foreigners come with an explicit desire to visit the main battlefields of World War II—Volgograd (Stalingrad), Kursk, Smolensk, Minsk, Brest, and so on—or the Borodino battlefield of 1812, so eloquently described in Tolstoy's *War and Peace*. For Russians, two other famous battlefields are Lake Chudskoe (Peipus) on the border with Estonia, where the Teutonic knights were

TABLE 16.1. Objects of World Natural (N) and Cultural (C) Heritage in the FSU, as Recognized by UNESCO

Armenia
- Monasteries of Haghpat and Sanahin (1996, 2000) (C)
- Cathedral and Churches of Echmiatsin and the Archaeological Site of Zvartnots (2000) (C)
- Monastery of Geghard and the Upper Azat Valley (2000) (C)

Azerbaijan
- Walled City of Baku with the Shirvanshah's Palace and Maiden Tower (2000) (C)
- Gobustan Rock Art Cultural Landscape (2007) (C)

Belarus
- Belovezhskaya Pushcha/Bialowieza Forest (1979, 1992) (N)
- Mir Castle Complex (2000) (C)
- Architectural, Residential, and Cultural Complex of the Radziwill Family at Nesvizh (2005) (C)
- Struve Geodetic Arc (2005) (C)

Estonia
- Historic Centre (Old Town) of Tallinn (1997) (C)
- Struve Geodetic Arc (2005) (C)

Georgia
- Bagrati Cathedral and Gelati Monastery (1994) (C)
- Historical Monuments of Mtskheta (1994) (C)
- Upper Svaneti (1996) (C)

Kazakhstan
- Mausoleum of Khoja Ahmed Yasawi (2003) (C)
- Petroglyphs within the Archaeological Landscape of Tamgaly (2004) (C)

Latvia
- Historic Centre of Riga (1997) (C)
- Struve Geodetic Arc (2005) (C)

Lithuania
- Vilnius Historic Centre (1994) (C)
- Curonian Spit (2000) (N)
- Kernav? Archaeological Site (Cultural Reserve of Kernav?) (2004) (C)
- Struve Geodetic Arc (2005) (C)

Moldova
- Struve Geodetic Arc (2005) (C)

Russia
- Historic Centre of St. Petersburg and Related Groups of Monuments (1990) (C)
- Kizhi Pogost (1990) (C)
- Kremlin and Red Square, Moscow (1990) (C)
- Cultural and Historic Ensemble of the Solovetsky Islands (1992) (C)
- Historic Monuments of Novgorod and Surroundings (1992) (C)
- White Monuments of Vladimir and Suzdal (1992) (C)
- Architectural Ensemble of the Trinity, Sergius Lavra in Sergiev Posad (1993) (C)
- Church of the Ascension, Kolomenskoye (1994) (C)
- Virgin Komi Forests (1995) (N)
- Lake Baikal (1996) (N)
- Volcanoes of Kamchatka (1996, 2001) (N)
- Golden Mountains of Altay (1998) (N)
- Western Caucasus (1999) (N)
- Curonian Spit (2000) (N)
- Ensemble of the Ferrapontov Monastery (2000) (C)
- Historic and Architectural Complex of the Kazan Kremlin (2000) (C)
- Central Sikhote-Alin (2001) (N)
- Citadel, Ancient City, and Fortress Buildings of Derbent (2003) (C)
- Uvs Nuur Basin (2003) (N)
- Ensemble of the Novodevichy Convent (2004) (C)
- Natural System of Wrangel Island Reserve (2004) (N)
- Historical Centre of the City of Yaroslavl (2005) (C)
- Struve Geodetic Arc (2005) (C)

Turkmenistan
- State Historical and Cultural Park "Ancient Merv" (1999) (C)
- Kunya-Urgench (2005)
- Parthian Fortresses of Nisa (2007)

Ukraine
- Kiev: Saint-Sophia Cathedral and Related Monastic Buildings; Kiev Caves Lavra (1990, 2005) (C)
- L'viv—the Ensemble of the Historic Centre (1998) (C)
- Struve Geodetic Arc (2005) (C)
- Primeval Beech Forests of the Carpathians (2007) (N)

Uzbekistan
- Itchan Kala (1990) (C)
- Historic Centre of Bukhara (1993) (C)
- Historic Centre of Shakhrisyabz (2000) (C)
- Samarkand—Crossroads of Cultures (2001) (C)

Note. Data from World Heritage Database online (*whc.unesco.org*).

defeated by Prince Alexander Nevsky in 1242, and the Kulikovo battlefield in Tula Oblast, the site of a decisive victorious battle led by Prince Dmitry Donskoy against the Tatars in 1380 (see Chapter 6, Table 6.1). Seasonal reenactments of battles occur in a few places in Russia and are very popular.

Religious Pilgrimages

The Russian Orthodox Church and Islam have a long-standing tradition of pilgrimages to holy sites, just as the Roman Catholic Church does (Chapter 14). "Pilgrimages" are defined as journeys to holy places to express devotion, to seek supernatural help, or to do penance. Among Christians, they became particularly popular in the Middle Ages as the cults of saints grew to become an important element of church life. Typical objects of pilgrimage include graves and churches with relics of saints; places where they lived and prayed (e.g., caves or monasteries); structures they built (cathedrals, churches); or places where they are believed to have appeared after their death. Pilgrimages are an established tradition in Russia. Since the fall of Communism, millions of people in the FSU have rediscovered it. Many pilgrimages are solo or small-group trips, but recently there have been also large public processions with relics or icons that go on for weeks and involve thousands of people. One of the largest takes place every summer in the Kirov region of northeastern European Russia and lasts for several days. Another pilgrimage to honor St. Seraphim of Sarov, the famous 19th-century monk and mystic, involves a few hundred participants walking from Kursk in southern Russia to Diveevo in Nizhniy Novgorod Oblast for 40 days, covering over 1,000 km.

Pilgrimages are important economic activities, because they generate revenue directly as donations to the communities that are visited, and indirectly by stimulating the development of services for pilgrims in otherwise poor locations. The town of Diveevo, for example, with its world-famous convent dedicated to the Holy Trinity, has a population of only 17,000 people (including over 800 nuns). On an average weekend in summer, over 3,000 pilgrims come to participate in church services, to venerate the relics of St. Seraphim of Sarov, to bathe in one of the holy springs, and so on. The majority of these people come on packaged tours organized by one of a few nationwide church service bureaus. Many of these are nonprofit ventures, using the proceeds from pilgrimages to support their own churches and communities in other parts of the country. Pilgrims to Diveevo stay in hotels or local people's homes. They also buy food, water, books, icons, candles, music, and other souvenirs. The impact of this economic activity on the small town is considerable. As an example, with local residents' monthly incomes from state pensions or salaries amounting to merely 5,000 or 6,000 rubles in Diveevo in 2006 (the average for Russia being 15,000), one group of pilgrims from Moscow staying overnight would pay 400–600 rubles to a local resident per night.

There is no definitive list of the top Orthodox pilgrimage sites in the FSU by numbers, but Diveevo and the monasteries listed in Chapter 14 are likely to be in the top 10. Ukraine, Belarus, Georgia, Moldova, and Armenia have many additional Christian sites. Over 20,000 Russians and Ukrainians participate in an annual pilgrimage to Jerusalem during Orthodox Holy Week. Muslim holy sites are found in the Volga region of Russia (especially in Tatarstan), Central Asia, Azerbaijan, and the northern Caucasus (especially Dagestan). And Muslims from the ex-Soviet states are common participants in the *hajj* to Mecca.

Active Tourism

The Russians are keen on active tourism. The most popular forms include hiking, mushroom and berry picking, game hunting, fishing, kayaking, downhill skiing and sledding, mountain climbing, horseback riding, and spelunking. Bicycling has always been popular in the Baltic republics and is gaining popularity elsewhere. There are some famous world explorers among the Russians—for example, Fedor Konyukhov (born in 1951), who has made over 40 record-breaking trips. He was the first person in Russian history to make a solo, nonstop circumnavigation of the globe; he has also climbed the top seven summits of the world (one on each continent) and made multiple trips across the Arctic and Antarctica on dogsled or on skis. With market reforms, new technologies and imported equipment have been introduced in most of Russia's heavily traveled areas. For example, the lower gorge of the Katun River in the Altay Republic of Russia now has over 50 companies offering whitewater rafting trips, using contemporary, well-built inflatable rafts. Similarly, snowmobiling, kayaking, tubing, scuba diving, windsurfing, and other outdoor endeavors have experienced phenomenal growth since 2000 in Russia.

Within Russia, the top outdoor destinations include Karelia (water tourism), the Caucasus and the Altay (mountain tourism; Vignette 16.1), the polar Urals, Lake Baikal, the Sikhote-Alin range, and (for the most adventurous) Kamchatka Peninsula. Other fine destinations include central European Russia, especially the glacial hills and lakes of Valdai National Park, the forests of Komi, the beaches of the Baltic coast, and the forests and rivers of the southern Urals (Bashkortostan and Chelyabinsk regions). In contrast, there are very few opportunities for active tourism for people living in the Russian steppe zone and western Siberia, where the land is flat and boggy, settlements are rare, and roads are poor or nonexistent. Even in these less optimal places, however, people go on outings to a local forest or a warm lake in summer, or go skiing across the frozen fields in winter. In fact, the so-called village tourism is enjoying high popularity everywhere in the region now because of its low cost, much as in the rest of Europe (Ostergren & Rice, 2004).

The official statistics on domestic active tourism in Russia are hard to come by. Nevertheless, Figure 16.3 provides an indirect way of assessing the popularity of various destinations, based on the trip reports filed online at the popular open-access Website *turizm.lib.ru*. Such reports are commonly used by others to gauge the suitability of a particular destination for their purposes. The most commonly reported destination is Karelia, which is close to St. Petersburg and Moscow; it provides unmatched opportunities for canoe and kayaking expeditions in the largely flat, lake- and river-dominated terrain. Karelia has postglacial scoured relief similar to that of the Boundary Waters Canoe Area in Minnesota, and attracts tens of thousands of tourists per year. The presence of national parks and preserves helps tourists to choose optimal routes, and the availability of a main railroad line provides easy access to many entry points. The second most commonly reported destination is the Moscow region. The actual total numbers of tourists here are likely to be higher than in Karelia, but most people probably spend much less time per trip. In Karelia, most trips are at least 7 days in duration; in the more accessible area near Moscow, weekend trips are common. The Moscow region has a few new skiing resorts that are attracting a wealthy clientele. It also offers diverse summer forms of recreation, such as mushroom hunting and berry picking. The third most commonly reported destination for adventure trips is the Altay Mountains. The Altay receives over 500,000 tourist visits per year, but only 10% of them are primarily outdoor oriented, including backpacking, horseback trips, and whitewater rafting. Most travelers to this area stay in established health resorts.

Currently in fourth place, the Russian Caucasus used to be the second most popular Soviet destination for mountain tourists, especially skiers, but its popularity has waned considerably because of the political instability there since 1991. Nevertheless, the skiing areas of Dombai and Baksan, and opportunities to climb Mt. Elbrus and other peaks taller than the Alps, continue to attract domestic and foreign visitors. New investments are also pouring in from the federal and regional governments. Two major international airports, Sochi/Adler and Mineralnye Vody, now provide access to the south and north slopes, respectively. The total number of developed ski resorts in the FSU in early 2009 was only 70; tiny Latvia had 27, Russia had 17, and Ukraine had 14. In comparison, Germany had 116, and Austria 275. In Russia, the busiest one is Baksan near Mt. Elbrus in Kabardino-Balkariya, with 11 lifts and 35 km of trails. Besides the Caucasus, big ski resorts exist in the Altay (Belokurikha), near Lake Baikal, and on Kola Peninsula (Kirovsk), as well as a few brand new ones near Moscow (Mtsensk, Volen, Sorochany). The latter ones use the newest snow-making machines and a combination of natural and artificial hills in the generally flat area.

Vignette 16.1. Touring the Golden Mountains of the Altay

In the Soviet period, the Altay (*altan* is Mongolian for "golden") was one of only five areas with a national reputation as a tourist region. The other four were the Baltic Sea coast, the Carpathians/Crimea, the Caucasus, and the mountains of Central Asia. Situated in the very heart of Eurasia, over 2,500 km from the nearest seacoast, the Altay is one of the most remote mountain ranges on earth. The closest big airport is in Barnaul, about 4 hours away by car. The Altay mountain system consists of over 30 separate ranges, rising in different directions away from the highest point, Mt. Belukha (4,506 m)—the highest mountain in Russian Siberia, and the second most popular destination in the Altay. About half of the Altay is in Russia, 30% is in Kazakhstan, and the remainder is divided between China and Mongolia. Its main uplift was caused by the India–Eurasia collision that started 50 million years ago, although many areas were already mountainous before that time, and others were elevated more recently (merely 5–8 million years ago). The relief consists of mountains of intermediate elevations (1,800–3,000 m) and extensive plateaus, with large intermountain depressions and deeply incised river valleys. The Mt. Belukha massif and the Northern and Southern Chuya ranges farther east contain over 1,500 glaciers, although they are now rapidly retreating due to global warming. The western Altay receives over 2,000 mm of precipitation a year, with the snow line running as low as 2,300 m above sea level. The eastern Altay is considerably drier because of the rain shadow effect; the snow line there is above 3,100 m, and it receives only 1,000 mm of precipitation. The Chuya Steppe depression, located in the rain shadow, receives less than 200 mm—a real semidesert.

The most interesting part of the Russian Altay is in the Altay Republic, attracting about half a million tourists per year. Additional 200,000 come to Altaysky Kray, with perhaps 100,000 visiting the Altay in Kazakhstan (East Kazakhstan Oblast). Portions of the Russian Altay were included in the World Heritage Program of UNESCO, including a number of wildlife preserves and the Altaysky Zapovednik.

The gateway community of Biysk, about 160 km south of Barnaul, is the last train station on the way to the mountains. Belokurikha is the largest resort in the Altay foothills, located about 1 hour south of Biysk. It received the heaviest resort investment in the region during the Soviet period, mainly sponsored by the Kuzbass and West Siberia metallurgy enterprises. Peaks of development occurred in the early 1960s and the mid-1980s. Today Belokurikha is a federal-level resort with over 20 sanatoria, a few ski lifts, numerous spas, upscale boutique shops, and otherwise good infrastructure (Figure 1). The warm mineral springs, clean mountain air, unusually warm microclimate, good ski slopes, and proximity to Barnaul and Biysk make this the most celebrated resort of south central Siberia. The majority of visitors come from Novosibirsk, the Kuzbass coal-mining area, and the Tyumen oil-producing region. The international segment is growing, but represents under 10% of all visitors at present. Other large resort complexes exist near Biysk, at Teletskoe Lake, at Biryuzovaya Katun and Chemal along the Katun River, and around Gorno-Altaysk.

Teletskoe Lake is the world's sixth deepest (average depth of 326 m). With a surface area of 231 km^2, crystal-clear water, and spectacular scenery rivaling that of Lake Baikal, it attracts about one-third of all visitors to the region, making it the top destination for tourists. Its eastern shores are entirely within Altaysky Zapovednik, which allows only limited tourism. Some boat tours and hikes to nearby waterfalls are offered at the gateway community of Artybash. The Chemal area of the Katun River gorge is the third most popular tourist destination after Teletskoe Lake and Mt. Belukha (Figure 2). The spectacular gorge provides ample whitewater rafting opportunities. Another local specialty is horse tourism along mountain trails. Caves are also ready for exploration. Over 100 possible back-country routes exist in the central Altay alone. However, recent overdevelopment—with over 50 resorts built near Mayma Lake (the only warm water lake in the area) and prospects for a huge federal casino complex—has led to many public debates over the best course of future development for the area.

(cont.)

FIGURE 1. Belokurikha is the most upscale resort of the Altay, recently visited by Vladimir Putin to proclaim a new chapter in the development of Siberian tourism. The main attractions here are mineral springs and ski slopes. *Photo:* Author.

FIGURE 2. The Katun River gorge attracts a quarter of all visitors to the Russian Altay. Whitewater rafting, horseback riding, and backpacking are offered here at more than 50 resorts. The area suffers from soil erosion, air pollution, plastic garbage, and lack of coordinated environmental planning, however. *Photo:* Author.

(cont.)

Nature and adventure tourism in the Altay includes hiking, backpacking, mountain bicycling, horse tourism, cave tourism, whitewater rafting, kayaking, cross-country and downhill skiing, paragliding, snowmobiling, fishing, hunting, and some extreme tourism. Also popular are spiritual retreats and archeology camps. The Altay is one of the oldest known areas outside of Africa with continuous human settlement (at least since 200,000 years ago). Also, many religious traditions of Buddhism and native Burkhanism suggest that the Altay is the cosmic gateway to the mystical, heavenly Shambala. The influential early-20th-century painter and philosopher Nikolai Rerikh in particular believed this, and he still has many followers in the area today.

Three big outside threats cloud the Altay's future tourism prospects:

- The possible construction of the Katun Hydropower Plant in the Katun River gorge, not far from Chemal. If this dam is built, the potential energy would be comparable to that produced by the largest hydropower installation in Russia, the Sayano-Shushenskaya GES on the Yenisei (6,000 MW), or about half of that produced by the Chinese Three Gorges Dam. The last major battle fought (and won) by the Soviet environmentalists was over this dam. It has not yet been built because of a well-organized protest campaign and the collapse of the socialist state, but recently there have been suggestions that it could be built after all, with some powerful backers from Moscow and Siberia.
- A newly proposed Gazprom natural gas pipeline from Russia to China. This would cut across the Ukok Plateau, a site of major archeological and biological importance listed as a World Heritage Site.
- Transboundary pollution from metal smelters at Ust-Kamenogorsk, Kazakhstan, as well as toxic fallout from Baikonur space launches.Ust-Kamenogorsk is located directly upwind, west of the Altay, and wind typically brings acid rain from it over the western slopes of the mountains over 150 days per year. The Baikonur rockets shed their parts all over western Altay; many contain highly toxic hydrazine fuel, a powerful carcinogenic and mutagenic agent.

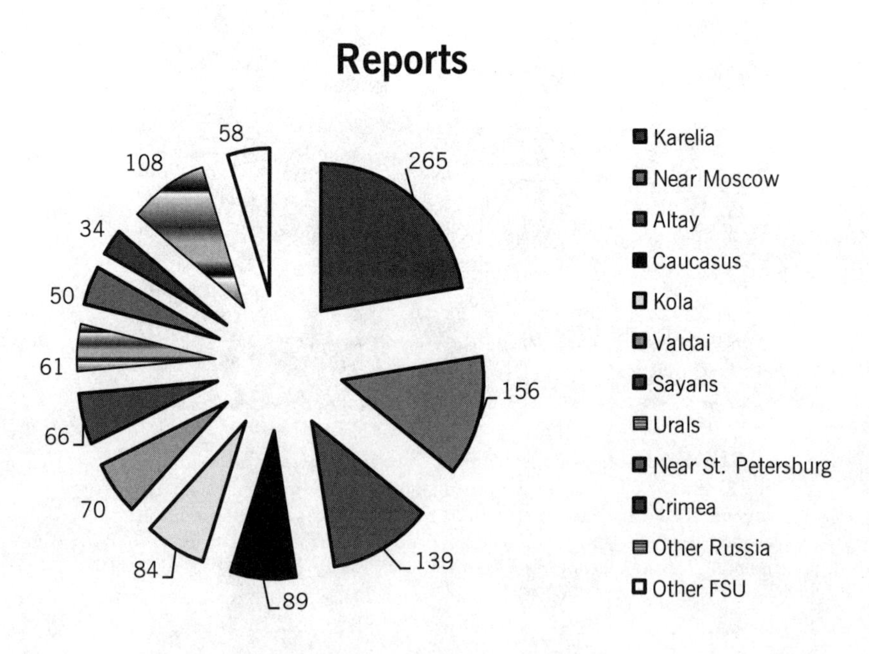

FIGURE 16.3. The number of outdoor trip reports published in 2007 on *turizm.lib.ru* by destination—an indirect measure of popularity of various destinations among outdoor-bound tourists in Russia. The regions are arranged clockwise, from Karelia at top right to other FSU destinations at top left.

Adventure Tourism

Adventure tourism is probably the fastest-growing form of tourism in the FSU. Many are extreme forms that attract thrill seekers, both domestic and foreign, to undertake risky activities outside the bounds of conventional behavior. Sometimes also called "shock tourism," it has evolved primarily among modern, upper-middle-income consumers who do not experience enough thrills in their mundane lives in big cities. The following opportunities may be mentioned:

- Taking rides in military fighter jets at supersonic speeds.
- Space and near-space tourism (e.g., being trained as if you were going to go to space, using Russian Star City centrifuge equipment, zero-gravity flights, etc.).
- Diving in lakes and rivers during winter.
- Extreme mountain biking down a steep slope with no trail.
- Paragliding.
- Exploring sewers and old bomb shelters in Moscow with so-called diggers.
- Bungee jumping into waterfalls or off cliffs.
- Walking a stretch of a former Siberian prison road with your feet in fetters.
- Wild caving in Central Russia, the Urals, or Siberia.
- Visiting abandoned mines or factories of the Soviet period.

There are now "extreme amusement parks" in the Altay and a few places in Central Russia, where one can spend a few days engaging in a variety of strenuous, sometimes life-threatening activities. Considering that many of these places also serve alcohol on the premises, accidents are common, including some fatalities.

Nature Tourism and Ecotourism

Broadly speaking, nature tourism involves the outdoors just as active and adventure tourism do, but the focus is on nature rather than on bodily exercise. True nature tourists use natural areas to observe, to heal, to learn, and to think. Ecotourism is similar to nature tourism, but it is not simply nature based; it must also be sustainable with respect to both nature and the local culture (Kolbovsky, 2008). Although western-style ecotours are just beginning to be developed in Russia, traditional Russian outdoor activities such as mushroom picking (mentioned above) may qualify as ecotourism when practiced sustainably.

Compared to Western Europe or North America, very few companies specialize in offering nature tours in Russia or the rest of the FSU. Moreover, most such companies cater to a very selective Western clientele, not to domestic tourists. Most Russian nature enthusiasts organize and outfit themselves. Since very few people in the U.S.S.R. owned a car, nature trips by necessity took place relatively close to home, in local natural areas that could be accessed via suburban train. Such stations in the Moscow region, for example, included Peredelkino, Bulatnikovo, Turist, and Opalikha, half an hour away from the city. More ambitious and better-prepared tourists would spend entire vacations hiking, backpacking, canoeing, or mountaineering in the remote corners of the Soviet Union, accessed via long-distance trains and plenty of walking with heavy backpacks.

There is much debate over how much nature tourism in Russia today qualifies as ecotourism. Although definitions of the latter vary, most would include two important provisions, as indicated above: the naturalness of the experience, and respect for/benefits to the local culture. For example, flying top executives around Sakhalin in a military helicopter shows them the beauty of the island's wild nature, perhaps, but the main mode of transportation is not natural; nor does it generate any revenue for the residents of the island. On the other hand, a kayaking trip down the famed Chuya River in the Altay Republic may very well be done in a natural way by using minimal-impact camping techniques, and it will benefit local communities if the party agrees to buy local food and supplies.

Few Russian travel agents have caught on to the importance of ecotourism, although some are learning quickly. Most assume that vacationing Russians are desperate only for the "three S's": sun, sea, and sand somewhere in the Mediterranean. In fact, you are more likely to find an ecotour guide from Russia advertising in English to Western clients than in Russian to local ones.

Nevertheless, hundreds of thousands of Russians do spend time on formal outdoor trips, many of which would qualify as ecotours. Protected natural areas, especially national and regional nature parks, provide interpretive trails and camps. Sometimes they are also the scenes of more questionable activities (e.g., private hunting trips for well-connected bureaucrats or wealthy foreign clients; this is a particularly common practice in Central Asia and the Caucasus).

Student groups from universities and schools organize longer nature expeditions, usually led by a dedicated teacher. Summer camps are held in different regions for underperforming or rural schoolchildren; these offer excellent environmental programs with elements of ecotourism in the curriculum (see Chapter 5, Vignette 5.1). Foreign clients demand more extensive ecotours; they cherish the wilderness experience that Northern Eurasia can still offer. Russia's zapovedniks and national parks (see Chapter 5) provide diverse opportunities for ecotourism. Additional possibilities exist in regional nature parks, in local natural monuments, in historical parks, and simply in undesignated wilderness. Only five countries worldwide have an amount of wilderness comparable to Russia's, and only Canada has a similar range of experiences. The other three are tropical Brazil, Australia, and China. In Brazil, tropical rain forests provide a radically different experience from Russia's taiga; in Australia, the main wilderness experience is that of a tropical desert; China has mainly arid mountains and subtropical forests. Although Canada may seem like an exact counterpart of Russia, in reality it is very different sociopolitically. For example, the provincial governments of Canada maintain a much tighter control over land use policies than local Russian jurisdictions do. Consequently, there are much stricter policies on back-country travel in Canadian provincial parks than in Russia, where pretty much "anything goes." Also, more land may be controlled by private logging companies in Canada, with restrictions placed on public access, than in Russia.

Besides Russia, outstanding opportunities for ecotourism exist in Kazakhstan, Tajikistan, and Kyrgyzstan, with their giant mountain ranges, wild steppes, and beautiful rivers and lakes. In the latter two, however, the political instability of the last 20 years has hampered tourism development. Kazakhstan is well positioned to create world-class ecotourism programs. It makes conscious efforts to attract tourists from China, with which it shares a long land border and a railroad link. Nevertheless, Kazakhstan lags behind Russia in the availability of outdoor tourist services. Ukraine also has good potential for rural tourism, with elements of ecotourism in the Crimea and the Carpathians. The Baltic states have well-run programs involving stays on farms, as well as excellent (albeit small) national parks. In Georgia and Armenia, opportunities for sustainable mountain tourism are likewise plentiful, but little known to outsiders due to political unrest and a lack of development.

Other forms of nature tourism are either being developed or likely to be developed in the FSU in the near future. One of these, scientific tourism, involves trips undertaken by researchers with a scientific goal in mind. For much of Soviet history, access to the Eurasian hinterland has been greatly restricted. Geophysicists studying the earth's magnetic fields, geologists interested in unique minerals, glaciologists pursuing remote glaciers and permafrost, hydrologists interested in the water balance of the Arctic Ocean, biologists looking for rare plants and animals, anthropologists and linguists studying endangered cultures, and archeologists searching for artifacts are some examples of specialists who visit Northern Eurasia on research grants these days. These trips are funded by Western taxpayers, are frequently conducted on tourist visas, and bring revenue to the FSU countries. Specialized bird-watching or whale-watching trips, archeological digs, and other types of nature tourism would fit into this category, as long as they are done by scientists with the purpose of obtaining new data.

Other Forms of Tourism

A growing segment of tourism is medical tourism, which involves long-distance travel to diagnose or treat a condition or disease. It is becoming better known, particularly with respect to dental services, Lasik eye surgery, cosmetic surgery, and cancer/heart disease diagnosis and treatment. Many such "tourists" are visitors to Russia from the poorer FSU republics, but some are also wealthy Westerners (especially former Russian residents who now live abroad) who are seeking cheaper treatments than those available where they currently live.

Organized shopping tours are less common in the FSU than in the United States or Europe, but they do occur. Moscow and St. Petersburg attract the lion's share of domestic and international shoppers, accounting for about 21% of the total retail activity in the country (Chapter 21). "Alcohol tours" to St. Petersburg attract busloads of tourists from Finland, a country where liquor is prohibitively expensive. Residents of cities within a day's journey of Moscow often come for weekend cultural tours that also include shopping. Shopping for electronics in border centers in nearby China is common for Kazakh tourists. Russians in the Far East visit South Korea or Japan to purchase cars or electronics.

Greatly underdeveloped is theme park tourism. There are still no equivalents of Disneyland or Sea World in the FSU. A few aquaparks and wildlife safari parks have been proposed for the Moscow and Leningrad areas or have been already built. The coastal areas have aquariums highlighting marine life. Veliky Ustyug in Vologda Oblast is the supposed home of Ded Moroz, the Russian Santa Claus, with associated heavily commercialized tourist attractions. The old Soviet amusement parks (e.g., Gorky Park in Moscow) are undergoing slow renovation.

Unfortunately, there is also sex tourism in many parts of the FSU (see Chapter 12). Many foreigners come to Russia and other Eastern European countries with the explicit purpose of procuring sex for money. Although such business is strictly illegal, economic realities and poor law enforcement make such sex relatively easy to obtain in much of the FSU, including the Baltic states. Strip clubs, escort services, massages, and dubious matchmaking agencies are common and operate with little hindrance in many cities. Western businessmen also procure sex workers for foreign markets in the FSU. Especially grave concerns exist about violations of women's rights and welfare in Moldova and Ukraine—both poor countries close to Europe that provide a large share of prostitutes to the European, Middle Eastern, and North American underground markets (Hughes, 2005). There are also concerns about pedophiles preying on children in the region—especially those in the poorest areas, but also wealthier urban kids online.

Some visitors come to the FSU for legitimate adoption of a child or for marriage, as a form of social or family tourism. Such prospective parents/bridegrooms are not tourists in the narrow sense, but they contribute to the rising fortunes of matchmaking companies (some of the same ones that also facilitate the "mail-order bride" business) and can be seen as a type of economic tourists as well.

To summarize, Northern Eurasia provides multiple opportunities for all kinds of tourism (Figure 16.4). However, major investment in infrastructure, advertisement, and planning is urgently needed to boost both domestic and international tourist numbers. The fastest-growing forms of tourism are nature and adventure tourism. Nature, however, is fragile, and much more must be done to make outdoor tourism sustainable over the long term. Also, Russia in particular is one of the world's top 10 suppliers of tourists to foreign destinations, and the numbers of Russian (and other ex-Soviet) tourists in Europe, Asia, and North America are expected to continue growing.

REVIEW QUESTIONS

1. Think of the top 5–10 destinations for conventional tourism in your country (check online resources or a library to determine the exact numbers of visitors). Which of these are primarily cultural sites? Which ones are natural sites? How many are wild areas as opposed to heavily developed areas? For every site on your list, try to come up with a similar site somewhere in Northern Eurasia.
2. Which countries are the main magnets for Russian tourists today? Why?
3. Name any five common outdoor activities that Russians enjoy. Are any of these available to you in the region where you live?
4. Who would travel more internationally—citizens of a small or a large country? Explain and back up with examples from the text and your own research.
5. Where do you think Russian students go on their winter break? (There usually isn't a spring break,

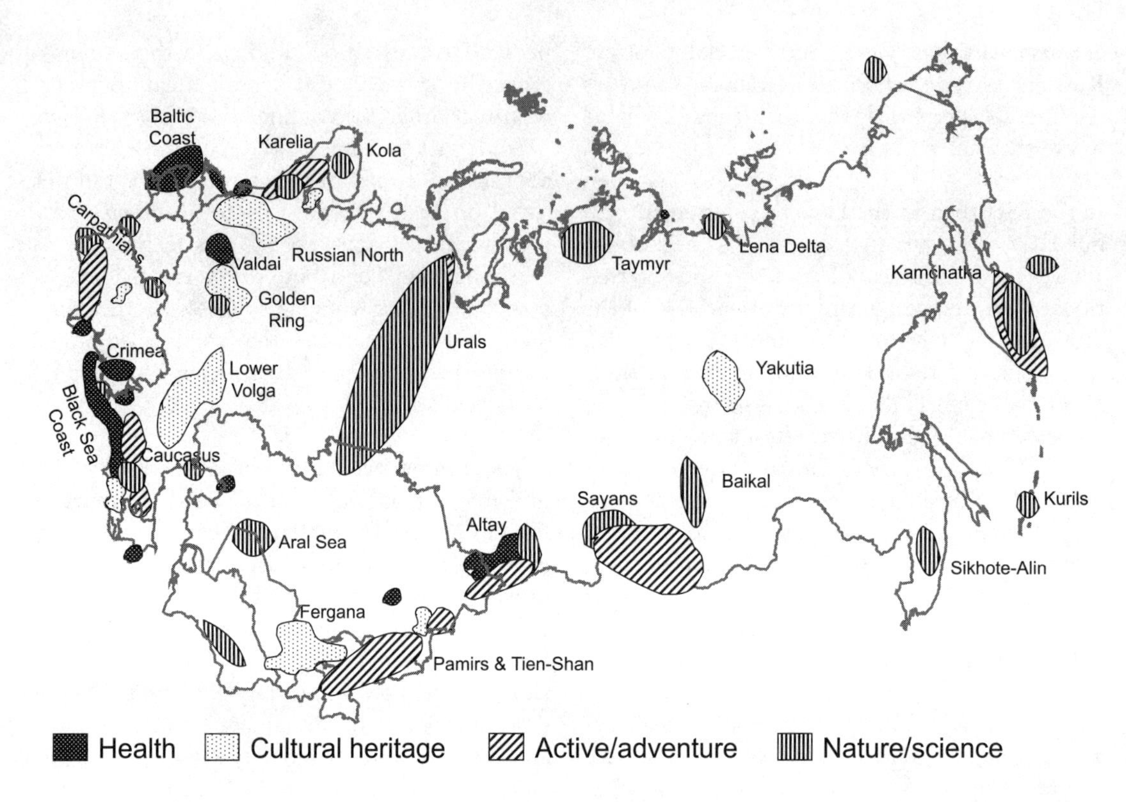

FIGURE 16.4. Areas with various forms of established tourism in Northern Eurasia (all forms may be present to a degree in each place).

because colleges operate on a two-semester system with no breaks at midsemester in Russia, but there is usually a long period of 4 weeks in January between the fall and spring semesters.)

6. Why do you think Moldova and Ukraine in particular have been dubbed the "mail-order bride capitals" of Europe? What are the geographic factors behind this phenomenon?

EXERCISES

1. The following countries were the most popular foreign destinations for Russian tourists in January and July 2006, as reported by the *www.tur-online.ru* travel agency. Try to explain the choice of these specific countries, using your sense of the world's regional geography. The numbers in parentheses are percentages of tourists out of the total:

January 2006	*July 2006*
Egypt (39%)	Turkey (44%)
Thailand (11%)	Egypt (9%)
UAE (8%)	Spain (6%)
Turkey (4%)	Greece (6%)
Czech Republic (3%)	Bulgaria (4%)
India (3%)	Croatia (4%)
Spain (2%)	Tunisia (3%)
Maldives (1%)	Cyprus (3%)

2. Try finding information on the top destinations for your country in the same 2 months (or at least in a recent year). What were these top destinations? How do those compare to the destinations visited by the Russians? Are some countries the same? Are there any countries on both lists that are very similar in terms of geography, but located in different hemispheres?
3. Investigate your options to purchase a basic tour of the Golden Ring (*Zolotoe Koltso*) of Russia. How many companies can you find that offer those? How many seem to be tour operators (as opposed to merely agents, who will sell you the package, but not operate the tour themselves)? What are the price ranges? How do those compare in both prices and amenities to a comparable tour in France? In an Eastern or Central European country? Which one seems like a better deal to you? Why do you think this is the case?

4. Stage a debate in your classroom on the pros and cons of developing (a) more conventional sanatoria, (b) more outdoor horse routes, (c) more highway-dependent motorized tourism, and (d) casinos in the Altay. Break into four teams, and let each one come up with positives for itself. Other teams should think ahead and come up with possible negatives for the others. Have fun trying to convince your instructor (or an investment board selected from among your classmates) to make the right choice.

Further Reading

Gorsuch, N. E., & Koneker, D. P. (Eds.). (2006). *Turizm: The Russian and East European tourist under capitalism and socialism*. Ithaca, NY: Cornell University Press.

Hughes, D. M. (2005). Supplying women for the sex industry: Trafficking from the Russian Federation. In A. Štulhofer & T. Sandfort (Eds.), *Sexuality and gender in postcommunist Eastern Europe and Russia* (pp. 209–230). Binghamton, NY: Haworth Press.

Jaakson, R. (2000). Supra-national spatial planning of the Baltic Sea region and competing narratives for tourism. *European Planning Studies, 8*(5), 565–579.

Kolbovsky, E. Y. (2008). *Ekologicheskiy turizm i ekologiya turizma*. Moscow: Academia.

Kosolapov, A. B. (2009). *Geografiya rossiyskogo vnutrennego turizma*. Moscow: KnoRus.

Meier, F. (1994). *Trekking in Russia and Central Asia*. Seattle, WA: The Mountaineers.

Mitchneck, B. (1998). The heritage industry Russian style. *Urban Affairs Review, 34*(1), 28–51.

Nikolaenko, D. V. (2003). *Rekreatsionnaya geografiya*. Moscow: Vlados.

Oko, D. (2007). Ripping with Borat. *Skiing, 60*, 2–3.

Ostergren, R. C., & Rice, J. G. (2004). *The Europeans: A geography of people, culture, and environment* (pp. 353–355). New York: Guilford Press.

Palmer, N. (2007). Ethnic equality, national identity and selective cultural representation in tourism promotion: Kyrgyzstan, Central Asia. *Journal of Sustainable Tourism, 15*(6), 645–663.

Pears, M. (2006). *Across the sleeping land: A journey through Russia*. Bloomington, IN: Trafford.

Pedersen, A. D., & Oliver, S. E. (1996). *The Lake Baikal guidebook*. Elizabethtown, NY: Ecologically Sustainable Development.

Rogers, P. (2003). Rent-a-MIG. *Bulletin of the Atomic Scientists, 59*(6), 6–7.

Schwartz, K. Z. S. (2005). Wild horses in a "European wilderness": Imagining sustainable development in the post-Communist countryside. *Cultural Geographies, 12*(3), 292–320.

Webster, P. (2003). Wild wild East. *Ecologist, 33*(1), 48–51.

Werner, C. (2003). The new silk road: Mediators and tourism development in Central Asia. *Ethnology, 42*(2), 141–160.

Lonely Planet has excellent guidebooks about Russia, Belarus, Central Asia, the trans-Caucasian states, Moscow, St. Petersburg, and the Trans-Siberian Railroad.

Websites

www.altaytravel.com—A tourism company specializing in the Altay.

baikal.travel—Lake Baikal travel.

www.crimea-portal.gov.ua/index.php?&f=us—Official site of the Crimea Peninsula in Ukraine.

www.ecotours.ru/en—Discussion of ecotour-related issues in Russia.

www.intourist.com—The original Soviet international tourism agency, still surviving today.

www.kamchatka.org.ru—A travel agency specializing in Kamchatka.

www.konyukhov.ru/eng—Fedor Konyukhov, a preeminent Russian explorer and adventurer.

www.russiatourism.ru/en—Official Website of the State Agency of Tourism of Russia.

www.sochi-travel.info—Information on the city of Sochi, located at the heart of the Russian Riviera.

www.waytorussia.net—Independent online guide to Russia.

www.zolotoe-koltso.ru—The Golden Ring, east of Moscow (the most famous area for heritage tourism). (In Russian only.)

PART IV

ECONOMICS

CHAPTER 17

Oil, Gas, and Other Energy Sources

Having considered the cultural and social geography of the former Soviet Union (FSU), let us turn our attention now to its "economic geography"—that is, its patterns of production, consumption, and trade. The post-Soviet reforms discussed in Chapter 8 have dramatically altered the economic geography of the entire FSU and of each post-Soviet state. Three of them, Estonia, Latvia, and Lithuania, are full European Union (EU) members about to join the "Eurozone." Ukraine, Georgia, and Moldova pursue fairly independent political pathways from Russia, but Ukraine and Moldova remain technically part of the Commonwealth of Independent States (CIS) and, in the case of Ukraine, continue to have strong economic ties to Russia. Belarus and most of the Central Asian republics (except Turkmenistan), in contrast, are much more engaged with Russia politically and especially economically. All are members of Evrazes, a new trade bloc set to encourage economic development, common customs, and better trade across Eurasia. Armenia and Azerbaijan have pragmatic, but more distant, economic relations with Russia. Turkmenistan is the most economically isolated country in the FSU, but is still engaged with Russia and Azerbaijan via the natural gas industry. Most of Part IV focuses on the economic geography of Russia, however. Several individual sectors of its economy are examined, starting with the most lucrative one, that of energy. As appropriate, examples from other FSU countries besides Russia are considered.

The Role of the Energy Sector in the Overall Russian Economy

Everywhere in modern Russia today, the energy industry has left its mark upon the landscape. Obvious signs include new office buildings, such as the Lukoil and Gazprom skyscrapers in southwestern Moscow; Western-style gas stations with colorful logos along rural highways; large, freshly painted railroad cars carrying petroleum products; and the names of new city streets, school buildings, and soccer stadiums. This is not an entirely new phenomenon. The U.S.S.R. was the largest producer of oil and natural gas in the world by the early 1980s, surpassing the United States and Saudi Arabia with production from the giant fields in western Siberia. (It remains the world leader in natural gas production and is currently second in petroleum production; see Table 17.1.) Back then, though, the energy sector consisted entirely of state-owned fossil fuels. In the new Russia, however, there are many private and semiprivate energy companies playing a

TABLE 17.1. Russia's Status as a World Energy Producer

	Russia	Global share	Rank worldwide
Coal production	341 mmt	5%	5th after China, United States, India, Australia
Petroleum production	9.87 mbl/day	12.5%	2nd after Saudi Arabia
Petroleum exports	5.08 mbl/day	8%	2nd after Saudi Arabia
Natural gas production	656 billion m3	20%	1st
Natural gas exports	182 billion m3	15%	1st
Electricity production	1 trillion kWh	5%	4th after United States, China, Japan
Per capita consumption	212 million BTU	NA	23rd

Note. Data from CIA World Factbook (estimates for 2007), except coal data (2006) and per capita energy consumption (2005) from U.S. Energy Information Administration. mmt, million metric tonnes; mbl, million barrels; kWh, kilowatt-houts; BTU, British thermal units.

much larger role in the overall economy than the state-owned energy industries played in the old one. These companies are highly visible, aggressive, profitable, and politically engaged. Table 17.2 lists the major petroleum companies.

Figure 17.1 illustrates the major role played by oil, gas, coal, and the rest of the energy sector in the new Russian economy. The share of this sector went up from only 12% of the total gross domestic product (GDP) in 1991 to 31% in 2002. Of course, the overall economy shrank by 50% over that period, so part of this relative change had more to do with the dramatic shrinking of other sectors. For example, light industry (e.g., textiles and clothing) contracted from 17% of the total output to only 1.5% (Chapter 19). However, there has been genuine growth in energy production in Russia since 1998. Production of petroleum, natural gas, and coal is up, while nuclear energy generation remains at levels similar to those of the late 1980s. The renewable sources of energy are expected to become more important now, because of high petroleum prices, pollution concerns over coal and nuclear energy, and a relative shortage of additional pipeline capacity for oil and natural gas. However, at present virtually all renewable energy in the Russian energy mix (a paltry 3%) is hydropower; most of this comes from old installations built in the 1950s, and few new dams are being planned. The largest hydropower facility, Sayano-Shushenskaya near Krasnoyarsk, suffered an unprecedented breakdown

TABLE 17.2. Russia's Major Petroleum Companies

Company	Market capitalization, mid-2007[a] (U.S. dollars, billions)	Production[b] (million barrels per year)
Rosneft (post-Yukos)	88	584 (2006)
Lukoil	68	654 (2007)
Surgutneftegaz	41	394 (2003)
TNK-BP	31	554 (2006)
Gazpromneft (Sibneft)	20	239 (2006)
Tatneft	11	185 (2006)
Slavneft	5	170 (2006)
For comparison:		
ExxonMobil	426	985 (2006)
Chevron	159	949 (2006)

[a]Data from RosBusinessConsulting (*www.rbc.ru*) for Russian companies, and from the *Fortune* 500 list for Chevron and Exxon Mobil.
[b]Data from companies' Websites.

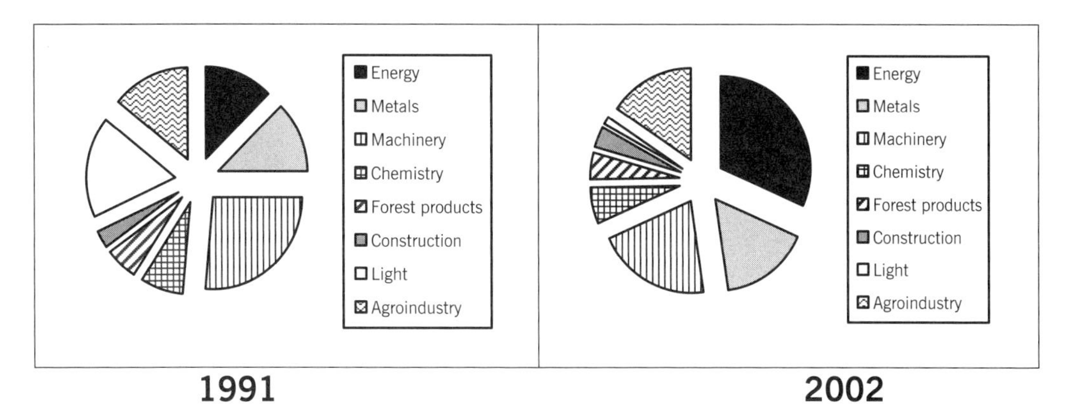

FIGURE 17.1. Relative shares of various industrial sectors in the Russian economic output, 1991 (before reforms) and 2002 (late reforms). Notice the rise in the relative importance of the energy sector at the expense of light industry. Data from the Federal Service of State Statistics, Russian Federation.

in the fall of 2009 that resulted in 75 casualties. The accident highlighted the fragility of the old equipment, as well as the top managers' lack of concern about safety. Although the Soviet Union built a few experimental geothermal (Kamchatka), solar (Uzbekistan and the Crimea), and tidal (Barents Sea coast) energy facilities, alternative energy remains virtually untapped in Russia today (Figure 17.2). When compared to that of the United States, Russia's energy mix is very high on oil and natural gas, and low on renewable energy, nuclear, or coal.

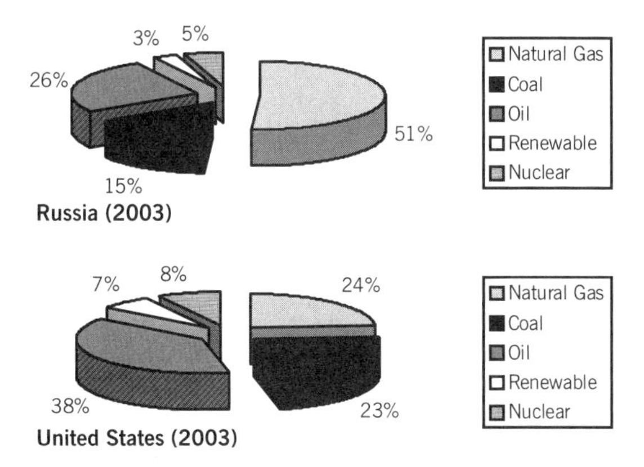

FIGURE 17.2. Energy mix of Russia and the United States by total primary energy supply, as measured in equivalence to tons of oil. Virtually the only renewable source in Russia is hydropower, whereas in the United States biofuels and wind play an increasingly important role. Data from U.S. Energy Information Administration.

The distribution of energy production in Russia is very uneven. The oil and gas fields in western Siberia produce 69% of all the petroleum and 91% of all the natural gas. Coal is mainly produced in the Kuzbass basin in central Siberia (47%). Most nuclear stations, by contrast, are located in the European part of the country, where the electricity need is greatest. Electricity production from all other sources is more evenly distributed, with each of the seven federal districts producing about an equal share; most of this production is unified in a national grid, except for the stations of the distant Far East. Coal and natural gas power stations dominate electricity generation (about two-thirds of the total), with 18% of electricity generated by hydropower, and 16% by nuclear energy. Finally, it must be kept in mind that the productivity of the Russian energy sector comes at a huge environmental cost: It accounts for 48% of all air pollution, about a third of all water pollution, and over 30% of all solid waste.

Petroleum

Petroleum is the most convenient and widely useful fuel known to humankind. It is liquid, so it can be piped long distances for a relatively low cost overland, or carried by tankers overseas.

It is much cleaner than coal and gives off more energy per unit of weight. It is extremely versatile: Hundreds of products can be made from it, including not only the obvious gasoline, jet fuel, diesel fuel, heating oil, and asphalt tar, but also wax crayons, pharmaceuticals, petroleum jelly, plastics, rubber, and even perfume. For much of the 20th century the oil supply was abundant, so it used to be cheap (except for a few brief periods in the 1970s). Petroleum is found in small quantities in most countries in the world, but in large quantities in only about 20 countries; these include the 12 Organization of Petroleum Exporting Countries (OPEC) members, as well as the United Kingdom, Norway, Mexico, the United States, Canada, and Russia.

The earliest commercial production of oil in the world started near Baku, Azerbaijan, in the Russian Empire, and in Titusville, Pennsylvania, in the United States, in the mid-19th century. The Soviet Union was constantly searching for more oil in the Caspian basin, and the important deposits of the Volga basin (Kuybyshev, Kazan) were discovered just before World War II. The Battle of Stalingrad was so strategically important because the Nazi government badly needed to get to the oil fields in these areas, and the Soviets were determined not to let this happen. During the war, new oil deposits were found in parts of eastern Siberia and on the remote Sakhalin Island in the Pacific. In the 1950s, a massive program of oil exploration was launched in the Arctic and in Siberia. Then, in the 1960s, extraordinary large oil deposits were found in the West Siberia economic region (now technically part of the Urals federal district; see Chapter 26) not far from Tyumen. For example, the Samotlor oil field, with initial reserves of 20 billion barrels of oil, was about 50% bigger than Prudhoe Bay in Alaska. These fields have been in constant production since the 1970s and remain the main producing areas in Russia today. Outside Russia and Azerbaijan, small deposits of petroleum are located in the Black Sea in Ukraine and a few large fields in western Kazakhstan.

How much petroleum does Russia have? For years, the amount of Soviet oil reserves was a closely guarded secret. The traditional Western sources of oil data, such as the U.S. Energy Information Administration or the experts of *Oil and Gas Journal*, would base estimates of Russian reserves on a number of assumptions and (frequently erroneous) rumors. Usually these estimates of Russia's oil would turn out to be on the low side. For example, a *National Geographic* feature article about the end of cheap oil (Appenzeller, 2004) used an estimate of 60 billion barrels of proven reserves for Russia, as compared to 23 billion barrels for the United States, 78 for Venezuela, 99 for Kuwait, and 261 for Saudi Arabia. There is strong evidence now that such estimates of oil reserves for Russia were too low, while those used for many OPEC nations were too high. In the past few years, more information has become available in Russia itself from newly privatized companies and from the government, and more realistic estimates of global oil reserves have emerged as independent experts have improved their assessments. A recent report from the German-based Energy Watch Group (2007) gives a much more realistic estimate of 105 billion barrels for Russia's oil reserves, as compared to 41 billion barrels for the United States, 35 for Kuwait, and 181 for Saudi Arabia. However, this same report observes that all oil-producing regions in the world, except Africa, are now past their peak of oil production and are expected to decline rapidly in the near future. The global oil crunch already seems to be well under way as reflected in skyrocketing prices in 2007–2008. In fact, because virtually all petroleum in Russia today still comes from the same fields developed during the Brezhnev period, it is very likely that Russian oil production will decline dramatically in the next 5–8 years (Gaddy, 2004). A top Russian government official acknowledged late in 2007 that the oil production had in fact dropped a few percent that year from the high level reached in 2006.

Meanwhile, the current rates of petroleum production in Russia are higher than they have been in the 1990s (about 9.5 million barrels per day), although Russia has not reached the all-time peak achieved in 1988 (12 million barrels per day). The latter figure will probably never be reached again. As noted earlier, Russia remains the second largest producer of oil on the planet after Saudi Arabia; in contrast to the latter, however, virtually all Russian petroleum comes from poorly accessible, extremely cold regions near the

Arctic Circle, and is piped to the consumers via an elaborate network of heated pipelines. Russian oil (the so-called Urals blend) is cheaper on the global markets than Saudi oil, because it has a higher sulfur content and is of lower quality. Unfortunately, it is also much more expensive to the producers: Instead of pumping up oil in a warm desert near the Persian Gulf, Russia must keep wells in year-round production thousands of kilometers away from the nearest seaport, in areas that stay below freezing for 8 months of the year. To make oil flow through the pipelines in the frozen tundra, some of it must be immediately spent on heating the pipes (this is also done in Alaska). Pipelines must be supported above the permafrost and require frequent and expensive repairs. In the late 1990s, a barrel of Saudi oil cost less than $1 to produce, while the same in Russia cost between $10 and $15. At present, very little of Russia's oil is produced offshore (less than 5%); by contrast, almost one-third is produced offshore in the United States, mostly in the Gulf of Mexico. However, massive investments have been made in the offshore oil fields near Sakhalin Island, and these are scheduled to start producing in the near future.

Russia is second only to the United States in the overall length of its pipelines, but Russian pipes are on average much larger (1,220 mm is the typical gauge) and move a lot more oil for longer distances. An average Russian pipeline moves oil at the speed of 10 km/hour, about half a billion barrels of oil per year. Although most Russian oil companies are now private or semiprivate, all pipelines belong to the state-run monopoly Transneft and its cousin Transneftproduct, thus ensuring the Kremlin's control over oil exports. The Russian state forces all oil exporters to pay a substantial access fee for the privilege of exporting oil; this fee was set at $340 per metric ton in 2009, which is about $45/barrel, or about 50% of the price of the oil on the world's markets. Most pipelines currently run from western Siberia through the middle Volga petroleum basin west to the central part of Russia, and beyond into western and southern Europe (Figure 17.3). New pipelines are now being built around Lake Baikal to the Pacific Coast and into China, and under the Baltic Sea from St. Petersburg to Germany (Chapter 21).

Russia has about 10 large oil companies, all of which are private, vertically integrated corporations except for state-owned Rosneft. A typical one is Lukoil, which has oil fields in western Siberia and in the Volga, Kaliningrad, and Timan-Pechora basins. It also owns a number of petroleum refineries and petrochemical complexes in Volgograd, Perm, and Nizhniy Novgorod, and many retail gasoline stations and other infrastructure abroad. Lukoil was among the first Russian companies to list its stock on international exchanges, and the first to break into the retail gasoline market in the United States with the purchase of 1,300 Getty Oil gas stations on the East Coast in 2000. It has an international presence in about a dozen countries, including Turkey, Iraq, Egypt, Romania, Bulgaria, and Kazakhstan. The brainchild of the career Soviet petroleum engineer turned oil tycoon Vagit Alekperov, Lukoil has managed to remain private despite recent pressures on the oil industry under Putin to renationalize. It maintains an impressive portfolio of accomplishments and is one of the most profitable businesses in post-Soviet Russia, with a net income of $9.5 billion in 2007. However—and this is not a secret—the only way Lukoil and other large companies could survive in the post-Yeltsin period was to distance themselves from open politics and forge alliances with the Kremlin. Companies that did not are no longer around (Vignette 17.1).

Despite improvements in the efficiency of petroleum extraction (e.g., modern methods of deep-field steam injection and better exploration techniques), very little new oil has been found in Russia in the past decade. The supposedly huge deposits of the Barents Sea, eastern Siberia, and well-explored Sakhalin Island will require enormous investments of capital in the very near future, if Russia wants to remain one of the top world oil producers. It is highly unlikely that this is going to happen soon enough to avert a massive downturn in the domestic oil business.

Another problem that has to be overcome is lack of high-quality refineries. Russia has 27 major refineries, all but one built during Soviet times. They remain very dirty and have much obsolete equipment, despite some recent updates. Most are located in the middle Volga basin (Volgograd, Saratov, Nizhniy Novgorod, and

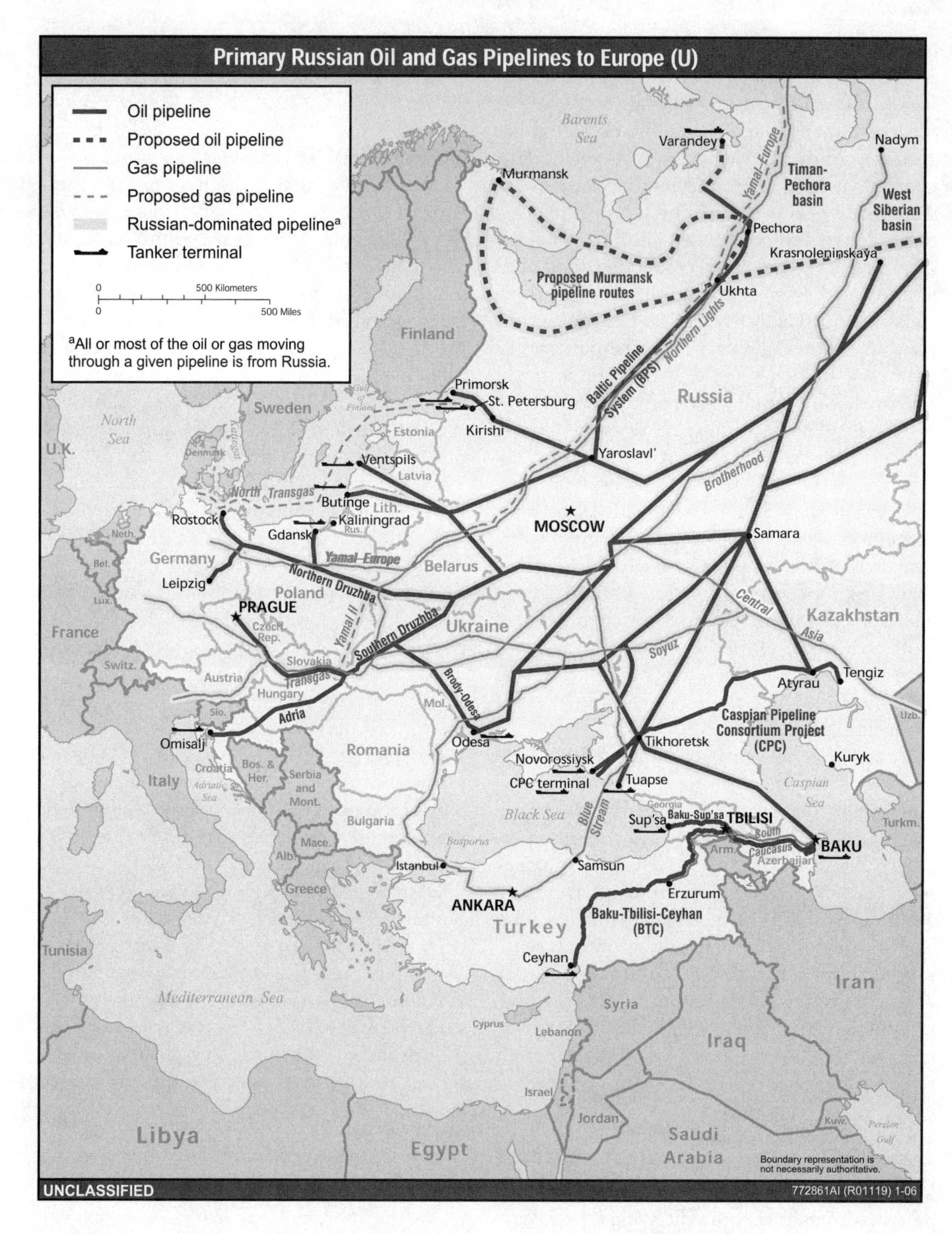

FIGURE 17.3. Selected oil and gas pipelines of Russia, Europe, and west Asia in 2006. From U.S. Department of Energy (*www.eia.doe.gov/emeu/cabs/Russia/images/772861%20%5BConverted%5D.pdf*).

Vignette 17.1. The Swan Song of Yukos

One of the most dramatic events of the post-Yeltsin period in Russia was the end of the Yukos oil company. In early 2003 it was the largest petroleum company in the country—poised to merge with the fourth largest, SibNeft, and eventually perhaps with a transnational giant (ChevronTexaco or ExxonMobil). Yukos was also one of the top five companies by market capitalization in Russia that year. In 2007, however, it went into bankruptcy because of a lengthy legal battle with the Russian authorities. Its key asset, the giant Yuganskneftegas oil field, is now part of the state-owned Rosneft company headed by a close friend of Vladimir Putin, Igor Sechin. The founder and chief architect of Yukos, Mikhail Khodorkovsky, is in a federal prison serving an 8-year sentence, along with a few other key company managers, after a trial for fraud and tax evasion. (A new trial is under way in Moscow as of the time of this writing.) Many other top managers left the country just in time to avoid prosecution. Whether the collapse of Yukos was a purely political case (as believed by many), or only a particularly egregious case of selective law enforcement in Russia today, the story is instructive in that it shows the extreme vulnerability of even the largest businesses in Russia—particularly those close to the top of the economic pyramid.

Yaroslavl), and also near the termini of the existing pipelines (Tuapse, Achinsk, Angarsk, Omsk). Refineries crack crude oil molecules into usable shorter ones to make gasoline, jet fuel, heating oil, and so on. At best, Russia is only able to refine about half of its crude oil. Most of the gasoline still sold in Russia would not pass European standards for quality and can actually damage Western car engines.

Of course, it makes economic sense to refine petroleum domestically, because the final products generate higher profits than export of crude oil does. Building new refineries, however, is very expensive, and few communities want one in their backyard. Interestingly, the United States likewise suffers from a lack of modern refineries for similar reasons. About three-quarters of the Russian-refined petroleum goes into producing heavy heating oil (*mazut*), which traditionally was used in power stations. Today, however, it makes more sense for power stations to use natural gas; what is needed instead is high-quality gasoline. Unfortunately, Russia largely lacks the domestic capacity for making decent gasoline. Consequently, domestic gasoline prices inside Russia are at the U.S. level (albeit much cheaper than in Europe), and much of the gasoline is of dubious quality. Current plans call for an increase in modernization of existing refineries and construction of a few new ones, so that by 2020 over 80% of all crude oil will be refined domestically.

The environmental impacts of the Russian oil industry are substantial (Chapter 5), although not very different from those in many other oil-producing regions. Exploration for and extraction of oil are rarely clean operations. Many wells must be drilled, but only a few become productive. Oil spills are common, both from the producing wells and during the transportation stage. Fires at refineries occur periodically. All the drilling equipment must be brought in and positioned over fragile wilderness (tundra and bogs in most places). Permafrost melt from oil production results in serious damage to the infrastructure above ground. Most of the natural gas that occurs along with oil in the same wells must be burned off for safety reasons. Passengers flying over western Siberia or Orenburg Oblast at night are treated to the ghastly spectacle of orange flames illuminating the night sky from dozens of wells—visible even from space.

Really large oil spills remain infrequent in Russia, but there are many cases of slow leaks that may go undiscovered for months. Russian pipelines are generally reliable, and with little Russian offshore oil drilling at present, there are as yet few chances for massive spills into the ocean. However, these may become a reality as the massive Sakhalin projects increase production; should spills happen there, not only the Russian but also the Japanese and Korean coastlines will be smothered with oil. To be fair, all

Russian oil companies have invested in more modern equipment and extracting/refining technologies in the past few years, so perhaps some of the worst environmental disasters can be averted. Unfortunately, corporate secrecy, greed, and lack of governmental accountability make environmental audits difficult. Of course, none of the new technologies remove the major problem of carbon dioxide emissions from the eventual burning of petroleum products.

Natural Gas

Natural gas is mainly methane that occurs in bogs. Fossil natural gas deposits exist in many of the same places where oil does. Basically, when oil is produced underground from the remains of the microscopic marine plankton, some of the shortest molecules escape and remain trapped underground in gaseous form right above the liquid oil. In cases where oil-producing rock has sunk very deep in the earth's crust and has been exposed to very hot temperatures, only gas is produced. Russia is exceptionally fortunate in having the largest gas deposits in the world. According to various sources, it has between 30 and 40% of all proven natural gas reserves on the planet, or about 48 trillion m^3.

Natural gas is the cleanest fossil fuel; when it burns, only carbon dioxide and water are produced. However, it can explode and therefore must be shipped with utmost caution. Usually it is shipped under pressure in pipelines, similar to oil. It must be shipped overseas in liquid form via specialized, very expensive tankers. This limits its applicability worldwide, because petroleum can be shipped anywhere in the world in regular (i.e., cheaper) tankers without liquefaction. Although gasoline or jet fuel cannot be made directly from natural gas, it can be used in its liquid form as a fuel for specially modified engines. Also, natural gas makes a great alternative to dirty coal in modern electricity-generating plants, and it is used widely for home heating and cooking. Natural gas must be liquefied and kept at low temperatures (at an energy cost) for long-term storage (Figure 17.4). So usually once it is released from underground, it goes immediately into the pipelines to the consumers. Because of this, its production and consumption are tightly intertwined. Even a short interruption in supply is immediately felt down the pipelines—something Europeans have learned to fear after several pricing disputes between Russia and Ukraine in recent years.

Just like the petroleum fields, most of the Russian gas fields are in western Siberia (60% by

FIGURE 17.4. Natural gas storage tanks near Syzran on the Volga. *Photo:* S. Blinnikov.

number of fields and 91% by volume of production). However, the gas fields there are generally located even farther north, near or on Yamal Peninsula (Urengoi, Yamburg, Zapolyarnoe). These are areas above the Arctic Circle, where summers are short and winters are long and bitterly cold. Additional large reserves exist in the Stockmann field in the Barents Sea, and in Orenburg Oblast in the Urals. Small deposits are found near the Caspian Sea. In the rest of the FSU, significant gas reserves are found only in Turkmenistan. According to the U.S. Energy Information Administration, in 2006 Russia produced 656 billion m^3 of natural gas, with the United States producing about 525 billion m^3. Three countries in Europe (the United Kingdom, Norway, and the Netherlands) collectively managed to produce about half of Russia's total output, and all of the Middle Eastern countries, including extremely gas-rich Qatar, produced about half of Russia's output as well.

Although the privatization of oil fields and refineries was allowed under Yeltsin, no privatization of the natural gas industry was allowed. Long-time Prime Minister Chernomyrdin (who had close ties to the gas industry) and a few other members of Yeltsin's inner circle ensured that the entire industry remained mostly in state hands, although partial private ownership of gas stock was allowed. Virtually all gas production in Russia today is controlled by one company: Gazprom, the country's largest corporation. Slightly over 50% of its stock is controlled by the state, and much of the rest is publicly traded. In 2007 Gazprom was ranked only 52nd in the *Fortune* Global 500 list by revenue ($81 billion)—somewhere along with Hitachi, Samsung, Nestlé, and Deutsche Telecom, but quite a bit behind such giants as Walmart, ExxonMobil, Royal Dutch Shell, or BP. In mid-2007, however, it was close to becoming one of the top five companies by market capitalization on the planet ($254 billion vs. $426 billion, for ExxonMobil and less than $200 billion for Walmart). The new president of Russia, Dmitry Medvedev, has expressed a wish to see it become the biggest company on the planet one day. Some political observers even joke that the true name of the ruling party, United Russia (the predecessor of which was Our House Russia), should instead be "Our House Gazprom" because of the heavy presence of Kremlin insiders on the company's board.

Gazprom has been active in maintaining existing gas pipelines and building new ones. The Blue Stream pipeline connected Krasnodarsky Kray in Russia and Turkey under the Black Sea in 2005. This gas line is the world's deepest (about half of it runs at depths close to 2000 m), and its underwater section is over 300 km long. A similar, but even longer and much more controversial, Northern Stream pipeline is being built from St. Petersburg to Germany under the Baltic Sea. Poland and other countries that are being bypassed criticize the project on both economic and environmental grounds, while Germany is naturally in favor of it. Its recently announced Southern Stream cousin will cross the Black Sea into the Balkans and will compete directly with the Western-backed Nabucco pipeline coming from Azerbaijan into Turkey. Another major international project is a future pipeline to connect Kovykta in Irkutsk Oblast to China. Yet another gas pipeline may be built across the Altay Mountains from central Siberia into western China across the scenic Ukok Plateau, despite the vocal protests of environmentalists.

Gas from Russia plays an increasingly important role in heating Europe in winter. Some European countries (e.g., Finland, Poland, Slovakia) are almost 100% dependent on Russian natural gas for their natural gas supplies, and Germany and Austria import about 30–50% of theirs from Russia. Even France and England receive small but important shares of Gazprom's bounty (Starobin, 2008). This of course creates political tensions over who is really in charge in Europe, as the energy prices continue to go up, while the Europeans have few options besides Russian gas for heating themselves in winter.

The transit of gas from Russia is an important geopolitical issue. Belarus and Ukraine have pipelines stretching across their territory from Russia to the EU member states, Russia's main global consumers. In the case of Ukraine, almost half of the gas actually comes from Turkmenistan—but technically it is still all Russian, because it enters the country from Russia and needs to be paid for through Gazprom-associated structures.

Both Belarus and Ukraine need some natural gas for themselves, but their primary role has been that of transit operators to Europe, for a fee. Gazprom, with the Kremlin behind it, has been keen on flexing its political pricing muscles; this has led to a few standoffs with Ukraine and less obvious complaints in Belarus over unfair pricing, both for the natural gas itself and for the transit of it. Although both countries continue to pay substantially less for natural gas from Russia than the "world price" of it (Belarus a lot less, and Ukraine about half of the world price), they want even more favorable rates, arguing that Russia has no other options for exporting its gas to Europe. At the same time, the Europeans are upset over the possibilities that Russia may shut down or restrict its gas flow because of pricing disputes with the transit countries. The "Orange Revolution" in Ukraine exacerbated the situation farther, because the more independent minded, pro-Western government of Viktor Yushchenko and Yulia Timoshenko immediately ran into problems with Gazprom that have made international news from time to time. It is interesting to note that the main source of Timoshenko's personal fortune in the past was, in fact, gas trading.

Natural gas extraction damages the sensitive tundra of the Yamal. In order to bring supplies in, all-terrain military vehicles with heavy tracks are commonly used. These tend to destroy fragile lichens and mosses for many kilometers around the wells, as each subsequent run must use a slightly different path, to avoid sinking forever into the liquefied mud. Such dirt "roads" a few kilometers wide can be visible on satellite images of the Yamal Peninsula. Natural gas is expected to continue being the top fossil fuel produced in Russia for at least another 25 years. As oil production drops, more and more natural gas will be required, and of course even the Siberian bounty will end at some point. The question remains what fuel will be available after that. In the FSU, outside Russia, major natural gas fields are found only in Turkmenistan and Azerbaijan. Recent independent audits in the former suggest that the country has much less gas than was previously believed. Kazakhstan and Uzbekistan have some as-yet-unrealized gas potential.

Coal

Coal was the first fossil fuel to be used by humans. It is also the most abundant, the cheapest, and by far the dirtiest. Coal industry has a long history in Russia. Some of the oldest mines in the FSU are over two centuries old; they are located in Ukraine today, mainly in the Donbass basin (Lugansk and Donetsk Oblasts), which partially extends into Russia's southwestern region of Rostov-on-Don. The old-fashioned underground mines of Ukraine and Russia are among the most dangerous in the world, with annual fatalities averaging a few hundred. These result from methane explosions and mine collapses. Modern open-pit and strip mines are common in Siberia, especially in the huge Kansk-Achinsk basin, which alone contains over 50% of Russia's coal. Over 65% of coal in Russia is produced now in open-pit mines. This method is safer for the miners, but results in much surface damage. In the Kuzbass and Donbass areas, however, mining continues primarily underground; this produces better-quality coal for making steel, but also leads to numerous accidents.

The total proven coal reserves of Russia are second in the world, after those of the United States—about 200 billion metric tonnes, with 114 billion in Kansk-Achinsk, 57 billion in the Kuznetsk basin of central Siberia, and 9 billion in the Pechora basin in the northeastern European part. The United States has about 240 billion metric tonnes in reserves, the largest on the planet. Only Ukraine and Kazakhstan have substantial coal reserves in the FSU besides Russia.

Traditionally coal was used in steel making, other metallurgy, the chemical and fertilizer industries, and electricity generation. However, since 1980 natural gas has largely replaced coal as the main fuel in Russia's power stations. In present-day Russia, only 15% of all energy comes from coal, as compared to 23% in the United States (Figure 17.2). Because of the shift toward natural gas, there has been an improvement in air quality around big industrial centers. The leading producer of coal in the country is the Kuzbass basin (over 160 million metric tonnes [mmt] in 2003). The second largest producer is the Kansk-Achinsk basin (34 mmt), with the

Pechora (13 mmt) and Russian Donbass (7 mmt) basins producing most of the balance. As in the United States, Russian coal companies maintain a lower profile than the petroleum producers do. Nevertheless, the largest coal mine in the country, Raspadskaya, had market capitalization of over $2 billion in mid-2007 and was ranked the 63rd largest company in the country. This mine alone can produce over 7.5 mmt of coal per year.

Many large coal-fired electric plants are located near the biggest open mines in Siberia (Surgut-2, Berezovo-1, Neryungrinskaya, Kharanorskaya, etc.). These stations each have a power generation capacity of between 4,000 and 6,000 megawatts (MW), comparable to that of the largest hydropower dams in the world. Each produces enough electricity to power a large city of a few million people, and millions of tons of carbon dioxide per year. RAO EES (United Energy Systems of Russia) was the monopoly that ran the country's electric grid until 2008 (now partially privatized and subdivided) and is one of the largest polluters in the world. It was the fifth largest company in Russia in 2008 in terms of market capitalization and is the largest unified electric grid in the world, stretching deep into most of the other FSU republics. It has since been reorganized, with independent energy companies created in different regions. Theoretically, this should lead to increased competition and lower tariffs for the consumers. So far, however, the results have not been very encouraging, with energy prices steadily rising between 10 and 15% per year. Russia produces about 6,000 kilowatt-hours per person per year; this equals the level of Germany or the United Kingdom, but it is quite a bit behind Australia (11,000) or the United States (14,500).

Russia ratified the Kyoto Protocol in 2004, after many years of dragging its feet. Because Russia's cumulative greenhouse emissions in the early 2000s were a lot lower than in 1990, owing to the economic downturn of the late 1990s, the country originally was in a position to gain a lot of money through the sale of carbon offsets to the EU. As of 2009, however, the country's emissions had climbed back almost to the 1990 level (Chapter 5), so in the very near future Russia must either lower its emissions (which is highly unlikely) or start paying other countries for permits as mandated by the Kyoto agreement. Heavy reliance on coal and natural gas for energy generation will not help in the short term. However, Russia also has over 30 nuclear stations and is likely to start building more in the future as a possible solution to the problem of carbon dioxide pollution.

Nuclear Energy

Russia was the first country in the world to start producing "peaceful" nuclear energy; the first small station, in Obninsk near Moscow, was opened in 1954 and is still operating today. Russia currently has 10 functioning stations with over 30 reactors. This is fewer than the United States has (105 reactors), but it still represents a lot of energy production capacity. In fact, Russia is the fourth largest nuclear energy producer in the world after the United States, France, and Japan. The largest stations are all concentrated in the European part: Kalininskaya in Tver Oblast (2,000 MW); Smolenskaya near Smolensk, west of Moscow (3,000 MW); Novovoronezhskaya near Voronezh, southeast of Moscow (1,800 MW); Kurskaya (4,000 MW); Leningradskaya (4,000 MW); and so on. There is even a small nuclear station at Bilibino in distant Chukotka to serve local electricity needs. It was built as an experimental facility in the Arctic in one of the coldest places on earth. Most stations in Russia now use water-cooled VVER-type reactors, not RBMK graphite reactors like the one that exploded in Chernobyl. Thus the nuclear stations of Russia are reasonably safe. Overall, Russia produces a surplus of electric energy, which is exported to its neighbors via power lines. (Vignette 17.2 describes an illegal but ingenious use of a small amount of this energy.)

Construction of new nuclear reactors has been put on hold since the Chernobyl disaster of 1986 (see Chapters 5 and 8), but is now being discussed again as the country struggles to meet its obligations under the Kyoto Protocol, and also as the energy from coal and natural gas gets more expensive. Overall, there is little opposition to new nuclear stations in Russia today; there is understandably more opposition in Ukraine, where

Vignette 17.2. Off the Energy Grid in a Russian Village

Victor N. is over 60. He lives alone in Berezovka, a village 200 km north of Moscow in the Kostroma region. (The names of the person and the village have been changed, but the story is real.) He is unemployed, a refugee from Belarus with no citizenship. The home he lives in is not his legal place of residence. It is a crooked two-room log cabin on 0.2 ha of land given to him as a temporary shelter by a friend, who sometimes comes from Moscow to stay with Victor for a few days in the summer. Victor stays in the cabin year-round; he has no other place to live. Because he is unemployed and not a legal resident, he has no salary or pension. He does odd jobs around the village, fixing things to earn a little cash. In his former life, he was a city engineer; he even patented a few inventions that allowed the city to do certain jobs faster. He can fix anything. His way of survival in Berezovka is living off the land, and to do this he needs energy.

If you come to Victor's cabin in winter, it is suspiciously warm inside, but there is no wood in the stove. He is too old to chop firewood. Also, firewood is not free any more: The local forestry enterprise is back in business after a period of bankruptcy, and its rangers zealously watch over the dwindling wood supplies. So what does Victor do? The answer is free power from an illegal wiretap placed on the main line that runs next to his house. Two electric heaters keep him warm all winter long for free. The electric power comes from a convenient nearby source: The Kalininskaya nuclear station, with four powerful nuclear reactors, is less than 50 km away. Incredibly, after 4 years of Victor's illegal siphoning of electric power, the local utility has yet to notice the power loss. Maybe someone has had to be bribed to look the other way. Victor is not the only one doing this in the village; a few homes have had fires because of poorly installed taps in recent years. Around Russia, about 20% of all electric power goes to such illegal users.

Victor also uses passive solar heating in his many greenhouses. He grows tomatoes, cucumbers, squash, and beans on the small plot of land behind the cabin from early spring to late fall. He also experiments with using wind power for draining water off his property, much of which is waterlogged from the nearby lake. To get around, he walks or bikes; he does not have a car, and he never needs gas. Ethanol fuel? Only for drinking. That comes from the village store.

Chernobyl is located. Although the Chernobyl accident is still painfully fresh in the public memory, many politicians now consider it to have been an extremely unfortunate accident that involved an obsolete reactor and was easily avoidable with proper precautions. It remains to be seen whether local opposition in Russia or other FSU nations will be able to stave off the return of the "peaceful atom." Environmental pollution from nuclear energy use is discussed in more detail in Chapter 5. A major issue is safe disposal of spent nuclear fuel waste. Russia must process its own, plus additional waste from Western Europe (especially France). The Russia–U.S. program of reprocessing old Soviet nuclear warheads is helping to generate electricity inside the United States, but does little to solve the domestic disposal of "peaceful" fuel in Russia.

Renewable Energy Sources

With the largest territory in the world, Russia should be able to capitalize on the use of wind, solar, and biomass power, which are more evenly distributed over large areas than coal, oil, or gas deposits are. However, at present the country is far behind the United States, Germany, or even Denmark in the use of renewables. According to the International Energy Agency (IEA, 2003), Russia could satisfy over 30% of its energy needs from renewables, but at present they account for less than 3% of the total energy mix, and less than 0.5% if hydropower is excluded. Compared to the United States, Russia generates 30 times fewer watts from alternative energy sources other than hydropower! Indeed, about the *only* renewable energy source generated in Russia is hydropower. Russia has 12% of the world's developed hydropower resources, and it is behind only the United States and Canada in hydropower generation worldwide.

Once a dam is built, hydropower costs less than either nuclear or thermal power per kilowatt produced. The Volga River has 11 big dams that produce cheap electricity. A major problem in the Volga basin, however, is the decline in sturgeon species because of the dams. Fish hatcheries help somewhat, but the large sturgeon are history. (Poaching of black caviar in the Caspian Sea is of course another reason for the sturgeon's decline.) Large Siberian rivers are also tapped, especially the Angara–Yenisei and the Ob–Irtysh systems. Recently dams were completed on the Zeya and the Bureya Rivers in the Amur basin and on some tributaries of the Lena in eastern Siberia. Relatively few dams exist in mountainous regions of Russia, however.

A site that has been repeatedly proposed for hydropower generation is the scenic Katun River gorge in the Altay (Figure 17.5; see also Chapter 16, Vignette 16.1). This, however, is a World Heritage Site and a popular tourist destination, so any future plans for building a dam and flooding the gorge are bound to generate massive opposition. To date, the most powerful dam in Russia is Sayano-Shushenskaya on the Yenisei (6,400 MW); this is about as powerful as the biggest hydropower plant in the United States, the Grand Coulee on the Columbia River in Washington State. However, it produces less than half the power of the Iguasu Dam on the Parana in Brazil, and only one-third of the Three Gorges' capacity. One of the geographic problems of using more hydropower in Russia remains its seasonal climate: An average dam produces only a fraction of the energy in winter than it does in summer, because of the much-reduced water flow under ice. Also problematic is flooding of large fertile floodplains. Construction of the Volga reservoirs in the 1950s destroyed hundreds of villages and a few historical towns, such as Kalyazin, with its famous old cathedral bell tower defiantly standing in the middle of the Rybinsk reservoir (Figure 17.6). Outside Russia, important hydropower facilities exist in Ukraine, Georgia, Kazakhstan, and Tajikistan. In the latter, United Energy Systems of Russia completed the new Sangtuda-1 hydropower plant in 2008. However, in the fall of 2009 the plant suspended its electricity sales

FIGURE 17.5. The Katun River gorge in Siberia is one of the possible sites for a large new dam. Power wires from a small dam built on the tributary can be seen. *Photo:* Author.

FIGURE 17.6. Impact of flooding of the Volga near Kalyazin: The St. Nicholas Cathedral bell tower is defiantly sticking out 50 years after the completion of the reservoir. *Photo:* S. Blinnikov.

to local users, citing payment delays. Additional hydropower installations in Tajikistan are being built by Chinese and Iranian interests.

Russia has a huge potential for using wind power, although Kazakhstan and Ukraine have an even better potential per square kilometer of territory. Coastal locations in the Far East; the Yamal Peninsula; mountain passes in the Urals and the Caucasus; and flat steppe areas show the greatest promise. So far, however, virtually no wind power has been utilized in any FSU country. The problem is lack of consistent wind over much of the Eastern European Plain, where the population concentration is the heaviest and the need for energy the greatest. Nevertheless, more wind generation is likely to be developed in the next 10–20 years; it is estimated that about 12 mmt of coal can be economically replaced by wind in the near future. About one-third of this power would come from European Russia, and two-thirds from Siberia and the Far East.

Russia has relatively low potential for solar power generation, despite its size. It is a northern country, and much of its territory experiences heavy year-round cloud cover. Better potential exists only in the extreme south (e.g., in the Kuban and Astrakhan regions) and in parts of eastern Siberia (Yakutia, Buryatia). However, even there the solar potential per square meter is much lower than in the Central Asian republics, especially Turkmenistan, as well as Armenia and southern Ukraine. About 46% of U.S. territory has good potential for solar power generation, but only 6% of the U.S.S.R. (Pryde, 1991). Nevertheless, the Soviet Union had experimental solar stations in the Crimea (Ukraine), Armenia, and Uzbekistan. There is continued interest in both solar heating stations and photovoltaics in the region. In the IEA (2003) report, the overall economic potential of solar power is estimated as the equivalent of 12.5 mmt of coal per year—about the same as for wind power. This, of course, is only an estimate of what is economically feasible in the near future, but it puts things in perspective: Russia mines well over 300 mmt of coal annually, so neither renewable option is likely to replace coal anytime soon.

The geothermal potential of Russia is virtually all concentrated in the Far East, with the exception of low-heat devices (heat pumps) that can be used anywhere. In the Far East, Kamchatka has 22 active volcanoes, and the Kuril Islands 21. Geysers and hot springs are found in Kamchatka and Chukotka. The economic potential of geothermal power, according to the IEA (2003) report, is about equivalent to 115 mmt of coal per year—a much higher figure than that for either wind or solar power, at least given the economic assumptions. However, almost all of this power will be coming from a very remote region unconnected to the national electric grid. Two small geothermal stations already exist in Kamchatka. Local uses of geothermal heat are also possible in parts of southern Siberia and in the Caucasus. Ukraine has limited potential for geothermal power development in the Carpathian Mountains and the Crimea.

Finally, Russia has a vast potential to produce ethanol, biodiesel, and energy from wood chips or hay (Figure 17.7). About 35 mmt of coal can be economically replaced with biofuels in the near future (IEA, 2003). Some agricultural regions (e.g., Kurgan, Altaysky Kray, Rostov)

FIGURE 17.7. Bales of hay represent biomass available as an alternative energy source. *Photo:* A. Fristad.

have already been using ethanol in tractors and combines. However, the efficiency of Russia's big farms is lower than that of U.S. farms, and consequently the biofuel production is more expensive. Also, because of Russia's vast petroleum reserves, the oil lobby is strong and does not want competition. Both ethanol and biodiesel are mainly made from warm-season crops (corn and soy, respectively), and only a few areas in the country have adequate climate for their production. There is also a concern that increasing the acreage dedicated to biofuels will use up some land available to food crops. Production of electricity from wood chips is a more feasible long-term option for Russia, given its huge forest reserves and plenty of waste available from the forestry sector. There is additional biofuel production potential in Ukraine, Belarus, Moldova, and Kazakhstan.

REVIEW QUESTIONS

1. What are the main energy sources in Russia? How do these compare to those of the United States in relative importance? Why?
2. Which areas of Russia produce the most fossil fuels? What types?
3. Is it true that the oil- and gas-producing regions of Russia are well-off now?
4. What regions of the world are connected to Russia via oil and gas pipelines?
5. What are the problems faced by the coal industry in Russia, Ukraine, and Kazakhstan?
6. Which alternative energy sources, in your opinion, should be the first priorities for development in the FSU?

EXERCISES

1. Compare and contrast five leading petroleum companies in Russia (see Table 17.2). Use their Websites to find answers to these questions: Where do they operate? How much of their operation is international? What are their main assets? For example, do they own refineries or only oil fields? Which of them seem to have better strategies for reaching out to global markets? Which one would you invest in, and why?
2. Analyze the energy resources of the five Central Asian states. Which of them seem to be most self-sufficient with respect to energy? Think of both conventional and alternative sources.
3. Investigate any recent international news story involving Gazprom. What happened, where, and why? Does this story provide positive or negative coverage of the company?
4. Speculate on the pros and cons of building a new petroleum pipeline from Angarsk to Asia. Explore two options: south to China, or east to the Pacific Ocean and then by tankers to Japan and other countries.

Further Reading

Appenzeller, T. (2004, June). The end of cheap oil. *National Geographic*, pp. 80–109.

Bradshaw, M. (2006). Observations on the geographical dimensions of Russia's resource abundance. *Eurasian Geography and Economics, 47*(6), 724–746.

Dienes, L. (2004). Observations on the problematic potential of Russian oil and the complexities of Siberia. *Eurasian Geography and Economics, 45*, 319-345.

Energy Watch Group. (2007). Crude oil: The supply outlook (EWG-Series 3). Retrieved November 20, 2007, from *www.energywatchgroup.org*

Feshbach, M., & Friendly, A., Jr. (1992). *Ecocide in the U.S.S.R.: Health and nature under siege.* New York: Basic Books.

Gaddy, C. G. (2004). Perspectives on the potential of Russian oil. *Eurasian Geography and Economics, 45*, 346–351.

International Energy Agency (IEA). (2003). *Renewables in Russia: From opportunity to reality.* Paris: Author.

Pryde, P. R. (1991). Renewable energy resources. In P. Pryde (Ed.), *Environmental management in the Soviet Union* (pp. 56–74). Cambridge, UK: Cambridge University Press.

Sagers, M. J. (2007). Developments in Russian gas production since 1998: Russia's evolving gas supply strategy. *Eurasian Geography and Economics, 48*(6), 651–698.

Starobin, P. (2008, June). Send me to Siberia: Oil transforms a Russian outpost. *National Geographic*, pp. 60–85.

Walker, M. (2007). Russia vs. Europe: The energy wars. *World Policy Journal, 24*(10), 1–8.

Websites

www.eia.doe.gov—U.S. Energy Information Administration (energy statistics from the U.S. government, including international data).

www.gazprom.ru—Gazprom, the Russian gas monopoly and largest company. (In Russian only.)

www.lukoil.com—Lukoil, the largest private Russian oil company.

www.oilru.com—Information about Russian oil. (In Russian only.)

CHAPTER 18

Heavy Industry and the Military Complex

Perhaps the heaviest legacy (both literally and figuratively) of the Soviet economy was its military–industrial complex, called in Russian the *voenno–promyshlenny kompleks* or VPK. Its presence was pervasive: Entire cities were built around steel mills, aluminum smelters, tank manufacturers, chemical factories, or nuclear weapons facilities. Over 50% of the country's industrial output in the 1980s was generated by this sector. The Soviet Union produced more tractors, tanks, missiles, turbines, and heavy military equipment than any other nation, including the United States. Not all of the machinery or industrial production was for national defense; many civilian products were made in the same factories as well. A common Soviet joke was that what looked one day like a baby stroller could be converted into a machine gun the next day. Because of the secrecy surrounding the details of the VPK's components, both in the Soviet past and in Russia today, it is not always possible to obtain exact data on the production or consumption of these products. Also, in the recent past there have been many changes in production patterns because of the ongoing, and still unfinished, conversion to the market economy. Nevertheless, it is possible to provide a broad overview of the main spatial trends of production, as well as to discuss the challenges facing this sector today.

I discuss the nuclear program first, as the most secretive and the most pivotal to the Soviet military machine. I then focus on other branches of heavy industry, both military and civilian: iron and steel, nonferrous metals, manufacturing of heavy machinery and equipment, and the chemical industry.

In Russia's industry today, the heavy machinery sector accounts for 17%, while metallurgy accounts for 16%. Combined, the two exceed the share of the energy complex discussed in Chapter 17. The geographic distribution of the VPK inside the country is as follows: 27% is concentrated in the Central federal district around Moscow, with 16% in the Urals, 14% in western Siberia, and 13% in the Volga region. St. Petersburg and the rest of the Northwest district account for 12%. Although the industry was hard hit by the reforms in the 1990s, the sector is now recovering strongly. The output decreased by more than half, to an all-time low of 47% of the 1990 level in 1998, but is now above 80% of the 1990 level. Bear in mind that because of much restructuring, many items produced today are different from what was made in the Soviet period. A large share of Russian heavy industry (over 80%) is now in private hands, typically in large-stock ventures. Some companies are nationally prominent, especially metal producers.

In 2007, 4 companies in the top 20 in Russia were engaged in metal production, heavy machinery production, or other heavy manufacturing: Nornickel in Norilsk, in 6th place (valued at $42 billion); the Novolipetsk metal combine, in 13th place ($19 billion); Severstal in Cherepovets, in 14th place ($17.5 billion); and the Magnitogorsk Metallurgical Combine, in 17th place ($14 billion). This last one, located in the city of Magnitogorsk, was built in 1929 as the first large industrial project of Stalinism. Today it employs 60,000 people and produces over 12 million tons of steel a year, about 16% of Russia's total. Other industrial giants farther down the list included TMK (a pipe manufacturer), Mechel (a metal and mining combine), GAZ and AutoVAZ (automakers), and many others. The sector as a whole employed almost 50% of industrial workers in Russia, over 7 million people.

FIGURE 18.1. A monument to the victims of the radioactive fallout from the Soviet atomic weapons testing near Semipalatinsk, Kazakhstan. The Nevada–Semipalatinsk movement was one of the first international nongovernmental organizations (registered in the late 1980s) to begin raising awareness of such radioactivity in Russia and the United States. *Photo:* Author.

Secret Nuclear Cities

The Soviet Union was the second country in the world to detonate a nuclear device in 1948, and the first to launch a civilian nuclear station in 1954. Much of the 1950s–1970s was spent on achieving nuclear parity with the United States during the Cold War. The Soviet Union had to build facilities for research and development (R&D) of nuclear weapons and intercontinental ballistic missiles (ICBMs). It needed uranium mines, enrichment facilities, plutonium-239 production facilities, and nuclear waste reprocessing facilities. It also had to build, transport, store, and test weapons of mass destruction: nuclear, chemical, and biological (Figure 18.1). According to some estimates, in the late Soviet period about one-quarter of all industrial workers in the country (5 million people) were employed by the VPK, including almost 1 million researchers at over 2,000 institutes and factories, and the sector accounted for almost 20% of the country's gross domestic product (GDP).

Hundreds of research labs, institutes, and factories were scattered over a few dozen small and medium-sized cities that did not appear on any maps (Figure 18.2). These secret nuclear (or otherwise military) cities constitute a fascinating subject for geographic research (Rowland, 1999). They were largely declassified, renamed, and finally put on maps by 2000. Most remain closed to casual visitors, however, and even Russia's residents (let alone foreigners) require special permits to enter. Some of this top-secret research also went on behind the facades of average office buildings in Moscow, Gorky (now Nizhniy Novgorod), Sverdlovsk (now Yekaterinburg), and Krasnoyarsk, hiding behind innocent names or simply post office box numbers. Such *yashchiki* (literally, "postal boxes") were good employers: The pay was better than usual, and the prospect of doing cutting-edge science research that had to be shrouded in secrecy added to the appeal. Workers were usually housed in nearby settlements with better-than-average clinics,

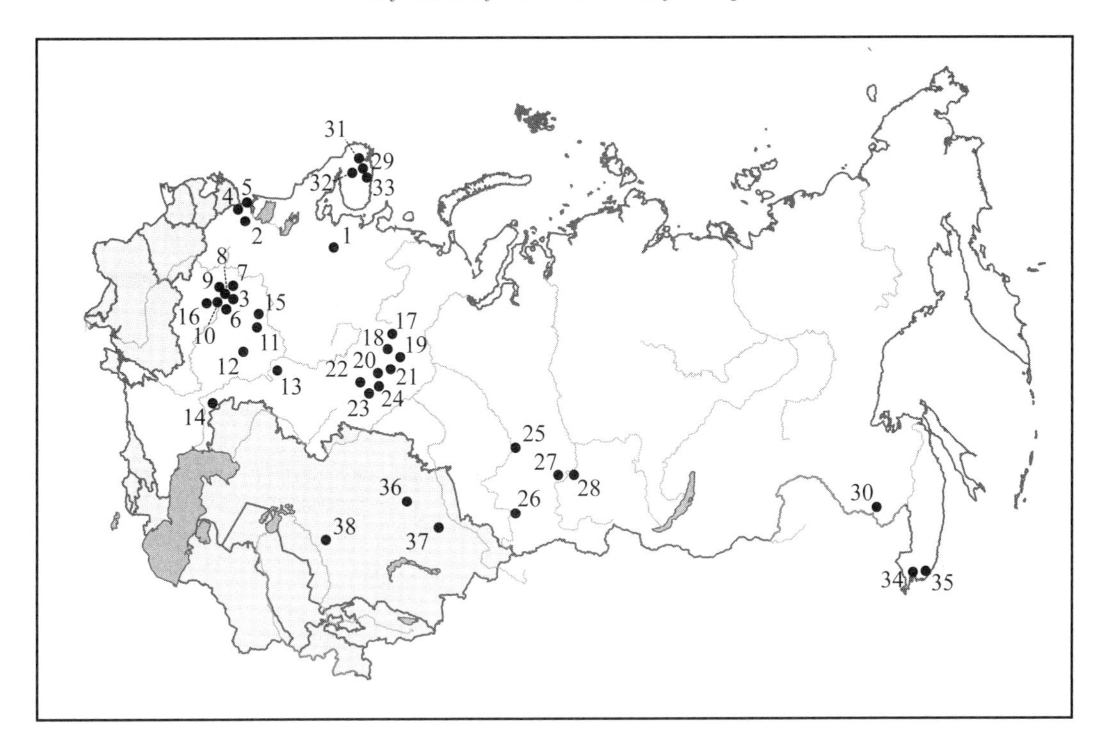

FIGURE 18.2. The main cities involved in the Soviet program of nuclear weapons, chemical weapons, and missile research, including secret cities (the names of these are given in parentheses): 1, Mirny; 2, Gatchina; 3, Korolev; 4, Sosnovy Bor; 5, Primorsk; 6, Zhukovsky; 7, Dubna; 8, Krasnoznamensk; 9, Troitsk; 10, Protvino; 11, Sarov (Arzamas-16); 12, Zarechny (Penza-19); 13, Dimitrovgrad; 14, Znamensk (Kapustin Yar-1); 15, Dzerzhink; 16, Obninsk; 17, Lesnoy; 18, Novouralsk (Sverdlovsk-44); 19, Yekaterinburg; 20, Ozersk (Chelyabinsk-65); 21, Snezhinsk (Chelyabinsk-70); 22, Ust-Katan; 23, Trekhgorny (Zlatoust-36); 24, Miass; 25, Seversk (Tomsk-7); 26, Biysk; 27, Zheleznogorsk (Krasnoyarsk-26); 28, Zelenogorsk (Krasnoyarsk-45); 29, Severomorsk; 30, Uglegorsk (Svobodny-18); 31, Polyarny; 32, Snezhnogorsk (Murmansk-60); 33, Zaozersk (Murmansk-150); 34, Fokino (Shkotovo-17); 35, Bolshoy Kamen; 36, Stepnogorsk; 37, Kurchatov (Semipalatinsk-21); 38, Baikonur. Data from Rowland (1999) and online research by author. Map: J. Torguson.

food stores, and schools available. In post-Soviet times, however, salaries at such facilities plummeted. Eventually, many employees quit their no longer satisfying jobs. My first job after college was with an environmental group that was renting two rooms for an office in one of the former "postal boxes" near the Airport metro station in Moscow. The linoleum floors were peeling, the washrooms stank, the guards were lenient, and our meager rent (paid out of Western grants) actually was one of the few sources of income for the almost bankrupt state facility!

Much Soviet-era research in economic geography was spent on optimizing "territorial production complexes" (called TPK in Russian)—that is, on aligning the locations of military factories with regional sources of fuels, metals, and water. The Soviet geographer N. Kolosovsky created a theoretical framework of TPK organization, proposing various types of solutions to the problems of optimal location, depending on the region. For example, for many industries dependent on coal (e.g., steel and chemical manufacturing), the most important factor in the location of enterprises was proximity to coal-mining districts. The location of other facilities followed a similar geographic logic. Warships and hydroplanes had to be produced near the sea (St. Petersburg, Taganrog, Komsomolsk-na-Amure, Severodvinsk). Nuclear facilities were hidden deep in the country's interior (the Urals, central Siberia) to protect them in case of an outside attack, as they are in the

United States. The Volga region, with its cheap electricity necessary for aluminum smelting, saw the development of the aerospace industry. The Urals traditionally were the center for production of lighter weapons (guns, rifles, grenades, and mines), as well as tanks and armored personnel carriers, because of the extraordinarily rich polymetallic ores available in the region.

The Soviet distribution of factories was determined by careful centralized planning unlike that in a market economy. For example, the top-secret Soviet city was arguably Arzamas-16 (historical Sarov), south of Gorky (Nizhniy Novgorod). Located in a beautiful pine forest in a former monastery town about 50 km south of the actual city of Arzamas, it is still one of the main centers of nuclear weapons research and production. The famous physicists Sakharov, Kapitsa, Tamm, Khariton, and others spent years living in the closed city in small but comfortable cabins in the woods, working on designs for some of the most powerful weapons ever built. The historical Sarov monastery was obliterated to make room for the nuclear center.

Other well-known towns and cities with nuclear facilities included Zarechny, radioelectronic center, southeast of Moscow; Obninsk, Dubna, and Protvino, near Moscow; Ozersk and Snezhinsk, near Chelyabinsk; Seversk, Zheleznogorsk, and Zelenogorsk, in central Siberia; and others. (Figure 18.2 shows the locations of all these, and the caption gives their secret names where applicable.) The Soviet Union had not only nuclear but also chemical and biological facilities at a few dozen sites. In addition, it engaged in production of space satellite equipment, antiaircraft ballistic missiles, cruise missiles, fighter jets, nuclear submarines, and of course ICBMs (e.g., in Votkinsk, Udmurtiya).

A typical "secret" city today has about 40,000 residents. Usually there is one main enterprise that gives the city its reason for existence. Sometimes there are two or three. The biggest such city is Seversk in Central Siberia, with over 116,000 residents in 1997 (Rowland, 1999); the second biggest is Novouralsk (91,000), near Yekaterinburg in the Urals. Rowland's list (which uses the pre-2000 economic region names) includes 11 settlements in the Urals, 9 near Moscow, 8 in the European North (mainly near Murmansk, where the nuclear submarine fleet is deployed), 7 in the Far East (submarine bases), 5 in Central Siberia near Krasnoyarsk, 3 in the Volga region, 2 in West Siberia, and 1 in the Far East. An additional 5 were located in Kazakhstan—for example, Kurchatov, near the Semipalatinsk bombing site (Figure 18.1), and Stepnogorsk, which was a center of uranium mining and of chemical and biological weapon production. There have been attempts to convert some of the former military factories for civilian use. For example, the first Russian computed tomography (CT) scanner was developed at Snezhinsk, the home of a leading thermonuclear bomb research facility. Many nonsecret cities around Moscow are also heavily involved with nuclear-weapons-related research.

Iron Ore and Steel

Russian iron ore production started in the 18th century. The earliest factories appeared in Tula (1712) and the Urals (the Demidov plant in Nizhniy Tagil, 1721), during the time of Peter the Great. The process of smelting iron ore requires a lot of cheap energy, and charcoal provided that in the Urals at first. When anthracite coal became available in the Donbass basin in the second half of the 19th century, the center of the iron industry shifted south. By the 1970s, the Soviet Union was the world's leading producer of iron ore, steel and pig iron, chromite, and manganese ores. Ironically, this was the time when the other leading world economies were shifting away from iron to titanium, plastics, and composite materials. Despite a very high volume of production, many Soviet ferrous metal technologies were energy-intensive, inefficient, and even obsolete. For instance, over half of all steel in the U.S.S.R. in 1988 was still produced with 19th century Siemens–Martin open-hearth furnaces—long before replaced with modern electric furnaces in the European countries and Japan. (Even more ironically, the electric method was in fact invented first . . . in the U.S.S.R.) In that same year, the Soviet Union lost more metal in production than was produced in Germany! Russia produced about half of all steel pipes in the U.S.S.R. Among other former Soviet Union (FSU) republics, Ukraine was by far the biggest producer. Once the reforms started, the industry entered a prolonged crisis throughout the FSU.

The overall production of marketable iron ore went down from 107 million metric tonnes (mmt) in 1990 to 78 mmt in 1995, but rebounded to 110 mmt by 2008. Russia was in fifth place worldwide in iron ore production in 2008, after China, Brazil, Australia, and India. The main area of production, yielding over half of all Russia's iron ore, is the Kursk Magnetic Anomaly in the Central district with four large ore deposits. Other iron mines exist in the Urals and on Kola Peninsula. Ukraine has very important iron ore deposits near Krivoy Rog in the south; its production is about 70% of Russia's total. Kazakhstan also has some significant iron ore deposits. Over 80% of all iron ore in Russia is mined via the open-pit method.

Steel production in Russia declined from almost 90 mmt in 1990 to a little over 74 mmt in 2008. Nevertheless, Russia remains a major producer of steel in the world (in 4th place after China, Japan, and the United States in 2008). The most important centers of steel production are in the European north (Cherepovets in Vologda Oblast, where Severstal is located), the European south (the Novolipetsk and Stary Oskol combines [Figure 18.3]), the Urals (Nizhniy Tagil, Chelyabinsk, Magnitogorsk, Novotroitsk), and Central Siberia (Novokuznetsk). The Urals produce about half of all steel, pipes, and other ferrous metal products in the country. The Urals region uses coal from Karaganda (Kazakhstan) and Kuzbass (Central Siberia) and iron ore from Kursk and Kazakhstan, all of which requires a lot of long-distance shipping. All of the largest plants there were built in the Soviet period, but have since been extensively modernized. A few major battles among different oligarchs and local mafia clans for control of these assets took place in the early 1990s.

FIGURE 18.3. The Stary Oskol steel combine (Belgorod Oblast) was one of the most modern in the late Soviet Union, using German equipment. It produces over 2.4 mmt of high-quality steel per year, 70% of which is exported. *Photo:* A. Shanin.

Today, Russia's iron and steel industry is a modern and powerful one and attracts worldwide investments. The Novolipetsk, Severstal, and Magnitogorsk combines were in the top 20 Russian companies by market capitalization in 2007, and their main owners are billionaires. Severstal recently purchased Rouge Steel in Dearborn, Michigan—the first acquisition by a Russian steel company inside the United States. At the same time, ArcelorMittal, the world's largest steel producer, has made some inroads into Russia (by buying coal mines) and especially into Ukraine (where it purchased Krivorozhstal, the biggest national steel maker).

Steel and iron production requires vast quantities of energy and is very polluting. For example, to produce 1 ton of pig iron, an average plant requires 1.2–1.5 tons of coal, 1.5 tons of iron ore, 0.5 ton of limestone, and about 30 tons of water. Private companies are well aware of this and are trying to improve efficiency and meet environmental standards by installing more efficient furnaces, adding better filters and scrubbers, and switching to new energy sources. For example, Severstal received the ISO 14001 certification in 2001 (this is the leading international standard of environmental quality). The company claims that its emissions of air pollutants declined by 70% in 10 years.

Nonferrous Metals

Nonferrous metals are called "colored metals" in Russia, as opposed to iron, which is "black metal." These metals can be divided into a few groups: heavy (copper, lead, zinc, tin, nickel), light (aluminum, magnesium, titanium, sodium, potassium), noble (gold, silver, platinum), rare (zirconium, indium, germanium, gallium), and high-temperature (tungsten, molybdenum). Many of these metals are found in mixed polymetallic ores and are mined together. For example, it is common to produce copper and molybdenum from one mine, or zinc and lead from another. This facilitates the use of large, diversified metal smelters. Most nonferrous ores contain only a small fraction of the useful metal, frequently less than 1%; therefore, much of the mined material has to be discarded. Also, energy consumption is very high, because many of these metals have to be produced with electrolysis. Typically smelters are located close to the cheapest sources of power (usually hydroelectric dams), and also near large deposits of ore.

Copper ores are concentrated in the Urals and eastern Siberia. The Udokan deposits in the latter are among the largest in the world (over 1.5 billion tons, with a copper content of 1.5%). Russia also imports copper ores from Kazakhstan (over 30%). The Urals produce the most copper at a few factories in Krasnouralsk, Revda, Karabash, Mednogorsk, Kyshtym, and others.

Lead and zinc are mined in a few areas in the Caucasus, in the Urals, in Kuzbass, and near Lake Baikal. Chelyabinsk is the leading center of lead and zinc metallurgy in Russia. Kazakhstan supplies additional lead; it accounted for 70% of the total Soviet production in the past, and has the fourth largest reserves of this metal on the planet.

Nickel and cobalt are mined on Kola Peninsula and near Norilsk in Krasnoyarsky Kray. Nornickel is the largest producer of nickel and platinum-group metals in the world. It was the sixth largest company in Russia by capitalization in mid-2007. Recently it purchased assets in Montana's Stillwater complex, and aspires to become another Russian company with a global reach. Norilsk itself is the largest city above the Arctic Circle in the world, with a population of 300,000, about 15% of whom are workers at Nornickel. This combine is also the largest air polluter in Russia, despite the company's recent investments in cleaner technologies.

Aluminum smelting is a big business in Russia. In 2008 the country produced about 6.4 mmt of bauxite ore—well behind Australia (63 mmt), Brazil, China, Guinea, Jamaica, or India, but still among the top 10 in the world. Bauxite ore is found in many areas of Russia: on Kola Peninsula, in the Komi Republic, and in eastern Siberia. Domestic sources, however, only cover about half of what is needed, making Russia a net bauxite importer. Kazakhstan was the biggest supplier of bauxite in the FSU, additional bauxite ore now comes from Ukraine, the Balkans, Venezuela, and other countries.

Russia is the second largest producer of finished aluminum after China. All aluminum smelting in Russia is concentrated in areas with cheap hydropower (Volgograd, Volkhov, Kandalaksha, Bratsk, Krasnoyarsk, and Sayanogorsk). For much of the Yeltsin period, aluminum production in Russia was concentrated in the hands of two competing companies, Rusal and Sual; however, they merged in 2007 to form the largest aluminum producer in the world, edging out the American giant Alcoa. The new Rusal employs 100,000 workers and is present in 19 countries; it produces 12% of the world's aluminum at 14 plants. It also controls four bauxite-producing mines and 10 plants that produce about 15% of the world's alumina (this is the enriched material needed to make pure aluminum). Some of this aluminum is sold to the United States, which produces virtually no domestic aluminum, but is one of the top world consumers.

Russia is one of the oldest producers of gold in the world; in 2008 it produced 165 tons and was in sixth place worldwide. The placer deposits of Siberia were the first to be mined, but are now largely depleted. However, there are still large lodes and complex polymetallic ores left, mainly in the Far Eastern Magadan Oblast (22 tons per year) and in Yakutia (30 tons). The Irkutsk and Krasnoyarsk areas also have substantial gold deposits. Much of the gold production in Russia is controlled by Polyus-Zoloto, a large private company with market capitalization of $8 billion in mid-2007.

The collapse of the Soviet Union hit the nonferrous metal industry hard. The output fell between 20 and 30% from 1990 to 1995 in Russia, and even more in some other republics. However, this sector was also among the first to recover, due to increasing demand from abroad and the relative ease of nonferrous metal production and sales (as compared to product manufacturing). By 2008, nonferrous production for metals was higher in Russia than before the reforms. The sector saw some of the worst criminal takeovers in the mid-1990s, as chaotic gangster wars erupted around aluminum smelters in Krasnoyarsk, for example. Nevertheless, the export orientation of the sector ensured a quick recovery; metal trade is one of the chief sources of foreign revenue. Metal production in Russia today is dominated by a few large, vertically integrated companies (e.g., Rusal and Nornickel are under the control of oligarchs with friendly connections to the Kremlin).

One of the biggest uncertainties at the moment seems to be skyrocketing energy prices. The nonferrous metal industry is particularly sensitive to the costs of electricity, coal, and shipping. Factories are idled when the electricity rates become unaffordable. Sales of scrap metal from idled and stripped factories were common on the black market in the chaotic 1990s. Rusal in particular was hard hit by the global recession of 2008–2009. The company made heavy debts to foreign lenders and was on the brink of default in the fall of 2009. In the first quarter of 2010, Rusal posted a modest profit of $247 million as compared with a loss of $638 million a year earlier. Overall, the company managed to avoid bankruptcy, but its future remains closely tied to the fate of the global metal markets.

Heavy Manufacturing

In Russia, the common term for heavy manufacturing is "machine building" (*machinostroenie*). It includes production of motors, boilers, tractors, agricultural and mining combines, industrial machines, manufacturing equipment, and their components. It also includes the manufacturing of some high-tech goods (e.g., electronic equipment, avionics and robots); some of these are discussed in Chapter 21. It includes some components of the military complex as well, such as construction of tanks, airplanes, and submarines, and their civilian counterparts (cars, trucks, passenger jets, and ships). Russia inherited about three-quarters of the Soviet Union's heavy industrial plants. Heavy manufacturing was the main priority of the Stalinist economy, because it was viewed as the primary source of the state's might, and a guarantee of its survival in a hostile world. About one-third of all industry in the late Soviet period was heavy manufacturing, especially military types. In relative terms, Russia had more of the heavy industries and other republics had more of the light industries. Two other republics that had major military machinery production were Ukraine and Belarus. Some heavy industrial production took place in Kazakhstan (mining and farm equipment) and the Baltic states (electronics and transportation), but very little was produced in the other four Central Asian republics, Moldova, or the Caucasus.

Much of the equipment used in heavy industry during the late Soviet period is still in use today, but is very old. Already in the early 1990s, the average lifespan of the equipment was approaching a quarter century; by 2006, 45% of the equipment used in manufacturing was over 20 years old. The overall output level of this sector fell by 50% in the first 5 years of reforms. Table 18.1 provides statistics for some specific types of machinery. Although the situation is slowly improving, even by 2008 Russia's heavy manufacturing output was only 85% of the 1990 level. However, the entire sector went through a major reorganization: While production of some items ceased or greatly declined, production of new items began. For example, the Soviet Union made over half a million heavy trucks per year, but it made no light pickups or vans. Now over 20 modifications of the Gazel van and the Sable pickup truck are produced by the GAZ plant in Nizhniy Novgorod (about 100,000 vehicles per year). The Soviet Union also did not build any railroad passenger cars, importing them from East Germany. Now at least two factories are making those domestically in Tver and Tikhvin.

Machinery building is the most complex of all industries. It requires coordination of production and a huge supply of parts. It is also heavily dependent on the availability of raw materials: energy; water; steel, aluminum, copper, nickel, and dozens of other metals; glass; and plastics. The main concentration of machinery building is in the Central federal district around Moscow (about 39% by production volume), but some industries are also located along the Volga (22%) and in the Urals (14%). Outside Russia, a very important machinery-building region is south central Ukraine, along the Dnieper River

TABLE 18.1. Amounts of Some Types of Machinery Produced in Russia in the Late Soviet Period and the Post-Soviet Period

Machinery	1990	1995	2000	2002	% of 1990
Diesel generators (thousands)	23.2	4.1	4.8	4.4	19
Mining combines (units)	406	128	93	82	20
Diesel train engines (sections)	46	12	21	23	50
Metal-cutting machines (thousands)	74.2	18	8.9	6.5	9
Trucks (thousands)	665	142	184	173	26
Passenger cars (thousands)	1,103	835	969	981	89
Combines (thousands)	66.7	6.2	5.2	7.5	11
Tractors (thousands)	214	21.2	19.2	9.2	4
Looms (thousands)	18.3	1.9	0.1	0.3	<2

Note. Data from the Federal Service of State Statistics, Russian Federation.

(Dniepropetrovsk, Dneprodzerzhinsk, Zaporozhye). Virtually all machinery building for civilian purposes has been privatized in Russia now. As in the other industrial sectors, a few large companies predominate; they are, however, much smaller than the petroleum, gas, or metal producers in terms of market capitalization (i.e., their stocks tend to be undervalued). The largest company by market value in this sector in 2007 was AutoVAZ, the manufacturer of Lada cars in Togliatti (about $5 billion), which was in 36th place among all Russian companies. KAMAZ (a truck manufacturer) was in 46th place, while GAZ was in 54th. One of the reasons for these low rankings is the low competitiveness of Russian products in the world markets. AutoVAZ entered into a complex negotiation process with Renault–Nissan in 2009 to avoid full bankruptcy; the company is expected to cease its production of the obsolete Lada line and switch primarily to the Renault brand by 2012.

The manufacturing of the heaviest machinery and equipment is mainly concentrated in the Urals (Yekaterinburg, Orsk, Chelyabinsk). Uralmash is the largest producer of heavy machinery in the region, making mining equipment and other machines for steel and nonferrous metallurgy, energy, and construction companies. The region is optimally located in the middle of the country, with a large pool of qualified labor; well-developed transportation networks; and major deposits of coal, iron, and nonferrous ores nearby. Many large factories were relocated to the Urals during World War II from European Russia and stayed there after the war. Large mining combines are produced in Siberia, where they are most needed (Irkutsk, Krasnoyarsk). Historically, another large machinery-building center has been Leningrad (St. Petersburg) and its vicinity, with proximity to the European markets and easy sea access; it is known for building ships, tractors, steam and hydraulic turbines, and nuclear reactors. Dnepropetrovsk in Ukraine is home to the large Yuzhny machinery plant, building aerospace equipment and more recently a wide array of consumer goods (e.g., bicycles and gym equipment).

Railroad car and engine building is an important branch of heavy industry in Russia, with its continued heavy use of railroads in shipping freight (40% by volume) and in passenger traffic (33%) (Chapter 21). Large diesel locomotives are built in Kolomna near Moscow. Novocherkassk in the northern Caucasus builds electrical locomotives. Although Russia holds the world record for the most powerful, fastest diesel-driven locomotives, only two ER-200 high-speed trains were built in Russia, and they were about only half as fast as a typical French TGV train. After the collapse of the socialist bloc in Europe and the loss of Ukraine, railroad cars also had to be built in Russia. Today freight train cars are built primarily in the European part of the country (Bryansk, Tver, St. Petersburg, Kaliningrad) and in the Urals (Nizhniy Tagil). Specialized subway trains and suburban commuter trains are built in

FIGURE 18.4. Electric commuter trains were built in Latvia during the Soviet period, but are now produced in Russia itself. This particular train connects Domodedovo Airport with downtown Moscow. *Photo:* Author.

a few cities around Moscow (Figure 18.4). Incidentally, the Nizhniy Tagil railroad car plant is also famous for designing and building the most massively produced Soviet tank (T-72), as well as the modern tank of the Russian Army (T-90), hundreds of which have been sold to India, Algeria, and other foreign countries. The T-90 still forms the backbone of the Russian tank force.

Ship building is another area where the Soviet Union had a great need for imports. Many cruise ships still working on Russian rivers were German-built. However, Russia still continues to build some of its own ships, unlike the United States, which ceased virtually all civilian ship building after World War II. About one-third of all ships in the world today are built in South Korea, with Japan, China, and Germany following suit. Compared to these giants, Russia's production is very small. Naturally, ocean-going ships are built near the sea (St. Petersburg, Vyborg, Severodvinsk, Astrakhan, Vladivostok), as are shelf drilling platforms and floating fish-processing facilities. Vessels for use on rivers are built along the Volga (Nizhniy Novgorod, Volgograd) and in some other areas (Blagoveshchensk on the Amur, Tyumen and Tobolsk in the Ob basin, etc.). The U.S.S.R. had well-developed river transport that included passenger commuter boats, hydrofoils, and cruise ships, as well as barges, tankers, and freighters. The Soviet Union also had a heavy presence on the world's seas, and its tankers, dry cargo ships, fishing boats, and icebreakers were all produced domestically.

Because the aerospace industry requires access to plenty of aluminum and energy, virtually all of it has been concentrated along the Volga, with its numerous dams near Kazan, Samara, Saratov, and Ulyanovsk (Figure 18.5). The region also has great research facilities, a highly skilled workforce, and a decent quality of life. Additional R&D for the industry occurs in Central Russia, with easy access to Moscow institutes and project bureaus (Korolev and Zelenograd). Some facilities for missile construction and testing are in Omsk and Krasnoyarsk.

The Soviet Union was making hundreds of large passenger jets per year in the 1970s (the most commonly known were the Il-62, Il-86,

FIGURE 18.5. Most of Russia's aerospace industry is concentrated along the middle Volga in Samara, Saratov, Ulyanovsk, Kazan, and Nizhniy Novgorod. The Progress factory in Samara builds the Soyuz spacecraft, one of which is shown here. *Photo:* S. Blinnikov.

Tu-134, Tu-154, and Yak-42 models). There were also a few reliable turboprop models (the An-2, An-12, and An-24) for shorter flights. Several were of excellent design and quality (Figure 18.6). For example, the Il-62, production of which began in 1962, was one of the most reliable long-distance planes ever built. It was able to cross the Atlantic Ocean from Moscow to New York without needing to refuel. By contrast, many late Soviet planes had unsurpassed aerodynamics, but noisy and thirsty engines. Perhaps the best example is the Il-96, one of which is still used as Russia's "Air Force One" to transport the president. It has a very smooth ride; however, its four engines use about twice as much fuel as their Western counterparts per kilometer of flight. Avionics were also lacking in quality.

During the Yeltsin period, production of passenger airplanes came to a halt, and production of military jets was greatly curtailed. One large plane manufacturer (Antonov) was left in Ukraine. Three others (Ilyushin, Tupolev, and Yakovlev) were struggling to continue production in Russia without adequate supplies or finances. The inability to sustain production during the reforms led to the virtual disappearance of modern Russian jets from the world's travel markets. Domestic airlines started switching to Airbus and Boeing models. To rectify the situation, Putin's government merged all existing plane producers into one consortium in 2006

FIGURE 18.6. The Tu-154 airplanes were the most commonly produced large jets in the Soviet Union. A few hundred were built, but they are now being replaced with Western aircraft. *Photo:* A. Fristad.

and provided new tax breaks and subsidies. Soviet fighter jets (the MIG and the Su-series) remain competitive on the world markets in flight performance, but generally lag behind U.S. and European models in pilot comfort and high-tech equipment. A few dozen of these fighters are still built per year and are significantly cheaper than American or French models. The Sukhoi Corporation also started making a civilian regional jet (the Superjet-100) in 2008, when prodded by the Kremlin. Primarily made out of foreign parts and not much different from the mass-produced Brazilian Embraer, it is nevertheless a source of much national pride.

Car and truck manufacturing is similarly concentrated in the middle Volga basin. Like the aviation industry, it was heavily militarized in the Soviet period. For example, UAZ and GAZ all-terrain four-wheel drive vehicles were used both by the military (like Humvees) and by civilians working in forestry, law enforcement, geology, and the like. GAZ also builds heavy-wheeled BTRs (armored personnel carriers) at its Arzamas plant in Nizhniy Novgorod Oblast. Many of the factories producing tanks or armored personnel carriers also make tractors or agricultural machinery. Trucks are mainly built by GAZ in Nizhniy Novgorod and KAMAZ in Naberezhnye Chelny. The historical ZIL plant in Moscow made hundreds of thousands of heavy trucks, and some limo cars for the Communist VIPs, but is now closed. Many of the long-distance trucks as well as the buses used in Russia today are imported, mainly from Europe (Volvo, Mercedes, etc.).

The biggest recent changes have occurred in the passenger car industry. The two giants of the Soviet period (AutoVAZ in Togliatti, making Ladas and Nivas, and GAZ in Nizhniy Novgorod, making Volgas) continue production as large private enterprises. Many new models have been introduced, some almost approaching Western standards of comfort . . . of about 20 years ago (Figure 18.7). The Moskvitch plant in Moscow did not survive Yeltsin's reforms and closed its doors indefinitely. However, the real revolution occurred when production of Western cars was allowed inside Russia. Although most of these factories assemble autos from parts manufactured outside Russia, their sheer presence makes Russian manufacturers try harder, while customers

FIGURE 18.7. This picture from Tomsk shows three AutoVAZ-built cars: a Lada 2105 sedan and Lada 2108 coupe in the front, and a Niva SUV on the other side of the street. *Photo:* A. Fristad.

benefit from many more choices. Employment for Russian workers is also a benefit. Foreign brands produced in Russia include Hyundai in Taganrog, Kia in Kaliningrad Oblast and Izhevsk, Ford in Vsevolozhsk near St. Petersburg, and a few others. GM, Toyota, Nissan, and Volkswagen either have limited production in Russia already or are planning to establish it in the near future. Another significant player is Sollers, a daughter enterprise of the steel-making giant Severstal, which now produces both foreign (Fiat, SsangYong, Isuzu) and Russian (UAZ) brands of cars and trucks.

In the Soviet period, most agricultural machinery was built in the breadbasket of the country—that is, in Ukraine (Kharkov), northern Kazakhstan (Pavlodar), Moldavia/Moldova (Kishinev), and Belorussia/Belarus (Minsk). Manufacturers in many of these republics depended on Russia and each other for parts, and with the breakup of the U.S.S.R. and the beginning of reforms, entered a deep crisis. Within Russia today, Rostov-on-Don and Taganrog, located near grain-rich Kuban, build giant combines (Rostselmash). Ryazan produces potato harvesters. The famous large Kirovets tractors were assembled in Leningrad, but today the plant produces mostly smaller machines for private farms. Incidentally, the same plant developed and produced hundreds of the late Soviet T-80 tanks. Volgograd, Lipetsk, Chelyabinsk, and Rubtsovsk produce tractors as well. Kurgan in the southern Urals produces armored personnel carriers (BMPs), thousands of which have been sold worldwide. Both John Deere and Caterpillar are present in Russia. Since 2000, Caterpillar has been making parts in its brand-new factory in Tosno near St. Petersburg; this is an attractive location because of easy connections with Western Europe and the presence of a highly qualified workforce. Their long-range plans involve building actual assembly lines for various types of tractors in Russia. A John Deere factory is present in the agricultural Orenburg area of Russia, where seeding equipment is assembled.

The Chemical Industry

Production of chemicals is critical for any economy. One of the earliest chemical industries to appear in Russia was the production of sulfuric and nitric acid, needed to make fertilizers and gunpowder. The production of potassium hydroxide for glass making was another early chemical industry. Some of Russia's earliest chemical factories were built in the early 1800s, mainly around Moscow, along the Volga, and in the Urals. During the Soviet period, much development occurred in the production of organic compounds from coal, petroleum, and natural gas, including plastics, fertilizer, paints, pesticides, detergents, and chemical weapons. Soviet chemists were at the forefront of research in many branches of modern chemistry (Chapter 15).

The geographic distribution of the chemical industry depends on the availability of the necessary raw materials, access to water, and cheap energy. Moreover, the production of many kinds of chemicals is highly polluting (Figure 18.8) and must be carefully located away from large settlements or fragile natural areas, but it also requires access to a highly skilled labor force, which is frequently problematic. There are five main types of chemical industries in Russia today: (1) mining and enrichment of raw materials (e.g., phosphates, potassium, and sulfur); (2) production of common acids, bases, and other feedstocks to be used in farther chemical processes; (3) basic organic synthesis (alcohols, ethers, formaldehyde, etc.); (4) advanced organic synthesis (e.g., plastics, rubber, and pesticides); and (5) other types, including biomedical, microbiological, and photochemical production.

The chemical industry of the U.S.S.R. was well developed and accounted for a bit less than 10% of all industrial output. In Russia today, chemical products account for about 5% of the total industrial output, but remain important both for domestic consumption and for exports. As in the rest of the industry, the early reforms of the 1990s hit the sector hard: Production of sulfuric acid, for example, decreased by about 40% from 12.8 mmt in 1990 to 8.5 in 2002. At the same time, production of synthetic fibers and paints dropped by 75%, production of tires dropped by 25%, and so on. Since 2002, there have been some increases in chemical production again. Russia remains one of the world's leaders in exporting fertilizers; it is also a large producer of plastics, rubber, and paints.

Within Russia, about 40% of the chemical industry today is concentrated in the Urals (if Permsky Kray is included, with its giant fertilizer operations) and another 20% in the Central federal district surrounding Moscow. Substantial concentration of chemical enterprises is also found along the Volga, especially in the middle part of the basin from Nizhniy Novgorod to

FIGURE 18.8. A factory near Saratov on the Volga is belching out fumes and polluted water. *Photo:* S. Blinnikov.

Kazan, Samara, and Saratov. Some types of chemistry production are localized in just a few places. For example, virtually all potassium fertilizer in Russia is produced in Permsky Kray, centered on the giant deposits of potash near Solikamsk and Berezniki. Russia is the second largest producer of potash in the world after Canada (6.3 mmt vs. 11 million in the latter in 2007), while Belarus is third (5.4 mmt). Nitrogen fertilizer is produced in many places where coal or natural gas is available. Russia is the second largest producer of ammonia in the world after China; much of it is produced using cheap natural gas. Ammonia is one of the main export items for Russia.

Table salt has traditionally been produced in the Urals and the lower Volga. Plastics are made in many places (e.g., Dzerzhinsk, Kazan, Volgograd, Yekaterinburg, Ufa, Salavat, Nizhniy Tagil, Tyumen, Kemerovo, Tomsk), but largely along the Volga (35%), in the Urals, and in Central Siberia. In contrast, production of synthetic fibers (e.g., polyester) is concentrated overwhelmingly (79%) in the Central district, near the large textile centers in Ivanovo, Shuya, Tver, Ryazan, and Kursk.

The Soviet Union was one of only five countries in the world known to stockpile chemical weapons (along with the United States, India, Libya, and Albania). Although new chemical weapons are supposedly no longer produced, there are some stashed away. About one-quarter of them were known to have been destroyed by 2007, in compliance with the international Chemical Weapons Convention. The chemical industry in Russia continues to produce many types of explosives at a few dozen factories, however. Solid and liquid fuels for missiles are widely manufactured as well.

Russia is a major producer of hundreds of medical drugs, including very sophisticated modern medicines developed either in the late Soviet period or since the fall of the Soviet Union; there are 340 pharmaceutical producers in Russia. Nevertheless, over half of all medical drugs are now imported. Only 2 Russian companies were in the top 20 suppliers of medical drugs in Russia in 2007: Farmstandard and Otechestvennye Lekarstva. The rest were well-known transnational corporations (e.g., Sanofi-Aventis, Berlin-Chemie, Gedeon Richter, Pfizer, Novartis, Bayer, and others from France, Germany, Switzerland, and the United States). The domestic pharmaceutical industry is also seeing increased competition from India. Little is being done to improve the investments in local production, and pharmacies are known for their corrupt business practices, resulting in high local costs.

Outside Russia, the largest chemical enterprises are found in Belarus, Ukraine, and Kazakhstan. All depend to some extent on raw materials from Russia as inputs or sell their products to Russian enterprises. Many of Russia's chemical exports to Europe are shipped via railroads going through Belarus or by trucks through Belarus, the Baltics, or Ukraine.

Overall, heavy industry remains the backbone of Russia's economy and provides a major share of the country's exports. It is also well represented in Ukraine, Belarus, and Kazakhstan, and the industries of these four countries remain integrated to a large extent. Light industry, considered in the following chapter, has fared less well.

REVIEW QUESTIONS

1. Which main industries are included in the VPK?
2. What are some common factors that influence the distribution of heavy industry?
3. Explain why the Central federal district and the Urals have such a concentration of heavy industry in Russia.
4. Use Table 18.1 to investigate which products showed the greatest decline after the fall of the Soviet Union. Try to explain why some were hit harder than others.
5. In your own country, find regions similar to the heavily industrialized regions of Russia or other FSU nations. What are some of the same economic challenges experienced in these regions?

EXERCISES

1. Look up pertinent data on production of any car or truck manufacturer in Russia (good ones to try are AutoVAZ, GAZ, UAZ, or KAMAZ). Where are they located? Does geography play a role in where these are located? How many types of cars/trucks do they make? How many models or modifications? Who are

their primary buyers? Do they sell vehicles outside Russia? How do their products compare to those of any major Western car manufacturer? You can have a class discussion and compare data for all of them, and additionally compare them to major Western manufacturers (e.g., Ford, GM).

2. Imagine that you work for the government of a rich Middle Eastern country. The boss wants you to prepare a comparative report that highlights the costs and benefits associated with the purchase of about 50 Russian tanks (T-90) or an equal number of Western ones (e.g., American or German). Make a recommendation. Make sure you explain your rationale and back it up with some concrete numbers on performance and price.
3. Visit your local pharmacy, and conduct an informal research of what you see on the shelves. What proportion of the over-the-counter drugs available are made in your country? What foreign manufacturers are represented? Are there any medicines that were made in developing countries (e.g., Brazil or India)? Why or why not?

Further Reading

Bond, A. R., & Levine, R. M. (2001). Norilsk nickel and Russian platinum-group metals production. *Post-Soviet Geography and Economics, 42*, 77–104.

Fortescue, S. (2006). The Russian aluminum industry in transition. *Eurasian Geography and Economics, 47*, 76–94.

Rowland, R. H. (1999). Secret cities of Russia and Kazakhstan in 1998. *Post-Soviet Geography and Economics, 40*, 281–304.

Shabad, T. (1969). *Basic industrial resources of the U.S.S.R.* New York: Columbia University Press.

Venables, M. (2006). The Russian invasion (automobile industry). *Manufacturing Engineer, 85*(4), 8–9.

ZumBrunnen, C., & Osleeb, J. (1986). *The Soviet iron and steel industry.* Totowa, NJ: Rowman & Allanheld.

Websites

eng.gazgroup.ru—GAZ Group, the leading manufacturer of passenger vans and pickup trucks in Russia.

www.lada-auto.ru—AutoVAZ, the largest auto manufacturer in Russia. (In Russian only.)

www.poly/engusgold.com—Polyus-Zoloto, the largest producer of gold in Russia.

www.rusal.ru/en—Rusal, the largest producer of aluminum in the world.

www.severstal.com—Severstal, one of the leading steelmaking companies in Russia.

www.sukhoi.org/eng—Sukhoi, the company that builds the Su-series of fighter jets and the new regional passenger jet, the Superjet-100.

www.uralmash.ru/eng—One of the largest machinery plants of the FSU.

CHAPTER 19

Light Industry and Consumer Goods

"Light industry" includes production of clothing, shoes, textiles, appliances, and other retail items. It was never a strong part of the Soviet economy. For example, although some clothing and footwear had to be produced, much of it was of abysmal quality. To get a pair of decent shoes, one had to shop at one of the hard-currency (*Beriozka*) stores, or beg a friend who went on a rare trip abroad to buy a pair. The Hungarian economist Janos Kornai developed a theory of socialism as a "shortage economy," explaining that in a socialist state shortages result not merely from "planning errors," as is commonly assumed; they primarily occur because enterprises exist to deliver goods, but not to make a profit. In other words, they can afford to lose money indefinitely without facing bankruptcy. All things being equal, Soviet managers had more incentives to produce heavy equipment, cement, or steel than to make consumer goods, because the Soviet Union reward system was heavily skewed toward the former rather than the latter. Thus less than a third of the total Soviet industrial output was in the consumer products sector.

Not only was light industry performing poorly before the breakup of the U.S.S.R.; it was also the one hardest hit by the recession resulting from the reforms of the 1990s. Some examples tell the story plainly: Russia produced only 11% of the shoes, 17% of the underwear, and 40% of the cotton fabric in 2002 that it produced in 1990. The sector's output in general fell a whopping 80%—an unprecedented decline, much more severe than that in the heavy industry. It is interesting to note that whereas the United States and other developed Western economies ceased production of their own shoes and clothing at the same time and shifted it overseas to Asia and Latin America in order to keep prices low for consumers, the production stopped or was greatly curtailed in the FSU when privatizing schemes failed. The eventual result was the same, however: Virtually all shoes and most clothes for sale in Russia today come from China and other Asian economies.

With respect to consumer electronics and other appliances, some common products (e.g., dryers or toaster ovens) were not made in the U.S.S.R. at all. Others (e.g., washing machines or tape recorders) were produced in small quantities and were of inferior quality. A handful of non-competing factories would produce refrigerators, stoves, TV sets, and so on. They were not interested in marketing their products to consumers or in improving quality and design. When Japanese, Korean, European, and U.S. brands began to flood Russia in the early 1990s, domestic production of appliances declined to a fraction of its former volume. Refrigerator production, for example, declined from 3.8 million units in 1990 to merely 1.3 million in 2000; camera production decreased from 1.9 million to 100,000; and so on. Imported goods have filled the shelves across Russia since the mid-1990s. Some new domestic production is beginning to occur, however. The products are not replications of the obsolete Soviet models, but contemporary products made to a large extent with foreign investment and technologies, and they have the potential to compete with foreign-made products. At the same time, Russian labor rates are higher than in most of Asia, and there is a lack of qualified workers in some regions of the country, making domestic production problematic.

Compared to the heavy industry, most of the light industry was located outside Russia during the Soviet era. For example, Latvia and Belarus assembled a lot of electronic products. With the devolution of the Soviet Union, the old factories in the other republics needed Russian parts, but these were no longer available because of the bankruptcies of many enterprises, the lack of credit, the collapsed banking system, and hyperinflation. In turn, Russian stores could not obtain finished goods from the other republics, because the newly independent states had lost their connection to Mother Russia. Within less than 10 years, a majority of the old Soviet factories in light industry either closed altogether or had to reinvent themselves with new sources of capital under new ownership. Although no one may regret the collapse of factories making obsolete models of TVs or washing machines, the decline in fabric manufacturing is very unfortunate. Soviet textiles were very durable and of high quality, and their loss is deplorable. Also, entire textile-manufacturing regions (e.g., Ivanovo) have lost numerous jobs and are financially depressed.

On the bright side, certain branches of light industry in Russia and other former Soviet Union (FSU) states today are genuinely booming. Here are some examples:

- Food processing (e.g., frozen dinners sold at supermarkets).
- Beer brewing (Figure 19.1).
- Cosmetics manufacturing.
- Book publishing.
- Manufacturing of construction materials (plywood, sheetrock, siding, etc.) and home improvement products (toilets, sinks, furniture, etc.).

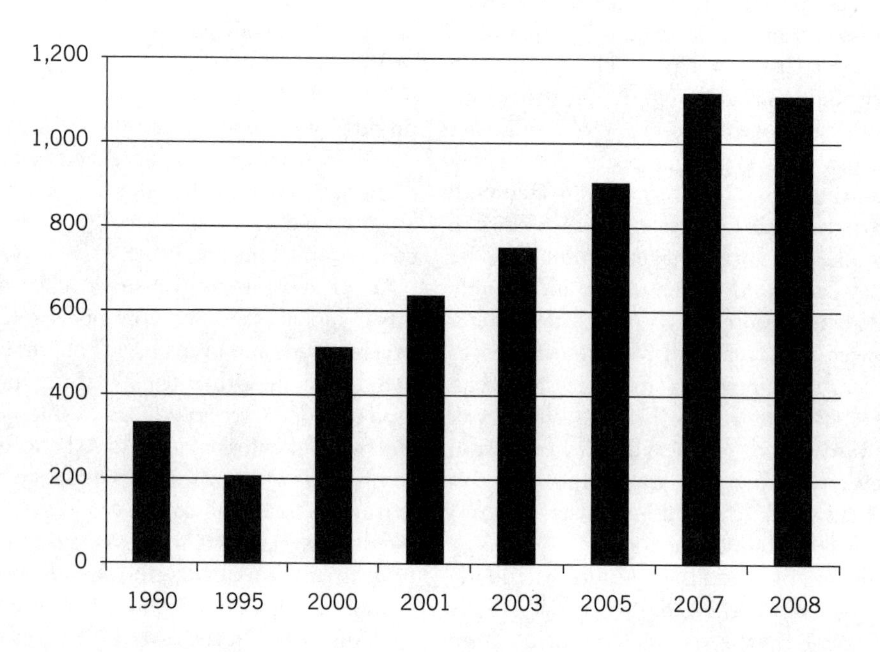

FIGURE 19.1. Beer production in Russia (in millions of decaliters). Data from the Federal Service of State Statistics, Russian Federation.

Note that many of these sectors were greatly underdeveloped in the Soviet Union. In this chapter, I focus on three representative industries: two growing ones (food processing and book publishing) and a struggling one (textile, shoe, and clothing manufacturing). Each illustrates typical challenges and opportunities in light industry of the post-Soviet period.

Food Processing

If you have seen the movie *Everything Is Illuminated*, you may recall the hilarious scene in which an unfriendly Ukrainian waitress brings out a lone potato on a plate to a vegetarian American tourist, who is played by Elijah Wood. The potato comes presumably baked and peeled, but without any dressing, gravy, salad, or side dish. Although this scene is admittedly exaggerated, it is based on real-life experiences with the restaurant industry in the FSU: The Soviet restaurant food was notoriously bad, and the staffers were unfriendly. The majority of Soviet citizens never ate out, except in their workplace cafeterias. All families cooked their own meals at home. Virtually no processed food was used; everything had to be made from scratch. Although much processed food is not healthy, in consumer societies it is heavily marketed as the sensible choice for busy people. Workers today have little time to cook, and recent innovations have made factory production of precooked food plentiful and cheap. The Soviet Union relied on some of its trade partners in Eastern Europe (Hungary, Poland, Bulgaria) for many canned and frozen food items, especially vegetables. It also produced certain types of processed food itself, mainly for the military. The most useful was *tushenka* canned meat—usually beef stew with plenty of fat and of dubious quality—which sustained many generations of Soviet geologists and students in summer camps. Although it was made mostly for the army (meat was a rare treat in a Soviet soldier's ration), much tushenka was sold on the black market. *Sguschenka* (sweet condensed milk) was another staple that was sometimes available. So, as far as processed and prepared food went, there were few choices. On the other hand, fresh bread of decent quality was almost always available everywhere, and so were many basic vegetables and grains. Therefore, at least three generations of Soviet women grew up with expectations of much cooking as part of their normal family life. Because nearly all of them also worked full-time, cooking was the most demanding of the household chores that Soviet women had to do, while men watched hockey on TV or read *Pravda*. When new business opportunities arose during the 1990s, one of the gaping holes in the economy begging to be filled was in the processed and ready-to-serve food industry.

Both Russian and foreign companies rushed in to fill the need. Among the foreigners came the transnational giants Nestlé and Kraft Foods, with full lines and dozens of brands of processed food products. Within a short time, both companies opened production facilities in Russia. Kraft Foods opened a coffee-packaging facility near St. Petersburg in 2000 to capitalize on the developing instant-coffee market, and it purchased a stake in the Russky Shokolad factory in the city of Pokrov to become the second largest chocolate maker in Russia. Nestlé purchased the famous Soviet ice cream factory in Zhukovsky near Moscow; it also made aggressive investments in baby food, coffee, candy, dry milk, and pet food production in the Vologda, Perm, Kostroma, Kuban, and Kaluga areas of European Russia, as well as in Barnaul in Central Siberia. Mars, Inc. brought in truckloads of Snickers, M&Ms, Skittles, and other candy products, and within a short period started manufacturing them in Russia. The French dairy giant Danone came in with offers of fresh yogurt, an undermarketed milk product during the Soviet period. Russians generally prefer *kefir* (which is a similar milk product, but fermented by a different culture) to yogurt. Danone quickly learned this and started making excellent kefir and farmers' cheese to suit Russian tastes.

McDonald's opened its first restaurant in Moscow in 1990. It enjoyed a runaway success, with the initial queues at the door exceeding those to Lenin's tomb. The company quickly learned, however, that in order to make decent fries it had to grow potatoes itself to its exact specifications, because no local suppliers of frozen fries could be found. As a form of geographic adaptation to local tastes, many foreign companies experimented with flavors unfamiliar to North Americans;

for instance, the black currant milkshake was my personal favorite at the first Moscow McDonald's for a while. Similarly, Skittles made in Russia contain beet sugar instead of corn syrup, and thus have a distinctly "European" taste, slightly different from that familiar to North American consumers.

Russian companies have slowly begun to respond to the challenge posed by outsiders. One of the early successes was the creation of the Wimm–Bill–Dann (WBD) Corporation in 1992—a major juice, milk, and baby food producer controlling about one-third of the Russian domestic market. Despite its vaguely European name (based on a cartoon character resembling a mouse with oversized ears), WBD is a distinctly domestic company owned by Russian capitalists. Today it has a presence in many parts of Russia, as well as in Georgia, Ukraine, and some Central Asian states, with 37 production plants. Many of its products' commercials bear direct references to the quality and naturalness supposedly common among Russian products. Cute brand names like Happy Milkman or Little House in the Village cause people to become nostalgic and buy more of the WBD products. However, the taste and quality of these products are indeed just as good as, or better than, those of WBD's Western competitors (Figure 19.2). This strategy of capitalizing on nostalgia for the Soviet past has been adopted by many other Russian manufacturers. For example, packaged Indian tea from the Moscow Tea Factory proudly bears the same elephant logo as its Soviet predecessors, and has the slogan on the package "That very tea, with the elephant" to distinguish itself from its numerous domestic and foreign competitors.

FIGURE 19.2. Russian-made cheese is of excellent quality and competes favorably with Dutch or Swiss cheese on the domestic market. *Photo:* A. Fristad.

Food processing has to be widely dispersed to be efficient, because transporting fresh, heavy, or frozen items over long distances is costly. Perishable items, like milk, must be consumed locally. The introduction of Western-style sterilized milk (e.g., Parmalat) did not meet with much success among Russians, who stubbornly prefer more natural-tasting alternatives. Some food processing must be done close to where the raw materials are harvested; for example, sugar or butter must be processed rapidly to avoid spoilage. Others must be produced close to the consumers—bread or pastry, which must be fresh, or bottled juices and beer, which are too heavy to ship far.

Speaking of beer, one of the biggest surprises of the 1990s was its emergence as the new national drink instead of vodka. Although vodka consumption remains high (Russia consumes half of all vodka produced in the world), beer now accounts for more alcohol consumed than vodka does within Russia. An average Russian in 2003 consumed 9.1 liters of pure alcohol, as compared to 15.4 liters in Luxembourg, 14.8 liters in France, and 8.3 liters in the United States. Although Russia still trailed 18 other developed countries in a 2005 rating by RosBusinessConsulting and the Organization for Economic Cooperation and Development, per capita consumption of alcohol in Russia was up about 65% since 1990. Most of this was due to an increase in beer, not hard liquor consumption. In fact, the Russian beer market is the third largest in the world, behind only the Chinese and the U.S. markets. Beer was not particularly popular or widely produced in the U.S.S.R.; the typical beer was soapy, sweet stuff of rather disagreeable quality. With privatization, unprecedented opportunities came for foreign and domestic investment in this arguably safer alternative to hard liquor, and beer production soared. Of particular note are early investments in obsolete breweries in the Urals by the Khadka family (the Sun Corporation from India), and more recent investments by Dutch, German, and other European beer makers in breweries in Moscow and especially St. Petersburg, the beer capital of Russia. The Danish giant Carlsberg controls production of Baltika beer (the top brand in Russia and the second biggest brand in Europe by volume sold), as well as of Arsenalnoe, Yarpivo, Nevskoe, and a few other brands. Several dozen beer brands are now available in the FSU, not including distinctly local microbrews. Also, beer commercials are easily one of the two most common types seen on TV, along with obnoxious cell phone advertisements.

Food processing in the other FSU republics has basically followed the Russian trends. The Baltics had the most advanced food-processing industry before the fall of the Soviet Union, but are now struggling because of the intense competition for food products within the European Union (EU). Ukraine and Belarus, and to a lesser extent Kazakhstan, Georgia, and Armenia, have sufficient domestic expertise with producing most common food items, as well as beverages. In all republics, the foreign presence is strongly felt (e.g., Coca-Cola, Pepsico, Nestlé, Unilever, and Kraft Foods), but many Russian companies are making their presence known as well. WBD, for example, routinely prints its food labels in the Russian, Ukrainian, and Kazakh languages, and has distribution networks in most of the 12 Commonwealth of Independent States (CIS) republics.

Textiles, Shoes, and Clothes

Soviet production of textiles focused on natural fabrics from domestic sources (flax and later cotton, hemp, wool, and silk), although some synthetics were also made. The textile industry was based primarily in the Ivanovo region northeast of Moscow, along the Volga River. This was the traditional area of flax production, although by the late Soviet period by far the most common fabric was made of imported cotton from Uzbekistan. Silk (Naro-Fominsk, Tver) and wool (Moscow) had to be produced mainly from imported raw materials as well. Wool, for example, would come from Kazakhstan, Tajikistan, Azerbaijan, and the Russian Caucasus, where sheep were extensively raised. The U.S.S.R. would even buy wool in Australia, New Zealand, and Uruguay. Wool fabrics were used in civilian clothing, army uniforms, blankets, and some industrial processes. Although Russian fabrics were relatively expensive to make, they were very durable. Socks, underwear, dresses, pants, and shirts were primarily produced in a few large factories in Moscow and nearby cities (e.g., Smolensk, Orel). Today the Central federal district accounts for 84% of all fabric and 47% of all shoe production in Russia. The St. Petersburg area, the Urals, and parts of Siberia also have many factories making fabrics for local consumption.

This sector suffered the greatest decline of any industry during the reform period. The volume of fabrics produced in 2005 was only 33% of the 1990 level, production of socks was at 32%, and production of shoes was at a mere 12%! Why did this happen? There were many reasons. First, as noted earlier, the Soviet state viewed light industry (especially clothing and shoe production) as a necessity, but not a priority. Therefore, few investments were made in it, and by the time of perestroika the industry was already in deep chronic crisis. Second, most of the raw materials for this industry (about 90%) came from agriculture—a sector that itself entered a period of deep crisis (Chapter 20). Third, imports of cheap materials from the other FSU republics were interrupted. A fourth big factor was heavy competition from cheap Asian imports, which flooded Russian markets in the early 1990s. Yet another factor was the relative expenditure involved in converting old equipment for modern production lines in this business, in which profit margins are low. For the oligarchs, it made much more sense to invest in lucrative petroleum and metals immediately available for exports, not in clothing or shoes for the poor domestic market. When the state ended most of its subsidies, the industry was left alone to struggle with the economic hard times. The recent modest increases in production in this sector have not made much of a difference: Russia imports most fabric, clothing, and shoes it needs (Figure 19.3). Russia still produces only 0.3% of all shoes in the world, despite the slight increase in production. This situation, by the way, is not very different from that in the United States, which was pushed to abandon domestic production of most textiles and especially

FIGURE 19.3. Most shoes for sale at this market stall in Novosibirsk are made in China. China accounts for 80% of all shoe imports into Russia, Turkey 9%, and Italy 2%. *Photo:* P. Safonov.

finished clothing and shoes in the 1990s by the globalization pressures from Mexico, China, Brazil, and other developing countries.

Some traditional domestic textile products made in Russia are highly regarded: Woolen scarves and shawls from Pavlovo Posad and Orenburg, and linen towels from Smolensk and Ivanovo, are considered superior to foreign alternatives. Special editions of these may have embroidery patterns designed by well-known artists and may retail for hundreds of dollars apiece. A quick walk through downtown Moscow, however, reveals hundreds of boutiques selling mainly overpriced Western goods: high-end shoes, leather jackets, business suits, dresses, and lingerie by the leading European, Asian, and North American designers and manufacturers. Given the very steep prices, it is astonishing to see so many stores with so many products actually in stock. Someone is obviously buying at least a few of these items. The majority of Muscovites, however, do not shop at these boutiques; they crowd the outdoor markets on the city periphery. One of the biggest of those is in Izmailovo, east of downtown, where Vietnamese and Chinese goods are peddled by Uzbek and Azerbaijani vendors in an enclosed area the size of 20 Red Squares (about 0.66 km^2). This is where the ultimate bargains can be found.

Book Publishing

Although it is not a commonly discussed area of light industry, book publishing, of course, is an industry dependent on raw materials (paper, glue, paint, energy), machinery (computers and printing presses), and high-tech services (computer layout and design, editing, marketing). This is one of the few sectors of the Russian economy today that is positively booming. Its remarkable rise coincided with the beginning of perestroika and merits a separate discussion.

The Soviet Union printed a lot of books; in 1988 the Russian Federation alone printed 1.8 billion volumes. Lenin's collected works alone were produced in millions of copies per year, for both domestic and international consumption. Works by the classic Russian authors (Pushkin, Tolstoy, Dostoevsky, Chekhov) were continuously in print. Some privileged Soviet writers saw their books printed in hundreds of thousands of copies. Children's books were also printed, both small and large. A typical press run of a single Soviet-era book would be an astonishing 100,000 copies—a figure that only a handful of bestsellers approach in the West. However, the selection of titles was limited. Many genres were not published at all, and the works of some authors (e.g., émigrés and dissidents such as Nekrasov, Solzhenitsyn, Aksenov, and Voinovich) were explicitly banned. Other great writers whose works were unconventional, satirical, or critical of the Soviet Union (e.g., Bulgakov, Platonov, Fadeev) would have only occasional books in print. The Bible was printed just a handful of times in the entire 70 years of Soviet rule, and then just a few thousand copies to be used by the clergy only. Romance novels, mysteries, thrillers, and other "light reading" were virtually absent. However the U.S.S.R. published a good deal of specialized scientific literature and a surprising number of some Western classics, especially by 19th-century authors (e.g., Dumas, Verne, Dickens). The total number of titles published per year approached 70,000. For comparison, the entire English-speaking publishing world (the United States, the United Kingdom, Canada, Australia, New Zealand) produced about 375,000 titles in 2004.

Overall, the Soviet population was very literate. Children memorized many poems in school, and most adults read quite a lot of serious literature, which in the absence of much TV was a common pastime. However, the bookstores were often scantily stocked, and finding good books could require much skill and haggling on the black market (book selling was not officially allowed in the streets, but this was done in a few areas where the authorities would turn a blind eye to the proceedings).

Gorbachev's glasnost allowed the production of many underrepresented genres to soar (listed here in no particular order):

- Contemporary Russian prose, including some excellent new authors (Chapter 13).
- Contemporary foreign fiction.
- Romance novels, science fiction, fantasy, and thrillers.
- Large-format art volumes.
- Encyclopedias of various sorts.
- Adult magazines and fiction.
- Travel books.
- How-to books for home, car, and pet owners.
- Textbooks, including many translations from Western sources, as well as original works.

Figure 19.4 depicts the relative numbers of books published in some of the categories above. The list could go on. What is significant to note is how quickly this happened. In perhaps 5 years, the Russian street kiosks and bookstores went from almost empty shelves to thousands of high-quality, glossy, colorful, exciting, and engaging books, most printed domestically (Figure 19.5). Hundreds of publishing houses were set up quickly. Some (Eksmo, AST, and Drofa) grew to be giants with millions of copies sold per year. Others struggled and went out of business, shifted from serious literature to printing calendars, or found a narrow professional niche.

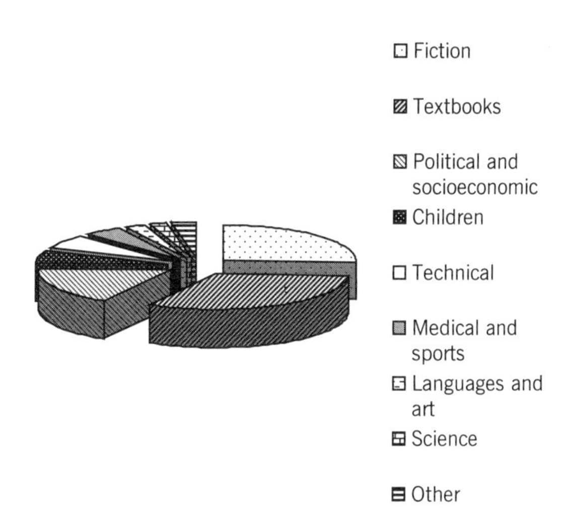

FIGURE 19.4. Textbooks and educational literature subsidized by the government predominate on the Russian book market; however, fiction and political and socioeconomic nonfiction are also massively printed. Data from Levina (2005).

A particularly Russian phenomenon of recent times is the emergence of "pseudotranslations." Publishers found out that, for example, a romance novel of 200 pages written by a certain Alice Smith and set in colonial India would sell many more copies than the same novel written by Nikolai Panov and set in the late 1970s suburbs of Petrozavodsk. So Panov (who might be a Moscow State University student with a major in philology) would publish the Indian novel under the pen name of Alice Smith, supposedly as a translation, and would make a few hundred dollars on the spot. The publisher would eventually pocket a large profit. Within the fiction segment, the top category now is domestic detective stories (about 15% of all titles), followed by foreign and "pseudotranslation" romance novels (7%). Serious novels and poetry are also popular. Overall, over 100,000 book titles were produced in Russia in 2004, although only 700 had press runs over 100,000 copies (Levina, 2005).

There was one big bottleneck in the Russian book-publishing business: While publishers proliferated beyond all initial expectations, the same printers were still operating in the same old Soviet factories. Typically found in the more forested parts of the Central district (Tver, Mozhaysk,

FIGURE 19.5. Street vending of books was particularly common during the Yeltsin period, but has since been mainly replaced with large bookstores. *Photo:* P. Safonov.

Yaroslavl), these printing monsters could print hundreds of thousands of copies per day, but only on cheap paper with inks of low quality. Books requiring color separations, high-quality bindings, or glossy paper had to be manufactured somewhere abroad—in Finland, Germany, Italy, or Singapore, countries specializing in high-end printing. In the early post-Soviet years, the Russian printers were also not equipped to deal with hundreds of individual customers and their individual projects; the resulting confusion added to the time and the cost. Gradually, however, the old printers found new investors, purchased better equipment, and reinvented themselves. Some smaller printing facilities offering more diversified services sprang up in other parts of the country. Russia's vast forest resources, relatively cheap labor, and huge customer base make it an ideal location for printing books. However, the cost of printing has been escalating quite a bit ahead of inflation, at 18–20% per year. By 2010 it is expected that the average book price in Russia will be equal to the average European price of 10 euros, while the average income even in big cities will still only be about one-third of the European level.

Moreover, although the publishing industry is booming as never before, the amount of serious reading in Russia has undoubtedly decreased (as it has in the rest of the world). Magazines, Sudoku, cable TV, MP3 players, and the Internet have all distracted Russian readers. Nevertheless, Russia remains a reading nation: It was in 25th place out of 81 countries in a United Nations Educational, Scientific, and Cultural Organization survey of the number of library books available per capita, ahead of the United Kingdom, Germany, or France. It is the 20th most literate country in the world out of 200 (99.4% literacy rate). Reading is the second most popular leisure activity for the Russians (14% do it, as opposed to 28% watching TV, in first place). However, about one-third of all Russians now claim that they never read—something unheard of in Soviet times.

One of the most successful Internet projects of the post-Soviet period is the Moshkov Library at *www.lib.ru*—a massive online library created, curiously, by a graduate of the very prestigious mathematics department of Moscow Sate University. The Website has over 5 gigabytes of text, including virtually all classical authors, many contemporary authors (foreign and domestic), and a wide variety of nonfiction (on tourism, computers, foreign travel, chess, etc.)—all available for free. Some titles that are available here may be copyrighted elsewhere, but there is as yet little perception in Russia today that publishing them online may be illegal. A large section of the online library contains thousands of self-published novels, ranging from very good to abysmally poor. Here is the ultimate freedom for the reading masses!

The FSU republics and the Russian diaspora across the world are two additional markets for Russian book publishers. Most major online bookstores (such as *amazon.com*) offer hundreds of Russian titles, and there are Russian retail bookstores all over the world in the biggest cities. Most Kazakhs, Ukrainians, and Georgians can still read Russian well, and many do so for business and pleasure. In these ways, publishing in the Russian language continues to sustain the global reach of Russian culture.

In summary, some sectors of light industry are in deep decline in Russia and throughout the FSU (e.g., clothing and textiles), while others are booming (e.g., publishing and food processing). Clearly, there has been a shift toward manufacturing consumer products that are locally in demand. Some of the new products are of high quality and reduce the need for more imports. At the same time, chronic underinvestment in traditional light industry has resulted in an employment slump in some regions. Overall, the light industry sector does not come close to rivaling the more established energy, metals, and heavy manufacturing sectors in Russia or the other FSU nations.

REVIEW QUESTIONS

1. What food products were not commonly available to Soviet consumers? Where were the greatest opportunities after the fall of Communism?
2. What are the main geographic factors that determine where textile industry may locate?
3. Why did the Ivanovo area become the leading producer of fabrics in Russia?
4. Why is Russia a good place for publishing books?

EXERCISES

1. Compare a leading food producer in Russia (e.g., WBD) with a similar producer in the United States. What are the commonalities in where and how they do business? What are the differences? Can you find any products from the Russian producer in your local food store(s)? Why or why not?
2. If you were asked for advice about starting a new publishing business in Russia, what would you suggest should be the focus? Where in Russia would you locate your business?

Further Reading

Levina, M. (2005). Knizhny rynok Rossii: Mify i realnost. *Otechestvennye Zapiski, 4*(24). Retrieved June 11, 2008, from *www.strana-oz.ru/?numid=25&article=1111.*

Tapilina, V. S. (2007). How much does Russia drink?: Volume, dynamics, and differentiation of alcohol consumption. *Russian Social Science Review, 48*(2), 79–94.

Vann, E. F. (2005). Domesticating consumer goods in the global economy: Examples from Vietnam and Russia. *Ethnos: Journal of Anthropology, 70*(4), 465–489.

Weir, F. (2006). Flax makes a comeback in Russia's mills. *Christian Science Monitor, 98*(51), 7.

Websites

www.baltikabeer.com—Baltika, the leading Russian beer brand.

www.eksmo.ru—Eksmo, the largest book publisher in Russia, with 11,600 new titles published in 2007 (15% of the Russian market). (In Russian only.)

www.nestle.ru—Nestlé's Russian division. (In Russian only.)

www.platki.ru—The Pavlovsky Posad factory, which has been making woolen and silk shawls and scarves since 1731. (In Russian only.)

www.wbd.com—Wimm-Bill-Dann (WBD) Corporation, the top producer of juice and dairy products in Russia.

CHAPTER 20

Fruits of the Earth

AGRICULTURE, HUNTING, FISHING, AND FORESTRY

Although only about 5% of Russia's gross domestic product (GDP) is produced by agriculture and another 5% by forestry, these two activities are strategic. These primary sectors of the economy are sometimes dismissed as "primitive" or even "irrelevant" by sophisticated postindustrial economists. Yet all of us need to eat. We need lumber and paper provided by forestry. In Russian society 100 years ago, 80% of the people were peasants. These people lived close to the land, growing food and cutting timber. Fishing and hunting supplemented protein from domestic meat sources. Today 15% of workers in Russia are employed in forestry or agriculture; this remains a much higher rate than in the West, where it is under 3%, but it is of course much lower than 100 years ago. Some discussion of village life and of settlement patterns in rural Russia and the U.S.S.R. has been provided in Chapter 11. Here I consider the impact of recent reforms on the current situation in agriculture, hunting, fishing, and timber harvesting.

Soviet Agriculture and the Post-Soviet Transition Period

Agriculture is one of the three main sectors of the economy, along with industry and services. It is indispensable for any country. Even if some food must be imported, it is always a good idea to rely on local sources for most staples—grain, milk, and meat. Agriculture includes farming and ranching, along with some less important areas, such as beekeeping and aquaculture. Ironically, the steady improvements in agriculture during the worldwide Green Revolution have put this sector at a disadvantage in all countries: As more and more efficient methods of growing food were introduced, fewer and fewer hands were needed to work on the land (Figure 20.1). This has resulted in the decline of the family farm, an exodus of cheap labor to the cities, and a sharp decline of the agricultural sector relative to the other two. This is of course a familiar scenario in the United States today, but it is also now happening in India, China, Mexico, Brazil, and even Africa.

The agriculture of the Soviet period was dominated by two forms of state farms: *kolkhozy* (Chapter 11) and *sovkhozy*. The former were collective farms made up of village farmers and, theoretically, cooperatively owned, and the latter were Soviet agricultural enterprises with state workers. Few actual distinctions existed between the two in the late Soviet period. Another time-honored tradition inherited from the Soviet period was the suburban *dacha* (Vignette 20.1).

FIGURE 20.1. Farmers in Arkhangelsk Oblast still harvest hay the same way they did for centuries, but this way of life is vanishing because of modern improvements in agriculture. *Photo:* A. Shanin.

Because the collective farming was notoriously inefficient, people were tacitly encouraged by the authorities to take care of themselves and to grow their own food. Small plots of land (averaging 0.06 ha) were grudgingly given out by the Soviet authorities to the urban residents, so that some food could be grown around cities. Vegetables and potatoes were most commonly produced, and sometimes apples or flowers (Figure 20.2). Villagers had slightly larger plots of land (usually 0.10–0.20 ha) immediately next to their houses to grow their own food. While all of this land was state owned, people were free to choose what to grow on their dachas, and they were doing it for themselves. This resulted in surprisingly high yields: A typical family of the late Soviet period would grow and can enough vegetables and fruit to last for about half of the winter. These tiny plots yielded an astonishing 30% of the total agricultural produce in the country in 1980, and yield even more today. There was not enough land, however, to grow wheat or corn on dachas. Livestock was also kept by the villagers close to home; in a collective farm setting with 2,000 state-owned cows, a few households would manage to keep a cow or two of their own. Pigs and goats could be raised at home as well. The urban dacha owners would typically not be able to keep big animals, because they could only be there during the summer vacation and on weekends, but even they sometimes managed to raise a few chickens or rabbits.

Because Soviet agriculture was so inefficient (Chapters 7 and 11), the Soviet Union had to import about one-fifth of its total calories by the early 1980s, making it the largest single importer of food on earth. The most common imports were wheat from Canada; sugar from Cuba; and vegetable, fruit, and meat from the Eastern European countries. In exchange, the U.S.S.R. sold grain, fossil fuels, timber, fertilizer, and metals on the world markets. In a sense, little has changed for Russia today. During the 1980s, the Soviet Union grain exports alone amounted to 30 million metric tonnes (mmt) *per year*, about as much as was grown in all of Turkey. The total grain production was then about 150 mmt per year, while the United States produced about 300 mmt. About one-quarter of all economic expenditures in the Soviet Union were on food.

Agricultural reform was certainly on Gorbachev's agenda: The short-lived Food Program attempted to boost domestic production in the mid-1980s, during his tenure as one of the secretaries of the Central Committee of the Communist Party. However, given the absence of economic incentives to produce more food, the

Vignette 20.1. Valentina's Dacha

Valentina is a busy woman. She raised five children and now has six grandchildren. She is retired, and although she lives in Moscow, she spends half of the year on her beloved dacha about 50 km east of Moscow. Beside the small cabin (which she and her husband expanded into four rooms from the original two), she has about 0.03 ha of land to farm. This is admittedly not much, but every square meter is diligently cultivated. Her husband holds three jobs in the city to keep the family above the poverty line. Nevertheless, *some* food must come from the tiny plot of land that she cultivates. She leaves the city in late spring, after the snowmelt in April, and returns in October. She sometimes travels to the city for shopping, but mostly she quietly spends time on her dacha planting, digging, hoeing, watering, weeding, and harvesting. She plants common crops: lettuce, cucumbers, carrots, beets, cabbage, herbs, strawberries for the grandkids, and other staples. The climate near Moscow is too cold to produce good tomatoes or eggplant, so she does not plant those; she does not have enough land for potatoes, either. She does like to have a few rows seeded in wild and garden flowers to keep the place pretty. She claims that her chores keep her healthy. Two of her grandchildren live in Moscow and spend most of their summers with their grandma. They also help with chores, although they are both preschool age. Figure 1 illustrates where things are grown on Valentina's dacha. Do not take the dacha lightly; half of all food produced in Russia is grown on plots like hers!

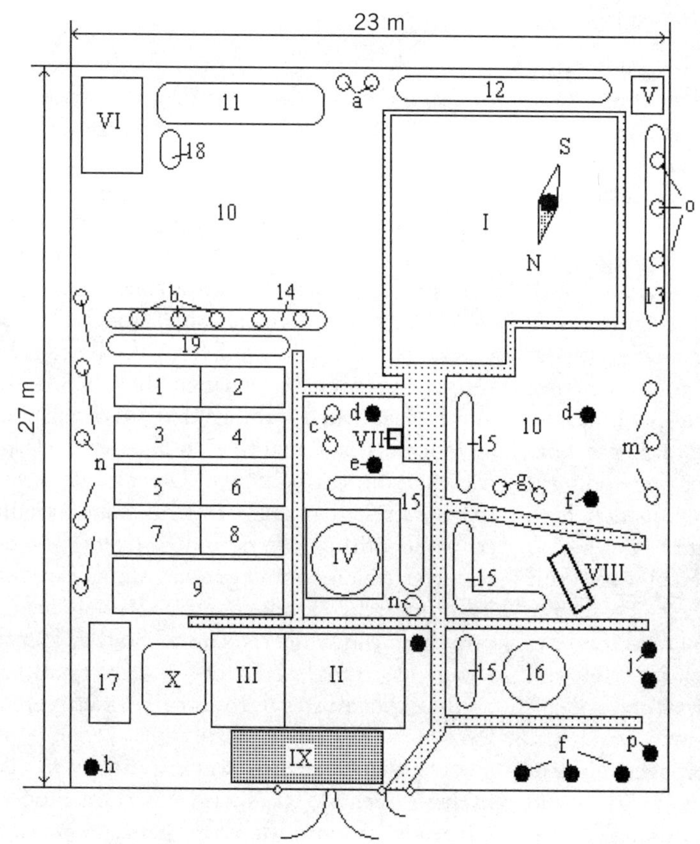

FIGURE 1. Dacha plot layout of 0.06 ha (about one-seventh of an acre). Buildings: I, house; II, sauna; III, storage shed; IV, sand pit for the grandkids; V, outhouse; VI, compost pile; VII, well; VIII, pond; IX, parking area; X, woodpile. Vegetables and flowers: 1, garlic; 2, radishes and cucumbers; 3, lettuce and dill; 4, beans; 5, turnips and cabbage; 6, peas; 7, carrots; 8, onions; 9, strawberries; 10, meadow; 11, raspberries; 12, lupins; 13, other annual flowers; 14, other annuals and spring wildflowers; 15, perennials; 16, alpine plants; 17, tulips; 18, zucchini; 19, parsley. Shrubs and trees: a, wild rose; b, black currants; c, red currants; d, apple tree; e, pear tree; f, juniper; g, Japanese quince; h, lilac; j, pine tree; m, hazelnut; n, jasmine; o, more lilacs; p, linden tree.

FIGURE 20.2. Dacha plot near Moscow. *Photo:* Author.

state farms were reluctant to grow more. Real agrarian reform would require many years of the post-Gorbachev period, and is by no means complete even today (Wegren, 2005). Although the reformers' initial hope was to quickly create millions of private farms, more or less U.S.-style, this effort has been largely unsuccessful to date. Russia had only 300,000 private farms in 2008, producing 9% of the total agricultural value. The largest producers in the country remain the former state farms, with over 50% of the total output. They control 76% of the land in cultivation and are now reorganized into "stock ventures," with much of the ownership concentrated in the hands of a few people who have ties to the farm directors (Wegren, 2005). Workers do own shares in the farms, although few are able to meaningfully exercise their stock owners' rights. In contrast, private farms control only 20% of the land, and the small private plots 4%. Table 20.1 details the output by the three main forms of ownership in 2008. It is interesting to note that despite some gains for the private farms and some decline for the big agricultural enterprises, the people growing *their own food* on dachas are the ones who continue to produce the bulk of Russia's potatoes and vegetables, and a large share of its meat.

The agricultural sector was hit hard by Yeltsin's reforms. The state subsidies were abruptly discontinued; competition from Western producers surged; and seed stock, fertilizer, pesticides, and equipment became prohibitively expensive. The overall agricultural output fell by 30% between 1990 and 1995, and an additional 10% by 2000. There was a proportional dropoff in the harvested acreage: Today Russia harvests only 65% of its former fields. The more recent period has seen a slight increase in production, at the rate of about 3–4% per year. This, however, is not yet enough to make a real difference in the trend. Russia has about 77 million ha under cultivation, a decline from the 118 million cultivated during the late Soviet period. Much of the abandoned land has been reverting to forest, as can be easily seen on satellite imagery collected over the past 15 years (Kuemmerle et al., 2008). Much of the farmland in the former Soviet Union (FSU) is also heavily polluted with pesticides (e.g., areas of the northern Caucasus, Moldova, and the "black soil" belt of Ukraine and Russia). Still, Russia has the third largest amount of arable land in the world—behind the United States and India, but ahead of Canada or China. There is 0.8 ha of arable land available per citizen, as compared to 0.6 ha in the United States or 0.09 ha in China.

TABLE 20.1. Structure of the Russian Agricultural Sector (%) with Respect to Ownership and Output Levels

	1995	2000	2005	2008
Grain				
Agricultural enterprises (former kolkozy)	94.4	86.9	80.2	78.1
Small private plots	0.9	0.9	1.5	0.5
Private family farms	4.7	8.4	18.3	21.0
Potatoes				
Agricultural enterprises (former kolkozy)	9.2	6.5	6.3	11.4
Small private plots	89.9	92.4	91.6	83.5
Private family farms	0.9	1.1	2.1	5.1
Vegetables				
Agricultural enterprises (former kolkozy)	25.3	19.9	14.0	19.2
Small private plots	73.4	77.9	80.3	70.7
Private family farms	1.3	2.2	5.7	10.1
Meat and poultry				
Agricultural enterprises (former kolkozy)	49.9	40.3	46.9	52.2
Small private plots	48.6	57.9	50.7	44.7
Private family farms	1.5	1.8	2.4	3.1

Note. Data from the Federal Service of State Statistics, Russian Federation (2009).

Patterns of Agricultural Production Today

This section describes the current production patterns in Russia (and, where relevant, other FSU nations) of the main agricultural products: grain; sugar and oil; potatoes; tobacco and tea; vegetables and fruits; and meat and poultry.

Grain

Before the Bolshevik Revolution, three grain crops were primarily grown in Russia: wheat, rye, and oats. Rye and oats can grow all the way to the Arctic Circle in European Russia; wheat can grow from Moscow to the Crimea. Barley began to be planted after World War II to provide cheap rations for the Soviet Army, and also for making small quantities of beer; it can grow even north of St. Petersburg. Corn and soy were introduced under Nikita Khrushchev in the late 1950s, after he returned from his famous U.S. tour. These warm crops cannot be grown in much of Russia because of the cold climate, and so they are limited to the extreme south of the country. Figure 20.3 illustrates Russia's total grain production.

Wheat remains the main grain (just a little under 50% of the total production by volume). Both the spring and summer varieties grow in Russia. Spring wheat has twice the yield of summer wheat, because it can use autumn rain and spring snowmelt, and is the one more widely planted. Overall, Russia harvested 45 million metric tonnes (mmt) of wheat in 2004—about half of China's level, and below that of India or the United States, but ahead of Australia or Canada. Spring wheat is primarily grown in western Russia, where winters are snowy but mild. Summer wheat is more common in the drier parts of the country farther south and east (e.g., in the lower Volga, southern Urals, and central Siberia). The largest wheat producers are the Kuban,

FIGURE 20.3. Russia's grain production. Each dot represents 100,000 metric tonnes harvested in 2005. Data from the Federal Service of State Statistics, Russian Federation.

Stavropol, and Rostov areas of southern European Russia, and Altaysky Kray in Siberia. Some of the best land near Kuban can produce 4 tonnes of wheat per hectare.

Barley is the second most common grain, grown on 20% of all acreage. It is a hardy grain that can grow far to the north (e.g., in Karelia) and in the dry south (e.g., in Kalmykia). Russia is the biggest producer of barley in the world, harvesting over 17 mmt per year. Canada and Germany harvest about 12 mmt each. Barley production is important for beer making and cheap cereals. Some barley flour is added to bread. Another common northern grain is oats. Russia is again the world's leading producer, harvesting about 5 mmt per year. Oatmeal is a popular cereal, and people in some rural areas still use horses for transportation and feed oats to the horses.

Rye is the oldest crop continuously cultivated in Europe. It can grow under the coldest conditions on poor and acidic soils. Because of these properties, it has been historically grown throughout northern Russia in the forest zone. Rye bread plays a large role in the traditional diet of Eastern Europeans. Some Russians claim that rye grain is the best source of alcohol for vodka. Russia is one of the three largest producers of rye in the world, at about 3 mmt per year, along with Germany and Poland. In comparison, the United States grows only 200,000 tonnes.

Corn (maize) can grow in Russia, but only in a limited area in the extreme south, where the vegetative season is at least 5 months long. Corn generally requires about 2,500 degree-days to develop, as compared to merely 1,000 for rye. Corn is an important source of livestock feed; it also can be processed for flour or syrup, or used directly in cooking. More recently, some regions in Russia have started processing corn for ethanol fuel. Whereas the United States produces almost 268 mmt of corn per year (almost 1 tonne per U.S. resident!), Russia only manages to produce 3.6 mmt, or about 1.5% of the U.S. level. The main corn-producing area of Russia is in the southern European part, in the Kuban and Stavropol areas. Ukraine is better suited for corn production, with about 6.3 mmt harvested in 2006. Corn chips are still uncommon in the FSU, partially because little corn of the right quality is available.

These southern areas are also best suited for soy production. The largest areas of soy production are concentrated in the Primorye region of the Far East, where the wet monsoonal climate helps its growth. Soy originates in China, so it is not surprising that it does so well in the Russian Far East along the Chinese border. Russia produces very little soy (about 0.8 mmt), as compared to the United States (producing 88 mmt per year) or Canada (producing 3.5 mmt). Thus soy must be imported. It is nevertheless one of the potentially lucrative new crops for Russia, and its production is expected to increase in the future. Soy is versatile; biodiesel and dozens of other products can be made from it. It also requires little fertilizer and needs less water than corn.

Rice is produced in a handful of places, mainly along the valleys of the large European rivers in southern Russia and in the Far East. Most rice must be imported, however, because there is simply not enough suitable land or high enough temperatures to grow adequate amounts of this essentially tropical crop.

Another grain that is culturally and economically notable in the FSU is buckwheat. It thrives in the northern forest–steppe zone of the European part. Unlike the other grains (except soy), which are wind-pollinated grasses, buckwheat is not a grass and is pollinated by bees. When bees are present, its yields are much higher. Unfortunately, the areas most suitable for buckwheat growth are also the zones where much of the chemical industry is located (Tula, Nizhniy Novgorod, Ryazan). Air pollution from chemical factories has a strong negative impact on bees; thus buckwheat yields are low in Russia. Russia still produces by far the most buckwheat in the world—about 850,000 tonnes per year of the total 2.2 mmt. The second biggest producer of buckwheat is Ukraine, with 200,000 tonnes.

Oil and Sugar

Oil-producing crops in the FSU include, first of all, sunflowers. Rapeseed (canola), corn, hemp, and a few other plants contribute small amounts of vegetable oil as well, but sunflowers are the traditional oil crop of Eastern Europe, providing over 80% of all vegetable oil. About 4.1 million ha in Russia are planted in sunflowers—mainly the southern part of European Russia along the border with Ukraine. Russia is the top sunflower producer in the world, accounting for nearly one-quarter, with Ukraine in second place. Collectively, the two countries produce about 8 mmt of sunflower seed out of 26 mmt harvested worldwide. Recently there has been an interest in producing biodiesel from all oil-rich plants, and their cultivation is expected to increase.

The main source of sugar in the FSU has traditionally been sugar beets. Most of this production takes place in the steppe and forest–steppe zones of the western European part. Sugar beets require a fair amount of heat (at least 2,000 growth-degree days), about a 5-month growing season, and plenty of moisture. Russia is in 4th place worldwide in sugar beet production, with about 21 mmt produced annually; it trails France, Germany, and the United States, which produce about 28 mmt each. Some additional sugar has to be imported from tropical countries, where it is made out of sugar cane. In the Soviet period, Cuba was the leading supplier of this type of sugar. Unlike the United States, Russia does not use corn syrup as a sweetener. This results in a distinctly different taste of Russian-made processed foods, pop, juices, and candy, as well as in a higher demand for sugar. Although the difference is hard to describe, a person who has tasted both will know the difference.

Potatoes

The all-important food in the region is potatoes (Figure 20.4). Introduced under Peter the Great, they have become the main Russian dietary staple. Potatoes are consumed boiled, baked, mashed, and fried. French fries and potato chips are becoming increasingly popular as well. Potatoes are an excellent source of starch and can also be processed for ethanol. Because potatoes originated in the Andes, they require a relatively short growing season, about 120 days. They prefer moderately cool summers and little water, so in fact much of European Russia has a perfect climate for them. In the 1970s over 4.4 million ha were planted in Russia with potatoes; today the acreage is smaller, about 3.1 million ha. About 11 tonnes per hectare are commonly produced, as compared to over 40 in the United Kingdom or the United States. About 40% of all Soviet potatoes came from large state farms, but today

FIGURE 20.4. Potatoes are the staple crop of Russia. *Photo:* I. Tarabrina.

only 11.4% do. Much of the Soviet large-scale production of potatoes was notoriously wasteful; in the fall, hordes of college freshmen and soldiers would be called into the fields for a few days of intensive harvesting. Well over 80% of all potatoes grown in Russia today are now grown by people on their small private plots. Worldwide, Russia is in second place with respect to potato production (after China), growing 39 mmt per year. Ukraine and Belarus together produce an additional 30 mmt, while the United States produces about 20 mmt.

Tobacco and Tea

Russia grows some of its own tobacco and tea in the northern Caucasus and along the Black Sea coast. Both are of relatively low quality, because these are subtropical crops that require plenty of heat during the growing season. Russia is a country of heavy smokers; 65% of its men smoke, as compared to 35% in France or 22% in the United States. Fewer Russian women smoke (about 10%), but their number is increasing (World Health Organization, 2007). Russia consumes over 250 billion cigarettes per year; it is in fourth place worldwide after China, the United States, and Japan in this regard, and must satisfy much of the demand for tobacco via imports. Russia is also one of the leading tea consumers in the world. It can satisfy less than 5% of this demand from domestic sources and must import the rest, primarily from Sri Lanka and India, where large plantations are leased by Russian companies.

Vegetables and Fruits

Russia grows a lot of vegetables—about 13 mmt per year on about 4.2 million ha. As with potatoes, most of these vegetables are grown on domestic plots. Fewer than 20% of all vegetables are produced on large farms. The main zones of vegetable production are located near big cities—for example, around Lake Nero (Rostov) east of Moscow, in the Oka and the Moscow valleys, and in the floodplains near St. Petersburg. The biggest production of tomatoes, cucumbers, watermelons, and other crops needing warm weather and a lot of water occurs in the lowest reaches of the Don and Volga, in the Stavropol, Rostov, and Krasnodar areas. Many essential vegetables are additionally imported from the European Union (EU) and other countries, including even dill and cucumbers, which could be produced domestically.

Russia grows a diverse array of fruits on about 1.2 million ha. Before the Revolution, there were over 200 varieties of apples alone. Unfortunately, the Soviet emphasis on mass production resulted in the disappearance of some of the tastiest ones. I. V. Michurin's (1855–1935) experiments resulted in a number of sturdy new Soviet varieties. He worked out a theoretical basis and practical means for hybridizing geographically distant plants, with good results. However, it is now known that in the process of overly zealous selection, he damaged or destroyed some of the good preexisting varieties. The other fruit staples widely grown in the FSU are pears, sweet and sour cherries, plums, black and red currants, gooseberries, raspberries, aronia, and a few others. In the warm valleys of Central Asia and in the northern Caucasus, apricots, peaches, quince, walnuts, and grapes are additionally grown.

Almaty, the former capital of Kazakhstan, literally means "[City] of Apples." Its legendary Aport apple variety produced fruit the size of a small football, weighing over 1 kg. Tragically, its commercial production collapsed along with the Soviet Union: Some of the best orchards were sold to developers of elite residential cottages, and others became the victims of neglect as their former owners left the country or switched to less demanding forms of agriculture. Overall, the biggest producer of fruits in Central Asia is Uzbekistan, because it has the largest share of the irrigated Fergana Valley. Some parts of central Siberia, especially the Russian Altay, have small areas where the microclimate is good for orchard crop production as well. The Russian Far East likewise has many unique varieties of orchard crops.

Virtually all Russian "viticulture" (cultivation of grapes for wine) is concentrated in the northern Caucasus, on about 72,000 ha. The Gorbachev antialcohol campaign of 1985 destroyed some of the best vineyards, however. Additional pressures came in the 1990s with the transitional period and increased competition from abroad (cheap imports from Australia, Chile, and Argentina, and not-so-cheap ones from France, Italy, and Spain). Besides Russia (Figure 20.5), notable wine production exists in the Crimea, in Moldova, and of course in Georgia. Although Russians only drink modest quantities of wine (7 L per person per year—about the same as in the United States, as compared to a whopping 55 L in France), its big population ensures it a spot among the top 10 wine-consuming nations worldwide. It produces about 310 million L of wine per year, and imports an additional 560 million L to satisfy domestic demand.

FIGURE 20.5. Sturdy varieties of vines can grow even in southern Siberia, as seen here in Biysk. All commercial Russian wine is produced along the Black Sea coast. *Photo:* Author.

Meat and Poultry

Let us now turn our attention to meat and poultry production. Russia is a Eurasian country, so its citizens are used to consuming a wide variety of meats—including chicken, beef, and pork, but also goat, lamb, horse, rabbit, duck, goose, and other animals in selected locales. An average Russian is accustomed to eating a greater variety of meats than an average American. The production of meat is a more complicated matter than production of grain, because it relies on so many factors—supply chains for feed, water, and vitamins; good shelter; and so on. Animals are thus expensive, and they are also slow to grow. They are the first agricultural products to fall victim to any economic perturbations.

As a result of poor management and lack of production incentives, the meat products of the Soviet period were substandard. The chickens produced by the state farms were notoriously skinny and tough (jokingly called "bluebirds"). The few types of beef, pork, or lamb sometimes available at state stores were mostly bones. One had to stand in long queues and depend on the mercy of the butcher to get a better slice. Prime rib, tenderloin, and other choice cuts were only available in the *nomenklatura* distribution centers (or for a bribe, if one personally knew a butcher).

Figure 20.6 illustrates the changes in the numbers of livestock and poultry in Russia between 1990 and 2008. During the difficult transition period after the fall of the U.S.S.R., animals were butchered and, for the most part, have not been replaced in adequate quantities ever since. In 2008 46% of all cattle were raised by large agricultural enterprises (the former kolkhozy), and 47.5% by individual villagers. Only 6% were raised by the modern Western-style farms. Most cattle are in the republics of Bashkortostan and Tatarstan in the Volga federal district, Altaysky Kray in Siberia (Figure 20.7), and Dagestan and Krasnodarsky Kray in southern Russia. Not surprisingly, all these regions have vast rangelands. Feedlots are rare; most cattle are free-ranging and grass-fed. It is interesting to note that in most areas in Russia, mixed dairy and beef production is the norm—unlike in the United States, Ar-

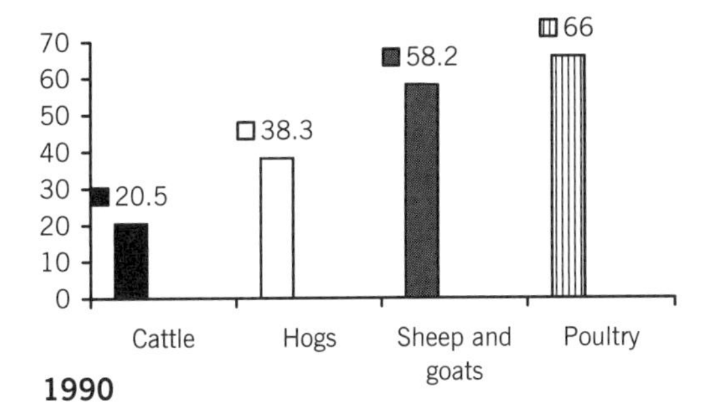

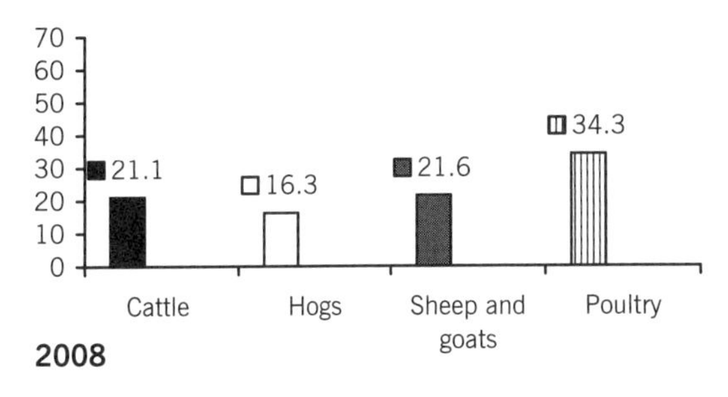

FIGURE 20.6. Data on livestock (in millions of live animals) and poultry (in tens of millions) in Russia in 1990 and 2008. Note that for most categories, the numbers were greatly reduced during the period of the reforms. Data from the Federal Service of State Statistics, Russian Federation (2009).

FIGURE 20.7. Cattle in the Altay. *Photo:* Author.

gentina, or Australia, where dairy cows are rarely raised in the same geographic areas as the beef cattle. One of the good places to observe Russia-like mixed cattle production in the United States is Minnesota, where dairy farms coexist with small-scale beef feedlots. The geography of hog production is similar to that of cattle, but most large farms are located in the South, especially in the Rostov-on-Don region.

Sheep and goats are most common in the republics of the northern Caucasus. Dagestan alone had over 3 million sheep and goats in 2008. Lamb meat and wool play an important role in the traditional Caucasian culture. Both fine-wool and coarse-wool breeds of sheep are raised. Central Russia's sheep specialty is the legendary Romanov breed from the Yaroslavl region, which has provided the best sheep hides for making winter clothing since the 18th century.

Outside Russia, the Central Asian and trans-Caucasus republics depend heavily on sheep and goats. Delicate wool fabrics are a local specialty and a source of great pride in Kyrgyzstan, for example. With about half of the country in high-elevation rangelands, wool production there makes perfect sense. Also, a meal in Central Asia is rarely served without lamb.

Ukraine and Belarus have many cattle and hog farms. Pig lard (*salo*) is an important component of the traditional Ukrainian diet. Some peoples of Central Asia, most notably in Kazakhstan and Kyrgyzstan, prefer horse meat (Chapter 14). Other unusual (to Westerners) forms of meat raised in the FSU include yak in Buryatia and Tyva, and elk in the Far East, Altay, and Khakassia. (The elk subspecies raised there is called *maral* and is famous not only for its meat, but for its antlers as well. Young antlers contain a lot of blood, which is used as an immune system booster in traditional medicine.) Reindeer meat is consumed locally by the peoples of Siberia and the north (the Nenets, Evenks, Evens, and others). About 2.2 million domestic reindeer browsed the Soviet Union tundras in the late 1980s. Today the numbers are lower, but remain high enough to sustain the native populations.

Poultry—chickens, ducks, geese, and turkeys—are common in the FSU. Since the 1960s, the national Soviet Ptitseprom program encouraged the creation of huge industrial complexes for breeding, raising, and processing chickens. These provided eggs and poultry meat to the Soviet masses. The top egg-producing region is Leningradskaya Oblast around St. Petersburg, with Krasnodarsky Kray second, and Sverdlovskaya Oblast around Yekaterinburg third. Clearly, egg production is geared toward consumers in the largest cities and must occur within a short distance from them. Moscow is supplied from about a dozen of the Central district's oblasts, with eggs coming from as much as 300–400 km away.

Although turkeys are raised in many places, the Eurasian goose was the traditional bird of choice for big feasts in the Slavic countries. Today only limited flocks of geese are found in peasant households, mainly in Ukraine. Ducks are raised throughout the FSU, but are rarely available in stores.

Food Imports and the Future of Russian Agriculture

Despite massive production of grains, oil, sugar, fiber, fruits, vegetables, meat, and poultry, Russia and most of the other FSU countries remain net food importers. In 2005 over $16 billion was spent by Russia to import food—almost 17% of all imports for the year. The cost went up to $35 billion by 2008, boosted by a national program (Wegren, 2009). Although for some African nations food constitutes one-third of all imports, for a typical European country food accounts for under 10% of imports (under 5% in the United States), and most of the imports are specialty items for which there are no domestic substitutes (e.g., deli cheeses, wine, or tropical fruit), not staples. In the FSU, only Armenia and Georgia have a higher proportion of food in their imports than Russia (about 18%). On the other hand, agriculture-heavy Moldova, Ukraine, and Kazakhstan spend only 10% on food imports and export some of their food. Kazakhstan is one of the leading wheat exporters in the world, while Ukraine exports wheat, sunflower oil, sugar beets, barley, buckwheat, and some fruit. Moldova is primarily a vegetable and fruit exporter, as well as a wine producer.

What foods does Russia import? About one-quarter of its food import budget was spent on tobacco and alcohol products, and 10% on tropical and subtropical fruits like bananas, mangoes, avocadoes, and citrus. Most of the remaining money (almost half, or about $15 billion per year) was spent on staples that Russia could grow in sufficient quantities itself: sugar, grains, non-tropical fruit, vegetables, and especially meat and poultry. Among the most popular import items from the United States are chicken drumsticks. Beef and pork are also imported in massive quantities from the EU or the United States. Butter, cheese, and some processed dairy products are significant import items as well. The top exporters of food to Russia overall are Ukraine, the United States, Germany, and France, as well as smaller EU countries with extra food capacity (Denmark, Finland, the Netherlands). In the Far East of Russia along the Pacific Coast, most food now comes from China, and some from Japan, South Korea, and Vietnam. A chronic inability to feed itself is a bad sign for any nation and has implications for national security. Neither Yeltsin's nor Putin's government was able to address this issue adequately. Recent politicalization of food imports has led to occasional sanctions against some exporting nations—supposedly for sanitary reasons, but frequently due to entirely political causes (Wegren, 2009).

Despite heavy imports, Russians today eat less protein than before the reforms. For example, an average Russian ate 69 kg of meat products a year in 1990, but only 41 kg in 2000; 20 kg of fish versus 10 kg; and 385 kg of dairy products versus 216 kg. Thus protein-rich products are not as easily obtainable now as before. Consumption of starchy foods (bread and potatoes) and junk food, on the other hand, has soared, raising fears of an imminent obesity epidemic. An average Russian now consumes 118 kg of potatoes per year, as compared to only 106 in 1990. For comparison, an average American consumes only 55 kg of potatoes, but 117 kg of meat per year.

What is likely to happen with Russian and other FSU agriculture in the future? In Russia's case, Ioffe et al. (2004) have documented profound depopulation of many old agricultural areas, especially in the European center and north. This process is likely to continue. However, it is not necessarily bad: Some of the most marginal farms are abandoned first, and transportation networks decline in the most disadvantaged areas (the so-called economic vacuum zones), which increases the concentration of food production near cities. Heavy investments in more intensive farming methods are likely to increase the overall production of food in the country eventually. The importance of food production on personal plots remains high; however, with better laws supporting private businesses, and national investment programs already in place, private family farms may receive a needed boost and are likely to continue to increase their share of the overall food production.

Many of the same processes are happening in other FSU republics. The Baltic farmers are facing increased opportunities, but also competition, within the EU framework. Ukraine is an agricultural giant with many unsolved internal political problems. Moldova and Belarus are agriculture-heavy economies as well; Moldova, however, is being increasingly marginalized as a trade partner by the Russian leadership. Most of the fruits, vegetables, and wine that Moldova produces can be purchased more cheaply or readily within the EU now, or even outside Eurasia. Few countries in Europe need Moldovan agricultural products, resulting in severe hardship imposed now on the republic.

The trans-Caucasian republics have an excellent climate for growing subtropical crops, such as tangerines and tea, but they must now compete with global producers (Spain, Morocco, Sri Lanka, etc.). They also have severe issues with transportation infrastructure and with ongoing military conflicts. The Central Asian states and Azerbaijan currently dominate Russia's fruit and flower markets, respectively. Internally in Central Asia, there is increasing competition between Uzbekistan and Kazakhstan for fruit and vegetable produce. Turkmenistan could produce more fruit and vegetables for export than it currently does, but is politically isolated. All of the Central Asian countries are heavy cotton producers. Their ranching is not at its best at the moment, but they all remain important producers of beef and mutton, for both domestic and Russian consumption.

Hunting and Freshwater Fishing

Domestic food production in Northern Eurasia is frequently supplemented by game and fish gathered in the wild. Siberia and the north have traditional hunter-gatherers and fishermen, for whom the game and fish of the taiga and tundra are their primary source of protein. Elsewhere, villagers hunt and fish to obtain extra protein because food from stores is expensive. There is a tradition of sports hunting among the political and business elite, as well as middle-class urban residents. Given the size of Russia's forests and steppe, and the number and extent of its rivers and lakes, both game and frreshwater fish are in plentiful supply.

The Russian tradition of hunting was well described by the classic writers of the 19th century. Leo Tolstoy's vivid description of hunting snipes in *Anna Karenina* is one of the best examples, along with the numerous hunting scenes from Ivan Turgenev's *Hunter's Sketches*. Wild game typical of Russia includes moose, elk, roe deer, brown bears, wild boars, capercaillie, grouse, quail, partridges, pheasants, and waterfowl. The peoples of the north also traditionally hunted walruses, seals, and whales, but today only the Chukchi and Inuit of the extreme northeast are allowed to do that.

A specialized form of hunting is trapping fur animals in the taiga. Russia is one of the top fur producers in the world, including such species as Arctic and other foxes, sable, hares, squirrels, mink, kolinsky, and muskrats. Although muskrats were only introduced in Eurasia from North America in 1928, Russia now has some of the largest and furriest muskrats in the world in the Selenga delta near Lake Baikal. In the steppes, bustards and other large running birds would be hunted historically, but they are now endangered. The Central Asian steppes have saiga antelope and wild kulan donkeys (both also now endangered). A specialized form of hunting in the Caucasus, Kazakhstan, and Kyrgyzstan is falconry, which brings in little food, but is a highly skilled form of hobby hunting.

Freshwater fishing is also common throughout the region. Over 50 species of fish were commercially harvested in Tsarist Russia. The short stories of N. Leskov and I. Shmelev have some vivid descriptions of the dozens of species available at noblemen's receptions. The most valued was sturgeon, including the huge beluga, weighing over a ton. Pike, perch, eel, and other fish were eaten boiled, fried, baked, smoked, dried, and in any other imaginable way. The traditional Siberian fish broth soup, *ukha*, is one of the fish specialties offered at many restaurants today. Much of the wild fishing declined in the 20th century because of water pollution, dams, and overfishing in the most developed parts of the region. The freshwaters of European North, Siberia, and the Far East remain remarkably productive, however, as even a brief visit to the fish counter at a local market in Russia will testify. The hardest hit were the Volga River sturgeon in the Caspian Sea basin, whose populations declined as a result of dam construction, water pollution, and caviar poaching.

Yet another form of wild harvesting is mushroom and berry hunting. A few Americans may hunt mushrooms on occasion, but many U.S. supermarkets will stock only the familiar button variety. Wild mushroom hunting is a national hobby bordering on an obsession in Russia, especially among middle-aged urban dwellers. Hunting mushrooms requires skill—knowledge of the correct places, the appropriate times, and edible kinds of mushrooms. Mushroom poisonings do sometimes happen (virtually all from consumption of the "destroying angel," *Amanita virosa*), but dozens of other varieties can be safely eaten. The king of the Russian woods is the white bolete, with caps sometimes reaching the size of a dinner plate. Mushrooms are consumed in soups and salads, and especially fried. Some people like to dry or can them for winter use. Wild strawberries, raspberries, dwarf blueberries, and lingonberries are also plentiful in most forests in Northern Eurasia. In Central Asia, there are forests where wild apples or plums can be found. The total amount of the wild mushroom and berry harvest is not known, but it may provide an important supplemental form of nutrition.

Marine Fisheries

Russia has one of the longest coastlines on the planet (about 37,000 km, mainly along the Arctic Ocean). However, its two main marine fishing areas are limited in extent: the Barents Sea in the European north, and the Bering and Okhotsk Seas of the Pacific. During the Soviet period, heavy investments were made in harvesting ocean fish: Salmon, cod, pollock, hake, sardines, herring, and many others were harvested from the coastal waters of the U.S.S.R., the inner seas (Caspian, Aral, Black, Baltic), and all over the world's oceans. In the late 1980s, the Soviet Union was the leading fishing nation on earth, surpassing Japan, Peru, and China in both the volume of catch and the size of its fleet. By 2004, however, Russia had dropped to sixth place in the amount of total seafood catch from marine fisheries—behind China, Peru, the United States, Japan, and Indonesia, but slightly ahead of Norway—with about 2.6 mmt caught. The main source of seafood in Russia is the Pacific coast of the Far East (about two-thirds of the total catch). Besides salmon and salmon eggs, the Far East is famous for its crab production, especially along the western shores of Kamchatka Peninsula. Russian king crab is available in U.S. markets, but it has been recently flagged by the environmental organizations as a poor consumer choice because of widespread poaching. Among other FSU republics, only Ukraine and the Baltic states have marine fishing of consequence according to the U.N. Food and Agriculture Organization (see **Websites**). One area of potential growth for all countries is aquaculture; very little of it is currently practiced, mainly in the form of raising carp in ponds.

Timber Production

Russia is about 50% forest-covered and has 20% of the world's timber supply. An average Russian has 5.2 ha of forests and 548 m^3 of timber available, as compared to the world's average of 0.9 ha (65 m^3) per person. Only Canadians have more forest acreage or timber per capita. Unfortunately, half of Russia's forest consists of larch, a hardy but scraggly species with low-quality wood. Practically speaking, Russia has few areas where timber production is profitable. Besides larch, the second most common tree is Scotch pine, followed by spruce and Siberian cedar pine. Overall, coniferous softwoods account for 82% of all standing timber. There are also large quantities of birch and alder (15%), but few hardwoods, such as oak, maple, ash, or basswood (<3%).

In the European part of the country, forestry is concentrated in Arkhangelsk Oblast and the Karelia and Komi Republics in the north. These areas are convenient sources of wood for domestic and European markets, especially Finland, Sweden, Austria, and Germany, where mills need logs and wood chips. Most timber harvesting in Siberia takes place in the central part, especially in Krasnoyarsky Kray near the Trans-Siberian Railroad and along the Yenisei. Much of this wood is consumed domestically, but increasingly large quantities are exported to China and Japan. In the Far East, production is concentrated in the Ussuri River basin and the Amur River watershed, close to Asian markets.

Until the new Forest Code went into effect in 2006, virtually all Russian forests (96%) were federally managed by the Ministry of Natural Resources. The rest were managed by the Ministries of Agriculture, Defense, and Education, as well as by local municipalities. Since the adoption of the new code, private ownership of forests finally became a reality in Russia. The new law allows limited outright ownership of forest land and encourages long-term leases, with minimal penalties for permanent degradation of the landscape. It is still unclear how much forest will become privatized, but potentially it represents one of the biggest business opportunities as well as environmental threats. So far, the most common scenario has been privatization of forests for suburban construction of recreational facilities, including housing. Technically speaking, forests in this case should not be cut, and no permanent structures should be allowed. In reality, construction results in major clearing of timber in the vicinity of big cities, and large houses placed deep in the woods. This is happening all around Moscow (Boentje & Blinnikov, 2007) and other major cities. Undoubtedly, large timber companies will also privatize some of the forest land now. Currently they are leasing over 80 million ha of forest lands (usually for 49 years), but many will want to get a permanent title now. At the same time, there is a strong public sentiment in Russia against private ownership of forests; this goes against a centuries-old tradition.

In 2004 Russia was in third place worldwide in timber production—behind the United States and Canada, but ahead of Brazil and China—with about 134 million m^3 of industrial roundwood (logs) produced. In terms of sawnwood (lumber) production, however, it was only in fourth place—again behind the United States and Canada, but also behind the lumber giant Sweden, and only barely ahead of Germany, Finland, and Japan—with about 21 million m^3 produced. Thus Russia processes only about 15% of all timber into lumber (Figure 20.8) as compared to 30% in Canada or 21% in the United States. This discrepancy reflects a history of underinvestment in the more labor-intensive, but also more lucrative, lumber mills. In the Soviet period, forestry was entirely state run, and its efficiency was low. The regional forestry units (*leskhozy*) in charge of inventorying, planting, and growing forests were supposed to grow healthy trees in sufficient quantities, but took no part in cutting or processing timber. The *lespromkhozy*, were the local logging units responsible for timber harvesting and processing, but their profits were fixed, regardless of quality or even quantity of lumber produced. Lumber-milling equipment was in short supply and of poor quality. Selling raw logs abroad brought less money than selling sawn lumber, but the Soviet state made do with it because roundwood brought an immediate profit in hard currency with little domestic

FIGURE 20.8. Only about 15% of all wood cut in Russia is processed into lumber. Shown is a lumber yard in Novosibirsk. *Photo:* Author.

investment. As the forestry sector went through privatization in the 1990s, there was a drastic reduction in the amount of timber cut, and almost no investments were made in new logging or lumber equipment for about a decade. Today Russia remains primarily an exporter of raw logs, not lumber. Swedish, Finnish, Japanese, and increasingly South Korean and Chinese lumber mills are the eager buyers of these logs and the real beneficiaries of this arrangement.

By 2000 the forestry sector in Russia was mostly privatized, although only 36% of all enterprises were completely free from partial state ownership. There were about 3,000 forestry companies registered in the country, but only a handful of big ones. The actual timber harvesting was done both by the state-run units (18%) and by private logging companies. The production of timber, lumber, and paper in Russia is dominated by a few very large companies; the biggest, Ilim Pulp Enterprise, accounted for 7 million m^3 of harvested timber (about 5% of the nation's total) in 2000. The largest timber processors are paper and pulp mills (PPMs) in the taiga zone, including the Arkhangelsk, Syktyvkar, Svetogorsk, Solikamsk, Kondopoga, Segezh, Ust-Ilimsk, Solombala, and Baikal PPMs, as well as others. Most are located in Arkhangelsk Oblast and the Karelia and Komi Republics in the European north, and in areas of central and eastern Siberia. Another important group includes large lumber processors (e.g., the Onega, Ust-Ilimsk, and Lesosibirsk lumber mills). Yet another group consists of former lespromkhozy that survived the reforms and are now organized in regional cooperative units (Karellesprom, Irkutsklesprom, etc.). This is geographically the most widespread group, with various production units of smaller size in most forest-covered regions of Russia (Karpachevsky, 2001).

Forestry in Russia remains dependent on a large pool of less skilled workers. It employs almost 1 million people. About two-thirds of all timber is exported, primarily to the EU, as well as to Japan, South Korea, and China. According to expert estimates, almost one-quarter of the timber exported in 2000 was sold illegally, without payment of export duties. The total loss to the treasury from this was conservatively estimated at $1 billion in U.S. dollars (Karpachevsky, 2001). The Russian forest industry also now has heavy foreign participation: In 2000 a few dozen large lumber mills, cardboard factories, or PPMs were controlled partially or wholly by foreign companies. The majority had mixed ownership; for example, one-fourth of the Syktyvkar lumber mill was owned by JPMorgan Chase Bank (U.S.), Burlington Investment Ltd. (U.S.), and Frantschach AG (Austria). International Paper (U.S.) wholly owned the Svetogorsk PPM in Leningrad Oblast near the Finnish border. Sixty-five percent of the huge Arkhangelsk PPM was owned by German, Dutch, and Austrian companies. Many enterprises were partially owned by the Finnish giants UPM-Kymmene and StoraEnso. Swedish and French companies were also prominent among the stockholders in Russian forestry enterprises.

Paper and pulp account for about 45% of the total revenue of the forestry sector, and 60% of all profits. The demand for paper and paper products domestically continues to soar, in part because of the increase in book publishing (Chapter 19) and in part because of the appearance of disposable products (tissue, diapers). Global paper consumption is likewise on the rise, and Russia is one of very few countries where production can grow to meet the rising demand.

There are about 2,800 furniture factories in Russia. Most are small local producers surviving from Soviet times, and many are emerging from bankruptcies caused by the economic restructuring. Collectively, they capture about 20% of all revenue in the forestry sector. Although 70% of the domestic furniture consumption is satisfied by them, the rest of the furniture has to be imported—mainly from European furniture manufacturers, both budget and luxury brands. The opening of the first Swedish IKEA store in Khimki, near Moscow, in 2000 greatly raised the competitive stakes for the domestic companies: IKEA alone reportedly sold $100 million worth of home and office furniture in the first year in Russia (about 8% of all sales for that year). More IKEA stores are being built near other big cities.

One critical issue facing the forestry sector is long-term sustainability. Soviet forestry was hardly sustainable, but it lacked the logging equipment required to inflict major damage on a large scale. Large intact areas of frontier forests still remain in the north of Russia, especially in the Komi and Karelia Republics, Murmansk Oblast, much of Siberia, the Altay, and the Sayans. Most of these are found away from the railroads or rivers, in the mountains, or within protected natural areas. Nevertheless, "the Russian forest is no longer a boundless belt of unbroken wilderness" (Aksenov et al., 2002, p. 10): Only 33% of Russian forests remain as intact, old-growth areas of sufficient size to permit uninterrupted natural cycles on the millennial scale (called "frontier forests"). Active reforestation is rare. Forest fires, about 80% of which are human-caused in Russia, devastate millions of hectares per year. With the improved economy, better equipment, and growing demand, the Russian companies are expected to increase logging greatly in the future. Areas along the Finnish and Chinese borders are already the most affected by clearcuts. Only a few timber producers in Russia are making efforts to get their wood production certified according to the ISO 14001 standard or to pursue Forest Stewardship Council certification; most of these are actually foreign-owned businesses, afraid of consumer backlash back home. Russian businesses must learn to be environmentally responsible too.

Among other FSU nations, only the Baltic republics and Ukraine have significant forestry industries. The Baltic states cut their own limited wood and reexport Russian wood and lumber to the global markets. They are well positioned geographically, with good seaports and easy railroad links to Russia (with which Estonia and Latvia share a border). Ukraine has a limited wood supply in the Carpathians and the Crimea, but many of those forests are in protected areas or are reserved for tourism. Kazakhstan and Kyrgyzstan have substantial forests in the mountains, but both suffer from lack of investment in processing of timber and from poor transportation linkages with China and other potential markets. The rest of the Central Asian and Caucasian republics have only small portions of forested territory and are wood importers.

REVIEW QUESTIONS

1. Name the main two types of agricultural enterprises during the Soviet period. How would they be different from a small family farm in the United States or France today? How would they differ from the large agribusinesses common in the West today?
2. What are the main grains grown in Russia? Where do they grow, and why are these specific ones grown?
3. Explain why U.S.-style family farms have had a hard time emerging in post-Soviet Russia.
4. What are the main forms of meat eaten in the FSU? How do these differ from those eaten in your home country? What might be the reasons?
5. What areas of the FSU have good freshwater fishing potential?
6. Name some important seafood harvested by Russia. Can you find any Russian seafood imports at your local store?
7. Explain what role hunting plays in the Russian culture and economy. What are the commonly hunted animals? Are they similar to the ones hunted in the area where you live?
8. What are the main timber-producing species in Russian forests?
9. Why are so few logs processed into lumber in Russia?
10. What needs to be done to improve the long-term sustainability of the Russian forest industry?

EXERCISES

1. Search FAOSTAT (see Websites) for any three agricultural products in any three FSU republics over a number of years (e.g., 1990, 1993, 1996, 1999, 2002, 2005). For instance, try wheat, watermelons, and cattle in Russia, Ukraine, and Kazakhstan. Graph the data and describe the patterns that you see. What may be the reasons for the similarities and differences?
2. If you live in an agricultural area, find out whether any local company or organization has done scientific or technological exchanges with Russia or other FSU nations (e.g., methods of low-impact tillage, organic farming, pest management programs, marketing, etc.). What were the results of such exchanges? Make suggestions on how to improve such exchanges in the future.

3. Use Google Earth to investigate the extent of clearcuts in the Karelia Republic of Russia. A good place to start would be areas west of Kondopoga. You can also try Arkhangelsk Oblast. Compare the patterns of timber cuts in terms of relative size, shape, and percentage of area cut to any area in Oregon, Washington, or Maine. What are the similarities? What are the differences? Which forests seem more affected—Russian or U.S.? Why?
4. Find out whether any companies you know carry products made from timber produced in the FSU. Are these products Forest Stewardship Council-certified? Would you buy them if they weren't?
5. Do you think the rise in foreign ownership of Russian timber processors is good or bad for the Russian forests? Should foreign companies be allowed to buy land in the country?
6. Investigate the Websites of a few large Russian lumber mills or PPMs mentioned in this chapter. How many of them have web pages in English? How many have a policy on sustainability? Would you invest in them? Why or why not?

Further Reading

Aksenov, D., Dobrynin, D., Dubinin, M., Egorov, A., Isaev, A., Karpachevsky, M., Laestadius, L., Potapov, P., Purekhovsky, A., Turubanova, S., & Yaroshenko, A. (2002). *Atlas of Russia's intact forest landscapes*. Moscow: Global Forest Watch Russia.

Boentje, J., & Blinnikov, M. (2007). Post-Soviet forest fragmentation and loss in the Green Belt around Moscow, Russia (1991–2001): A remote sensing perspective. *Landscape and Urban Planning, 82*, 208–221.

Field, N. C. (1968). Environmental quality and land productivity: A comparison of the agricultural land base of the U.S.S.R. and North America. *Canadian Geographer, 12*(1), 1–14.

Gregory, P., & Stuart, R. (2001). *Russian and Soviet economic performance and structure*. Boston: Addison Wesley Longman.

Ioffe, G., Nefedova, T., & Zaslavsky, I. (2004). From spatial continuity to fragmentation: The case of Russian farming. *Annals of the Association of American Geographers, 94*(4), 913–943.

Karpachevsky, M. (2001). *Khozyaeva Rossijskogo Lesa*. Moscow: Biodiversity Conservation Center.

Kuemmerle, T., Hostert, P., Radeloff, V. C., Perzanowski, K., & Kruhlov, I. (2008). Post-socialist farmland abandonment in the Carpathians. *Ecosystems, 11*, 614–628.

Wegren, S. K. (2005). Russian agriculture during Putin's first term and beyond. *Eurasian Geography and Economics, 46*(3), 224–244.

Wegren, S. K. (2009). Russian agriculture in 2009: Continuity or change? *Eurasian Geography and Economics, 50*(4), 464–479.

World Health Organization. (2007). *The tobacco atlas*. Geneva: Author.

Websites

faostat.fao.org—FAOSTAT, the statistical branch of the U.N. Food and Agriculture Organization.

www.forest.ru/eng—The NGO Russian Forest Club, covering all aspects of use and misuse of Russian forests.

CHAPTER 21

Infrastructure and Services

After industry and agriculture, the service sector is the next step in our survey of post-Soviet economic geography. In the developed countries of the West, services account for over 60% of the gross domestic product (GDP), because with high productivity and mechanization fewer people are needed for production in industry and agriculture. In Russia's case, the sector accounts for a little less (58% in 2008), but it is nevertheless the biggest sector of the economy. In Tajikistan, however, it is merely 50%. Only 15 years ago, the service sector constituted under 30% of Russia's economy. The explosive growth in services was caused both by the relative decline of industry and agriculture as a result of economic restructuring, and also by the genuine increase in many services as the demand for those soared (Table 21.1).

TABLE 21.1. Structure of Russian Service Sector (%) with Respect to Types of Paid Services Provided

	1995		2002	
	Russia	Moscow	Russia	Moscow
Household services (e.g., dry cleaning, laundromats)	19	9	13	11
Passenger transport	28	35	25	31
Communications (mail, phone)	8	8	12	10
Housing and utilities	19	22	22	18
Education	3	1	7	1.5
Medical care	3	1.5	5	2.5
Recreational (camps, sanatoria, tourism)	5	2.5	4.5	6.5
Legal services	8	5	5	7
Other	6	15	4.5	10.5
Total	100	100	100	100

Note. Moscow (the city and the oblast) accounted for almost one-third of all paid services in the country in the mid-2000s, and the proportion of service types there is very different from that in the rest of the country. Data from Plisetsky (2004).

"Services" is a diverse category, and economists disagree on what exactly should be included in them. Generally speaking, services are nonmaterial elements of an economy (e.g., health care, banking, information, housing, law enforcement, and education). Some aspects of Russia's service sector have already been covered in Chapter 15. I have chosen to include infrastructure along with services in this chapter, although it could also have been included in chapter 18 on heavy industry. For example, transportation of people and goods is both a service and a production process. Other services highlighted in this chapter include information and leisure services.

The service sector was greatly underdeveloped in the Soviet Union, because the government always gave the highest priority to heavy industry. Although mass transit was well developed, other services lagged far behind Western norms. After World War II only 10% of all workers were in the service sector, and by 1990 only 25%, as compared to over 70% in the United States at that time. Clearly, recent years have seen a massive increase in the relative importance of services, with Russia's emergence as a new consumer society. This chapter begins, however, with the most tangible, material form of services—transportation.

Transport Near and Far

It takes over 10 hours in a passenger jet to cross Russia's airspace from west to east. The famous Rossiya train takes about 6 days to travel the length of the Trans-Siberian Railroad from Moscow to Vladivostok. Moving people and freight has always been one of the biggest challenges and top priorities for the Russian government. Ukraine and especially Kazakhstan are also very large countries with many transportation needs. Long distances put the former Soviet Union (FSU) at a competitive disadvantage worldwide, relative to small, compact regions near seacoasts (e.g., Southeast Asia, where goods go from coastal factories directly to ports). Within the FSU, the most advantaged nations in regard to transport are the Baltic states, with their easy access to the Baltic Sea; the least advantaged are the landlocked Armenia, Kyrgyzstan, and Tajikistan.

Russia has some geographic advantages, as well as disadvantages. Its exclave of Kaliningrad on the Baltic Sea is a convenient coastal locale, with a deep seaport that never freezes, a mild climate, and the European Union (EU) all around. Russia spans 11 time zones and can take advantage of its vast size when it comes to energy generation across the entire country (Chapter 17). It also can charge foreign nations for access to its vast airspace or Arctic territorial waters; it can provide reliable international freight shipping via the Trans-Siberian Railroad from Europe to Asia; and it furnishes steady employment for hundreds of thousands of transportation workers. In fact, transportation services account for 10% of Russia's GDP and 6% of all its workers.

The main forms of transport in Russia and other FSU republics are railroads; automotive transport (cars, trucks, buses); air transport; marine and river shipping; and pipelines. The first four can be additionally divided into passenger and freight, as well as local and long-distance. Pipelines are used primarily for movement of liquid fossil fuels and are all long-distance. Additional forms of infrastructure that could be discussed along with transport are power lines, ground telecommunications, and (more recently) wireless telecommunications and satellites; telecommunications are discussed separately later in this chapter.

Railroads

The Russian railroads are products of the Tsarist and Soviet periods. The Soviet Union had a unified railroad network with a standard wide gauge. The Soviet system was designed for a much larger country than present-day Russia and for far greater capacity: It assumed unimpeded travel across the interior republics' borders, and it was built for a militarized, industry-heavy economy almost twice the size of Russia's economy today. Thus we cannot ignore the legacy of the Soviet period and must pay close attention to the impact it has had on today's transport development. (Wireless and Internet infrastructures, on the other hand, were developed after the collapse of the Soviet Union and do not have the same limitations as the older infrastructure. Geography still plays a role here, however: Kazakhstan's connection to European servers, for example, is most easily accomplished via Russia and Belarus.)

Railroads in Russia are second only to pipelines in the volume of freight shipped and are first in passenger volume. They account for 40% of all freight shipped by weight and 33% of all passengers moved (Figure 21.1). Russia is in second place in the world after the United States in the overall length of its railways (over 87,000 km), and the first in the total length of electrified railways. Russia also has the widest commercial gauge in the world, at 1,520 mm (the U.S. gauge is 1,465 mm, and the European one is 1,435 mm). With a wider gauge, more cargo per car can be moved, and the car ride is more stable given the same car height. For a country that relies so heavily on railroads, these are important advantages. Russia's railroad network is dense (about 5 km of railways per 1,000 km^2 of territory); this is about the same density as China's or about one-quarter of the U.S. level.

Russian railroads are heavily used and are generally in good shape. An overnight ride on a passenger train in Russia is faster, cheaper, and much smoother than a ride on Amtrak, in my personal experience. In the United States, Amtrak has to beg private freight railroads to borrow tracks one train at a time, which results in delays. In Russia, the state-run monopoly Russian Railroads controls all traffic and always gives priority to the passenger trains, which therefore generally run on schedule. In this sense, Russia's railroads work similarly to the French SNCF. In contrast to France, however, Russia does not yet have true high-speed trains. One type of train, the ER-200, achieving speeds of only 200 km/hour (kmh), has been running intermittently between Moscow and St. Petersburg since the late 1980s; a new German-built Sapsan replaced it in 2009.

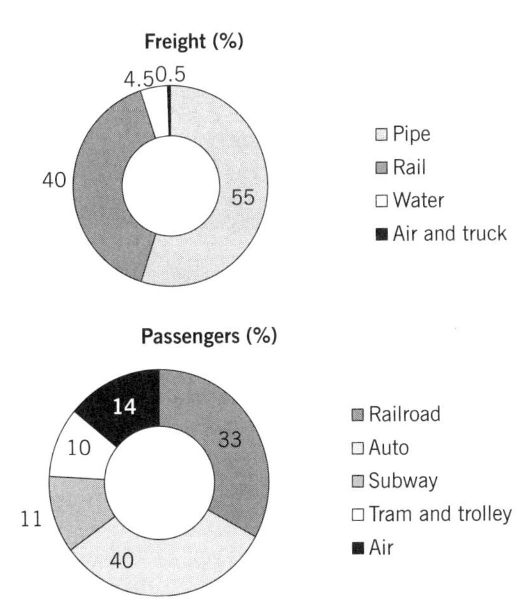

FIGURE 21.1. Freight (weight x kilometers) and passenger turnover (passengers x kilometers) in Russia in 2002. Data from the Federal Service of State Statistics, Russian Federation (2006).

The first railway in Russia was built in 1837 for the tsars; it connected their summer residence, Tsarskoe Selo, to St. Petersburg. By 1851 Russia had the first public railroad from St. Petersburg to Moscow. Railroads were later built along existing roads like spokes in a wheel, from Moscow to the important regional centers Yaroslavl, Nizhniy Novgorod, Saratov, Simferopol, Novorossiysk, Riga, Kiev, and Warsaw. This is clearly visible on a railroad map of Russia (Figure 21.2). Moscow has nine major stations, each providing services in a particular direction (Paris, with basically the same general railroad layout, has six). The Trans-Siberian Railroad (Figure 21.3) was mostly finished by 1898 (Montaigne, 1998), but a small segment around Lake Baikal was not completed until World War I. Most construction during the Soviet period focused on expanding the capacity and electrification of existing lines, building a few critical connectors among them, and accessing some new coal- and ore-mining areas in northern European Russia and in Siberia. Important breakthroughs of the Soviet period included construction of Turksib and other Central Asian railroads connecting Almaty and Tashkent in three directions with Barnaul to the northeast, Orenburg to the north, and Astrakhan to the northwest in Russia. Another major new railroad was built to connect Central Russia with the Vorkuta coal mines and labor camps in the Komi Republic, northeast of Moscow. The Baikal–Amur Mainline (BAM)—running parallel to the Trans-Siberian Railroad, but much farther to the north—was started with prison labor in the 1930s, and was eventually finished by the 1990s after a long hiatus. It was built partially in

FIGURE 21.2. Main railroads of the FSU. Data from ESRI ArcAtlas (1996).

FIGURE 21.3. The Novosibirsk–Moscow train stops at Barabinsk along the Trans-Siberian Railroad. The Moscow-to-Vladivostok trip takes 6 days to cover over 9,000 km. *Photo:* Author.

response to the Soviet leadership's concerns over potential invasion from China.

The northernmost functioning railroad in the world remains the short Dudinka–Norilsk–Taynakh line, running along the Yenisei River at 70°N latitude. There is a proposal to connect that line with the Vorkuta–Moscow line in Europe, and even to extend it all the way east to the Bering Strait. There is also a proposal to build a line from Yakutsk toward Chukotka and eventually to North America under the strait. The proposed tunnel would span 109 km and run under water up to 55 m deep. Plans include construction of a corridor for car travel, a rail line, and electric and fiber-optic cables within the tunnel. Some believe that up to 3% of the world's cargo could eventually be moved through the tunnel. Construction of a tunnel underneath the Bering Strait is within today's technological reach, and if completed, such a tunnel would allow uninterrupted railroad and car travel from Europe or Africa to South America. The project would not be feasible without a massive increase in trade between Russia's Pacific Rim and the United States, however, which is unlikely at present.

Although the long-distance trains of Russia have a glamorous reputation because of the Trans-Siberian Railroad, the majority of passengers on a daily basis use commuter trains (*elektrichki*), which run for 100–150 km within and around major cities. An average train in Moscow takes about 5 minutes between stops within the city, and consists of 8–12 designed for about 200 passengers each. These trains complement the *metro* (subway). Outside the city limits, stations are spaced about 10 minutes apart, offering easy access to the dachas. A lot of people who live outside the city also use the trains for their daily commute.

Automotive Transport

Russia's automotive transport lags far behind that of Western Europe, North America, or most of developed Asia. Russia has about one car per six people at the moment, whereas most Western European countries have one per two. Although Russia has the seventh largest number of vehicles (over 24 million), the road network is not adequate for much intercity or inner-city movement and is only in eighth place worldwide in total length (871,000 km). The United States, by comparison, has over 6.4 *million* km of highways. The Soviet planners emphasized mass transit in their city plans (Chapter 11): trains for between-city travel, and buses and subways for within-city travel. The road network therefore received less attention.

Another area where Russia lags behind is in road quality. Until just a few years ago, there were no highways in Russia approaching the U.S. interstate system's capacity or possible speed of travel. At this writing, the Moscow beltway and segments of a few highways to the south and west are the only fully divided, multilane highways with controlled access and no traffic lights in the entire country. The Moscow–St. Petersburg highway is reminiscent of average American county roads in places. A typical Russian long-distance national highway is a two-lane asphalt road with potholes, an uneven grade, narrow shoulders, and a practical speed limit of about 80 kmh (<50 mph). In more rural areas, roads even between important regional centers lack asphalt altogether (Figure 21.4). Incredibly, until 2004 there wasn't a highway connecting Irkutsk and Chita Oblasts in eastern Siberia, making a cross-country car trip from Moscow to Vladivostok impossible. Despite all this, the car ownership rate in Russia has almost tripled since the fall of the U.S.S.R. and is expected to grow farther. There is frantic road construction throughout the country. Still, the 19th-century saying attributed to Nikolai Gogol rings true today: "Of all the problems, the two main ones in Russia are fools and roads."

Besides the dramatic rise in private car usage, there has been also an increase in long-distance bus service in many parts of the country—both to compensate for high rail prices and less convenient train schedules, and to reach more destinations. For example, from Novosibirsk one can travel by long-distance buses to Tomsk and Barnaul (5 hours each), Kemerovo and Biysk (7 hours), Semipalatinsk (8 hours), and even Almaty (20+ hours). Although trains are available to all of these destinations, buses are cheaper and have more flexible schedules. On the local level, inner-city buses, trolleybuses, and tramways serve millions of commuters daily. In fact, about 80% of all those commuting to work in Russian cities do so by bus—the reverse of the U.S. pattern, where over 90% commute by private car. Tramways (old-fashioned street cars) are generally being phased out, as they increasingly compete with cars for the same congested streets (Figure 21.5). St. Petersburg has the most extensive tram system in Russia. Electric trolleybuses are more popular because they do not require rails. Buses move over 90% of all city passengers in Russia among above-ground mass transit forms. The majority of those bus trips are intraurban, 15% are suburban, and <1% are long-distance. An increasingly popular option is the *marshrutka* (minibus service). Typically private, these minibuses run along the same routes as regular city buses, but stop only on demand and provide faster service for the same price.

Although long-distance freight is mainly carried by railroads, if measured by cargo moved

FIGURE 21.4. Typical rural Russian roads: paved (above) and unpaved (below). *Photo:* Author.

FIGURE 21.5. Trams are electric and pollution-free. However, they occupy the street median and are the slowest mass transit on average. *Photo:* Author.

per kilometer (Figure 21.1), trucks and pickups are increasingly commonly used for shorter trips. In fact, trucks are definitely visible in Russia today, with hundreds of them hauling freight along all the major highways, much as they do in the United States. The true long-distance trucks, however, are not nearly as prevalent as in the United States; most trucks are used on local delivery trips of 10–50 km. Long-distance truck traffic is heaviest in western European Russia between Poland and Moscow and St. Petersburg and Moscow. Trucks are also common along the Trans-Siberian corridor. Trucks mostly move high-value import items in Russia (electronics or refrigerated food), with heavier and cheaper freight (e.g., construction materials or lumber) being hauled almost exclusively by the railroads. Compared to the United States, though, trucks move a minuscule amount of freight in Russia (less than 1%, vs. 67% for U.S. trucks)!

Air Travel

Air travel is the most common and familiar long-distance travel option in North America. Although travel by car is almost always cheaper even cross-country, the high speed of air travel more than compensates for the price in our fast-paced world, and flying is chosen by most U.S. passengers who travel over 1,000 km each way. In the mid-1990s, U.S. domestic air passengers accounted for about 10% of all passenger-miles traveled, whereas rail and bus travel together accounted for only 3.5%. The Europeans and Japanese tend to fly less (e.g., the EU's original 15 members' proportion of airline passenger-miles is only 6.5% of the total) because of the competitiveness of high-speed trains. In Russia only 14% of travel happens by plane, as compared to 40% by automobile and 33% by train. The proportion of air travel is higher than in the United States because a lot fewer people travel by private car in Russia (under 10% of all passenger-kilometers, as opposed to almost 85% in the United States). With respect to freight, virtually nothing travels by air in Russia except express mail, high-value perishable goods, and military equipment.

The airline industry can be best thought of as consisting of a few key interlinked parts: aircraft, airlines, and airports. All were developed in the Soviet Union for interurban and especially interregional traffic. Russia was not the first nation in the world to have commercial air flights, but by World War II there was a regular passenger air service from Moscow to Kamchatka and back, with a few stops in between. Mass production of war planes helped to improve civilian aviation after the war was over. Small cities were served by An-2 biplanes, and in Siberia and the north by Mi-8 helicopters. In 1956, the Tu-104 became one of the first passenger jets flown commercially in the world. In the 1960s, large passenger jets (Tu-154 and Il-62) began providing service on long-distance routes, and hundreds of airports were opened. The Soviet Union's Aeroflot had the largest airline fleet in the world and served 87 countries. It flew exclusively Soviet-built planes and was the only domestic carrier in the country at the time.

Since the late 1980s there has been a dramatic decline in the production of new planes due to the economic slowdown (Chapter 18). With reforms in 1992 came the near-collapse of the air travel system because of increased fuel prices and safety concerns, as well as shrinkage of the available pool of pilots, mechanics, and airport staff. A few dozen private airlines were created out of the old Aeroflot regional units (e.g., Sibir, KrasAir, Pulkovo Airlines, UTair, and DalAvia). This did not help to create competition, as was initially thought, because each airline remained a virtual monopolist in its own region. For example, over 80% of all flights out of Tolmachevo Airport in Novosibirsk are carried by the locally based Sibir (S7) airline. A few airlines were created from scratch and have tried to follow Western business models using Western airplanes (e.g., Transaero in Russia and Air Astana in Kazakhstan). However, despite an increase in competition and attempts to improve air safety, little else has been done to increase air travel.

In terms of safety, a big problem is the obsolete fleet. The Tu-154 was a very reliable airplane when it started flying in the 1970s, but today these planes are at least 20 years old. Maintenance is spotty, and some spare parts installed on planes are known to be counterfeits. Although foreign planes are now gradually replacing the old Soviet models, they are not necessarily much safer either. The imported airplanes flown by

Russian carriers are typically old Boeing and Airbus models, discarded after many years of leased service in South Africa, Poland, or Brazil. Moreover, Russian maintenance facilities are not always able to service foreign models properly, and there is also a shortage of pilots trained to fly Western airplanes. Aeroflot and the flagship airlines of some FSU republics are flying new leased aircraft and can be considered more reliable than most regional carriers. Of all major airlines in Russia today, only about five can be recommended as reliable for domestic flights. Aeroflot remains the leading carrier in Russia, with 12% of domestic and 45% of international flights, and aims to increase its domestic share to 25% in the next few years.

To put anyone at ease who is planning a trip to Russia, most of the crashes that have made headlines in recent years there were actually quite unrelated to the mechanical soundness of the planes. Two airliners were downed by Chechen suicide bombers in 2004, highlighting the reality of terrorism in the country, and two crashed in 2006 because of preventable pilot errors. The FSU is certainly not the safest place in the world to fly, but it is not the worst either. If the entire airspace of the FSU is considered, between 1990 and 2007 there were only 23 fatal accidents involving commercial airplanes, of which 18 occurred in Russia. A few more happened abroad, mainly on charter flights. Over the same period of time, 10 accidents occurred in India, 15 in China, 16 in Latin America, and over 30 in Africa. North America had 30 accidents over the same period of time, including the four airplanes lost on September 11, 2001. However, the United States has over 8.5 million airplane departures per year, as compared to only 330,000 in Russia, so the U.S. accident rate per passenger is of course much lower than Russia's.

Moscow is by far the busiest air hub in Russia (Table 21.2). The busiest three airports in Russia surround Moscow: Domodedovo to the southeast, Sheremetyevo to the northwest, and Vnukovo to the southwest. Domodedovo International Airport is the busiest in Russia, with over 20 million passengers per year, while Sheremetyevo gets 15 million and Vnukovo 8 million; thus the total for Moscow is 43 million air passengers per year. By contrast, the busiest passenger airport in the world, Hartsfield–Jackson International in Atlanta, Georgia, served 90 million passengers in 2008. So, despite Moscow's having over 11 million residents, the total number of air passengers in and out of the city was only slightly more than at the Minneapolis–St. Paul (MSP) International Airport, which serves a region of only about 4 million. All three Moscow airports are now connected to the city center with express trains and convenient bus shuttles.

The fourth largest airport in Russia is Pulkovo near St. Petersburg, serving about 7 million passengers per year. Other important air hubs include Tyumen, Omsk, Novosibirsk, Krasnoyarsk, Irkutsk, Yakutsk, Khabarovsk, and Vladivostok in Siberia, as well as Samara, Yekaterinburg, and Nizhniy Novgorod in European Russia, and the vacation destinations Sochi, Anapa, Krasnodar, and Mineralnye Vody in the Caucasus region (Table 21.2). In other FSU countries, Kiev's airport is the busiest, with about 6 million passengers per year; the airports of Almaty, Tashkent, and Baku are also prominent. In contrast, Minsk and Ashgabat see very few air passengers.

Whereas about 30 of the biggest airports remain busy in Russia, about 600 airports serving medium-sized cities have been shut down or have had only occasional charter service since the 1990s. In the European part of the country, buses and trains can serve these cities well, but in Siberia and the distant north, the disappearance of small airports may spell doom to the local economy. The most active smaller airports are located in oil-rich western Siberia, where even cities of 20,000–50,000 residents have modern airports to serve the oil crews. In other parts of the country, even cities with 200,000–300,000 people may not have a functioning airport any more. Clearly, there is an ample business opportunity waiting here. A few national initiatives in 2007 and 2008 were seeking to provide federal incentives for additional airport development, but their impact remains to be seen.

Russian air travel tends to be expensive, due to lack of competition and high domestic fuel prices (Derudder et al., 2007). Russian airports are also among the most remote on the planet from the biggest global hubs; Zook and Brunn

TABLE 21.2. Main FSU Airport Hubs and the Airlines Associated with Them

Airport	Passengers in 2007 (millions)	Hub for airline(s)
Russia[a]		
Domodedovo, Moscow	18.7	Transaero, VIM Avia, Domodedovskie Avialinii (AiRUnion), many foreign airlines
Sheremetyevo, Moscow	14.0	Aeroflot
Vnukovo, Moscow	6.7	VIM Avia, Atlant-Soyuz, Rossiya
Pulkovo, St. Petersburg	6.1	Pulkovo Airlines (Rossiya)
Yemelyanovo, Krasnoyarsk	2.3	KrasAir (AiRUnion)
Koltsovo, Yekaterinburg	2.3	Uralskie Avialinii
Tolmachevo, Novosibirsk	1.8	Sibir (S7), Novosibirsk Avia
Adler, Sochi	1.6	NA
Kurumoch, Samara	1.4	Samara (AiRUnion)
Pashkovsky, Krasnodar	1.4	Avialinii Kubani
Ufa	1.2	Ufimskie Avialinii
Irkutsk	1.0	Sibir (S7)
Rostov-on-Don	1.0	Aeroflot-Don
Vladivostok	0.9	Vladivostokavia
Khrabrovo, Kaliningrad	0.8	KD Avia
Mineralnye Vody	0.7	Kavminvodyavia
Roshchnino, Tyumen	0.6	UTair
Yakutsk	0.6	Yakutskie Avialinii
Vityazevo, Anapa	0.6	NA
Kazan	0.6	Tatarstan
Khomutovo, Yuzhno-Sakhalinsk	0.6	Sakhalinskie Aviatrassy
Bolshoe Savino, Perm	0.5	Permskie Avialinii
Omsk	0.5	Omskavia (AiRUnion)
Khabarovsk	0.4	DaLavia
Surgut	0.36	UTair
Other FSU		
Tallinn, Estonia	1.7	Estonian Air
Riga, Latvia	3.2	airBaltic
Vilnius, Lithuania	1.7	flyLAL
Boryspil, Kiev, Ukraine	6	Ukraine International, AeroSvit
Donetsk, Ukraine	1.5 (est.)	Donbass Avia
Dnepropetrovsk, Ukraine	1.0 (est.)	DniproAvia
Minsk, Belarus	0.5	Belavia
Kishinev, Moldova	0.7	Air Moldova
Tbilisi, Georgia	1.5 (est.)	Georgian Airways
Zvartnots, Yerevan, Armenia	1.1	Armavia
Heydar Aliev, Baku, Azerbaijan	2.0 (est.)	AZAL
Astana, Kazakhstan	0.8	Air Astana
Almaty, Kazakhstan	2.0	Air Astana
Manas, Bishkek, Kyrgyzstan	0.6	Kyrgyzstan Airlines, Itek Air
Tashkent, Uzbekistan	2.0	Uzbekistan Airways
Ashgabat, Turkmenistan	0.3 (est.)	Turkmenistan Airlines
Dushanbe, Tajikistan	0.5 (est.)	Tajik Air

Note. Data from online research by author.
[a]Russian airports are ranked by volume of traffic, in descending order.

(2006) found that Novosibirsk is about as remote as Majuro, Marshall Islands, for example. Air travel in Russia is less expensive than in Africa, but more expensive than in Europe, Asia, or North America. A project in my Geography of Russia class surveyed airfares between the top 25 cities in Russia in spring 2007. The average one-way airfare was about $300. The most expensive destinations were naturally the most remote ones, with little competition among the air carriers: A trip from Moscow to Yuzhno-Sakhalinsk would cost $667 one way, for example. The distance traveled in this case is about the same as between Boston and Anchorage (6,500 km), a trip that would cost about half of that price in the same year. (Bear in mind that Russia's incomes on average are only one-quarter of the U.S. level, so the relative expense of flying in Russia is much greater.) Russia remains poorly connected to North America in particular; only Aeroflot, Transaero, Delta, United Airlines, and Air Canada had regular service in and out of Russia as of summer 2010.

The FSU's airspace is not exclusively populated by travelers in and out of Russia. Every FSU republic now has its own national carrier, and some have additional private airlines. One of the most successful non-Russian carriers today is Air Astana, the flagship carrier of Kazakhstan. It took over the bankrupt old government airline, hired a Western-born chief executive officer, and completely changed to a Western model of doing business. It flies to about 30 destinations worldwide, including Moscow, London, Beijing, Seoul, Bangkok, and Amsterdam, using primarily new A-320 jets. Its service is good, and its pricing is competitive. Almaty remains the largest city in Kazakhstan, with a newly opened international airport terminal. However, because the city is no longer the capital, it receives fewer government travelers than the much smaller new political capital, Astana. Ukraine has a few airlines based in Kiev. Like Aeroflot, they largely inherited the old Soviet aircraft, which is now in urgent need of replacement. Other than Russia, Ukraine was the only republic that assembled airplanes in the Soviet period—in particular, Antonov turboprops and some military models. Hundreds of these are still flying around the world, especially in Africa, where they are frequently flown by Ukrainian crews.

Water Transport

Water transport was formerly well developed in Northern Eurasia, but is slowly dying out. Ocean-going vessels remain important: Russia is in 10th place worldwide as ranked by marine tonnage and has almost 4,000 marine vessels in civilian use, not counting the navy. The most common ocean-going ships are fishing boats (trawlers, seiners, etc.). There are also 720 general-purpose cargo ships and 215 petroleum tankers in use. However, Russia's lack of refrigerator, container, roll-on/roll-off, and other specialized modern ships hampers freight shipping. For example, only 14 container ships were operated out of Russia in 2007, as compared to 82 in the United States. The number of ocean-going vessels continues to drop. Over 60% of all Russian ships are sailing under flags of other countries to avoid registration taxes, so the statistics are not easy to obtain. The most common types of marine cargo are petroleum products, coal, timber, grain, sand, gravel, fertilizer, and metal ores.

The busiest area of marine shipping for the Soviet Union was traditionally the Black Sea, which allows trade with the Middle East and Europe via the Bosporus Strait. Novorossiysk currently is the busiest port in Russia, shipping about 50 million metric tonnes (mmt) per year. By comparison, the busiest port in France, Marseille, handles over 100 mmt per year, and the busiest in the world, Singapore, handles over 340 mmt. Other important seaports on the Black Sea include Tuapse, Rostov-on-Don, and Taganrog in Russia, and Odessa, Kerch, Sevastopol, Kherson, and Nikolaev in Ukraine. Georgia has important ports at Batumi and Poti.

St. Petersburg is the second busiest port in Russia, with about 20 mmt moved per year. New terminals are being built north and south of St. Petersburg to accommodate the increase in petroleum exports across the Baltic Sea (e.g., at Primorsk). Kaliningrad is a strategically important port between Poland and Lithuania. Baltic ports are Tallinn and Parnu in Estonia; Riga, Ventspils, and Liepaja in Latvia; and Klaipeda in Lithuania. Murmansk and Arkhangelsk in European Russia account for over three-quarters of all cargo shipped in the Russian Arctic. Murmansk is the only port open year-round in the north, because of the warm Norwegian current. It is a multipurpose port with a significant navy presence. Arkhangelsk specializes in shipping timber. Along the northern shores of Russia, Dixon, Dudinka, Igarka, Tiksi, and Pevek receive oceanic traffic a few months per year along the famous Great Northern Route. Even with nuclear-powered icebreakers, there is too much ice in winter there at present to keep them operating year-round east of Dudinka. This may be changing with global warming within the next few decades, but at present Russia must build new icebreakers to replace the aging ones that have been in service since 1959.

In the Far East, the largest cargo ports are all in the south: Vostochny (13 mmt), Nakhodka (11 mmt), Vladivostok (5.5 mmt), and Vanino (5 mmt). Their specialties are fishing, timber and ore shipping, and importing Asian consumer goods. Petropavlovsk, Magadan, and Yuzhno-Sakhalinsk have strategic importance for the Russian Navy, which keeps its nuclear submarines there. The inner Caspian Sea basin has two major Russian ports, Makhachkala and Astrakhan. They are mostly used for fishing and shipping petroleum products, and account for less than 1% of all sea shipping in the country. Other ports on the Caspian include Baku in Azerbaijan; Atyrau and Aqtau in Kazakhstan; and Turkmenbashi in Turkmenistan.

Virtually no passengers travel by sea any more in the FSU. Some European cruise lines call on St. Petersburg in their "seven capitals" voyages around the Baltic Sea. Limited ferry lines and suburban commuter boat rides existed during the Soviet period in the Baltics, on the Crimea Peninsula, and along the Black Sea coast, but most of these are history now. A handful of boats continue to operate pleasure cruises along the shorelines. River passenger traffic has suffered a similar fate: Scheduled ferries no longer serve the major European rivers in Russia (e.g., the Oka, Volga, or Neva). Most boats that you see on rivers now are either small private motorboats or cruise ships. Freighters and barges are still plentiful, although not nearly as common as in the late Soviet period. Back then, an observer along the Oka River between Ryazan and Murom would have counted over a dozen ships per hour traveling in either direction. Today an observer will be lucky to see one. A typical Russian river-going ship is a self-propelled barge with a capacity of 2,000–3,000 metric tonnes. Some of these vessels can sail into the open sea, and most carry loose material (e.g., gravel, sand, stone, grain, or fertilizer). Petroleum products and timber are also commonly transported by river.

The busiest river watershed in Russia remains the Volga–Kama system, moving about half of all in-country river cargo. The second busiest is the Ob system, which extends upstream into Kazakhstan via the Irtysh and downstream all the way to the Arctic Ocean. Some parts of Siberia and the North have rivers as their only connections to the world at large, given the lack of roads and airports. Overall, Russia has over 100,000 km of navigable interior waterways, about three-quarters of which are in the European part of the country. The extensive canals permit travel from the Black Sea up the Don into the Volga, and then all the way to either St. Petersburg or the White Sea coast via Moscow, earning Russia's capital the nickname "port of five seas."

Pipelines

Russia is the queen of pipelines. It has more of them than any country except the United States by overall length, and carries more products via pipelines than any other country. The first pipeline in Russia was an 853-km kerosene pipeline from Baku to Batumi in the Caucasus, built in 1907. The U.S.S.R. had fewer than 2,000 km of petroleum pipelines built before World War II, but over 66,000 km by 1990. The peak of pipeline construction occurred in the 1980s, when some of the key components of both petroleum and gas pipeline networks were laid out between Eastern Europe and the western Siberian oil and gas fields (Starobin, 2008). For example, the Druzhba (Friendship) pipeline connected the Almetyevsk refinery center in Tatarstan with Samara, Mozyr, and eventually Brest in Belarus. This remains the top export line for Russian oil today. Its celebrated gas counterpart is the Urengoy–Pomary–Uzhgorod line, which is the world's longest at 4,451 km. Russia had over 44,000 km of petroleum pipelines and over 150,000 km of gas pipelines in 2008.

Although less glamorous than trains or planes, pipelines move more freight, about 55% of the total (Figure 21.1). Of these, 59% move natural gas and 41% move petroleum. Gas pipelines are operated by Gazprom, and petroleum pipelines by the Transneft monopolies. Individual oil companies, which are numerous in Russia, have to purchase transit rights from the state-run Transneft, thus ensuring the Kremlin's control over a strategic resource. The majority of pipelines run from northeast to southwest—from Yamal Peninsula and the Western Siberian Lowland across the Urals to the refineries of the Volga basin, and beyond to consumers in Central Russia, Ukraine, Belarus, and the EU. A pipeline recently completed from Baku to Tbilisi, Georgia, to Ceyhan, Turkey (BTC), with heavy American and European involvement, completely bypasses Russia and thus seriously undermines the Putin–Medvedev administration's attempts to control all petroleum flows in and out of the Caspian basin. A large pipeline like the BTC can carry up to 1 million barrels of oil per day. If you recall that Russia produces "only" 9 million barrels per day, all it needs is about nine large lines. However, many pipelines are much smaller and serve local markets.

A new petroleum pipeline is being built east from Angarsk near Lake Baikal to China and Japan. Transneft's original plan to route the line around the northern end of Lake Baikal met with unprecedented opposition from the public and required Putin's direct involvement. Lake Baikal is located in a seismically active zone, and even a mild spill into the lake would prove disastrous to its ecology. The current plans call for the pipeline to bypass the lake over 100 km to the north. Additional pipelines are being built in European Russia to the areas near St. Petersburg and to the Barents Sea. When these are completed, more oil from Russia will flow to European and even American consumers, bypassing the politically hostile regimes of Eastern Europe.

The first natural gas pipeline was built between Moscow and Saratov in 1940. One of the most important remains the Urengoy–Pomary–Uzhgorod export system, which ushered in the current era of Russia's major exports of natural gas to Central and Western Europe. Another one, the Soyuz pipeline, connects Orenburg with Uzhgorod in western Ukraine. Turkmenistan, a major natural gas producer, is connected to Europe via Russia-operated pipelines. However, the Nabucco company is now proposing a new gas line under the Caspian Sea to Baku and Turkey, to avoid Russia altogether. Most of Russia's and Turkmenistan's gas is exported to Europe via Ukraine, whose leadership has recently been at odds with Russia's government-controlled Gazprom over the price for transshipment. To counterbalance the proposed Nabucco pipeline and avoid issues with Ukraine, Russia operates the Blue Stream pipeline under water from Russia to Turkey and is planning a new Southern Stream line from Tuapse into Bulgaria, Romania, Serbia, and Slovenia. Yet another gas pipeline is being built under the Baltic Sea from St. Petersburg to Germany (the Northern Stream). This project is favored by Germany and Russia, but is opposed by Estonia and Poland on both political and environmental grounds.

Building and running pipelines are expensive. The Southern Stream construction alone is estimated to cost between $10 and $14 billion. Gas leaks are especially dangerous, because they may lead to explosions. In the cold climate of Siberia, petroleum must be heated to flow through the pipelines. The possibility of terrorists' sabotaging a pipeline also exists in Russia. Nevertheless, the current political regime clearly has the strategic goal of supplying more and more of the world with fossil fuels from Russia. The question remains how much of these fuels will last. If recent estimates by the World Energy Group (see Chapter 17) are correct, Russia has less than 10 years' worth of petroleum left in its existing oil fields, and little will be available after that unless a dramatic increase in exploration takes place immediately, which is very unlikely. Gas is more plentiful and will probably be still available 25 years from now.

As the preceding discussion has made clear, Russia has major achievements but also major problems with its transport infrastructure. What about the less tangible infrastructure, such as telecommunications and the Internet? This brings us to the broader question of high technology in Russia.

High-Tech Russia

A cartoon from the 1990s depicted a Russian *bogatyr* (ancient warrior) on horseback, incredulously poking with his spear at a computer keyboard in front of Baba Yaga's (a witch-like Slavic folklore character's) log house. The keys on the keyboard were made out of tree stumps, cut to various sizes. A sign on the house proudly said, "The first ancient Russian computer." Such may be the popular image of high-tech Russia. Could there be a real high-tech Russia—a Russia of cutting-edge technology, not of tree stumps? We need to remember that although Russia's recent history has been one of deconstruction and humiliation, it remains a country producing many sophisticated weapons and spacecraft, with millions of highly talented and well educated scientists and engineers. Nevertheless. the investments in high-tech enterprises have been lacking over much of the post-Soviet period. Investments in research and development (R&D) in Russia are trailing those in virtually all developed countries—not only in absolute amounts, but even in proportion to the GDP (Chapter 15). The total spending in Russia on scientific research in the mid-2000s was 13% of the U.S. level, even with GDP adjusted for purchasing power parity. Russia at present has a hard time even holding on to the areas in which for a while it enjoyed a leading position in the world (e.g., laser development, nuclear physics, organic chemistry, microbiology, and mathematics).

The following high-technology industries, services, and research areas are nevertheless being developed in Russia: electronic hardware design; computer software design; telecommunications (cable and satellite TV and the Internet); avionics; geospatial technologies; "smart" weapons design; advanced biotechnology; genetic engineering and molecular biology; industrial chemistry and pharmaceuticals; new medical devices; and nanotechnology. Not all of these have a strong domestic innovative component, however. Although you can get a computed tomography (CT) scan in a Russian clinic now, until just a few years ago no CT machines were made in Russia; all were imported from Germany or the United States. Similarly, all personal computers (PCs) are either imported whole or assembled in Russia from parts made in Taiwan, Singapore, Malaysia, and increasingly China.

However, some areas are positively booming. Arguably the most noticeable is the proliferation of cell phones and wireless gadgets in just the past 5 years. With some of the lowest rates in the world, unlimited pay-as-you-go call plans, and at least three major companies vying for customers coast to coast, Russia is a cell phone buyer's market. Almost 20 years ago, fewer than half of Russia's households had even a land phone, and automated intercity dialing was not available in many areas of the country. Even for big-city residents, no call waiting, caller ID, or collect calling was available. How much has changed in just 15 years! Instead of investing in obsolete landlines, the newly founded private wireless companies provided everyone with cell phone coverage, no matter how remote the location, within 10 years. There are more registered cell phone users in Russia now than there are people (over 150 million and counting). Many customers keep multiple phones to get the best rate, depending on their exact calling locations. The coverage for the largest country in the world is steadily improving. I had better reception on my cell phone in the middle of the Altay Mountains 10 km from a nearby village than near the entrance to Yellowstone National Park within 3 km of Gardiner, Wyoming, in 2007. A phone call within Russia usually costs less than 10 U.S. cents per minute, and since 2007 all incoming calls have been free. There are no credit checks to get a cell phone in Russia, and prepaid minutes do not expire. Some areas remain poorly served; still, cell phone service is now expected and makes good money for its providers. The top two, MTS and Bee-Line, were in the 10th and 12th spots, respectively, on the Russia Top 100 list of companies in 2007. Each was worth about $20 billion—just below the largest oil and gas producers, and comfortably ahead of car manufacturers, retailers, and most banks.

More digital phone lines and fiber-optic connectors are being put in annually. Cross-country digital trunk phone lines now run from St. Petersburg to Khabarovsk and from Moscow to Novorossiysk. The telephone systems in 60 regional capitals have modern digital infrastructures, many using U.S.-manufactured equipment and software. In rural areas, however, the telephone services are still outdated, inadequate, and very low-density compared to Europe's.

Another important high-tech system developed in the late Soviet period was GLONASS satellite navigation. Currently there are plans to bring it back to full operation with a few satellite launches per year. This is one of only two alternatives to the U.S.-controlled global positioning system (GPS); the other one is the Galileo system, which is still being deployed by the EU. The regular GPS receivers available to Western consumers cannot use the GLONASS signal; a special receiver is required. Some GLONASS units are manufactured in Russia, mainly for the military market, but more and more are now available from Asian manufacturers. Russia is second only to the United States in the number of communication satellites in orbit: Besides GLONASS, Russia maintains dozens of TV, radio, telephone, and military satellites. In 2006, Kazakhstan became the second FSU country to launch its own satellite, KazSat 1, although it was built with Russia's involvement.

Russia lags far behind the West in robot and computer development. Massive mainframe computers were built in the Soviet Union for military and civilian use, but even then they were not quite as good as American or Japanese models. Production of computer equipment almost stopped with the launch of economic reforms. For years the U.S. Congress banned exports of sensitive computer technology to Russia. Many Soviet computers were essentially clones of existing IBM systems that were built with stolen blueprints, but some (e.g., the BESM-6) were original domestic creations rivaling their U.S. analogues. However, with the advent of PCs, Russia followed the rest of the world in adopting American computer technology for home and business use (mainly Windows, although Mac and Linux are both available). Importing Western PCs was a common early path to wealth for the Russian entrepreneurs of the 1990s. Much of Russia's civilian computer manufacturing today consists of assembling laptops and PCs from prefabricated parts, using screwdrivers.

The Internet is yet another infrastructure development that is growing in leaps and bounds in Eurasia (Warf, 2009). The Russian segment of the Internet ("Runet") is adding millions of users per year, with the number of users nearly doubling every year for the past 7 years (Vignette 21.1). The domain zone *.ru* had almost 138,000 registered domains (i.e., Websites) as of December 2009—behind France and Poland, but ahead of Sweden and Chile. The Baltic republics had about 15,000 domains registered per country, and there are a few thousand each in most other FSU republics. The Russian language is among the top 15 languages encountered on the World Wide Web, although not yet in the top 10. About 27% of the Russian population had online access in 2008 (38 million users), as compared to about 71% (35 million) of South Koreans or 57% (33 million) of Italians. There were 62 million users in the entire FSU in 2008. Internet access is about as common in Russia now as it is in Turkey or Brazil, but not nearly as common as in developed Asia or Europe. In addition to accessing the Internet from home or work, many people use Internet cafés (Figure 21.6), which are available in all big and medium-sized cities (Warf, 2009). Many also now have Web-enabled cell phones or personal digital assistants. Broadband connections are growing fast; few Russian homes have cable TV, so these are mostly digital subscriber lines at present, though wireless broadband is available in Moscow and other big cities. Golden Wi-Fi in Moscow is one of the largest citywide Wi-Fi systems in the world. Early in 2008, the Russian Ministry of Information announced a new federal program that would make broadband affordable to all urban users (75% of the population) within the next few years. In a vast country such as Russia, connections across long distances are of ultimate importance, so the Internet penetration is expected to grow. Because of the central government's current tight control over TV and newspapers, the Internet is the source of the least censored information in Russia at present, and ideally it will remain so.

Where are high-tech services found in Russia? Primarily in the biggest cities, of course, especially in and near Moscow, St. Petersburg, Nizhniy Novgorod, Yekaterinburg, and other cities with populations over 1 million, heavy industry, good universities, and a decent social life. Novosibirsk in particular is the high-tech hub of Siberia; it

Vignette 21.1. Runet Takes Off

Runet is the segment of the Internet written in Russian. In the narrow sense of the word, it encompasses only the Websites of the *.ru* domain. However, many Websites in Ukraine, Belarus, and Kazakhstan, as well as in Israel, Germany, and the United States, are written in the Russian language. In a science fiction novel *Year 4338*, written in the 1830s, the Russian philosopher Vladimir Odoevsky presciently outlined a "magnetic telegraph" that would allow people to talk to each other from far away, as well as "household journals" in which people would discuss their lives "for all to see." Indeed, this all became reality with the Internet. Runet took off in the mid-1990s, along with the rest of the Internet.

The first Internet lines from the U.S.S.R. to the West were put in place in the late 1980s. The first Internet providers appeared in 1992–1993. GlasNet was one of the best known, providing Internet access to numerous small grassroots organizations and using Western (largely American) sources of funding. In 1994 *.ru* and other republican domains were established in response to the devolution of the Soviet Union; Azerbaijan, Estonia, Georgia, Latvia, Lithuania, and Ukraine had already received theirs in 1992–1993. In 1995 a number of Internet "firsts" happened: The first online store was opened; the first Website design studio appeared; and the first Internet news agency was started. In 1996 the first Internet café in the country—called Tetris, to honor the famous Russian computer game—opened in St. Petersburg. The first Russian search engine, Yandex, appeared in 1997. In 1998 the first free e-mail service, *mail.ru*, began. The first Russian blog appeared in 1999. Russian Wikipedia was started in 2001; it featured over 466,000 articles in 2009 (the 10th most of all languages). The first free video hosting was started by Rambler, one of the largest Internet providers, in 2004. Google opened its Moscow office in 2006. (One of the two cofounders of Google, Sergey Brin, is Russian-born.) In October 2007, then Vice-Prime Minister Dmitry Medvedev announced that all 59,000 Russian schools had been connected to the Internet. Also in 2007, the Russian company SUP purchased the software platform for the famous blog site LiveJournal.com from the American company SixApart. Russia is second only to the United States in the number of blogs on LiveJournal.com, with about 500,000 people in 2007. The actual number of Russians on the site is probably higher, because many Russian-language bloggers live outside Russia.

FIGURE 21.6. An Internet café in Altaysky Kray. *Photo:* A. Fristad.

has great universities, diverse companies, and excellent international connections via the airport and the railroad. Other Siberian cities with high-tech R&D presence include Omsk, Tomsk, Krasnoyarsk, and Irkutsk. There are also a few dozen small academic towns in Central Russia, the Urals, and Siberia (Chapters 11 and 18) where much high-tech development takes place.

Other FSU republics have various degrees of high-tech development. If Internet usage is taken as an indicator, Tajikistan is the least high-tech country, with merely 0.3% of the population having online access. Other low-usage countries are Turkmenistan (1%), Kyrgyzstan, and Armenia (5% each). Estonia tops the list of high-tech FSU nations, with 58% of its population using the Internet in 2007, followed by Belarus (56%), Latvia (47%), and Lithuania (30%). Rates for Estonia and Belarus are comparable to the U.K. level. The U.S. and Japanese rates were 69% in the same year.

Retail and Leisure Services

Besides transportation and telecommunications, services include medical care, education, retail, and leisure. Medical care has already been covered in Chapter 12, and science and education in Chapter 15. Tourism-related leisure services have been covered in Chapter 16. Here I discuss retail and some other leisure services.

The Soviet-era retail sector was dreadful: There were not enough stores, few available goods, and plenty of rude clerks. All stores were state run. Shop clerks were, like everybody else, state employees. Their salaries were fixed. They did not care whether customers liked their service or not, because no alternative stores existed (except the closed shops for the top Communist elite). Being a store clerk was well regarded as a cozy job that gave one easy access to *deficit* (scarce goods). When scarce goods were available, long queues would form even though prices were low, because there wasn't enough quantity to satisfy the demand. The economic planners never could guess accurately how much of anything was needed, especially with respect to consumer goods. In short, shopping in the Soviet Union was a bleak experience. In one of the funniest cultural portrayals of the period, the late-Soviet comedy *Blonde around the Corner* featured a grocery store employee (a woman) falling in love with an astrophysicist (a man). This description suggests a considerable economic tension between the two. It existed, but not in the way Westerners would expect: The clerk, not the researcher, was considerably better off in the Soviet Union. When they had their first date at the man's apartment (where else, in the absence of decent restaurants?), the woman suggested making a pizza from "whatever leftovers are in the fridge," and then discovered that the fridge was empty—so she had to call her brother to deliver a carload of groceries straight from her store.

Not only were goods not necessarily available at the Soviet shops, but entire categories of stores simply did not exist. For example, there were no shopping malls with brand-name stores, because there were no brands; all clothing was made by the state, with minimal differences among the available models. There were no craft stores, no car dealerships, and no home improvement stores. The main retail forms that did exist were these:

- Specialized food stores, selling baked goods, candy, meat and poultry, or dairy products.
- General grocery stores, including supermarkets from the 1970s on. These had far fewer goods available than the average U.S. or European supermarket, but they were basically organized according to the familiar self-service model (Dries et al., 2004).
- Specialized hardware stores, electronic stores, pet stores, or bookstores. These existed in only a handful of places and had few goods on the shelves.
- Liquor stores.
- Large department stores. The two most famous were GUM (Hilton, 2004) and TsUM in Moscow.
- Street kiosks selling newspapers, ice cream, or tobacco products.
- Second-hand stores, called *komissionnye*.
- Farmers' markets. This was the only retail form that allowed negotiation over prices. Only fresh farm produce and small crafts were allowed to be sold in these markets.

In addition there were two types of stores not accessible to average citizens:

- Hard-currency (*beriozka*) stores, where the fortunate few who could travel abroad were able to spend their earnings. These stores also catered to the handful of foreigners in the country.
- Distribution centers for *nomenklatura* party members.

Clearly, Soviet retail needed a major overhaul, which came in the 1990s with the reforms. The first new retail form that emerged in that period was the most archaic one: that of street-corner vendors (Chapter 8). Yeltsin issued a decree in January 1992 that allowed "private citizens to sell things in any place of their convenience" without registration, in an attempt to rapidly boost the availability of goods everywhere without the cumbersome government regulations. So millions of common people took to the streets and resold whatever they could find at the state stores with a small markup. Many were elderly women supplementing their meager pensions. The street retailers would sell their wares at more convenient times and locations than the state stores (e.g., near subway station exits during the evening rush hour). Street kiosks proliferated a bit later, selling everything from bubble gum to bread, beer, cigarettes, condoms, and VHS tapes. Many were vandalized by racketeers or by street gangs.

The next stage was the development of large outdoor markets selling not only fresh produce, but also cheap imported goods from other parts of Russia, as well as from Turkey, China, Vietnam, and other countries. The prices at these markets were considerably lower than at stores, reflecting their low overhead.

At about the same time, the first private stores appeared. They were boutiques charging exorbitant prices for brand-name Western luxury goods. Soon thereafter, cheaper private stores opened: bakeries, bookstores, pharmacies, and so on. Many of these were privatized former Soviet establishments, but some were brand-new ventures. Eventually some stores grew larger and spawned chains and franchises. Some evolved into giant retail empires, including Seventh Continent, Perekrestok, and Pyaterochka. They are big in Russia, but not in the global sense. In a recent ranking by Deloitte and Touche, only two Russian retailers made it to the top 250 worldwide: Group X5 (which includes Pyaterochka and Perekrestok) in 191st place, and Euroset (the main cell phone retailer in Russia) in 229th place. Group X5's sales volume was a meager $3.5 billion in 2006, as compared to $348 billion for Walmart, $66 billion for Kroger, and $40 billion for Safeway in the same year.

The French Auchan and German Metro chains are competing with Russian supermarkets now (Roberts, 2005); and Walmart is expected to enter the country in the next few years. Most foreign retailers are from Europe and Asia, not from North America. American retailers were slow in coming. Starbucks, for example, had early plans to enter Russia, but then bailed out; it even lost its trademark registration to a Russian brand squatter, after Starbucks failed to open a single store in 5 years. It finally arrived, however, with a few dozen stores open in 2008. The European and Asian competitors of the U.S. companies were less picky about legal niceties and so managed to fare better. I have already mentioned the triumph of the Swedish furniture retailer IKEA in the Moscow market (Chapter 20). One of the main obstacles all foreign chains face, besides local competition, is the lack of large plots of land suitable for big-box construction in the densely packed Russian cities. Most chains chose locations along the periphery, but doing so excludes many potential customers who do not own cars. The most successful future retailers in Russia will have to learn what appeals to the most people, not to the selected clientele of Gucci and Armani boutiques (Karpova et al., 2007). Walmart may be just what Russia needs; however, its global expansion at the moment seems to be focused more on China, a much larger market.

In recent years, both domestic and foreign retailers have shown steady improvements in customer service. Many now have loyalty discount cards; virtually all accept bank-issued credit and debit cards for payment; most have regular sales events, free parking, usually free bagging, and (increasingly) consumer credit with favorable interest rates. In short, the shopping experience in the biggest cities in Russia is not much different from that in North America, Europe, or Japan. In the provincial towns, however, relatively little has changed. One may run into an occasional unfriendly shopkeeper, the floors of the store may be dirty, and the range of products on the shelves will be more limited.

The Russian restaurant business is also booming. There are restaurants of every imaginable kind, both foreign and domestic (Figure 21.7). Of the major U.S. restaurant chains, only McDonald's is well represented, with almost 200 restaurants in 37 cities serving half a million customers per day. Pizza Hut had some early successes, but later faltered. There is no shortage of pizzerias, though, mainly of the domestic variety. Some of the best restaurants in Russia are the creations of Arkady Novikov, chef and entrepreneur, who conceived a few widely popular chains (Yolki-Palki, Kish-Mesh, Little Japan, and others). His main idea was to provide high-quality food with a regional theme—Russian, Central Asian, or East Asian, for example—at reasonable prices (Baker & Glasser, 2005). He also started a number of

FIGURE 21.7. T.G.I. Friday's, anyone? *Photo:* A. Fristad.

high-end restaurants (Vanil and Cantinetta Antinori in Moscow, for example), but most people are not likely ever to dine in these places. An average dinner bill at some of the better restaurants in Moscow will set you back about $100, not $15 or $20 as at democratic Yolki-Palki. In recent years the top-rated cuisines have been Japanese, Thai, French, Central Asian, and Mediterranean. Ukrainian, Georgian, and of course Russian foods are also well represented.

Leisure services include all occupations and organizations dealing with leisure activities, from tourism to therapeutic recreation to parks. Generally these are services that people use voluntarily for pleasure, although other motives (e.g., the pursuit of health) may also be involved. Some of the services within this category have already been discussed in Chapters 15 and 16 (arts and tourism, respectively). Here I focus on other types. As Russians' disposable incomes grow, so does the demand for leisure services. As everywhere in the world, the biggest cities have the most creature comforts available to their residents. Moscow, St. Petersburg, Novosibirsk, and 15 or so of the other biggest cities in Russia have virtually every imaginable leisure service available: multiplex cinemas, aqua-parks, disco clubs, dancing halls, bowling alleys, tanning salons, fitness centers, spas, hairdresser shops, tattoo parlors, oxygen therapy bars, and dozens of other types of ventures. Some (e.g., swimming pools, cinemas, and discotheques) had existed in the Soviet period but are now newer, bigger, and better. Others, like aqua-parks, are newly imported concepts.

The majority of leisure services available to city residents cater to the same crowd: young people with extra money, typically between the ages of 18 and 35, and usually either students or professional employees in business or government. This group is supposed to represent the new "Russian middle class," although the exact definition of this varies (Chapters 10 and 12). Some high-end establishments (e.g., private clubs) are open to members only and cater to a considerably older, really well-off segment of the population. There are relatively few options available for the elderly, since that segment has small incomes. Facilities for children, on the other hand, are numerous, with playgrounds and children's parks being the two most obvious ones.

Some of the newest forms of entertainment are very expensive. The market demand is growing fast, and the supply does not catch up quickly enough. In a recent marketing survey, an average annual membership in a Moscow fitness club cost $1,600, while one in Japan cost only $500. Only about 1.5% of the Russian population has a

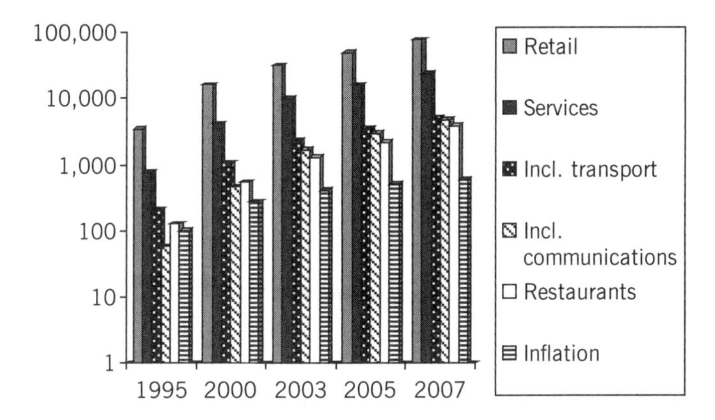

FIGURE 21.8. Service sector indicators (thousands of rubles per capita, not adjusted for inflation) in Russia. Inflation is shown as percent per year relative to 1995 (= 100). Growth in services has surpassed inflation in all categories. Data from the Federal Service of State Statistics, Russian Federation (2009).

membership in any fitness club. Needless to say, for those people it represents a major acquisition, a boost to their self-image.

Although many of the new services are commercialized, most of the municipal services are still free—libraries and city parks, for example. In nice weather, city boulevards are crowded with strolling pedestrians. Older people play chess or dominoes; younger kids play soccer or tag; moms promenade with strollers; young couples hug affectionately. Options in smaller towns and especially villages are more limited. Typically, small cities have at least a city park, a cinema, a theater, and a stadium. Village life is even more basic, with perhaps a single building (the so-called club) dedicated to multiple recreational purposes. In some corners of Russia life has hardly changed at all, with people's main leisure choices limited to TV and gossip about the neighbors. Figure 21.8 summarizes the growth in retail services in Russia between 1995 and 2007.

QUESTIONS

1. Why do you think intercity train and bus service are so underdeveloped in the United States? Why are they so well developed in Russia and other FSU republics?
2. If you had a chance to travel from Moscow to Vladivostok, what mode of transportation would you choose? Explain your choice. What if you had to travel from St. Petersburg to Murmansk? From Kazan to Sochi? From Moscow to Almaty?
3. Which of the following countries is likely to have the fewest fitness clubs per capita: Russia, Ukraine, Armenia, or Tajikistan? Explain your rationale.
4. Which countries of the FSU would be most attractive to large retailers from the United States? Explain your choice.

EXERCISES

1. Look at the map of pipelines in Chapter 17 (Figure 17.4). Propose three alternative routes for exporting Russian oil and gas to East Asia, and two to Western Europe, to the ones that are currently being used.
2. Propose three priority routes for high-speed train development in the FSU. Use the map of existing train routes (Figure 21.1), a population map, and any other maps that may help guide your decision. Defend your plan in a class presentation.
3. Research any combination of airlines available to you to travel from your nearest metropolitan area to Moscow, St. Petersburg, Kiev, Yerevan, Astana, and Bishkek. How many options involve a carrier from the FSU? How many involve only major Western airlines? What are the price differences? Which option would you choose if you had to travel yourself? Why?
4. Do a quick search on the numbers of Internet pages in your country. Compare them to the numbers in Russia.
5. What are other common non-English Internet languages?

6. Make a list of all types of leisure services available in your city or town. Which of these would you expect to find in Russia? Which ones would you not expect to find? Why might this be the case?

Further Reading

Baker, P., & Glasser, S. (2005). *Kremlin rising: Vladimir Putin's Russia and the end of revolution.* New York: Scribner.

Derudder, B., Devriendt, L., & Witlox, F. (2007). An empirical analysis of former Soviet cities in transnational airline networks. *Eurasian Geography and Economics, 48*(1), 95–110.

Dries, L., Reardon, T., & Swinnen, J. F. M. (2004). The rapid rise of supermarkets in Central and Eastern Europe: Implications for the agrifood sector and rural development. *Development Policy Review, 22*(5), 525–556.

Hilton, M. L. (2004). Retailing the revolution: The state department store (GUM) and Soviet society in the 1920s. *Journal of Social History, 37*(4), 939–964.

Karpova, E., Nelson-Hodges, N., & Tullar, W. (2007). Making sense of the market: An exploration of apparel consumption practices of the Russian consumer. *Journal of Fashion Marketing and Management, 11*(1), 106–121.

Larimo, J., & Larimo, A. (2007). Internationalization of the biggest Finnish and Swedish retailers in the Baltic states and Russia. *Journal of East–West Business, 13*(1), 63–82.

Lemarchand, N. (2001). Le développement du commerce en Russie: La place de Moscou. *Bulletin d'Association de Geographes Francais, 2001*(4), 363–372.

Lentz, S. (2000). Die transformation des stadtzentrums von Moskau. *Geographische Rundschau, 52*(7–8), 11–18.

Lorentz, H., & Hilmolla, O.-P. (2008). Supply chain management in emerging market economies: A review of the literature and analysis of the Russian grocery retail sector. *International Journal of Integrated Supply Management, 4*(2), 201–229.

Montaigne, F. (1998, June). Russia's iron road. *National Geographic, 193*(6), 2–33.

Panfilov, V. S., & Chernovolov, M. P. (2007). The condition and trends of the modern retail market system in Russia: Macroeconomic and financial aspects. *Studies on Russian Economic Development, 18*(4), 417–427.

Plisetsky, E. L. (2004). *Sotsialno-ekonomicheskaya geografiya Rossii. Spavochnoe posobie.* Moscow: Drofa.

Roberts, G. H. (2005). Auchan's entry into Russia: Prospects and research implications. *International Journal of Retail and Distribution Management, 33*, 49–68.

Starobin, P. (2008, June). Send me to Siberia: Oil transforms a Russian outpost. *National Geographic*, pp. 2–27.

Warf, B. (2009). The rapidly evolving geographies of the Eurasian Internet. *Eurasian Geography and Economics, 50*(5), 564–580.

Zook, M. A., & Brunn, S. D. (2006). From podes to antipodes: Positionalities and global airlines geographies. *Annals of the Association of American Geographers, 96*(3), 471–490.

Websites

www.airastana.com—Air Astana, the main Kazakhstan airline.

www.domodedovo.ru/en—Domodedovo Airport, the biggest and most modern in Russia.

www.internetworldstats.com—A source for current statistics on Internet usage in the FSU.

www.novikovgroup.ru—Restaurants of Arkady Novikov, including Yolki-Palki. (In Russian only.)

www.riverfleet.ru—Photos and descriptions of many models of the river fleet used in Russia. (In Russian only.)

eng.rzd.ru—Russian Railways (still a federal monopoly).

PART V

Regional Geography of Russia and Other FSU States

CHAPTER 22

Central Russia

THE HEART OF THE COUNTRY

This chapter begins this book's section on the regional geography of Northern Eurasia. "Regions" in this context have been defined as "human constructs ... of considerable size, that have substantial internal unity or homogeneity, and that differ in significant respects from adjoining areas" (Hobbs, 2009, p. 4). In the United States, examples of regions include the Midwest and the South; in Europe, they include Scandinavia and the Mediterranean. In Northern Eurasia or the former Soviet Union (FSU), there are 15 countries in four groups: the Baltic states; Russia, Belarus, Ukraine, and Moldova; the three states in the trans-Caucasus; and the five states of Central Asia. Russia is presently divided into seven regions, distinguished on the basis of political units.

Chapter 8 details the number of regions within the Russian Federation during the Soviet and immediate post-Soviet periods, when it was common to divide Russia into 11 economic regions and almost 90 oblasts and autonomous republics. It also describes the new scheme of 7 federal districts and 83 subjects of federation (see Chapter 8, Table 8.3 and Figure 8.3). It takes a while to learn the names and geographic peculiarities of even a few of these units (see Table 22.1 and Figure 22.1).

TABLE 22.1. Comparative Characteristics of the Seven Federal Districts of Russia

District	Administrative center	Subjects	% of area	% of population	% of GDP	Areas of economic specialization
Central	Moscow	18	4	26	31.5	Machinery, banking, retail
Northwest	St. Petersburg	11	10	10	10	Machinery, forestry, fishing
Volga	Nizhniy Novgorod	14	6	21	16.5	Oil and gas, machinery, agriculture
South	Rostov-on-Don	13	3.5	16	7.5	Agriculture, recreation
Urals	Yekaterinburg	6	10.5	8.5	18	Oil and gas, metallurgy, defense
Siberia	Novosibirsk	13	30	14	11.5	Coal mining, metals, forestry
Far East	Khabarovsk	9	36	4.5	5	Fishing, defense

Note. Data from the Federal Service of State Statistics, Russian Federation (2006).

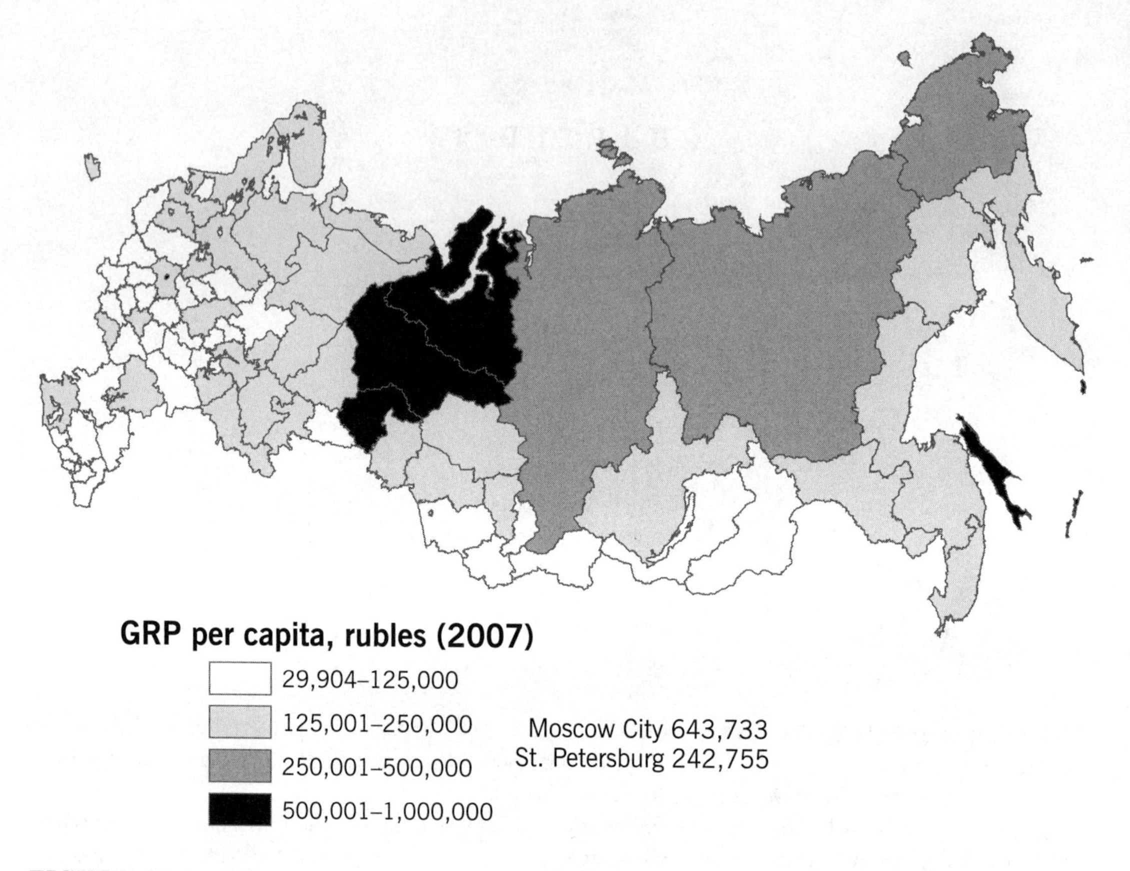

FIGURE 22.1. Gross regional products (GRPs) of Russia's subjects of federation in 2007. Data from the Federal Service of State Statistics, Russian Federation (2009).

The region of Russia centered on Moscow (Figure 22.2) was formally defined during the Soviet period as consisting of Moscow itself and the 12 oblasts immediately surrounding it. An "oblast" is broadly analogous to a U.S. state or a Chinese or Canadian province. However, oblasts have not been as stable as the U.S. states; their borders and names have frequently changed over the last 150 years. Most have remained unchanged for the last 30 years, with the exception of some recent mergers of the small ethnic autonomous okrugs in Siberia with larger Russian-populated krays. Moscow and the 12 oblasts around it constituted the so-called Central economic region, which was used by Soviet geographers to report statistical data. This region was analogous to the mid-Atlantic states in the United States, centered on the national capital with high population density and heavy dependence on government jobs. Since 2000, the new Central federal district created by the Putin administration has included an additional five oblasts to the south of Moscow, which were previously classified as part of the Chernozemny economic region adjacent to Ukraine (see Table 8.3). Because the most recent governmental statistics pertain to this new federal district, it makes sense to discuss it here as one unit. Therefore, this chapter focuses on the 17 oblasts in the middle of European Russia and the federal city of Moscow (Table 22.2).

The Central region/district thus defined (0.7 million km^2) accounts for only 4% of Russia's territory but over 25% of its population (38 million people), and is 80% urban. The population density is 58/km^2, which is seven times Russia's average. However, the population distribution is uneven: The region's center is the city of Moscow, a giant with over 11 million people, and an additional 5 million in the surrounding Moscow Oblast. There is no exact analogue to Moscow

FIGURE 22.2. The Central federal district of Russia. Map: J. Torguson.

City or Oblast in the United States: Moscow combines the functions of the U.S. political (Washington, D.C.), financial (New York), and historical (Philadelphia) capitals. A better analogue would be Paris, the primate city of France, and the Île-de-France region surrounding it.

The highways, railroads, and power lines converge on the national capital like spokes in a wheel (see, e.g., the rail map in Chapter 21, Figure 21.2). At the same time, few people live over large swaths of rural territory north of Moscow—for example, in northeastern Kostroma Oblast or northwestern Tver Oblast, where the population densities approach Siberian levels (12/km^2). The region's population is growing because of immigration from other parts of the country and from other parts of the FSU. The natural growth rate, however, is negative.

TABLE 22.2. Subjects of Federation of the Central Federal District

Subject	Administrative center	Population (2006) (thousands)	% ethnic Russian	GRP/capita (rubles in 2006)	Top products and areas of economic specialization within Russia (% of nation's total production for some)
City of Moscow	Moscow	10,425	85	493,190	Machine building, food processing (30%), banking, education, research
Moscow Oblast	Moscow	6,628	91	141,397	Large train engines (100%), tea and jams (42%), drugs (17%), paints (16%)
Tver Oblast	Tver	1,407	93	89,784	Railroad cars (63%), compressors and excavators (43%), flax
Yaroslavl Oblast	Yaroslavl	1,328	95	118,187	Air conditioning engines (33%), paints (20%), auto engines (15%), tires (13%)
Kostroma Oblast	Kostroma	709	96	75.154	Spinning machines (100%), plywood (11%)
Ivanovo Oblast	Ivanovo	1,100	94	47,950	Fabrics (57%), truck-mounted cranes (42%), flax
Vladimir Oblast	Vladimir	1,473	95	76,328	Wheelchairs (35%), AC engines (33%), tractors (17%), TV sets, glass
Ryazan Oblast	Ryazan	1,182	95	87,651	Potato harvesters (36%), metal cutting machines (12%), oil refining
Tula Oblast	Tula	1,600	95	90,123	Sock-making machines (100%), milk separators (82%), cranes, fibers
Kaluga Oblast	Kaluga	1,014	94	83,817	Cash registers (39%), matches (37%), locomotives, strand board (6%)
Bryansk Oblast	Bryansk	1,331	96	61,888	Diesel locomotives (63%), radiators (28%), cranes (14%), ethanol, cheese
Smolensk Oblast	Smolensk	1,006	94	79,254	Large electric engines (25%), canned milk (18%), butter (2%)
Orel Oblast	Orel	834	95	75,222	Centrifuge pumps (43%), city equipment (14%), tile (13%), loaders (9%)
Kursk Oblast	Kursk	1,184	96	85,350	Rubber belts (48%), batteries (44%), cash registers (44%), iron ore (18%)
Lipetsk Oblast	Lipetsk	1,181	96	159,468	Refrigerators and freezers (44%), fruit preserves (28%), steel (15%), sugar
Tambov Oblast	Tambov	1,130	97	69,840	Synthetic paints (41%), sugar (10%), ethanol (6%), potato starch
Belgorod Oblast	Belgorod	1,511	93	119,673	Iron ore (36%), asbestos roofing (16%), sugar (13%), washers (9%)
Voronezh Oblast	Voronezh	2,314	94	70,849	Winnowers (86%), ore enrichment machines (30%), synthetic rubber (19%), sugar

Note. Data from the Federal Service of State Statistics, Russian Federation (2009). To convert gross regional product (GRP) to U.S. dollars, divide GRP by 27.19.

Physical Geography

Much of the Central region fits in the watersheds of the Volga and Dnieper rivers. The relief is flat, slightly undulating plain covered with thick glacial deposits, at about 300 m above sea level. There are few valuable mineral deposits, except the massive deposits of iron ore (55% of Russia's total) near Kursk. Peat (about one-quarter of Russia's total), construction stone, gravel, and sand industries are well developed. Slow-flowing rivers, numerous lakes, and many wetlands provide for excellent navigation, fishing, and easy access to many natural areas in the region. Some small rivers, such as the Klyazma (Figure 22.3), Moskva, and Upa Rivers, are heavily tapped for municipal needs.

The climate is mildly continental, with about 4 months of winter (average January temperature = –10.8°C) and four distinct seasons. Summers are warm and moderately wet (average July temperature = +18.4°C). There is a well-developed gradient of annual precipitation across the region from the northwest (600 mm) to the southeast (420 mm). Soils are primarily poor podzols (spodosols) in the north, and considerably richer gray forest soils (alfisols) and chernozems (mollisols) in the south. Because of the moisture gradient, the northern half of the region is in the coniferous and mixed forest zone, while the southern half is in the forest–steppe and true steppe zones. In the northern half (Tver, Kostroma), forestry is common; in the southern half (Lipetsk, Tambov, Belgorod), agriculture is more important. Moscow Oblast is about 40% forest covered, similar to Minnesota. The Central region accounts for about 19% of Russia's arable land, and the chernozem soils in Kursk, Belgorod, and Tambov Oblasts are among the most productive on the planet. A few world-class nature parks are also located in the Central region. These include Priokssko-Terrasny Zapovednik, where one can observe the endangered Eurasian wood bison (*zubr*); Ugra National Park in Kaluga Oblast and Bryansky Les Zapovednik in Bryansk Oblast, both well known for strong natural interpretation programs and a good blend of cultural and natural elements on their trails; and Valdaysky National Park on the border with Novgorod Oblast in the Northwest region, providing an authentic experience of glacial lakes and fishing in a southern taiga setting.

FIGURE 22.3. The Klyazma River. *Photo:* Author.

Cultural and Historical Features

Culturally, the region is overwhelmingly Russian (Table 22.2), with some Uralic minorities living in the forested east and north (Mordvinians and Mari) and Ukrainians in the south. The presence of cosmopolitan Moscow also makes this region the most visited by foreigners. There are an estimated 2 million migrant workers and refugees from near and far abroad in Moscow City and Oblast. The largest minorities in Moscow include Tatars, Jews, and Ukrainians, who have lived here for centuries, but also Azerbaijanis, Chechens, Georgians, Moldovans, Chinese, Vietnamese, and a host of Western nationals (Americans, Germans, French, British, etc.). The region has the best universities in the country. It also has the greatest number of theaters, museums, sports events, cultural sites, and other cultural landmarks, although the vast majority of those are located in Moscow City, as discussed in Chapter 15. Outside Moscow, the most important historical and cultural places are located along the Golden Ring northeast of Moscow, a tourist route of international importance that includes many sites on the World Heritage list (see Chapter 16, Table 16.1): Vladimir, Suzdal, Rostov, Kostroma, Yaroslavl (Figure 22.4), Uglich, Rybinsk, and others. Many of the cities along the way predate Mos-

FIGURE 22.4. Historical Yaroslavl, part of the Golden Ring tourist itinerary. *Photo:* S. Blinnikov.

cow, and contain important religious and cultural artifacts, kremlins, cathedrals, museums, and newly reopened monasteries. These are also among the most visited places in Russia, popular with large tour groups.

Good alternatives to the heavily traveled Golden Ring cities are the historical cities of Kasimov in Ryazan Oblast (Figure 22.5) and Murom in Vladimir Oblast. Located about 4 hours east of Moscow by car, they provide excellent examples of well-preserved old city centers with important Uralic (Murom) and Tatar (Kasimov) influences. Each boasts many churches, some monasteries, typical pre-Soviet merchant-class districts, and pretty riverfronts along the Oka. Both also have a fair share of Soviet-era-built apartment blocks and large factories, and are thus also very typical (rather than exceptional) cities. Murom is astonishingly old; its official founding date is 862 A.D., which is about three centuries older than Moscow and about 1,000 years older than many U.S. or Canadian cities. With a population of only 126,000, Murom is not large, although it is advantageously located on the lower Oka River, the largest right tributary of the Volga. The deep forests and thick wetlands of the Meshchera Lowland between Kasimov and Murom are legendary for their pristine wilderness.

Another cultural specialty of the Central region is artistic landscapes. Many poets, writers, composers, and artists had their estates in villages located around Moscow. Of particular note are Yasnaya Polyana (Leo Tolstoy) in Tula Oblast; Spasskoe-Lutovinovo (Ivan Turgenev) in Orel Oblast; and Melikhovo (Anton Chekhov), Abramtsevo (many famous artists worked there), Zvenigorod (Mikhail Prishvin), and Klin (Pyotr Tchaikovsky), all in Moscow Oblast. Right next to Moscow, the village of Peredelkino attracts thousands of people to the grave of Boris Pasternak, the author of *Dr. Zhivago* and one of the best Russian poets of all time. Tarusa, about halfway between Kaluga and Moscow, has sites associated with the life of another famous poet, Pasternak's friend and contemporary Marina Tsvetaeva, as well as with Anton Chekhov, Konstantin Paustovsky, Nikolai Zabolotsky, and others. The painter Vasily Polenov lived and worked near Tarusa as well.

The areas west and north of Moscow are famous for historical sites associated with many wars. Dear to every Russian heart is the Borodino battlefield, so aptly described in *War and Peace*, the site of the epic battle between Napoleon's and Kutuzov's armies in 1812. Borodino hosts an annual reenactment of the battle in September—probably the largest such event in Russia, involving tens of thousands of people—and has numerous monuments and historical landmarks. Sites associated with the much bloodier

FIGURE 22.5. Panorama of Kasimov on the Oka, Ryazan Oblast. *Photo:* Author.

and more recent World War II also abound in the same area, between Moscow and Smolensk. Smolensk, Vyazma, and Kaluga saw some of the fiercest battles in 1941–1943, as the Nazis were advancing toward and then retreating from Moscow. It is still not uncommon to discover unexploded mines buried deep in the woods. Far to the south, the battle of Kursk (summer of 1943) was the largest tank battle in history, involving over 5,000 tanks and millions of men. The battle unfolded from north-central Ukraine to Kursk and Belgorod Oblasts.

Economics

The Central region is the heart of Russia not only historically and culturally, but also economically, accounting for almost a third of the nation's GDP. The main economic strengths of the region are manufacturing and services. The presence of the best universities and research centers, as well as a large consumer base, makes the Central region an attractive place for domestic and foreign investments. In 2006 it attracted about one-quarter of all domestic investments in Russia and 54% of all foreign investments.

Manufacturing remains the key economic activity, accounting for about one-quarter of the region's total industrial output. Some of the earliest factories in Russia emerged here in the 18th century (flax mills in Ivanovo, gun factories in Tula, ceramics and crystal glass manufacturers in Ryazan and Vladimir, etc.). The second most important industry is food processing, to satisfy the appetites of the large population. Construction materials, energy generation (mainly from coal, natural gas, and nuclear power), and chemicals are also well-represented industries.

Among the machinery-building centers, besides Moscow, a few medium-sized cities can be mentioned. Kolomna, southeast of Moscow, is the leading producer of large diesel train engines in Russia. Railroad cars are built in Tver and Bryansk. Kostroma and Rybinsk have shipbuilding facilities. Zhukovsky, near Moscow, is the leading development and testing site for the aerospace industry in Russia. Ivanovo produces construction cranes. Vladimir makes tractors. A number of military factories are also located in the region (Zelenograd, Kaluga, Ryazan, Korolev, and Dubna), producing handguns, ammunition, short- and long-range missiles, radioelectronic equipment, lasers, space satellites, and other items. One of the dirtiest industries of the Central region is its chemical industry, mainly concentrated east and south of Moscow in Vladimir, Orekhovo-Zuevo, Shchekino, and Novomoskovsk. Plastics, fertilizer, shampoos, creams, gels, detergents, and so on are made here. The textile industries and associated clothing manufacturers are concentrated in the opposite end of the region, north and west of Moscow in the areas historically suitable for growing flax. Today, however, most textiles and clothing are made from imported cotton in Ivanovo, Smolensk, Tver, Kostroma, and Vladimir Oblasts. Ivanovo Oblast alone makes 57% of all textiles in the country.

The construction industry is booming in and around Moscow. The region accounts for about half of all cement produced in Russia; it also accounts for one-third of all housing units built in recent years. There is a lot of individual housing construction in the suburbs, as well as busy reconstruction of old city buildings. The region is able to satisfy most of its needs for construction materials from local sources. Cement, plywood, glass, brick, metal and wood farming, flooring, siding, and roofing materials are all produced here. Finnish, German, and Canadian firms predominate in making contemporary composite materials for siding and roofing. The production of construction materials is concentrated in Podolsk (Moscow Oblast), Tula, Bryansk, Voskresensk, and Ryazan.

Kursk, Belgorod, Tambov, and Lipetsk Oblasts are heavily farmed, with wheat as the main crop grown on the excellent chernozem soils (similar to the mollisols of the U.S. Midwest). They also contain the largest facilities for mining and processing iron ore in Russia, in the area known as the Kursk Magnetic Anomaly (KMA). The KMA contains so much iron ore that a magnetic compass cannot be used here, because the needle points down instead of north! It contains about 31 billion metric tonnes (mmt) of iron ore in reserves. (This is about twice as much as is left in all of the United States, mostly in Michigan and Minnesota.) Much of the KMA mining takes place in open pits. Belgorod Oblast has the best ores; however, they must be dug up from deep underground. Unfortunately, the KMA lacks local energy sources, so these must be imported. Nearby Lipetsk Oblast has one of the largest steel-processing plants in the country, the Novolipetsky Combine, as well as a large steel pipe plant and a refrigerator plant. Stary Oskol has another large, modern steel plant. Lipetsk has some of the highest salaries in Russia for workers outside Moscow and western Siberia.

With respect to agriculture, the Central region has diverse specializations. Near Moscow are mainly potato, vegetable, and fruit farms, as well as dairy production. North and west of Moscow are areas of flax cultivation, while wheat and potato farming take place in the "black earth" belt to the south. As explained in Chapter 20, much of this farming actually takes place on tiny private plots next to dachas. For example, 93% of the potatoes in the region come from such plots, and less than 5% from large agricultural enterprises. However, egg and dairy production are dominated by large poultry farms. Despite heavy production, not enough grain is grown in the Central region (only about 5–6 mmt), so additional grain must be imported.

The biggest problem with agriculture in the region is the lack of young people in rural areas. The farming population is rapidly aging; the average villager's age is over 50 years, versus 38 years countrywide. Young people are leaving for jobs in the cities, and no one is left to replace them. Some communities manage to attract immigrants from other FSU republics and from Siberia, but in most places the need for laborers is great.

Whereas the immediate vicinities of large cities are doing well because of their easy access to urban markets, much of the region's periphery can be characterized as "economic black holes." Such areas are remote from potential markets and cannot support themselves over the long term (Chapter 20); they include western Tver, southwestern Kaluga, eastern Ryazan, northeastern Ivanovo, and eastern Kostroma Oblasts. Over a third of all arable land has been abandoned in these areas since the collapse of socialism.

The infrastructure of the Central region is well but unevenly developed. Moscow, of course, is the giant hub of communications and transportation networks. It is a true primate city, surpassing the next three biggest cities combined (St. Petersburg, Novosibirsk, and Nizhniy Novgorod). It is served by 11 major railways, three large airports, two ferry terminals on the Moscow River, and a dozen federal highways radiating in all directions. Moscow is surrounded by new cargo terminals, storage warehouses, and customs facilities that serve not only the city, but other areas of the country as well. The density of roads and railways decreases dramatically as one leaves the city. At 50 and 100 km outside the city center, there are ring roads (beltways). Both survive from the Cold War period, when they were built to deploy antiaircraft and missile defense rapidly around Moscow in the event of a NATO attack; they were military roads not shown on maps. Today both are being renovated. The one at the 50-km mark is going to be partially replaced by a private tollway, which would permit improved freight traffic circulation around the city. The Central region is also crisscrossed by a number of large oil and gas pipelines stretching from Siberia and the Volga to Europe, and by electric power lines from nuclear, hydropower, and thermal power plants to the cities. The region is a net exporter of electricity, but a major importer of fossil fuels, petroleum products, and industrial chemicals.

The heaviest concentrations of population and industrial centers occur east (Vladimir and Ivanovo), south (Podolsk, Serpukhov, and Tula), and southeast (Moscow Oblast, Kolomna, and Ryazan) of Moscow itself. Moscow is surrounded by immediate satellite cities (Khimki, Mytischi, Lyubertsy, Krasnogorsk) right next to the 50-km beltway, and more distant satellites (Ramenskoe, Podolsk, Zelenograd, Zvenigorod) about 30 km away. The periphery of Moscow Oblast has a number of medium-sized cities (Serpukhov, Dmitrov, Kolomna, Mozhaysk) located 100 km away from the city. Many of the latter are historical cities over 10 centuries old. Others (Dubna, Obninsk, Pushchino) are Soviet-era towns built explicitly for scientific or weapons research.

The western and northern parts of Moscow Oblast are less attractive for industry or agriculture, but more attractive for tourism and recreation. For example, Istra Rayon, located west (i.e., upwind) of Moscow City, is attractive for year-round outdoor enthusiasts. Old Soviet sanatoria and new private lodges receive hundreds of thousands of visitors per year. Summer camps for children abound. Hunting, fishing, biking, cross-country skiing, and downhill skiing are all well developed here. A new development will create "Moscow Switzerland" on a vast tract of land in the Central district in the near future, complete with artificial ski slopes, ski lifts, and glitzy restaurants and bars. For those wishing a wilder experience, large tracts of forests in the Egoryevsk, Dmitrov, and Taldom areas allow for plenty of hunting. The Zavidovo reserve in the extreme north of Moscow Oblast is the traditional hunting area for members of the Kremlin administration, Duma deputies, and important foreign guests.

The surrounding oblasts have certain specializations as well. Clockwise from the top of Figure 22.2, these oblasts (and their capital cities) are Tver, Yaroslavl, Kostroma, Ivanovo, Vladimir, Ryazan, Tula, Kaluga, and Smolensk. South of Tula and Kaluga are Orel Oblast and five oblasts in the "black earth" belt: Bryansk, Lipetsk, Voronezh, Kursk, and Belgorod. The oblast centers are typically located about 200 km from Moscow and are fairly large cities, ranging from almost 1 million (Voronezh) to about 500,000 (Yaroslavl, Ryazan, Tula, Lipetsk) to about 300,000 people (Kaluga, Orel, Kostroma). Most of these cities date back eight or nine centuries, and a few (Vladimir, established 1108 A.D.; Yaroslavl, established 1010 A.D.) predate Moscow. Each oblast center has its own legislature and governor's office; at least one large university; and numerous factories, hospitals, shopping malls, transportation facilities, and so on. The majority of oblasts have a unicentric structure, with the oblast capital dwarfing all other cities. However, Rybinsk in Yaroslavl Oblast and Kovrov and Murom in Vladimir Oblast are cities just a little smaller than the capitals. The largest oblast by area, Tver, is also the least densely settled. Generally, the northern oblasts (Tver, Kostroma, Yaroslavl, Ivanovo) specialize in timber and textile processing, while the southern ones (Kursk, Belgorod, Tambov) specialize in agriculture. The oblasts in the middle are most heavily industrialized (Tula, Ryazan, Kaluga, Vladimir) or have an intermediate agricultural–industrial profile (Bryansk, Voronezh, Smolensk).

Challenges and Opportunities in Central Russia

The Central region is uniquely positioned in Russia to take full advantage of new economic opportunities: It is the most centrally located, richest, best educated, and most globally connected part of the country. However, some of its peripheral units, especially in the north and south, will need a lot more time to catch up with the rest of the area economically. For example, the average gross regional product (GRP) per capita in 2004 in the poorest three oblasts (43,000 rubles) was half of that in the three richest (Moscow, Yaroslavl, and Lipetsk Oblasts at 97,000). The biggest uncertainty at present involves the future of agriculture: After the passage of the new Land Code, it has become possible to purchase agricultural land. This may help farmers on the one hand, but will lead to massive land speculation on the other. Already large areas of Kaluga and Ryazan Oblasts are experiencing speculative increases in land prices. Many suburban zones of oblast centers are being converted into residential housing for the rich, to the detriment of the local environment and social fabric.

In addition, the future of the region is uncertain because of the lack of funding and political goodwill for technological innovations, infrastructure improvements, or local land policy change. For example, the old academic cities near Moscow (Obninsk, Chernogolovka, Pushchino, Dubna) have suffered from years of neglect of even the most basic infrastructure, due to the drop in funding from the Russian Academy of Sciences. Many years of private and public investments are needed to reverse the trend. Likewise, some historical small cities (Klin, Torzhok, Kozelsk) look like ghosts of their former past, with many vacant lots, weed patches, broken pavement, and Soviet-era apartment blocks in serious need of repair. They could play a new role as tourism magnets, given the proper regional planning and adequate investments.

At the other end of the economic spectrum, Moscow and a few other booming regional centers are trying to cope with the influx of migrant workers, rising incomes, and unbridled consumption. The time spent in traffic jams alone has increased by a factor of two over the past few years in Moscow. The air gets dirtier every year; water supplies are inadequate; and the electricity grid runs at close to 100% capacity in the summer with the increased use of air conditioners. Constraining the growth of the Moscow metropolis, and encouraging development of its satellite cities and especially other oblast capitals, may be the top challenges facing the Central region's planners. One speculative, but attractive, idea being discussed is moving some governmental functions to St. Petersburg, Nizhniy Novgorod, or even an entirely new city, to take some of the pressure off Moscow. This remains merely a speculation at the moment, although the Constitutional Court did get moved to St. Petersburg late in Putin's presidency.

EXERCISES

1. Use a map or atlas to research and develop a 3-day bus excursion itinerary for the following groups of tourists:

   a. Those interested primarily in the historical battlefields of Central Russia.
   b. Those interested in its religious heritage.
   c. Those interested in cultural, especially literary landscapes.
   d. Those interested in ecotourism and nature tourism.

   This activity may be done as a take-home exercise in your class, with different groups of students working on different topics and later presenting their itineraries to the entire class.
2. Use Table 22.2 and any additional sources that you can find to compare and contrast the economic production of selected oblasts in Central Russia. For example, it may be interesting to compare a northern (Tver, Kostroma) with a southern (Tambov, Belgorod) oblast, to see differences in the relative importance of agricultural crops or types of industries.

Further Reading

Ahlberg, R. (2000). Economic development, civil society and democratic orientation: A study of the Russian regions. *Journal of Communist Studies and Transition Politics, 16*(3), 21–38.

Benini, R., & Czyzewski, A. (2007). Regional disparities and economic growth in Russia: New growth patterns and catching up. *Economic Change and Restructuring, 40*, 91–135.

Blinnikov, M., Shanin, A., Sobolev, N., & Volkova, L. (2006). Gated communities of the Moscow green belt: Newly segregated landscapes and the suburban Russian environment. *GeoJournal, 66*, 65–81.

Boentje, J., & Blinnikov, M. (2007). Post-Soviet forest fragmentation and loss in the green belt around Moscow, Russia (1991–2001): A remote sensing perspective. *Landscape and Urban Planning, 82*, 208–221.

Hobbs, J. J. (2009). *World regional geography* (6th ed.). Belmont, CA: Brooks/Cole.

Hough, J. F. (1998). The political geography of European Russia: Republics and oblasts. *Post-Soviet Geography and Economics, 39*, 63–95.

Ioffe, G., & Nefedova, T. (2001). Land use changes in the environs of Moscow. *Area, 33*(3), 273–286.

Iwasaki, I., & Suganuma, K. (2005). Regional distribution of foreign direct investment in Russia. *Post-Communist Economies, 17*, 153–172.

Mitchneck, B. (1998). The heritage industry Russian style. *Urban Affairs Review, 34*, 28–52.

Mitchneck, B. (2007). Governance and land decision-making in Russian cities and regions. *Europe–Asia Studies, 59*, 735–760.

Pallot, J., & Nefedova, T. (2003). Trajectories in people's farming in Moscow Oblast during the post-socialist transformation. *Journal of Rural Studies, 19*, 345–362.

Rodoman, B., & Sigalov, M. R. (2007). *Tsentralnaya Rossija: Geografiya, istoriya, kultura.* Moscow: Gelios APB.

Wegren, S. K. (2008). Land reform in Russia: What went wrong? *Post-Soviet Affairs, 24*, 121–148.

CHAPTER 23

Russia's Northwest

FISHING, TIMBER, AND CULTURE

The United States has its own Pacific Northwest. Russia's Northwest borders the seas of the Atlantic Ocean and is much farther to the north, but it does have some similarities to its American counterpart; for example, both have a maritime climate, are highly dependent on timber and fishing, and house large navy fleets. Russia's Northwest, however, has St. Petersburg—the second largest city in Russia, its former capital, and one of Russia's top three seaports by tonnage. St. Petersburg is also the unofficial cultural capital of Russia, with its numerous museums, theaters, and famous historical sites. Visitors from abroad arrive in this region through St. Petersburg's Pulkovo Airport; on a train from Moscow; or via one of its three main seaports (St. Petersburg, Murmansk, or Arkhangelsk). Some also come overland from Finland. For Russians, the region is shrouded in the nostalgic imperial past because of St. Petersburg. It is also a perpetual frontier, with a history of border conflicts with the Swedes, Finns, Poles, and Baltic peoples going back over 1,000 years.

This region as discussed here coincides with the Northwest federal district (as defined in 2000) and includes 11 subjects of federation: the Karelia and Komi Republics, Nenetsky Autonomous Okrug, seven oblasts, and the city of St. Petersburg (Tables 8.3 and 23.1). The oblast that surrounds St. Petersburg is an independent subject of federation and is still known by its Soviet name, Leningradskaya. Note that Kaliningradskaya oblast is an isolated exclave on the Baltic Sea; it borders Lithuania and Poland and is surrounded by the European Union (EU) (see Chapter 9, Vignette 9.1). The federal district includes two old Soviet economic regions: the Northwest proper, with four oblasts near the Baltic Sea (Leningradskaya, Pskovskaya, Novgorodskaya, and Kaliningradskaya); and the North, with the other subjects of federation (Figure 23.1).

The Northwest region/district thus defined (1.7 million km^2) accounts for 10% of Russia's area and 10% of its population (14 million people). It is the most heavily urbanized of all Russia's regions, with an 82% urbanization rate. The population density here is exactly Russia's average, 8.3/km^2. However, the population distribution is very uneven: The region's most important city, St. Petersburg, accounts for 5 million people (about a third of the total). Nenetsky Autonomous Okrug, on the other hand, contains merely 42,000 people spread over 177,000 km^2 of territory, roughly the size of Oklahoma. Northwest Russia is entirely in Europe. It borders Finland, Poland, and the Baltic states; the Baltic, Barents, White, and Kara Seas; the Ural Mountains; and the Urals and Central federal districts. The re-

TABLE 23.1. Subjects of Federation of the Northwest Federal District

Subject	Administrative center	Population (2006) thousands	% ethnic Russian	GRP/capita (rubles in 2006)	Top products and areas of economic specialization within Russia (% of nation's total production for some)
Murmanskaya Oblast	Murmansk	865	85	181,488	Phosphate fertilizer (100%), seafood (18%), iron ore (10%), Navy services
Arkhangelsk Oblast	Arkhangelsk	1,291	94	160,530	Pulp (34%), cardboard (25%), lumber (10%), seafood (4%)
Karelia Republic	Petrozavodsk	698	77	124,260	Paper (23%), iron ore (10%), pulp (8%), wood (6%), tourism
Komi Republic	Syktyvkar	985	60	216,296	Paper (15%), plywood (13%), coal (4%)
Nenetsky Autonomous Okrug	Naryan-Mar	42	62	NA	Paper, gas, reindeer herding
City of St. Petersburg	St. Petersburg	4,581	85	177,387	Turbines (90%), cigarettes (21%), tractors (11%), soda (9%), education
Leningradskaya Oblast	St. Petersburg	1,644	90	161,752	Tea (45%), cigarettes (15%), paper (12%), oil refining (9%), cars (3%)
Pskov Oblast	Pskov	725	94	68,713	Small electric engines (27%), dairy, flax
Novgorod Oblast	Novgorod	665	94	110,666	Ammonia (9%), plywood (7%), TV sets (5%), dairy
Vologda Oblast	Vologda	1,235	97	168,772	Steel rolling (18%), flax fabric (10%), dairy
Kaliningrad Oblast	Kaliningrad	940	82	106,422	TV sets (66%), fish canning (33%), fish and seafood (9%)

Note. Data from the Federal Service of State Statistics, Russian Federation (2009). To convert gross regional product (GRP) to U.S. dollars, divide GRP by 27.19. NA, not available.

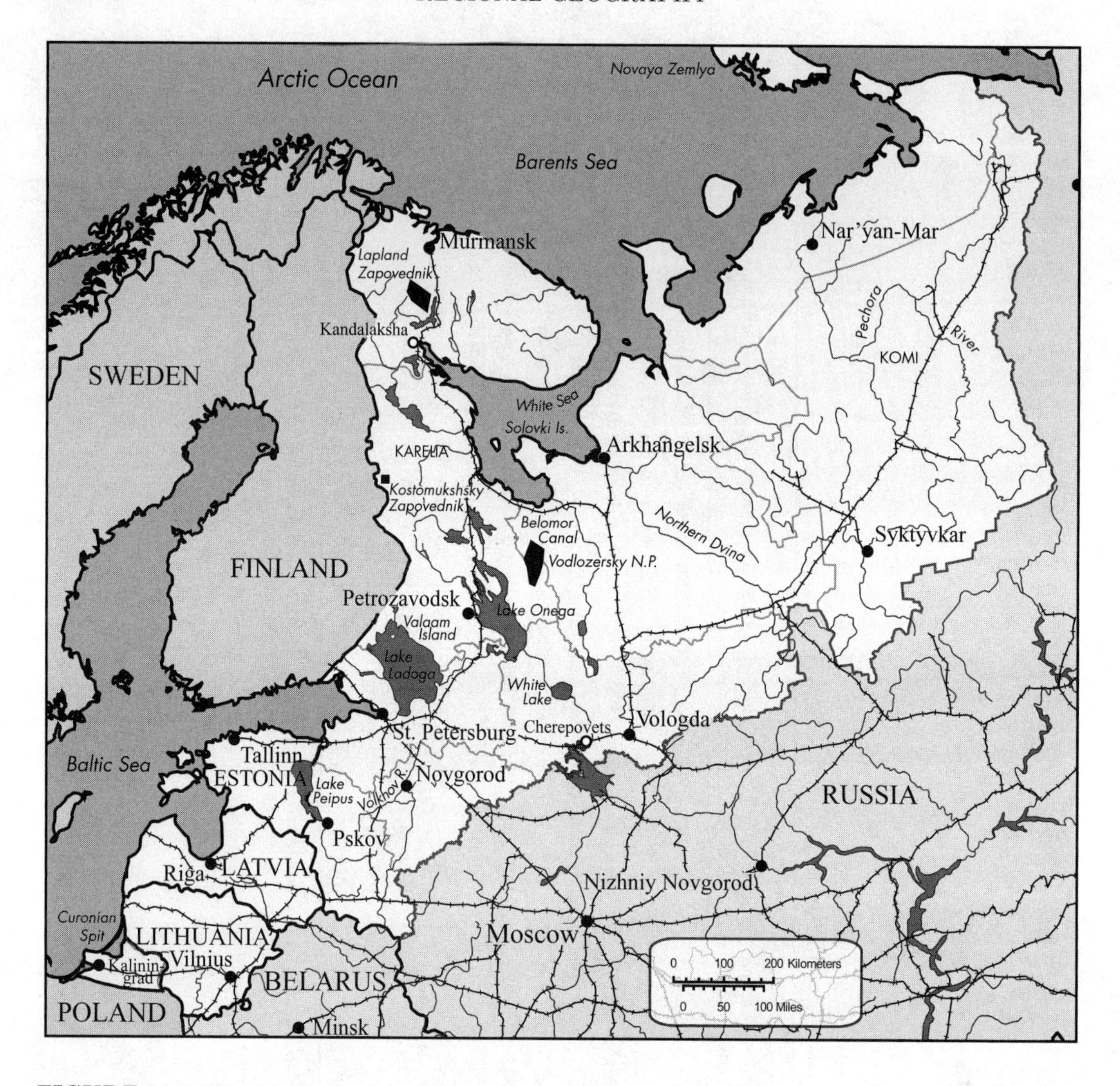

FIGURE 23.1. The Northwest federal district of Russia. (The Baltic states, Lithuania, Latvia, and Estonia, are included at bottom left; see Chapter 29.) Map: J. Torguson.

gion has a lower-than-average birth rate and a higher-than-average death rate. St. Petersburg and Leningradskaya Oblast together have the lowest fertility rate in Russia—1.2 children per woman, as opposed to about 1.4 in Russia as a whole in 2008. With high mortality and low fertility comes rapid depopulation (–0.7% per year in 2005). Only the Far East region's population is declining at a similar rate. Despite considerable immigration from other parts of Russia (Siberia, the Far East, and the Caucasus), as well as from other former Soviet Union (FSU) nations, the region is shrinking by 100,000 people per year. This is equivalent to the disappearance of a sizable city.

Physical Geography

At the height of the last glaciation, much of the Northwest region was buried under the Scandinavian ice shield. As the ice melted, some of the oldest rocks on the planet were exposed in parts of the Kola Peninsula and in Karelia; these are similar in age and composition to the rocks found in eastern Canada and northern Minnesota (over 3 billion years old). The world-record-breaking research shaft near Murmansk is a hole deeper than 15,000 m drilled into crystalline bedrock by geologists to enable them to study the composition and structure of the earth's crust. It is still only about halfway down to the earth's mantle. Thousands of glacial lakes dot the landscape, including large lakes Ladoga and Onega, as well as Lake Chudskoe (Peipus), Ilmen Lake, and White Lake. The biggest river of the region is the Northern Dvina, which enters the White Sea near Arkhangelsk. The short and powerful Svir connects Lake Onega and Ladoga, continuing as the Neva River into St. Petersburg on the Gulf of Finland. One of the best ways to see much of the region is to take a boat cruise along the Volga–Baltic canal. The Belomorcanal stretches from Lake Onega north to the White Sea. Both canals were built with prison labor in the Soviet period. The relief of the region is mostly flat in the south and west, and hilly in the north. The tundra-covered Khibiny Mountains in the Kola Peninsula rise just to 1,000 m above sea level and are treeless. In the extreme northeast of the region, Mt. Narodnaya in the northern Urals rises to 1,895 m.

The climate is humid continental (the Dfb type as defined by the Köppen system) in the southern half, and subarctic (Dfc) in the north. The shores of the Arctic Ocean are in the polar climate type (ET). Winters are over 5 months long, and insolation (solar radiation) is low (Figure 23.2). St. Petersburg is located at 60°N, the same latitude as Stockholm, Oslo, or Anchorage, Alaska. If in June one can read outside at 2:00 in the morning "without a lamp" (as Alexander Pushkin famously said), in winter one has only 5–6 hours of daylight, and winter-caused depression is common. Murmansk, above the Arctic Circle (68°50'N), experiences real polar night: No sun is visible at all for about 45 days from early December until mid-January. Because of twilight it is not absolutely dark outside for a few hours per day even then, but local residents have to cope with the long period of darkness. Coastal locations in the region experience a damp maritime climate. St. Petersburg is infamously soggy; rain is common any time during the summer, and a lot of heavy wet snow is common in winter. The coldest temperature recorded in St. Petersburg is –36°C; the hottest is +33°C. The coldest temperatures inland from Arkhangelsk may plunge to –51°C, approaching the coldest recorded in the lower 48 U.S. states. Bioclimatically,

FIGURE 23.2. Winter in Arkhangelsk Oblast. *Photo:* A. Shanin.

the region is not warm enough for most crops to grow. In the middle of the region, near Petrozavodsk, Karelia, the vegetative season is not long enough even for wheat to grow. Only rye, barley, oats, some hardy potato varieties, green peas, radishes, turnips, carrots, and flax can ripen here. Soils are primarily poor podzols or peat soils (histosols) in the north and richer turf podzolic soils in the south.

Shrubby and mossy tundra covers much of Nenetsky Autonomous Okrug and the Kola Peninsula's shores. Much of the region, however, is covered in boreal forests (taiga). This region is home to the largest surviving fragments of taiga old growth in European Russia (Chapter 4, Figure 4.6), especially in areas of the Karelia and Komi Republics away from the railroads. About half of the forests in the region are Scotch pine stands; the other half are mixed forests of birch, aspen, fir, and spruce. Generally, the vegetation resembles northern Minnesota and parts of Ontario (albeit with fewer species of trees), and is almost identical to nearby Finland. The Northern Dvina valley is covered in azonal grasslands, which make excellent pastures.

The region has many zapovedniks and national parks. Laplandsky Zapovednik on Kola Peninsula (established 1930), covering 278,000 ha, preserves the last wild herds of reindeer in Europe and unique landscapes of fragile northern alpine tundras of the Khibiny Mountains. It also contains archeological sites of the Lapps. Vodlozersky National Park, on the border of Karelia and Arkhangelsk Oblast, is one of the best places to experience the middle taiga and associated waterways. The main focus here is on horse and *baidarka* (kayak) tourism. Kandalakshsky Zapovednik protects the littoral zone of the White Sea, with a host of marine organisms. Rare eider ducks nest here that provide the warmest down insulation known to humankind. The largest parkland of all, Virgin Forests of Komi (a World Heritage Site; see Chapter 16, Table 16.1), includes Pechoro-Ilych Biosphere Preserve and Yugyd-Va National Park in the eastern Komi Republic along the Ural Mountains on over 2.5 million ha, which is about double the size of Yellowstone National Park. This area is home to the largest intact boreal forests of Europe.

Cultural and Historical Features

Culturally, the Northwest region is primarily Russian, with important Uralic minorities of Karelians and Komi in their respective republics, as well as Lapps and Nenets tribes in the extreme north. The forces of assimilation were strong, especially during the Soviet period, and few of the natives continue to speak their languages or practice traditional lifestyles. The cities of Novgorod (established 860 A.D.) and Pskov (established 903 A.D.) are among the oldest Russian cities as recorded in the *Primary Chronicle* (see Chapter 6). In the first three centuries of the Russian state, they rivaled Kiev as major centers of crafts and trade. Merchants in Novgorod and Pskov had easy access to the Baltic Sea and helped explore and settle the inhospitable and distant shores of the White Sea. The old sections of both cities are included on the list of World Heritage sites (see Table 16.1). St. Sophia Cathedral is the oldest church of the Russian north, built between 1045 and 1050 A.D. under Prince Vladimir, son of Yaroslav the Wise. Novgorod also boasts an impressive kremlin and over 50 churches, many of which were badly damaged during World War II but have now been restored.

The city of St. Petersburg was established in 1703 by Peter the Great as the imperial capital of Russia. It is second only to Moscow in the number of its theaters, museums, universities, and sport facilities. It is, however, first in per capita visits to cultural sites, earning it the nickname of the "cultural capital of Russia." St. Petersburg is home to the Mariinsky Theater for opera and ballet; the largest art museum in the world, the State Hermitage Museum; the excellent Russian Museum, exhibiting Russian art (Figure 23.3); the magnificent Peter and Paul Fortress; and the Alexandro-Nevsky Lavra monastery. (Chapter 15 has mentioned several of these facilities.) It also has dozens of sites associated with the lives of poets and writers (Pushkin, Lermontov, Gogol, Dostoevsky, Nabokov, Blok, Akhmatova, Brodsky, and many more). Many sculptors, painters, composers, and musicians lived in the city in the 19th and 20th centuries as well.

St. Petersburg has a number of historical sites associated with the two revolutions of 1917, including the battleship *Aurora* (which signaled the

FIGURE 23.3. The Russian Museum in St. Petersburg houses a fine collection of Russian art. *Photo:* A. Shanin.

beginning of the October revolt) and Smolny Institute (the site of the first Bolshevik government). In the Soviet period, Leningrad was famous for its heroic 900-day resistance to the Nazi siege, when about 1.2 million of its residents died but did not give up the city. Piskarevsky Memorial Cemetery is one of the largest memorials in the world, dedicated to those who died in World War II. Surrounding the city are a few well-preserved royal palaces and estates, which receive millions of tourists per year: Tsarskoe Selo, Petrodvorets (Figure 23.4), Lomonosov, Pushkin, and Pavlovsk. The city is also surrounded by wide sandy beaches and dozens of resorts stretched along the Gulf of Finland.

Other heavily visited sites in Northwest Russia include the 1,000-year-old Valaam Monastery on a group of islands at the north end of Lake Ladoga; Kizhi wooden churches on Lake Onega; sites in Pskov Oblast associated with the life of Pushkin; and small towns in Vologda Oblast, the Komi Republic, and Arkhangelsk Oblast with their unique wooden log churches (Veliky Ustyug, Kargopol).

Economics

The Northwest region is also an economic powerhouse of Russia. In 2006 the region attracted about 13% of all investments in the country, slightly higher than its share of the population.

FIGURE 23.4. The Petrodvorets estate, west of St. Petersburg, rivals Versailles in its royal splendor. *Photo:* A. Shanin.

The region accounts for 8% of Russia's petroleum production, 10% of its hydropower, 18% of its diamond mining, 26% of its peat extraction, 27% of its bauxite mining, 44% of its oil shale production, and 55% of its phosphate fertilizer production. Large copper, nickel, and phosphate mines are found on Kola Peninsula, and the Pechora basin is an important source of coal. The region's main specialty is forestry, however, accounting for 35% of all timber produced in Russia and 60% of all cardboard, paper, and pulp. The region is second only to the Far East in commercial fishing, accounting for about 35% of the nation's total catch.

Manufacturing remains a major activity, accounting for 10% of all industrial output. Although the region lacks iron ore and petroleum, and has only modest quantities of coal and nonferrous metals, the early Soviet program of industrial development favored large steel, aluminum, and machine-building factories in this region (especially in Leningrad and Cherepovets). The machinery building focuses on three very different types of items: (1) high-tech radioelectronic, testing, and medical equipment; (2) heavy machines (e.g., power plant turbines, printing presses, and road construction machinery); and (3) ships. As in the Central region, construction materials, energy generation, and the chemical industry are well represented. Use of the Baltic Sea coast's oil shale is an unusual specialty of the region; numerous facilities exist near St. Petersburg to process this shale into usable petrochemicals. The region is also one of the top fertilizer producers in the country, due to a high concentration of phosphate deposits on the Kola Peninsula. Finally, because of the proximity to Europe, since 2000 car manufacturers from other parts of the world have opened factories in the Northwest for assembling vehicles from prefabricated parts. For example, there is now a Kia factory in Kaliningrad, and a Ford plant in Vsevolozhsk near St. Petersburg now makes the Focus model for the Russian market.

Northwest Russia is a great place to grow trees and hunt game, but not to grow food. As noted above, the climate is cool and soggy, and the soils are poor. The region produces negligible amounts of grain or sugar beets, and only modest amounts of vegetables, meat, and milk (about 7–9% of the national total in each of the latter three categories). Much food must be imported from other regions of Russia and from Western Europe. Flax production is regionally important. As in Central Russia, agricultural efficiency is very low: Farms use a lot of obsolete equipment, and the rural population is elderly.

The infrastructure of the Northwest focuses on St. Petersburg as the main hub. Its proximity to Western European markets, its good Internet and phone connections, and its highly educated workforce make it especially attractive for export and import of consumer goods, as well as for retail and financial services. Connections to Finland, Germany, and Sweden are particularly strong. As described above, St. Petersburg is also a cultural magnet for foreign tourists. As the most "European" of all large Russian cities, it welcomes people interested primarily in the rich imperial history of the Romanov period. Its tricentennial in 2003 and a more recent Group of Eight (G8) summit were attended by thousands from around the world. To make these events possible, a major renovation of the old city core required massive investments of federal funds (Bater, 2006). As in Moscow, Gazprom and the major petroleum companies have a conspicuous presence in the city. Gazprom's controversial Okhta Center tower, to be finished by 2012 for about $2.5 billion, will be the tallest building in northern Europe, reaching 300 m. The height is symbolic; it celebrates 300 years of the city's history. The controversy revolves mainly about the visual impact of a huge, ultramodern skyscraper on the historical low-rise city center (Blinnikov & Dixon, 2010).

Transport connections across the vast Northwest region include the following:

- The Baltic–White Sea waterway from St. Petersburg to Murmansk, across Lakes Ladoga and Onega and the Belomorcanal.
- The Northern Dvina River to Arkhangelsk.
- A number of federal highways, especially the St. Petersburg–Vyborg, St. Petersburg–Petrozavodsk–Murmansk, St. Petersburg–Cherepovets–Vologda, and Vologda–Arkhangelsk routes.
- A few railroads (generally alongside the highways), including the lines to Vorkuta and Murmansk that reach above the Arctic Circle.

A unique corner of Russia is Kaliningrad Oblast, the only exclave of the country completely surrounded by EU territory (see Vignette 9.1). Established in 1946 on the territory formerly known as eastern Prussia, the oblast is centered on the old port city of Koenigsberg, renamed Kaliningrad after one of Stalin's figurehead ministers. After World War II, the victorious U.S.S.R. chose to keep the territory to itself, both as one of several war reparations from Germany and as a convenient gateway to Europe. When Lithuania was Soviet, the U.S.S.R. saw no problem with keeping a sliver of the Baltic Coast for Russia proper. However, after the end of the Soviet Union and the withdrawal of Soviet troops from the Baltic republics, questions arose about the future of the isolated Russian oblast, which is entirely surrounded by EU territory now that Poland and Lithuania have joined NATO and the EU. The biggest issue is getting the Russian population in and out, since they now need EU visas if they travel by car or by most trains. As a partial solution to the problem, a new express train takes Russian citizens from Belarus to Kaliningrad Oblast without stopping inside Lithuania. The city of Kaliningrad has 430,000 people, and the oblast is densely populated (62 people/km^2). It is an important center of heavy industry (shipbuilding, small engines, paper and pulp), transportation, and fishing. The Kurshskaya Kosa sandspit, covered in dunes, is an international park shared by Russia and Lithuania and is very attractive for swimmers and sun bathers.

The future of the Northwest region depends on a few key factors:

- The continued development of St. Petersburg must be thoughtfully addressed. At present the city lags far behind Moscow in per capita investments or income, but it is likely to gain on the capital in these areas over time. St. Petersburg is particularly attractive to Western companies because of its coastal location, proximity to Europe, and large, educated workforce. It is also arguably a better place to live than the congested and suffocating Moscow; in fact, St. Petersburg covers twice as large an area as Moscow, but has only half of Moscow's population. It has a well-developed urban transit system and has many attractive coastal suburban communities where the new upper middle class can live. Its cultural sites, numerous parks, and relatively low real estate prices also make the city a good choice for living. At the same time, its dark, damp winters and the city's bad reputation as a high crime area (its official murder rate is about 50% higher than Moscow's) make it a poor choice, and solutions to the crime problem must be found.
- The demographics of the region are worse than Russia's average. Some of the biggest cities in the region also have among the top HIV infection rates in the country. Also, the incidence of drug use is rising, and so are interethnic tensions. For example, in 2007 Karelia witnessed spontaneous outbursts of anti-Chechen (and broadly anti-Caucasus) mob violence; this was fueled by high unemployment rates for the local Russians, as well as by the outsiders' economic success (which was perceived as unfair) in retail and restaurant businesses.
- Improvements in the agriculture and tourism infrastructure of the traditionally poor Novgorod and Pskov Oblasts need to be made soon. These areas have the warmest climate in the region, as well as plenty of agricultural land, which has been sitting idle since the end of Soviet rule. With global warming, the region is expected to gain more farmland farther to the north, while the water supplies should remain adequate for successful farming even in much warmer temperatures.
- Federal support for the aging Northern Fleet of the Russian Navy, with a heavy concentration of nuclear submarines in a few closed settlements near Murmansk, must continue; otherwise, these areas will see a dramatic decline in population and living standards. At the same time, the problem of sea pollution from dumped nuclear waste must be addressed. Recall that the Novaya Zemlya islands are among the most radionuclide-polluted areas on earth (Chapter 5).
- The long-term sustainability of the forestry sector is questionable. Many areas of the Komi and Karelia Republics are severely overcut. New paper and pulp mills may be built and will require more logs in the near future. Some areas of Arkhangelsk Oblast experience water and air pollution from giant paper and pulp mills located there.
- The problems associated with development of coal, oil, and gas fields in the extreme north-

east of the region must be addressed. The needs of traditional cultures and environmentally sensitive tundras frequently clash with the insatiable demands of fossil fuel extraction.

EXERCISES

1. Find supporting evidence that the Northwest region of Russia is similar to the U.S. Northwest (usually defined as the states of Washington, Oregon, and Idaho). You can use physical, cultural, or economic elements in your analysis of each region's position in its country.
2. Suggest reasons why car manufacturers have been attracted to Northwest Russia, especially Leningradskaya Oblast.
3. What strategic arguments can be made for and against the idea of moving the Russian capital back from Moscow to St. Petersburg? Have a classroom discussion of pros and cons of such a move.

Further Reading

Bater, J. H. (2006). Central St. Petersburg: Continuity and change in privilege and place. *Eurasian Geography and Economics, 47*, 4–27.

Blinnikov, M. S., & Dixon, M. L. (2010). Megaengineering projects in Russia: Examples from Moscow and St. Petersburg. In S. D. Brunn (Ed.), *Engineering earth: The impacts of megaengineering projects*. Dordrecht, The Netherlands: Springer.

Crate, S., & Nuttal, M. (2004). The Russian north in circumpolar context. *Polar Geography, 27*, 85–96.

Hanell, T. (1998). Facts in figures: The Baltic States and Northwest Russia at a glance. *North, 9*(2–3), 9–16.

Round, J. (2005). Rescaling Russia's geography: The challenges of depopulating the northern periphery. *Europe–Asia Studies, 57*, 705–727.

Stanley, W. R. (2001). Russia's Kaliningrad: Report on the transformation of a former German landscape. *Pennsylvania Geographer, 39*, 18–37.

CHAPTER 24

The Volga

Cars, Food, and Energy

If you are familiar with U.S. geography, it may be helpful to think of the Volga region as Russia's Midwest. The region is located in the middle of the country, along the longest river in Europe, the Volga. It is rich in agricultural lands and hydropower resources. It is also home to some of the largest factories of the former Soviet Union (FSU), including most of the automotive and aerospace factories. It is also pretty much "average" in its demographics, economics, and politics, just like the Midwest.

The region discussed here coincides with the Volga federal district (as defined in 2000) and includes portions of the old Volga-Vyatka, Povolzhye, and Urals economic regions (Table 8.3 and Figure 24.1). The district now contains 14 subjects of federation, including six autonomous republics, one kray, and seven oblasts (Table 24.1). Note that Astrakhan and Volgograd Oblasts and the Republic of Kalmykia, which were included in the Povolzhye economic region, are not considered here; they are included in the South federal district now. The Volga district as currently defined and as discussed in this chapter occupies 1 million km^2 (about the area of Bolivia) and is home to 31 million people (a little less than California's population). The population is 71% urban, which is about Russia's average. Its population density of 30/km^2 makes it the third most densely populated district in the country after the Central and South districts. The region is home to many important cities, including five over 1 million (Nizhniy Novgorod, Samara, Ufa, Kazan, and Perm) and seven between 500,000 and 1 million (Saratov, Togliatti, Ulyanovsk, Izhevsk, Penza, Orenburg, and Naberezhnye Chelny). Most of these are capitals of their respective subjects of federation. The largest city, Nizhniy Novgorod (called Gorky in the Soviet period), is the fourth biggest city in the country and is one of Russia's top manufacturing centers and consumer markets.

The region has an excellent position with respect to transportation networks. The Volga River, with its tributaries the Oka and Kama, connects the region with Moscow and St. Petersburg, as well as with the Black, Baltic, and Caspian Seas. Numerous railroads, highways, oil and gas pipelines, and airports provide additional infrastructure. The Volga district's location between the Central and Urals districts ensures robust trade links with the rest of the country. In the south, the district borders now-independent Kazakhstan. The Northwest district to the north provides coal and timber resources. In short, this is one of the most advantageously located areas in the entire country.

The Volga River, the longest in Europe at 3,690 km, gives the region its name. A cascade of hydropower plants—Gorkovskaya (580 megawatts

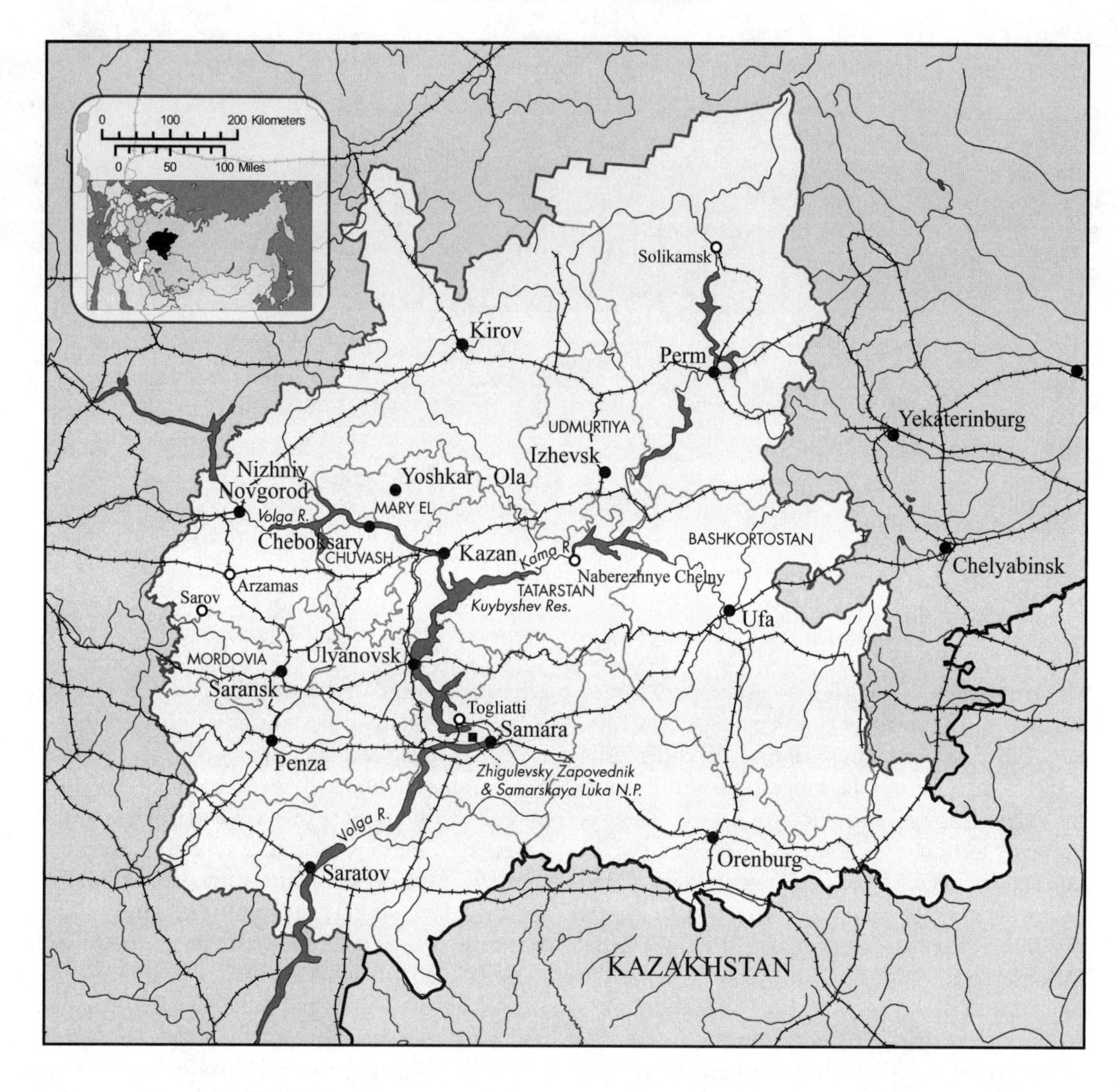

FIGURE 24.1. The Volga federal district of Russia. Map: J. Torguson.

[MW]), Cheboksarkaya (1,379 MW), Saratovskaya (1,290 MW), Volzhskaya (2,300 MW), and Nizhnekamskaya (1,080 MW)—on the Volga provides cheap electricity; however, it also creates disruptions for migrating sturgeon and other fish, and a lot of fertile land has been taken out of production because it is flooded by the reservoirs. The Volga is navigable from about mid-March to November in the middle reaches. It is one of the most heavily used rivers in the world, both with respect to the amount of water used for industry and irrigation and in terms of total freight shipping. It is also one of the most polluted rivers of the FSU. However, swimming in many stretches away from the big cities is safe, and fishing for pike and sturgeon is common.

Physical Geography

The relief is flat, slightly undulating plain, covered with thick glacial deposits in the northern half. A slight hilly plateau 350 m high extends from north to south along the right (west) bank of the Volga. Interesting caves are found in the foothills of the Urals (e.g., the Kungur ice cave in Perm kray is over 6 km long, and the Kapova cave in Bashkortostan is 2 km long); they are natural

TABLE 24.1. Subjects of Federation of the Volga Federal District

Subject	Administrative center	Population (2006) (thousands)	% ethnic Russian	GRP/capita (rubles in 2006)	Top products and areas of economic specialization within Russia (% of nation's total production for some)
Kirov Oblast	Kirov	1,443	91	67,185	Wood-processing equipment (19%), tires (16%), OS board (14%), defense
Permsky Kray	Perm	2,748	85	143,570	Turbo drills (100%), mineral fertilizer (42%), gas ranges (29%), defense
Udmurt Republic	Izhevsk	1,544	60	105,767	Small engines (11%), cars (4%), trucks (3%), military, beef, dairy
Nizhegorodskaya Oblast	Nizhniy Novgorod	3,411	95	112,162	Buses (65%), trucks (62%), thermoplastics (39%), food processing (17%), cars
Mariy El Republic	Yoshkar-Ola	712	48	59,964	Refrigerators (19%), grains, potatoes
Chuvash Republic	Cheboksary	1,292	27	71,242	Looms (76%), bulldozers (27%), beef, dairy
Mordoviya Republic	Saransk	857	61	66,573	Power transformers (75%), cistern cars (51%), light bulbs (27%), beef, dairy
Tatarstan Republic	Kazan	3,762	40	161,013	Synthetic rubber (30%), tires (28%), trucks (16%), plastics (14%), agriculture
Bashkortostan Republic	Ufa	4,064	36	124,647	Baking soda (55%), wheelchairs (47%), light bulbs (29%), rubber, oil
Penza Oblast	Penza	1,408	86	64,962	Seed combines (23%), pipeline equipment, sugar (5%), grain
Ulyanovsk Oblast	Ulyanovsk	1,336	73	76,244	Buses (26%), trucks (10%), beef, dairy, grain
Samara Oblast	Samara	3,189	84	153,967	Cars (75%), synthetic rubber (21%), oil refining, beverages, plastics, aerospace
Saratov Oblast	Saratov	2,608	86	77,148	Trolleybuses (49%), synthetic fibers (29%), glass (26%), tractors, refrigerators
Orenburg Oblast	Orenburg	2,138	74	140,586	Nonferrous metallurgy equipment (34%), salt (19%), oil, gas

Note. Data from the Federal Service of State Statistics, Russian Federation (2009). To convert gross regional product (GRP) to U.S. dollars, divide GRP by 27.19.

wonders of the region. There are relatively few valuable mineral deposits. Nationally significant are petroleum (9% of Russia's reserves), natural gas (6%), oil shale (51%), and sulfur deposits. Petroleum is produced at over 150 sites, with a few large oil fields dominating the production, mainly in Tatarstan Republic and Samara Oblast. Proximity to the iron ore deposits of Kursk, the coal of northern European Russia, and the nonferrous metals of the Urals makes shipping of raw materials into the region an easy task. Additional iron and copper ore deposits are found in Orenburg Oblast within the region. Most of the forest resources are concentrated in the north of the region, especially in Nizhniy Novgorod and Kirov Oblasts, which are over 50% forest covered. Average forest cover in the southern part of the region is only 8%.

The climate is more continental here than near Moscow. Snow cover lasts for at least 5 months, with the average temperature in January reaching –16°C. Summers can be warm in the north and hot in the south. The average July temperature in Nizhniy Novgorod is +18°C, and near Saratov is +20°C. The absolute temperature records in Kazan were –47°C for winter and +38°C for summer. Precipitation values in the north of the region are about 500–600 mm/year, but in the south only about 400 mm/year. The overall climate here is virtually identical to that in the upper U.S. Midwest (i.e., Minnesota and Wisconsin). Soils are primarily gray forest soils (alfisols) in the north, and chernozems (mollisols) and chestnut semidesert soils (aridisols) in the south. The northern third of the region is in the coniferous and mixed forest zone, while the southern two-thirds are in the forest–steppe and true steppe zones. Some rare plants and vegetation types occur on the west bank of the Volga on chalk cliffs (Figure 24.2).

The region is rich in protected areas, with 14 zapovedniks, 8 national parks, and 4 federal wildlife refuges. Volga–Kama Zapovednik occupies about 11,000 ha on the left bank of the Volga in Tatarstan; it protects undisturbed forest–steppe areas of the middle Volga. Zhigulevsky Zapovednik, surrounded by Samarskaya Luka National Park near Samara, protects basswood forests and forest–steppe areas with many "relict" species (i.e., species surviving from earlier ages) of plants along the high right bank of the Volga. Some of the relicts date from the Pliocene

FIGURE 24.2. Chalk cliffs along the Volga at Samarskaya Luka National Park. *Photo:* S. Blinnikov.

era—typically, remnants of the warmer period when steppes were more widespread. Orenburgsky Zapovednik protects steppes in the transition zone between Europe and Asia south of the Ural Mountains.

Cultural and Historical Features

Culturally, the Volga region is one of only two in Russia with a heavy presence of many minorities from the Uralic (Finnish) and Altaic (Turkic) families (Chapter 13). Most of the Uralic peoples are Orthodox Christian, while most of the Altaic peoples are Muslims. The Mari people have not been thoroughly converted one way or the other; many retain their traditional animist/shamanist beliefs. The Chuvash people are distinct in that they are of mixed Tatar and Uralic ancestry, and they are predominantly Orthodox Christians (Figure 24.3). The Uralic peoples (Maris, Mordvinians, Komi Permyaks, and Udmurts) live in the north and west of the region, while the Altaic peoples (Tatars, Bashkirs, Chuvashs) live mainly in the center and east of the region. Each of the groups has its own autonomous republic, where they are either a majority (e.g., the Tatars represent 53% of the population in Tatarstan, and the Chuvashs 70% in Chuvashia) or a large minority (e.g., the Mordvinians represent 35% of Mordovia's population, and the Maris 43% of Mari El). The Tatars, Bashkirs, Chuvashs, and other cultures created their own literature, folk and fine arts, styles of dress, and architecture. For example, Mari and Mordvinian peasant women in traditional dress have different patterns on their headgear. Some of the traditions are shared by many ethnic groups, thus establishing a general "Volga" culture that crosses ethnic/linguistic boundaries. The Russian language serves as a common communication tool; virtually all people in the region can speak it, and most can also read and write Russian. One of the Russian cultural specialties in the region is a distinct style of hand-painted decoration, used to ornament wooden spoons and bowls in the village of Khokhloma since the 17th century. Intermarriages between Russians and non-Russians, and between different groups, of non-Russians, are very common in the region. Several of the well-known contemporary artists in Russia from the region have multicultural backgrounds—for example, the rock singers Zemphira Ramazanova (an ethnic Tatar born in Ufa, Bashkortostan) and Yuri Shevchuk (a Ukrainian–Russian raised in Ufa). Other celebrities from Ufa include the Russian painter Mikhail Nesterov and the Tatar dancer Rudolf Nureyev. The pop singer Alsou comes from Bugulma, Tatarstan, and is an ethnic Tatar.

FIGURE 24.3. Chuvash traditional dress on display at Cheboksary. An offering of bread and salt is a traditional form of hospitality throughout Russia, not only among the Slavs. *Photo:* S. Blinnikov.

Nizhniy Novgorod and Samara are the largest cities in the region, each with over 1 million people. These two cities have given many celebrities to the world as well. For example, Nizhniy Novgorod was the birthplace of the Soviet writer Maxim Gorky, and the city bore the writer's name in Soviet times. Other famous people from Nizhniy include Nikon, the great 17th-century patriarch of the Russian Orthodox Church; inventor Pyotr Kulibin (d. 1818); mathematician Nikolai Lobachevsky (d. 1856); and composer Mikhail Balakirev (d. 1910). The legendary Soviet aviator Valery Chkalov, who was the first in the world to fly nonstop over the North Pole in 1935, was born in Vasilevo (now Chkalovsk) in Nizhniy Novgorod Oblast. Nizhniy also gave the world the microbiologist Irina Blokhina, the pianist and conductor Vladimir Ashkenazy, and the supermodel Natalia Vodianova.

Samara, well known from the Russian folk song "Samara Gorodok," has also produced a few well-known personalities, including Eldar Ryazanov, one of the best movie directors; the actress Ershova; and the spacecraft designer Kozlov. Kazan is a typical Tatar city and another important cultural center of the region (Vignette 24.1).

Despite the region's high cultural diversity, the demographic situation in the Volga district mirrors that for Russia as a whole: Its population is declining (the average rate of decline was –0.6% in 2005). In the economically depressed Kirov and Ulyanovsk Oblasts and in Mordovia, the decline exceeded 1%. On the other hand, in the relatively booming Tatarstan, the decline was merely –0.2%. The Volga region is one of the three best in Russia with respect to average life expectancy: In 2005 it was 65.26 years (58.64 for men and 72.59 for women). This reflects good nutrition and the economic prosperity of the region, as well as lower-than-average alcoholism rates (especially in the Muslim subjects of federation).

Economics

The Volga region is an economic giant in Russia, especially with respect to machine building, car and aircraft manufacturing, and the space industry. In 2004 it accounted for 16.5% of Russia's gross domestic product (GDP) and 23% of all industrial output. Two of its wealthiest areas, Tatarstan Republic and Samara Oblast (Figure 24.4), have higher-than-average gross regional products (GRPs) in Russia; these were about $4,150 per capita in 2004. The two poorest, Mari El Republic and Penza Oblast, had GRPs of only $1,654 per capita in the same year. The region's richest subjects of federation are the ones that produce or refine petroleum, have a strong machinery-building base, and/or have well-developed and diversified agriculture. In 2004 Tatarstan was in 4th place among the 88 subjects of federation at that time in the amount of money made in mineral extraction. Overall, the Volga region accounts for 14% of Russia's petroleum production, and about 2% of its natural gas. It is, however, responsible for about one-quarter of all petroleum refining.

With respect to machinery production, the region is second only to Central Russia and accounts for about 25% in the country. Until a few years ago, when new car factories appeared in Kaliningrad and Leningradskaya Oblasts in the Northwest, the Volga region was the only one in the country to assemble passenger cars. There are automobile factories in Togliatti (AutoVAZ), Nizhniy Novgorod (GAZ and PAZ), Izhevsk (Izhmash), Ulyanovsk (UAZ), and ZMA (Naberezhnye Chelny). Now they make a total of about 1 million cars per year, or about 80% of Russia's total. The KAMAZ plant in Naberezhnye Chelny is one of the largest truck assembly plants in the world. The company makes about 67,000 heavy trucks (over 14 tonnes) per year and is the largest truck manufacturer in Russia, although only 11th worldwide. Another regional specialty, the aerospace industry, is well represented in Kazan (helicopters), Samara (Tu-154 airplanes, Soyuz and Progress spacecraft, civilian and military satellites), Ulyanovsk (Tu-204 passenger airplanes and An-124 cargo airplanes), Nizhniy Novgorod (MIG-29 and MIG-31 fighter jets), and Saratov (Yak airplanes and some helicopters). Many factories in Mari El, Mordovia, Chuvashia, and Udmurtiya specialize in producing radioelectronics and testing equipment for civilian and military uses. Nizhniy Novgorod Oblast is the giant of the Russian shipbuilding industry, producing over half of all ships for river traffic and many ocean-going vessels.

Petroleum refinery and petrochemicals constitute the second leading industrial specialty of the region. All of Russia's major petroleum companies have some presence in the region, including Lukoil, TNK-BP, Tatneft, and Rosneft. The Volga region produces the lion's share of Russia's motor oil, and leads the country in production of plastics and fertilizer. Production of potassium fertilizer from the extremely rich Solikamsk and Berezniki deposits in Permsky Kray accounts for 42% of the country's total! There is a large nitrogen fertilizer factory in Togliatti, and phosphate production is heavy in Salavat, Bashkortostan. Yoshkar-Ola (Mari El), Saransk (Mordovia), and Penza are major pharmaceutical producers. Kirov specializes in making shoes, leather jackets, fur coats, and toys (Figure 24.5), as well as, ironically, weapons.

The Volga region is also an agricultural giant, producing about 25% of the grain and sunflowers, 15% of the sugar beets, and 12% of the potatoes

Vignette 24.1. Profile of Kazan

Kazan (established 1177 A.D.; population 1,105,000) is the most important Tatar city in Russia. The city was founded by the Volga Bulgars, who had become part of the Golden Horde under Genghis Khan, and later became its own khanate in 1438. Kazan Khanate was conquered by Ivan the Terrible's troops in 1552. Today Kazan is a large industrial center on the banks of the Volga River near the mouth of the Kama, the largest left tributary of the Volga. There are 150 industrial enterprises in the city, including the Gorbunov factory, which builds Tu-214 jet planes; a helicopter factory; medical and instrumental equipment manufacturers; an organic synthesis plant; a few petrochemical factories; and cosmetics and food producers. Kazan State, Kazan State Technical, and Kazan Architecture Universities are among the best in Russia. The city recently opened its own subway system, which is the 7th largest in Russia and 15th largest in the FSU. Its center is undergoing massive renovation (Figure 1), with Tatar nationalism playing a role in developing the new city identity. There are nine theaters, including the renowned Kachalov's Russian Drama Theater and Galiaskar Kamal's Tatar Academic Theater. There are about 20 museums, including a national museum of Tatarstan, a museum of fine arts, science museums of Kazan State University, various ethnographic museums, and museums devoted to famous persons (e.g., Lenin, Gorky, and the famous opera singer Feodor Chaliapin).

FIGURE 1. Large-scale reconstruction of Kazan is under way to provide room for more commercial development in the congested city core. *Photo:* S. Blinnikov.

The historical core of Kazan, with a beautiful kremlin, has unfortunately been greatly remodeled in the last 10 years. Entire neighborhoods of mid-19th-century merchant homes have been demolished. Nevertheless, there are many sights to see, including the kremlin (a World Heritage Site), many mosques and churches, the Suyumbike leaning tower from the 17th century (Figure 2), old streets for pedestrians only, and so on. Kazan has strong ice hockey (Ak Bars) and soccer (Rubin) teams. It is a diverse, multicultural city, with Turkic and Slavic cultures coexisting and enriching each other. The recent rise in Tatar and Russian nationalism sometimes makes this harmony less than ideal, however: Both ethnicities have supremacist groups in Kazan, which have been involved in repeated street clashes with each other.

(cont.)

FIGURE 2. Suyumbike was built in the late 17th century in Kazan, probably as a guard tower. Its tilt from the vertical is 1.8 m near the top. *Photo:* S. Blinnikov.

and vegetables in the country. It also accounts for about 15% of meat and milk production. The region leads the nation in arable land—18.7% of the territory, for a total of 45 million ha. Typical crops grown in the north include barley, rye, oats, and winter wheat. In the south production of summer wheat is important, along with buckwheat, millet, hemp, hops, sugar beets, and mustard. The main zones of agriculture are located along the right (west) bank of the Volga, which has a milder climate. The main livestock production is concentrated in Bashkortostan and Tatarstan. Together these two republics account for 13% of the cattle, 11% of the milk, 9% of the hogs, and 7% of the eggs produced in Russia.

The infrastructure of the Volga region is multimodal; it is mainly centered on Nizhniy Novgorod and Kazan in the middle, and on Samara and Saratov in the south. The cities along the Volga tend to be very long, but narrow, stretching parallel to the river for 20–30 km. Kirov and Perm (Figure 24.6) are located along the northern branch of the Trans-Siberian Railroad, while Kazan, Ulyanovsk, and Samara lie along its southern branch. All major cities in the region have good connections to each other and the rest of the country via railroads and paved highways. The Volga is, of course, the main waterway. Numerous oil and gas pipelines run through or originate in the region, with the most important nodes being Alemetyevsk (the start of the Druzhba pipeline to central Europe), Samara, and Nizhniy Novgorod. Nizhniy Novgorod's airport is used as a reliever for Moscow in poor weather. Samara, Saratov, Kazan, and Perm have large airports as well.

FIGURE 24.4. New housing in Samara for people with higher-than-average incomes. *Photo:* S. Blinnikov.

FIGURE 24.5. Locally made toys are sold directly from the factory to the train passengers at the Trans-Siberian Railroad station in Kirov. *Photo:* Author.

Challenges and Opportunities in the Volga Region

The Volga region is advantageously located in Russia between north and east, south and west, along a major waterway. It is also a culturally diverse place, where different cultures have generally enriched each other, although interethnic tensions are on the rise. The region is less developed over much of the heavily forested, cold Kirov Oblast, and in the poor rural Mordovia Republic and Penza Oblast. However, all units have at least one big city with strong industry, education, and cultural services. The region has enough qualified workers for the available employment. The current boom in the oil and gas industry has helped the region, which specializes in both production and refining of these fuels. Also, production of cars and airplanes is growing again, including some joint ventures with foreign participation. Future challenges include conversion of large military facilities to civilian uses, continued modernization of the old Soviet factories, investment in better infrastructure and education, and retention of the best young workers instead of losing them to Moscow or St. Petersburg. The region is a net importer of energy, despite the availability of large hydropower dams on the Volga, and may need new nuclear and wind power stations in the near future. Lack of water for irrigation in the warming world may present challenges for the region's agriculture, especially in the already semiarid south. Tourism in the region has a lot of potential for growth; the main focus of foreign tourism at the moment is on the Volga itself, as seen on week-long

FIGURE 24.6. The city of Perm—a large industrial and military hub in the northeastern part of the Volga region. *Photo:* Author.

ship cruises from Moscow or Nizhniy Novgorod. More inland itineraries could be developed that capitalize on the district's rich cultural and natural heritage.

EXERCISES

1. Investigate the hydropower potential of dams on the Volga River, and compare those with dams on another large river (e.g., the Nile, Colorado, or Missouri). What are the similarities and differences in major issues surrounding the hydropower resource?
2. Discuss the relative strengths and weaknesses of the GAZ and AutoVAZ automakers. Are there any compelling geographic factors responsible for the success of these two enterprises in capturing the lion's share of Russia's light truck and car markets, respectively?
3. Develop a ranking of all the Volga region's subjects of federation, based on their potential for interethnic violence. What were the main factors used in your index? Compare it to those developed by your classmates.
4. Study existing itineraries for the Volga River cruises. What can be improved?

Further Reading

Backman, C. A. (1997). The forest sector of European Russia. *Polar Geography, 21*, 272–296.

Cashaback, D. (2008). Assessing asymmetrical federal design in the Russian Federation: A case study of language policy in Tatarstan. *Europe–Asia Studies, 60*, 249–275.

Demin, A. P., & Ismaiylov, G. K. (2003). Water consumption and drainage in the Volga basin. *Water Resources, 30*, 333–346.

Hanson, P. (1998). Samara: A preliminary profile of a Russian region. *Europe–Asia Studies, 49*, 407–430.

Le Calloc'h, B. (1999). The Finno-Ugrian peoples of Russia: The Mordvines. *Acta Geographica, 120*, 17–33.

Matsuzato, K. (2006). The regional context of Islam in Russia: Diversities along the Volga. *Eurasian Geography and Economics, 47*, 449–461.

Romanov, P., & Tartakovskaya, I. (1998). Samara Oblast: A governor and his *guberniya*. *Communist Economies and Economic Transformation, 10*, 341–361.

Schmidt-Kallert, E. (1994). New settlements of Soviet–German families on the Volga. *Geographische Rundschau, 46*, 141–147.

CHAPTER 25

The Caucasus

Cultural Diversity and Political Instability

The Caucasus is located at various crossroads: those between Europe and Asia, north and south, east and west. The Black and Caspian Seas are separated by 500 km of high mountains (Figures 2.1 and 25.1). To the north is European Russia; to the south is Asia. The main Caucasus range is the highest in Russia. It provides a natural barrier to cold air masses from the north, as well as to travel between the Russian northern Caucasus and the independent trans-Caucasian republics to the south—Georgia, Armenia, and Azerbaijan. Both sides of the mountain range are considered here, making this the first chapter in Part V to discuss some non-Russian republics of the former Soviet Union (FSU). The reason for this is geographic: The Caucasus is both a physical and a cultural region that is best discussed as a whole. Doing so will make it much easier to understand the multiple conflicts taking place there.

The Russian Caucasus is included in the South federal district, which occupies 600,000 km^2 and contains 23 million people in 13 subjects of federation (Tables 8.3 and 25.1). The northernmost subjects are not in the mountains: Kalmykia, Volgograd, and Astrakhan Oblasts were formerly included in the Povolzhye economic region and mainly contain flat steppes. On the other hand, Krasnodarsky and Stavropolsky Krays and the eight autonomous republics are partially or wholly in the mountains. Again, the Caucasus main range provides a convenient natural boundary, with the mountains running at about 2,000–3,000 m continuously from northwest to southeast. Only one highway that is drivable year-round connects Russia with Georgia, via Ossetiya. Georgia, Armenia, and Azerbaijan together account for 186,000 km^2 (a territory 25% smaller than the United Kingdom), and have a combined population of only 15.7 million.

With respect to population, the South district of Russia is the second most densely populated territory after the Central district, with an average density of 40/km^2. It is also the least urbanized region, with only 58% of its population living in cities. It leads the country in fertility (only –0.1% natural decrease, as compared to –0.6% for the country as a whole in 2005), and it is also the poorest region among the seven federal districts, with only half of Russia's average gross regional product (GRP) per capita. If we were to seek an analogous region in the United States based on economic and social characteristics, this would be also in the South (Alabama,

FIGURE 25.1. The South federal district of Russia and the independent FSU republics of the Caucasus. Map: J. Torguson.

Mississippi, Louisiana). Most of the demographic and economic peculiarities of the Russian Caucasus have to do with the ethnic republics, not with the predominantly Russian oblasts and krays.

The urbanization rate for the three independent trans-Caucasian republics is even lower (53%), and because of the high fertility in Azerbaijan, their overall natural growth rate is positive at 0.4%. The population is actually decreasing slightly in Armenia, while Georgia's population level is stable. The republics of the Caucasus, particularly Georgia and Dagestan, lead the FSU in average life expectancy. Some attribute this to clean mountain air, plenty of exercise, and a balanced natural diet (Karny, 2000).

TABLE 25.1. The Caucasus: Subjects of Federation of Russia's South Federal District, plus Georgia, Armenia, and Azerbaijan

Subject or nation	Administrative center	Population (2006) (thousands)	% ethnic Russian	GRP/capita (rubles in 2006)	Top products and areas of economic specialization (% of Russia's total production for some)
Volgograd Oblast	Volgograd	2,636	89	94,812	Tractors (26%), gas ranges and steel pipes (18%), fibers, oil refining
Astrakhan Oblast	Astrakhan	994	70	85,172	Vegetables, watermelons, rice, sturgeon (caviar), oil, gas
Kalmykiya Republic	Elista	289	34	42,728	Sheep and cattle ranching
Rostov Oblast	Rostov-on-Don	4,304	89	78,328	Electric locomotives (100%), anthracite (76%), combines (75%), vegetable oil
Krasnodarsky Kray	Krasnodar	5,097	87	91,369	Cement, shipping, tourism, sugar (34%), wine (26%), vegetable oil (22%)
Stavropolsky Kray	Stavropol	2,710	82	66,058	Plows (28%), brandy (19%), mineral water (12%), wine (10%), plastics
Adygeya Republic	Maykop	443	65	47,373	Food industry, wheat, sunflower, orchard crops
Karachaevo-Cherkesskaya Republic	Cherkessk	432	34	52,715	Tourism, food industry, mineral water (11%), paints (2%), sheep ranching
Kabardino-Balkarskaya Republic	Nalchik	894	25	46,813	Tourism, diamond-cutting equipment (41%), wine (7%), liquor, mineral water, fruit
North Ossetiya Republic	Vladikavkaz	702	23	60,979	Wine (14%), liquor (6%), finished metal products
Republic of Ingushetiya	Magas	487	1	17,471	Vegetables, fruit, sheep/cattle ranching, nonmetal minerals
Republic of Chechnya	Grozny	1,163	4	24,920	Vegetables, fruit, sheep ranching, oil refining, cement
Dagestan Republic	Makhachkala	2,641	5	44,644	Brandy (19%), milk separators (13%), hydropower, petroleum
Georgia	Tbilisi	4,616	2	65,256	Steel, copper, manganese, wine, tea, tangerines, wood products, tourism
Armenia	Yerevan	2,967	0.5	89,074	Machine tools, electric motors, tires, knitwear, jewelry, brandy
Azerbaijan	Baku	8,239	2	114,198	Petroleum, steel, petrochemicals, cotton, rice, grapes, tobacco, livestock

Note. Data for Russia from the Federal Service of State Statistics, Russian Federation (2009). To convert gross regional product (GRP) to U.S. dollars, divide GRP by 27.19. Data for Georgia, Armenia, and Azerbaijan from CIA World Factbook and calculations by author.

Physical Geography

The physical environment of the Caucasus is remarkable. This is the warmest part of the FSU, with some subtropical vegetation present, especially along the Black Sea coast (see Chapter 3, Figure 3.6). The mountains, on the other hand, have snow-capped peaks and a number of substantial glaciers, especially near Mt. Elbrus, Dombai, and Kazbegi. The highest point, Mt. Elbrus, is the tallest peak in Europe at 5,642 m (see Chapter 2, Figure 2.3). Mt. Aragats in Armenia is slightly over 4,000 m. The Volga flows into the Caspian Sea in the northeast of the region, resulting in a massive delta rich in plant and animal life; the delta is actually located below sea level, at –28 m. The Volga is connected by a canal to the Don, which empties into the Sea of Azov and allows shipping to the Black and Mediterranean Seas beyond. The main rivers of the Caucasus (the Kuban, Terek, Rioni, and Kura) are short but powerful, and are heavily tapped for hydropower. Lake Sevan is an important body of water in central Armenia and is the largest lake in the region, if one does not include the Caspian Sea. Large reservoirs are located in Kalmykia Republic and Volgograd Oblast.

The Caucasus is relatively poor in minerals, but is rich in agricultural and bioclimatic resources. Azerbaijan and parts of the Russian Caspian Sea coast are rich in petroleum and have some natural gas. Georgia has significant deposits of manganese ore. Armenia has substantial reserves of construction stone. Despite the rich petroleum reserves of Baku, the region overall is energy deficient and has to import electricity and fossil fuels from other parts of the world, especially from the Volga region in Russia and from Turkmenistan. The Donbass coal basin is partially located in Rostov Oblast, on the border with Ukraine. Armenia has a nuclear power station inherited from the Soviet period, while all countries also have numerous hydropower installations.

The region has the mildest climate in the FSU. The growing season lasts over 6 months, about 1½ months longer than in Central Russia. Sochi has an annual temperature of +14°C and remains frost-free all year. It receives 1,570 mm of precipitation, mostly in winter, and has a Mediterranean-like climate. Batumi, farther south along the coast in Georgia, is even warmer. It was the only subtropical part of the Soviet Union. Armenia has a dry mountainous climate that varies with elevation. Azerbaijan on the Caspian Sea has a mild climate, although quite a bit drier than on the Black Sea coast. In the mountains the tree line occurs at about 2,200–2,500 m, with subalpine and alpine meadows extending to the snowline at around 3,000 m.

The South district of Russia is about 10% forest covered and provides little wood for the country, but what it has are the most valuable hardwood varieties—beech and oak. Moreover, it has about 80 million ha of arable land; consequently, it accounts for about 20% of all agricultural production, including some subtropical crops that cannot be grown anyplace else in Russia. The flat lands north of the mountains are used for raising wheat, corn, and (in a few places) rice. Kalmykia Republic and Volgograd Oblast have large sheep- and cattle-ranching areas. Astrakhan Oblast's specialty is producing watermelons in the Volga floodplain. The Kuban River watershed produces all types of agricultural products. Farms in the region are among the largest and wealthiest in Russia. A strong Cossack culture ensures social stability and a good work ethic (Figure 25.2). Overall, the South region of Russia accounts for two-thirds of the sunflowers, one-third of the grain, and one-quarter of the sugar beets and fruit harvested in the country.

The mountainous areas on the northern slopes of the Caucasus range provide opportunities for sheep ranching (over 60% of Russia's total), as well as for orchard crop production (apples, plums, cherries, and apricots) similar to that of the Yakima River basin in Washington State (Figure 25.3). The strip of the Black Sea coast between Anapa and Sochi and into Abkhazia is the only area in the FSU where tea and citrus fruit can be grown. This is also the main viticulture area of Russia. Georgian wines are legendary in quality and are produced from a few varieties of grapes with unique taste (e.g., the dark red *saperavi*). Armenia has limited arable land, but produces some highland grains, mutton, and brandy. Azerbaijan grows rice, tobacco, tea, and a variety of vegetables and fruit. It is the lead-

FIGURE 25.2. Cossack monument in Krasnodar. *Photo:* A. Pugach.

FIGURE 25.3. A variety of fruits at a farmer's market in Krasnodar. *Photo:* A. Pugach.

ing exporter of flowers to Russia among the FSU countries.

The South district of Russia has a few large protected areas, including the mountainous Caucasus and Teberda biosphere reserves, a unique yew–box tree grove in Khosta near Sochi, and two national parks surrounding Mt. Elbrus and Sochi. In the steppe-dominated north of the region, the very interesting Black Earth Zapovednik protects some of the driest ecosystems of Europe, including fragments of a real desert; it also includes Manych-Gudilo Lake, of international importance, with a host of bird species (pelicans, herons, geese, swans, and shorebirds). The Astrakhan Zapovednik in the Volga Delta protects a pivotal wetland of the northern Caspian basin, with heron and ibis rookeries and sturgeon spawning grounds. The three independent trans-Caucasian republics had a few zapovedniks in Soviet times, but their state is uncertain now, with the ongoing conflicts in the region and a lack of state funding. Funding for Georgian preserves has improved under the new leadership of Mikheil Saakashvili. Poaching of animals and illegal logging of wood remain common in much of the Caucasus, both inside and outside Russia.

Cultural and Historical Features

The Caucasus is the first place in Asia where human remains are recorded, and the Fertile Crescent dated back to 1.8 million years ago. Anatomically modern humans lived in the region 250,000 years ago. In more recent times, the Caucasus was the easternmost fringe of the Roman Empire, and then was contested by the Persian, Ottoman, and Russian Empires. It remains a cauldron of ethnic conflicts today (see many references in **Further Reading**); those in Chechnya, Dagestan, Ingushetiya, Ossetiya, Abkhazia, and Nagorno-Karabakh can be mentioned. (Vignette 25.1 describes the Chechen conflict in some detail.) Most revolve around the issue of land control in the aftermath of the Soviet period, but many have also taken on religious or cultural overtones. This being said, the region presents remarkable opportunities for cultural studies and international cooperation.

Vignette 25.1. The Chechen Conflict: A Black Hole, or a Light at the End of the Tunnel?

Of all the conflicts in the Caucasus, the Chechen war in Russia has been the worst in terms of the number of casualties, the number of high-profile terrorist acts on Russian territory, its intensity, its length, and the economic impact on the Caucasus region. Numerous books have been written about the details of the conflict (see **Further Reading**), but some of these are written from decidedly partisan perspectives, so one must exercise caution in interpreting the existing literature. Here I patch together a brief geographic account of the conflict, to help you navigate the pertinent literature and appreciate the spatial scope of its impact.

The Chechens are one of the two Vainakh people (the other are the Ingush), self-named the Nokhchi Cho. The Vainakhs are indigenous mountainous ethnic groups of the eastern Caucasus, mainly living on the northern slope. Unlike most of the Caucasian peoples, who are shepherds, the Chechens have been primarily hunters. The eagle and the wolf are their totems. Because of their area's harsh environment, the Chechens practice very little agriculture. When game was scarce in the Middle Ages, they were known to stage raids on people's dwellings in the foothills to obtain food and loot. They were among the last in the Caucasus to adopt Islam in the 18th century, and originally that was of the contemplative Sufi variety. They have always been known as fiercely independent people and strong fighters. They forged an anti-Russian alliance with the Avars from Dagestan during the prolonged and bloody Russo-Caucasian war in 1816–1864 (Baddeley, 1969).

The Chechens are split into 20 *teips* (tribes), which frequently do not get along with one another. In particular, the teips from the plains are thought to be of mixed ancestry and are looked down upon by the mountain teips. Understanding the teip structure is important, because much contemporary politics in the region still depends on it. A dramatic legacy of the past is the blood feud, which is common among many groups in the Caucasus, but is especially developed among the Chechens. When a member of the family is killed by a member of another teip, the whole family must seek an opportunity to avenge the blood with blood (i.e., to kill a member of the offending clan). Although blood feuds were illegal in the Soviet Union (and also are not allowed under Islam), the tradition did not disappear, and has come to haunt the region in the post-Soviet era.

During World War II, the entire Chechen nation (about 400,000 people) was rounded up and deported on the orders of Joseph Stalin to Kazakhstan in 1944. Stalin, himself a Georgian/Ossetian, accused Chechens of collaboration with the advancing German army. An entire generation passed before the Chechens were allowed to return in the late 1950s. In fact, much of the Chechen leadership of the 1990s (including their first president, Dzhokhar Dudayev), had been born in Kazakhstan. There was little development in rural Chechnya during the Soviet period. The city of Grozny (a Russian 19th-century fortress whose name means "Fearful") was developed into a typical Soviet industrial city, mainly centered on petroleum refining and the cement industry.

Toward the end of the Soviet Union, Chechen volunteer fighters helped their distant kin the Abkhazy across the mountains in their conflict with the Georgians in 1989. Busy with fighting the remnants of the Communist Party and seeking support among regional elites, Yeltsin made a few poorly conceived promises to the interior ethnic republics, including an opportunity for "as much sovereignty as they could swallow." Tatarstan, Yakutia, and Chechnya were among the first to claim more tax breaks and less political control from Moscow. The situation in Chechnya very quickly took a dramatic turn for the worse. In late 1991 it unilaterally declared its independence from Moscow; this went almost unnoticed, as the entire Soviet Union was falling apart at the time. However, the Chechen declaration was not an empty statement. Dzhokhar Dudayev, a former Soviet Air Force general, was determined to lead his people to an independent future (or at least a more independent one, depending on what Moscow was willing to concede). Although he was not radical at first, he eventually aligned himself with the broader Islamist movement in Asia and grew increasingly uncompromising in his dealings with Moscow. Cash, weapons, and volunteer jihadis poured into Chechnya from the larger Muslim world,

(cont.)

especially the radical groups of the Middle East (including Al-Qaida and the Taliban). Dudayev was sometimes supported and sometimes opposed by Shamil Basayev, an even more radical freedom fighter turned terrorist.

Between 1991 and 1994, the de facto independent Chechnya gradually gained strength. Some Soviet weapons were left behind by the retreating Russian Army in 1991 and fell into the hands of the Chechen leaders. Chechnya was still nominally part of Russia, which allowed major sums of money to be laundered off by its leaders through legitimate banking channels. A number of financial scams involving major Russian banks were perpetrated through their Chechen branches, involving billions of rubles. Petroleum shipments through the territory of the republic were confiscated by the Chechens and disappeared without a trace or were resold to Russian vendors. Kidnapping of aid workers and journalists became another source of easy cash. All the while, Yeltsin's government was too busy with economic reforms back home to pay much attention.

Late in 1994, the situation finally exploded: On December 11, Russia moved in troops to regain control of the defiant province. Some hardliners close to Yeltsin (especially his minister of defense, Pavel Grachev) were involved in pushing the president to make that fateful decision. The Russian troops quickly regained Grozny, but they lost hundreds of men and much equipment in the process. They were also unable to finish Dudayev's government off in "three days of special operations," as some Russian generals had boasted; instead a real war broke out on the Russian southern fringe, raising the specter of an all-out civil war engulfing the entire Caucasus. The fighting had subsided somewhat by May 1995, but many thousands of Chechen and Russian refugees had left the republic, and a few thousand were dead. Most Chechen fighters had been driven into the forests and the mountains. Nevertheless, Basayev's raid on Budennovsk (300 km north of Chechnya) in June surprised everyone. With about 20 other Chechens, he took 1,600 hostages at a regional hospital deep inside Russia's territory. Russia was forced to sign a truce agreement. Soon after the release of the surviving hostages (about 100 had been killed in the skirmish), the Russian troops started a negotiated pullout. Renamed the independent Republic of Ichkeriya, Chechnya now had its long-sought independence, although it was not recognized by any world government except the like-minded Taliban in Afghanistan.

Unfortunately for Dudayev, he got too closely involved with some dangerous people, including Wahhabi fighters from Saudi Arabia, Jordan, Pakistan, and Afghanistan with direct links to Osama bin Laden. Financing for Ichkeriya would now come from the Muslim diaspora abroad, but strings were attached. Strict sharia law, head coverings for women, no schooling for girls and very little for boys, mass public executions, and lashings of the "enemies of the state" became part of daily life for the remaining Chechnya residents. Life was hardest for the Russian residents of Grozny, who were trapped in a bombed-out city in a hostile society, with no relatives in the countryside to provide even basic food or shelter. Thousands of Chechen refugees dispersed themselves throughout the FSU, mainly ending up in Moscow and some other large Russian cities; others went to Ukraine, including the off-limits zone around Chernobyl, where empty houses were available.

Meanwhile the kidnappings near Chechnya continued, including a particularly gruesome case in which four BBC journalists were beheaded by Chechen warlords in 1998. Russia could not tolerate this situation for long: Not only had it suffered a public, humiliating defeat, but it needed to stabilize the rest of the region, since five other autonomous ethnic republics were being infiltrated with potential jihadis from Chechnya. It could not afford to allow another hostage crisis to happen.

Nevertheless, in 1999, Basayev with his men managed to infiltrate neighboring Dagestan and stir up some action in the western mountains. The situation deteriorated to such an extent that another war seemed imminent in the Caucasus. Then, in a few fateful days in August and September, explosions rocked apartment buildings in Moscow, Rostov-on-Don, and Buynaksk. A few hundred innocent civilians died at night, crushed in their own beds. The brand-new prime minister Vladimir Putin and the rest of the Federal Security Service squarely blamed the Caucasian terrorists, specifically Basayev. Some sources suggested that Russian authorities had *allowed* this to happen as a pretext to send the army back to Chechnya. This contention has never been adequately proven, but in any case, the Russian Army was back in Chechnya in the fall of 1999—and this time it was there to stay.

(cont.)

Many detailed accounts of the next few years exist. Eventually Dudayev, Maskhadov (his replacement), and Basayev were all killed by the Russian special security forces. A few major terrorist attacks occurred in Russia after 1999, most notably the Nord-Ost theater siege in Moscow in 2003, the bombing of two airliners in August 2004, and finally the bloody Beslan school siege in September 2004. The president of the Chechnya Republic at this writing is Ramzan Kadyrov, the son of the former warlord Akhmad Kadyrov, who had become a pro-Russian Chechen leader. (The elder Kadyrov was killed in a mine blast at a soccer stadium in Grozny in 2004, apparently by pro-Basayev forces who viewed him as a traitor.)

Chechnya did have a constitutional referendum in 2003 that allowed it to stay within the Russian Federation. Although the referendum's objectivity has been questioned, it is hard to deny that for the vast majority of Chechnya's population, life is more stable now. A functioning multiparty parliament was elected in 2005; however, the president has sweeping powers and is only minimally controlled from Moscow. About 50,000 Russian troops are still located in the republic, but these are gradually being withdrawn. Only about 40,000 Russians continue to live in Chechnya, as compared to over 100,000 before the war. Moscow is again spending billions of rubles on restoration of the destroyed industry and infrastructure in the republic, but much of the money is being lost to corruption. Although the major conflict is over and is unlikely to resurface in the same dramatic ways, Chechnya is years away from being a prosperous and stable society, and this is one area in Russia where travel is not advisable.

Culturally, the Caucasus is the most diverse region in the FSU, if measured in number of languages per unit of area. Dagestan alone has five main languages (Avartsy, Dargintsy, Kumyks, Lezgin, and Lakhs) and two dozen secondary languages spoken in different parts, although the republic is only about the size of Costa Rica. Besides the Russians, important groups in the northern Caucasus include the Circassians, Vainakhs, Ossetians, and Turkic-speaking Karachai and Balkars (Chapter 13). Most of these groups (except Ossetians, who are Orthodox Christians) accepted Islam in the 17th and 18th centuries. The Georgians (Kartli, in their own language) have four subethnic groups and are indigenous inhabitants of the southern Caucasus with no immediate relatives. They are also Orthodox Christians and have a highly developed old culture (their golden period was in the 12th century under King David II, followed by Queen Tamar), with a unique alphabet, poetry, music, architecture, and dress style. The Armenians are Indo-European people of Asia Minor, with their own distinct language and alphabet, and a culture spanning two and a half millennia. Almost 2 million Armenians live in diaspora, from the Middle East to the United States. The Azerbaijanis are closely related to the Turks linguistically, but are Shiite Muslims by religion. More Azerbaijanis live in Iran than in Azerbaijan. The Naxcivean region of Azerbaijan is a separate exclave, with Armenia, Turkey, and Iran between it and the rest of the country.

Economics

Economically, the South district lags behind much of the rest of Russia. Although Volgograd and Rostov are large industrial centers, and Sochi is a Riviera with booming real estate, much of the region has below-average incomes. The poorest three republics in Russia are war-torn Chechnya (GRP unknown) and its neighbors Ingushetiya (about 15% of the national average) and Dagestan (about one-third of the national average). These are also the areas with the highest unemployment (24%), highest poverty rate, and highest fertility (2.15 children per woman, as compared to 1.40 for Russia as a whole). Even the richest subject, Krasnodarsky Kray, has only two-thirds of the national average GRP per capita. Among the three non-Russian states, Azerbaijan is the richest ($9,500 GRP per capita in 2008), and Georgia is the poorest ($4,700), with Armenia slightly above Georgia's level ($6,300). Most of the wealth in Azerbaijan comes from exports of petroleum, but it is spread across the population very unevenly. Armenia is heavily reliant on remittances from the large diaspora living abroad, especially in the United States, Lebanon, and France; in this sense, it is similar to El Salvador and other Central American economies. It also has the friendliest relations with Russia of the three countries.

Georgia is currently led by a strongly pro-Western president, Mikheil Saakashvili, who seems eager to push for closer relations with NATO and is known to have received Western cash support for his package of reforms. Nevertheless, the main foreign sources of income for Georgia remain exports of fruit, wine, mineral water, and a limited amount of minerals. Russia's politically motivated embargo on Georgian wine exports since 2006 has hurt the Georgian economy somewhat. At the same time, the Baku–Tbilisi–Ceyhan (BTC) pipeline completed in 2006 now allows Georgia to receive substantial payments for transshipment of Caspian Sea oil to the West. In August 2008, a vicious military conflict between pro-Russian South Ossetiya and Georgia escalated into a real war when the Georgian government shelled the Ossetian capital, Tskhinvali, in the middle of the night in a bid to retake the lost territory. The Russian-backed military response was unexpectedly strong, and Georgia lost after a few days of fighting what was the most significant conflict for Russia outside its proper territory in the entire post-Soviet period. The situation in South Ossetiya and Abkhazia in the aftermath of the conflict remains uncertain, with both appealing to the international community to recognize them as independent countries. Both like to cite Kosovo as a precedent in Europe, but so far they have been recognized only by a handful of U.N. members, with Russia being the most prominent.

The two largest cities within the South region of Russia are Rostov-on-Don and Volgograd (the Stalingrad of World War II), with about 1 million people each. The former specializes in coal processing, production of synthetic fibers and plastics, agricultural machinery building, banking, and educational services. It is also a large port on the Don, connected to both the Azov Sea and the Volga via a canal, and is thus a major transportation hub. Rostov Oblast gave the world the writers Anton Chekhov (author of *The Seagull, Three Sisters*, other plays, and numerous short stories) and Mikhail Sholokhov (*And Quiet Flows the Don*). Volgograd is one of the longest cities in Russia, stretching for 100 km along the west bank of the Volga. It has a large tractor plant producing about one-quarter of all tractors in Russia. It also has a number of steel and petrochemical enterprises, an aluminum smelter, a ball-bearing factory, and many others. Both Rostov and Volgograd Oblasts are important agricultural areas. Rostov Oblast, for example, is the second biggest producer of hogs in the country and produces a lot of vegetable oil.

The city of Krasnodar (population 646,000) is a major industrial center of the region as well. Its main specialties are petroleum processing, agricultural machinery building, and light industry (clothing, shoes, etc.). Krasnodarsky Kray is one of the top five agricultural producers in the nation. Of particular note are rice, tea, wine, and tobacco production. There are over 100 sanatoria and 75 large tourist camps along the Black Sea coast from Novorossiysk to Adler (Figure 25.4). Sochi is the future home of the 2014 Winter Olympics and is one of the most expensive places to own a home in Russia. North of Sochi, the city of Tuapse has a major oil refinery and is the terminal for the future underwater pipeline into Turkey. Farther north, an underwater gas pipeline was completed into Turkey from Novorossiysk in 2004. Novorossiysk is Russia's biggest

FIGURE 25.4. Beachgoers in Krasnodarsky Kray, the leading provider of seacoast tourism services in Russia. *Photo:* A. Pugach.

seaport and is especially important for shipping petroleum across the Black Sea.

The small republics on the northern slope of the Caucasus have relatively underdeveloped cities and industries. Many specialize in mining and ranching. There are some small factories making cement, plywood, electric equipment, furniture, clothing, and food, mainly for local consumption. Karachaevo-Cherkessiya and Kabardino-Balkariya have developed downhill ski resorts and backpacker routes. The mineral spa areas of Pyatigorsk and Kislovodsk are a big draw for people from Central Russia and are well described in Mikhail Lermontov's poems from the 19th century. Over 80 resorts existed here in the late Soviet period, but fewer are open now.

The economic strengths of Georgia are few. Good climate and beautiful scenery do not make a country particularly rich or enable it to count on many tourists unless it is also politically stable, which Georgia at the moment is not. However, Georgia does have small-scale mining of manganese and other ores in the mountains, some hydropower production, and limited light manufacturing and food industries. The capital, Tbilisi (population 1,090,000), is a big, ancient, and beautiful city undergoing much-needed renovation and restoration (Figure 25.5). The poet Lermontov painted a small picture of Tbilisi in 1837 that transmits the city's charm really well. Tbilisi now is a much larger city, of course; it even has its own subway. It is the main political, cultural, educational, and economic center of Georgia. Other cities include Gori, the birthplace of Joseph Stalin (Figure 25.6); Batumi and Poti, ports on the Black Sea; and Rustavi, Kutaisi, and Zugdidi. Georgia's future prospects depend on developing the Caspian oil transshipment and perhaps refining; forming closer ties with the European Union (EU) and NATO; and especially improving its currently strained relations with Russia, still the biggest trade partner and most influential political player in the Caucasus region. Casting an uneasy shadow on the situation are the two regional conflicts in Abkhazia (capital, Soukhumi) and South Ossetiya (capital, Tskhinvali).

FIGURE 25.5. Old Tbilisi. *Photo:* K. Van Assche.

FIGURE 25.6. At the museum of Joseph Stalin in Gori, Georgia. *Photo:* K. Van Assche.

Armenia is the only landlocked country of the trans-Caucasus and is the smallest in area. Although it is slightly wealthier than Georgia in 2008, partially due to a strong flow of remittances from abroad, it is geographically much less advantaged. Two other nations—its historical archenemy, Turkey, perpetrator of the hotly debated Armenian genocide in the late 19th and early 20th centuries; and its more recent antagonist, Azerbaijan—prevent shipping of any supplies in or out of Armenia from the west and the east, respectively. Political negotiations with Turkey in 2009 seem to have eased the situation somewhat. Limited trade exists with Georgia and Iran (soon to be connected by a railroad), but its biggest trade partner is Russia. Armenia is an arid high-mountain country, with virtually all of its land above 1,000 m elevation (Figure 25.7). It has well-developed industry and formerly had a very well educated workforce, including world-famous doctors and engineers; its musicians and chess players were also renowned. Many of these people, unfortunately, were forced to leave the country in the turbulent 1990s or earlier, and are unlikely to come back soon.

FIGURE 25.7. Armenia is full of dry mountains and beautiful old churches. *Photo:* K. Van Assche.

Energy independence is a big goal for Armenia. It recently restarted its Soviet-era nuclear power plant to provide much-needed electricity. Armenia has limited production of gold and gems, as well as plenty of construction materials, but virtually no fossil fuels. It has endured a prolonged conflict with Azerbaijan over the control of the Nagorno-Karabakh exclave, east of Armenia proper. That Armenian-populated region was placed by Stalin inside Azerbaijan as an autonomous region, but was long sought by the Armenians. The smaller but more motivated Armenian army, along with paramilitary groups, drove the Azerbaijanis out in 1994, and a cease-fire was declared. Many of the Azerbaijani refugees from that war ended up living in Baku. Most of the industries in Armenia now are scaled-down versions of those from Soviet times, when it used to produce some high-quality washing machines, electrical equipment, clothing, and tools. Pig iron and nonferrous metals are currently Armenia's most valuable exports. Yerevan, with over 1 million people, is the biggest cultural and economic center of the country (Figure 25.8). Other important cities include Sevan near the famous lake of the same name, Vanadzor, and Gyumri in the country's north.

Azerbaijan is the wealthiest of the three independent Caucasian republics, but much of its wealth belongs to a small minority of businesspeople with close ties to the autocratic president, Ilham Aliyev. Azerbaijan's economy is tied to oil and gas development; the oldest continuously producing oil fields in the world are located on the Apsheron Peninsula. The country is thought to have between 9 and 14 billion barrels of oil left in reserves, comparable to the reserves of Algeria or Norway. Virtually all of the oil is produced offshore, but the Caspian Sea is shallow and warm, and with the construction of the Baku–Tbilisi–Ceyhan pipeline the oil can now quickly reach customers in Europe (Balat, 2006). Some of Azerbaijan's oil (out of a total of 1.1 million barrels per day) still flows to Novorossiysk in Russia, where it is put on tankers to be shipped across the Black Sea. The Apsheron Peninsula is badly polluted with petroleum and heavy metals. In 2006 Sumgait, north of Baku, was listed as one of the 10 most polluted places on earth by the U.S.-based Blacksmith Institute.

The second most important source of income is agriculture; the climate of Azerbaijan along the coast is warm enough to grow cotton, rice, tobacco, and a variety of vegetables and fruit. Fresh-

FIGURE 25.8. A street in Yerevan. *Photo:* K. Van Assche.

FIGURE 25.9. A famous landmark in downtown Baku, the Qiz Galasi, or Maidens' Tower. *Photo:* S. O'Lear.

cut flowers are supplied to Moscow markets from early spring onward. The biggest industrial hub is the capital, Baku (Figure 25.9). Baku is a large metropolitan area with a population of 2 million. It has a mixture of old and new architecture and is located on the south side of the Apsheron Peninsula in a scenic bay. Most of the country's wealth and power is concentrated here. Ganja is the second biggest city in the far western part of the country, with a rich historical past. The famed 13th-century poet Nizami is buried in a mausoleum here.

Challenges and Opportunities in the Caucasus

The future of the Caucasus is very important for the rest of the FSU and for the entire world. First of all, the Caspian Sea basin represents one of the last relatively large remaining sources of petroleum in the world, and at present the main way to take that oil out is via either Russia or Georgia. Second, the Caucasus has the mildest climate in the entire FSU. It used to receive up to 10 million tourists from the rest of the U.S.S.R. per year in the 1980s, but has only half as many now. Tea and citrus plantations, vineyards, and a variety of natural areas, sanatoria, spas, camps, and of course beaches are amply represented here. The Caucasus is also a well-watered place, which is important in the warming and drying world of today. Third, the Caucasus is an amazing melting pot of cultures and traditions. It is a haven for anthropologists, archeologists, linguists, and other researchers (Catford, 1977). Fourth, the Caucasus is one of only 25 global "biodiversity hotspots," according to Conservation International (2010). Its entire flora, for example, is over 6,000 species of vascular plants—about the same as California's. Many rare animals make their homes in the mountains. There are still large tracts of wilderness available for maintaining ecological balance, for primitive recreation, and for education. Finally and significantly, the Caucasus is located on the Asian fringe of Europe. Whether Georgia, Armenia, and Azerbaijan align themselves most closely with the EU, Russia, or the Middle East has major implications for the political balance of power in the region and for the prospects of peace on the planet (Goltz, 2006; Huttenbach, 1996; King, 2008).

EXERCISES

1. Explore any personal connections with the Caucasus and its cultures in your family, among friends, and/or in the community where you live. Which cultures, languages, or traditions do you have access to?
2. Propose a new investment strategy for a wealthy Western investor in the Caucasus. What countries or industries would be good choices? Which ones would be poor choices? Present your findings in class.
3. Do an online search of nature-oriented tour offerings in the greater Caucasus. Which areas are heavily marketed and why?

Further Reading

Baddeley, J. F. (1969). *The Russian conquest of the Caucasus.* New York: Russel & Russel.

Balat, M. (2006). The case of Baku–Tbilisi–Ceyhan oil pipeline system: A review. *Energy Sources, Part B: Economics, Planning and Policy, 1*, 117–126.

Catford, J. C. (1977). Mountain of tongues: The languages of the Caucasus. *Annual Review of Anthropology, 6*, 283–315.

Conservation International. (2010). The biodiversity hotspots. Retrieved April 1, 2010, from *www.conservation.org/explore/priority_areas/hotspots*

Frolova, M. (2006). The landscapes of the Caucasus in Russian geography: Between the scientific model and the sociocultural representation. *Cuadernos Geograficos, 38*, 7–29.

Gachechiladze, R. (1995). *The new Georgia: Space, society, politics*. College Station: Texas A&M University Press.

Gall, C., & DeWaal, T. (1998). *Chechnya: Calamity in the Caucasus*. New York: New York University Press.

Goltz, T. (2006). *Georgia diary: A chronicle of war and political chaos in the post-Soviet Caucasus*. Armonk, NY: Sharpe.

Hahn, G. M. (2005). The rise of Islamist extremism in Kabardino-Balkaria. *Demokratizatsiya, 13*, 543–595.

Herzig, E. (1999). *The new Caucasus: Georgia, Armenia, Azerbaijan*. London: Pinter.

Hunter, S. T. (1994). *Transcaucasia in transition: Nation-building and conflict*. Boulder, CO: Westview Press.

Huttenbach, H. (1996). *The Caucasus: A region in crisis*. Boulder, CO: Westview Press.

Karny, Y. (2000). *Highlanders: A journey to the Caucasus in quest of memory*. New York: Farrar, Straus & Giroux.

King, C. (2008). *The ghost of freedom: The history of the Caucasus*. New York: Oxford University Press.

Knezys, S., & Sedlickas, R. (1999). *The war in Chechnya*. College Station: Texas A&M University Press.

Meier, A. (2005). *Chechnya: To the heart of a conflict*. New York: Norton.

Miller, D. E., & Miller, L. T. (2003). *Armenia: Portraits of survival and hope*. Berkeley: University of California Press.

Nation, R. C. (2007). *Russia, the United States and the Caucasus*. Carlisle, PA: Strategic Studies Institute.

O'Lear, S., & Gray, A. (2006). Asking the right questions: Environmental conflict in the case of Azerbaijan. *Area, 38*, 390–402.

O'Loughlin, J., Kolossov, V., & Radvanyi, J. (2007). The Caucasus in a time of conflict, demographic transition, and economic change. *Eurasian Geography and Economics, 48*, 135–157. (See also other articles in this special issue.)

Pereira, M. (1973). *Across the Caucasus*. London: Bles.

Plunkett, R., & Masters, T. (2004). *Georgia, Armenia and Azerbaijan*. Hawthorn, Victoria, Australia: Lonely Planet.

Rybak, E. A., Rybak, O. O., & Zasedatelev, Y. V. (1994). Complex geographical analysis of the Greater Sochi region on the Black Sea coast. *GeoJournal, 34*(4), 507–513.

Shami, S. (1998). Circassian encounters: The self as other and the production of the homeland in the north Caucasus. *Development and Change, 29*(4), 617–646.

Therborn, G. (2007). Transcaucasian triptych. *New Left Review, 46*, 68–89.

Ware, R. B. (1998). Conflict in the Caucasus: An historical context and a prospect for peace. *Central Asian Survey, 17*(2), 337–352.

CHAPTER 26

The Urals

Metallurgy, Machinery, and Fossil Fuels

The Urals economic region of the Soviet Union included four subjects of federation (Bashkortostan and Udmurtiya Republics, Permsky Kray, and Orenburg Oblast) that have already been discussed in Chapter 24. It also included Sverdlovsk Oblast around Yekaterinburg, as well as Chelyabinsk and Kurgan Oblasts, all of which are discussed here. The federal districting scheme of 2000 has added Tyumen Oblast, and the Khanty-Mansi and Yamalo-Nenets Autonomous Okrugs to the Urals as well, although they were previously part of the West Siberia economic region (Table 8.3, Figure 26.1, and Table 26.1).

As currently defined, the Urals federal district has 1.8 million km^2 populated by 12 million people. It is 81% urban, but with a low population density of 6/km^2. It is hard to find an analogue to this region within the United States. On the one hand, the Urals resembles the industrial lower Midwest, with its emphasis on steel mills and manufacturing. On the other, the region resembles Texas and parts of the American South, with the emphasis on fossil fuels. Finally, because of the scenic low continental mountains in the middle, it resembles the Appalachian states.

Physical Geography

The Ural Mountains run almost perfectly north–south, dividing Europe from Asia. The average elevations near Chelyabinsk are only 1,000 m, with the highest point, Yaman-Tau, reaching 1,638 m. Mt. Narodnaya in the polar Urals reaches 1,895 m. The Western Siberian Lowland, in the eastern part of the region, extends to the Arctic Ocean (see Chapter 2, Figure 2.1) and is one of the largest wetlands in the world. During the late glacial period, a vast, shallow lake covered the area. In the more distant past, the entire Western Siberian Lowland was a tropical sea that left massive deposits of shale and limestone. The Urals are located along an old, inactive tectonic boundary uniting the Western Siberian and European platforms. The mountains are about 230–300 million years old; they were formed in the Permian period of the late Paleozoic age. The Permian period's name comes from the city of Perm, west of the Urals. The Urals are seriously worn down by water and wind erosion. East of the mountains, the Irtysh and the Ob form the largest river basin in Russia (see Chapter 2). The western slopes give rise

FIGURE 26.1. The Urals federal district of Russia. Map: J. Torguson.

to the Kama and the Ural Rivers in the south and the Pechora in the north. The Ural Mountains are a treasure trove of resources: coal, iron ore, manganese, titanium, chromium, gold, copper, nickel, vanadium, marble, and many other minerals. This is the richest area in all of Russia with respect to nonferrous metals and gemstones. Over 1,000 minerals are found in the Urals, including some with names derived from local landmarks: uralite, ilmenite, and others. Now that Tyumen Oblast and the two autonomous okrugs are included in the Urals district, the region has also become by far the richest area in Russia with respect to petroleum and natural gas, accounting for over 70% of all Russia's oil and more than 80% of its natural gas reserves.

The climate is strongly continental, mainly of the Dfb (humid continental) and Dfc (subarctic) Köppen types. The coldest temperature ever recorded in Yekaterinburg, which is in the middle of the region, is –48°C, and the warmest is +38°C. The mean annual temperature is +2.8°C, which is almost 4° colder than Minneapolis, Minnesota. The rivers freeze in late November and thaw in late April. The western slopes of the Urals may receive over 800 mm of precipitation, while the dryer southeastern slopes receive less than 400 mm per year. Podzolic, gray, and brown forest soils are common in the Urals proper, and gley-

TABLE 26.1. Subjects of Federation of the Urals Federal District

Subject	Administrative center	Population (2006) (thousands)	% ethnic Russian	GRP/capita (rubles in 2006)	Top products and areas of economic specialization within Russia (% of nation's total production for some)
Tyumen Oblast	Tyumen	3,323	72	782,429	Petroleum and gas production and refining, electric energy
Khanty-Mansi Autonomous Okrug	Khanty-Mansiysk	1,478	66	NA	Petroleum production (57%), electric energy, fur-bearing mammal farms
Yamalo-Nenet Autonomous Okrug	Salekhard	531	59	NA	Natural gas (87%), petroleum (11%), reindeer farms
Sverdlovskaya Oblast	Yekaterinburg	4,410	89	148,710	Steel rolling (66%), railroad cars (55%), pipes (31%), turbines, defense
Chelyabinsk Oblast	Chelyabinsk	3,531	82	125,558	Bulldozers (73%), steel rolling (26%), tractors (19%), jewelry (41%)
Kurgan Oblast	Kurgan	980	92	68,208	Automated milking equipment (81%), agricultural machinery, wheat

Note. Data from the Federal Service of State Statistics, Russian Federation (2009). To convert gross regional product (GRP) to U.S. dollars, divide GRP by 27.19. NA, not available.

podzolic and peat soils in the former West Siberia. North of Tyumen, permafrost is common, but not very thick.

Most of the region is located in the southern taiga and mixed forest zones, with small areas of steppe in the south, many peat bogs in the east, and diverse forest and alpine communities in the mountains (Figure 26.2). About 40% of the district is forest covered. The main area of timber harvesting is in Sverdlovskaya Oblast, providing 60% of the regional total of 6.2 million m^3 of timber per year, or about 6% of the national total. Agriculture is underdeveloped, due to the harsh climate and poor soils, but still accounts for 7% of Russia's total agricultural output; most of the agricultural activity is concentrated in Chelyabinsk and Kurgan Oblasts because of their warmer climate.

Protected natural areas include nine zapovedniks, three national parks, and nine federal wildlife refuges. Denezhkin Kamen Zapovednik in northern Sverdlovsk Oblast protects about 70,000 ha of mountainous taiga in the northern Urals. The preserve straddles the boundary between Europe and Asia, which ensures high biological diversity of its flora and fauna. The preserve has healthy wildflower, wolverine, sable, bear, wolf, lynx, moose, and wild boar populations. Pripyshminsky Bory National Park, also in Sverdlovsk Oblast, focuses on outdoor recreation and preservation of unique pine forests along the ancient terraces above the Tobol and Tura Rivers. Hiking, cycling, horseback riding, and whitewater rafting routes are being developed. Kurgansky Wildlife Refuge, located on the border with Kazakhstan in Kurgan Oblast, protects rare steppe birds, including bustards and red-breasted geese.

Cultural and Historical Features

Culturally, the Urals region is overwhelmingly Russian, with small Uralic minorities in the forested east—the Mansi people west of the Ob, and the closely related Khanty east of the Ob. Siberian Tatars near Tyumen, and Bashkirs in the south near Chelyabinsk, form two important Turkic minorities. One of the oldest Russian settlements in the region is Tobolsk, the former

FIGURE 26.2. The low and old Ural Mountains are forest covered, except at higher elevations near the Arctic Circle. *Photo:* A. Shanin.

capital of Siberia, established at the confluence of the Tobol and Irtysh Rivers in 1587. Its stone kremlin is Siberia's first. The city lost much of its significance when the Trans-Siberian Railroad was completed a few hundred kilometers to the south in the late 19th century. Yekaterinburg and Chelyabinsk have over 1 million people each, while Tyumen has over 500,000. Yekaterinburg was founded in the 18th century, when the first iron smelters were established in the Urals. It is the main cultural and business center of the region today, with world-class universities, museums, theaters, and even its own movie studio. Magnitogorsk, south of Chelyabinsk, and Nizhniy Tagil, north of Yekaterinburg, are among the largest steel producers in Russia. Miass and Zlatoust, west of Chelyabinsk, are also important industrial centers with about 200,000 residents each. The central part of the Urals Mountains is the most urbanized, whereas Kurgan Oblast is the least urbanized. The city of Kurgan itself (population 345,000) is an important agricultural and weapons-building center.

FIGURE 26.3. Yekaterinburg's new All Saints on the Blood cathedral marks the site of the former Ipatiev house, where the family of the last tsar was murdered by the Bolsheviks in 1918. *Photo:* I. Tarabrina.

The Urals cannot match Central Russia in terms of cultural or historical sites worth visiting. Yekaterinburg was the place where the last tsar and tsarina of Russia, Nicholas II and Alexandra, were executed with their family and servants in 1918 by the Bolsheviks. Today the murder site is adorned with a magnificent cathedral, All Saints on the Blood (Figure 26.3). Alapaevsk is visited as the site where Elizabeth Romanov, Alexandra's sister, was murdered with her associates. Verkhoturye is another heavily visited religious site in the Urals, associated with St. Symeon, a 17th-century saint. Not too far away is the village of Pokrovskoe in Tyumen Oblast—the birthplace of Grigory Rasputin, the controversial friend and spiritual associate of the last royal family. D. Mamin-Sibiryak is a well-known late-19th-century writer who grew up in Yekaterinburg Oblast. Another well-known literary figure from the Urals is Pavel Bazhov, whose fairy tales were based on local folk legends about the masters of the gemstone underworld. Other well-known figures associated with Yekaterinburg include Vladislow Krapivin, a popular Soviet children's writer; movie director S. Govorukhin; Boris Yeltsin, the first president of the Russian Federation; Alexey Yashin, a legendary ice hockey player; sculptor Ernest Neizvestny; mathematician Nickolai Krasovsky; and Olympic swimming champion Alexandr Popov. Yekaterinburg also gave Russia three very popular rock bands: Nautilus Pompilius, Agata Kristie, and Chaif.

Economics

Economically, the Urals region is a mining and metallurgy giant, accounting for about 19% of all industrial output in Russia. As noted earlier, the region now accounts for over 70% of all petroleum in Russia (most of it in Khanty-Mansi Autonomous Okrug) and over 80% of all natural gas reserves (mostly in Yamalo-Nenets Autonomous Okrug). The Urals district also leads the country in production of bauxite, and is second only to the Siberia district in production of copper ore and to the Central district in production of iron ore. Sverdlovsk and Chelyabinsk Oblasts together account for 38% of all steel made in Russia. Only the Central district produces more machinery.

The oil and natural gas fields of what was then the West Siberia economic region were discovered in the 1960s and developed in the 1970s. In

1965 this area produced only 1 million metric tonnes (mmt) of petroleum, but by 1985 it was already 400 mmt, or about 2.9 billion barrels per year! To put this in perspective, the entire United States uses about 5.5 billion barrels per year. The production of oil in this area dropped dramatically in the 1990s because of the economic downturn, to about 200 mmt per year in 1995, but has since risen to about 320 mmt. This number is unlikely to increase farther, because the oil fields are rapidly being depleted. Tyumen is a large historical center in the former West Siberia, on the banks of the Tura River. It is a railroad hub and a major river port. Oil and gas processing and shipbuilding are two major activities. Courtesy of the energy boom in recent years, Tyumen Oblast leads the nation in gross regional product (GRP) per capita—about $28,800 in 2006 (Table 26.1).

The largest oil fields are concentrated along the middle reaches of the Ob River, in the central and eastern parts of Khanty-Mansi Autonomous Okrug. There are four major concentrations of extracting areas near cities, each associated with one or two of the major oil companies of Russia: Surgut (Surgutneftegaz), Nizhnevartovsk (BP-TNK and Lukoil), Nefteyugansk (formerly Yukos, now Rosneft), and Kogalym (Lukoil). Gazpromneft (formerly Sibneft) controls the very large Noyabrsk oil and gas field northeast of Surgut, and the new Priobskoe oil field near Khanty-Mansiysk.

Farther to the north, there is primarily natural gas production in the giant fields near the base of the Yamal Peninsula. Most of the gas production is controlled by Gazprom; however, some gas is also produced by TNK-BP, Surgutneftegaz, and Novatek. The earliest gas fields to be developed, near Urengoy (just below the Arctic Circle), started producing in 1964. In 2005, 557 billion m^3 of gas was produced in Yamal alone. The Urengoy fields are connected to Tyumen and the Urals proper via a railroad link and multiple oil and gas pipelines. The more recently discovered gas fields are located farther north, above the Arctic Circle: Yamburgskoe on Gydansk Peninsula, and Bovanenkovskoe on Yamal Peninsula. A new railroad link from Labytnagi to Yamal is being built. Overall, Russia is thought to have about 48 trillion m^3 of gas in proven reserves, more than in all of North and South America, Europe, and Africa combined. Only some Middle Eastern countries, especially Qatar and Iran, have comparable reserves.

Steel production in the Urals is concentrated in a few very large factories dating back to Soviet times—in particular, the Magnitogorsk, Nizhniy Tagil, Chelyabinsk, and Novotroitsk steel combines. A few dozen smaller factories specialize in rolling steel, making pipes, and manufacturing precise parts. Also associated with the large steel-making combines are factories producing cement, gypsum panels, nitrogen fertilizer, and plastics. Historically, the fuel used in steel production was locally produced charcoal. In the past 30 years there has been a definite shift toward using more imported fossil coal, especially from central Kazakhstan, which is conveniently connected to the Urals by a few railroads. Another metallurgical specialty of the Urals is copper production (the Krasnouralsk, Revda, and Karabash plants) and refining (the Kyshtym and Verkhnepyshminsky plants). Nickel, aluminum, zinc, titanium, and magnesium production are also well developed. Sverdlovsk and Chelyabinsk Oblasts have a number of so-called secret cities producing components for nuclear weapons (Novouralsk and Lesnoy in the former oblast, and Snezhinsk, Ozersk, and Trehgorny in the latter; see Chapter 18). The vicinity of the Mayak factory in Chelyabinsk Oblast was the scene of a few nuclear accidents in the late 1950s and early 1960s, and is heavily damaged by radioactive waste (Chapter 5).

The machinery-building sector in the Urals focuses on industrial machinery and turbines (Uralmash, Uralelektrotyazhmash, Yuzhuralmash), tractors (Chelyabinsk), trucks (Novouralsk, Miass, Kurgan), and railroad cars and tanks (Nizhniy Tagil). Motorcycles are built in Irbit. The construction industry is also well developed, based on local nonmetal mineral mining, cement, and plywood production. Finally, Yekaterinburg and Chelyabinsk are major producers of consumer goods, including leather jackets, cotton shirts, shoes, radioelectronics, and appliances.

As noted earlier, most of the Urals' farming areas are in the south, mainly in Kurgan and Chelyabinsk Oblasts (80% of the total). Grains are planted here, especially wheat, as well as sunflowers. Sverdlovsk Oblast leads the region in production of potatoes and vegetables; as in the rest of Russia, these are mainly grown on small dacha plots by city residents.

The infrastructure of the Urals has two main hubs in the middle (Yekaterinburg and Chelyabinsk) and one in the east (Tyumen). All three cities are on branches of the Trans-Siberian Railroad. The northern line (Yekaterinburg–Tyumen–Omsk–Novosibirsk) runs entirely within Russia. The southern line, however, runs through Chelyabinsk and Kurgan to Petropavlovsk in Kazakhstan and then to Novosibirsk. In the Soviet period, that route was the faster of the two, but it is slower now because the double delays on the international borders add 5–6 hours to the length of the trip. The regional line from Tyumen northeast to Surgut and Urengoy is very important for moving people and freight in and out of the oil- and gas-producing districts. Construction of highways, railways, and pipelines in the fragile tundras and bogs of western Siberia is hampered by the presence of extensive wetlands, and in some areas, permafrost. Some Yamal "highways" are little more than 2- or 3-km-wide muddy ruts left by tractors. Paved highways largely run parallel to the railroads. A newly proposed federal highway will connect Yekaterinburg and St. Petersburg via Perm and Kirov. The Ob, with its tributaries, serves as an important shipping lane. However, about half of the year it is ice-bound in the north, so navigation only occurs during the warmer half of the year.

Challenges and Opportunities in the Urals Region

The biggest future challenge for the region is reducing its dependence on heavy machinery, militarized enterprises, petroleum, and natural gas. Diversification of the local economy is progressing slowly. The large influx of capital due to fossil fuel extraction and processing development should be wisely invested to improve infrastructure and social life. Climatically, the region may experience some improvement with global warming, but it is certainly not an attractive place to live (especially in its eastern and northern parts). However, the Ural Mountains themselves are well positioned for development of tourism and recreation, especially whitewater rafting, cave tourism, backpacking, and horse tourism. Heritage and religious tourism are also popular and could be made more so. Heavy air pollution and large areas of nuclear and other industrial contamination pose some serious challenges for future successful development.

EXERCISES

1. In groups, conduct research on the geography of oil production of one of the Russian oil majors. As of 2009, suitable companies included Lukoil, Rosneft, Gazpromneft, Surgutneftegaz, and TNK-BP. Find out where their oil fields are, where their production facilities and refineries are concentrated, and what their recent discoveries and investment trends have been. Give in-class presentations highlighting the accomplishments of each company. Try to convince an independent panel of judges (including your instructor) that your company has the best potential to attract new investments.
2. Write a report on any traditional culture of the Urals (Khanty, Mansi, Siberian Tatars, etc.). Compare and contrast their lifestyle with that of any other traditional culture from around the world that you are familiar with.
3. Investigate educational options available to a foreign student in either Yekaterinburg or Chelyabinsk.
4. Read any of Pavel Bazhov's tales about the people and gemstones of the Urals. What are some of the key cultural elements in them that are related to the natural riches of the region?

Further Reading

Aarkrog, A., Dahlgaard, H., Nielsen, S. P., Trapeznikov, A. V., Molchanova, I. V., Pozolotina, V. N., Karavaeva, E. N., Yushkov, P. I., & Polikarpov, G. G. (1998). Radioactive inventories from the Kyshtym and Karachay accidents: Estimates based on soil samples collected in the south Urals (1990–1995). *Science of the Total Environment, 201*(2), 137–154.

Bazhov, P. P. (1955). *Malachite casket: Tales from the Urals.* Moscow: Foreign Languages Publishing House.

Considine, J. I., & Kerr, W. A. (2002). *The Russian oil economy.* Northampton, MA: Elgar.

Dienes, L. (2004). Observations on the problematic potential of Russian oil and the complexities of Si-

beria. *Eurasian Geography and Economics, 45*, 319–345.
Grace, J. D. (2005). *Russian oil supply: Performance and prospects.* New York: Oxford University Press.
Harris, J. R. (1999). *The Great Urals: Regionalism and the evolution of the Soviet system.* Ithaca, NY: Cornell University Press.
Hassmann, H. (1953). *Oil in the Soviet Union: History, geography, problems.* Princeton, NJ: Princeton University Press.
Lane, D. S. (Ed.). (1999). *The political economy of Russian oil.* Lanham, MD: Rowman & Littlefield.
Medvedev, Z. A. (1979). *Nuclear disaster in the Urals.* New York: Norton.
Peterson, J. A., & Clarke, J. W. (1983). *Petroleum geology and resources of the Volga–Ural province, U.S.S.R.* Alexandria, VA: U.S. Geological Survey.
Radzinsky, E. (2000). *The Rasputin file.* New York: Nan A. Talese.
Rodgers, A. (1974). The locational dynamics of Soviet industry. *Annals of the Association of American Geographers, 64*(2), 226–241.
Ryabinin, B. S. (1973). *Across the Urals.* Moscow: Progress.
Scott, J. (1989). *Behind the Urals: An American worker in Russia's city of steel.* Bloomington: Indiana University Press.
Smit, P. J. (1996). *Magnitogorsk: Forging a new man* [Video]. New York: First Run/Icarus Films.
Tarasov, P. E., Volkova, V. S., Webb, T., Guiot, J., Andreev, A. A., Bezusko, L. G., Bezusko, T. V., Bykova, G. V., Dorofeyuk, N. I., Kvavadze, E. V., Osipova, I. M., Panova, N. K., & Sevastyanov, D. V. (2000). Last glacial maximum biomes reconstructed from pollen and plant macrofossil data from Northern Eurasia. *Journal of Biogeography, 27*(3), 609–621.
Zimin, D. (2007). The role of Russian big business in local development. *Eurasian Geography and Economics, 48*(3), 358–377.

CHAPTER 27

Siberia

Great Land

The greatest Russian scientist of the 18th century, Mikhail Lomonosov, famously said that "Russia will increase through [the use of] Siberia." In the early 21st century, his words have proven prophetic: Siberia is pivotal to Russia's economic might. It is part of Asiatic Russia and is usually defined as the land east of the Urals and west of the Lena River, sometimes including the entire watershed of the Lena. Thus the territory west of Siberia is European Russia, and the land east of it is the Far East, also called the Russian Pacific. The origin of the word "Siberia" is unclear, but it probably came from a Turkic word meaning "clean land" or "magic land" (or, alternatively, from a Mongolian word for "swampland"). In the 15th century a powerful Sibir khanate existed in western Siberia, populated by a mix of Turkic-speaking Siberian Tatars and some Uralic tribes. As defined in the districting scheme of 2000, the Siberia federal district does not include the oil- and gas-rich Tyumen Oblast, which is now part of the Urals federal district (Chapter 26). It includes four republics, two krays, and six oblasts located in western and central Siberia (Table 8.3, Figure 27.1, and Table 27.1). It also does not include Yakutia (Sakha), which is part of the Far East district, although Yakutia is quintessentially Siberian in its nature and culture. Four areas that were previously autonomous okrugs within Siberia were merged with oblasts or krays in 2007–2008; some statistics may still be reported for those separately, however. The district center and the informal capital of Siberia is its largest city, Novosibirsk, with over 1.4 million residents on the Ob River.

Siberia thus defined (5.1 million km^2) is just a little smaller than the largest (Far East) federal district, and is bigger than the European Union (EU) in size. Although it accounts for about one-third of Russia's territory, it has only 20 million residents, giving it an average population density of only 3.9 people/km^2. The population is 71% urban. In many ways the region is analogous to west-central Canada in North America: It has few people, plenty of natural resources, and a very cold continental climate. Like the rest of Russia, Siberia is losing population fast—in part because of the usual demographic imbalance between fertility and death rates, but also in Siberia's case because of substantial emigration to the warmer European part of the country. The overall decline is about –0.6% per year, among the fastest in Russia. As a result, the shortage of workers is acute: In 2006 Krasnoyarsky Kray alone officially received 5,000 migrant workers from China (38%), Ukraine (27%), North Korea (10%), and Kyrgyzstan and other Central Asian states. The actual need in the kray was close to

FIGURE 27.1. The Siberia federal district of Russia. Source: J. Torguson.

70,000 workers per year, but only 40,000 were available from the domestic pool.

Physical Geography

Physically, Siberia has a vast lowland in the west and an upland in the east. The Western Siberian Lowland, partially discussed in Chapter 25, is a bed of an ancient sea, slightly smaller than the Amazon Lowland in size. It is drained by the slow, north-flowing Ob–Irtysh river system. Western Siberia forms a large saucer sloping very gradually to the north; the city of Omsk, for example, is located at only 94 m above sea level, although it is 1,500 km south of the Arctic Ocean. Eastern Siberia is drained by the Yenisei and is a plateau with elevations reaching 1,000 m (and, in Putorana, 1,700 m). The highest point in Siberia is Mt. Belukha in the Altay Mountains in the south, with an elevation of 4,506 m. The Sayans range farther east is a bit lower, reaching 3,000 m.

Siberia is famous for its lakes, including Lake Baikal, as well as Lake Taymyr in the extreme north, Khantayskoe near Norilsk, Teletskoe in the Altay, and saline Lake Chany in the steppe south of Barabinsk. As noted in earlier chapters, Lake Baikal is the deepest and oldest lake in the world and has the greatest volume of any freshwater lake (23,000 km^3, or about 20% of all freshwater on the planet). Baikal is 636 km

TABLE 27.1. Subjects of Federation of the Siberia Federal District

Subject	Administrative center	Population (2006) (thousands)	% ethnic Russian	GRP/capita (rubles in 2006)	Top products and areas of economic specialization within Russia (% of nation's total production for some)
Omsk Oblast	Omsk	2,035	84	121,934	Tires (12%), oil refining, defense, food industry
Tomsk Oblast	Tomsk	1,034	91	180,441	AC electric engines (12%), plastics (12%), food industry, education
Novosibirsk Oblast	Novosibirsk	2,650	93	108,454	Coke and mining equipment, grain, dairy, food, high-tech, education
Kemerovo Oblast	Kemerovo	2,839	92	119,124	Coal mining (55%), pig iron (13%), steel (13%)
Altaysky Kray	Barnaul	2,543	92	66,275	Steam boilers (68%), cereal (15%), flour (11%), wheat, defense
Republic of Altay	Gorno-Altaysk	205	57	54,398	Food industry, sheep/cattle/elk ranching, tourism
Krasnoyarsky Kray	Krasnoyarsk	2,906	89	202,030	Grain combines (26%), refrigerators (21%), coal, electricity, timber, defense
Khakassiya Republic	Abakan	538	80	94,950	Electricity (14%), railroad cars (5%), coal, aluminum
Tyva Republic	Kyzyl	309	20	47,968	Food industry, textiles and clothing, coal, cobalt, sheep ranching
Irkutsk Oblast	Irkutsk	2,527	90	128,277	Pulp (27%), timber and lumber (13%), salt (14%), electricity (6%), aluminum
Buryatiya Republic	Ulan-Ude	963	68	94,169	Coal, graphite, machine building, textiles, pulp and cardboard, sheep, furs
Chita Oblast	Chita	1,128	90	77,899	Coal, molybdenum, tin, uranium, machine building, wool, ranching

Note. Data from the Federal Service of State Statistics, Russian Federation (2009). To convert gross regional product (GRP) to U.S. dollars, divide GRP by 27.19.

long and up to 80 km wide, and is about 25 million years old. It has exceptionally clear water, with visibility 40 m down. Lake Baikal is home to 1,500 aquatic species, 80% of them endemic to the lake, including 255 species of shrimp-like amphipods and 80 species of flatworms. Its most famous endemic is the freshwater Baikal seal (*Phoca sibirica*). There are also a few large human-made lakes in Siberia (Figure 27.2), such as the large Obskoe reservoir upriver from Novosibirsk on the Ob, the Krasnoyarsk and Sayano-Shushenskoe reservoirs on the Yenisei, and the Ust-Ilimsk and Bratsk reservoirs on the Angara. More dams could be built elsewhere, but all of these projects are being challenged by environmentalists (especially controversial is the Katun Gorge project, discussed in Chapters 16 and 17).

Siberia is also famous for its mineral riches. Its settlement by the Russians in the 16th and 17th centuries was mostly driven by the insatiable demand for sable and mink furs in Europe. Later quests for gold and diamonds brought in miners instead of trappers. Coal, oil, and natural gas, as well as many metallic ores and uranium mining, brought in even more people in the 20th century. Both the tsarist and Communist governments encouraged voluntary settlement of Siberia. They, especially the Communists, also sent millions of unwilling prisoners to the remotest and least hospitable parts of the region. Many mines and timber-cutting areas were developed first as tsarist-era penal colonies and later as Soviet GULAGs. Today Siberia produces 80% of Russia's coal, but only 3% of its petroleum and natural gas, if Tyumen Oblast and the two autonomous okrugs that are now part of the Urals district (Chapter 26) are excluded. Nevertheless, large yet untapped deposits of both oil and gas exist in eastern Siberia, in the Lena basin. Siberia also accounts for about 30% of Russia's timber production, from about 300 million ha of forest. Eastern Siberia is 57% forest covered and western Siberia about 37%, although much of this land is occupied by unproductive larch forests in the Lena basin and only moderately productive fir and spruce forests along the Ob. The Yenisei and the Ob together account for one-quarter of all river runoff in Russia and a similar share of its hydropower potential. Fishing and hunting provide much-needed protein to the local population and are very popular subsistence and recreational activities.

Siberia also has a strong agricultural base, principally in the south (Altaysky Kray, Omsk and Novosibirsk Oblasts, and Tyva Republic). The region has 45 million ha of arable land, almost double what the Volga region has, but less than 40% of it is currently cultivated. Only very

FIGURE 27.2. One of many dams in Siberia: Chemal GES in the Altay. *Photo:* Author.

hardy varieties of barley, rye, oats and wheat can be grown in the south of the region and along the river floodplains. The rest of the agricultural land consists of scattered pastures and small vegetable plots. In drier or colder areas, sheep and cattle ranching are very common (Figure 27.3).

The Siberian climate is strongly continental. The bulk of the region is in the Köppen Dfc climate type, meaning that it is subarctic with enough humidity year-round, comparable to Yukon and central Alaska. The eastern parts of Siberia are in the Dw type, characterized by dry winters with little snow cover and severe frost; no exact matches for this exist in North America. The coldest temperature recorded near Novosibirsk is –50°C, and in Chita –54°C. Summers can be as hot as +37°C, but they are short: Over much of the region, only 4 months of the year have average temperatures above freezing. Parts of Taymyr Peninsula have some of the thickest permafrost in the world, exceeding 500 m. Future warming of the Arctic is likely to have a huge impact on this region. The shores of the Arctic Ocean are at present tundra covered, but most of Siberia is forest covered, with forest–steppe and true steppe in Altaysky Kray and southern Omsk and Novosibirsk Oblasts.

The mountains in the south have distinct climate and vegetation belts that vary with the slopes' elevation and orientation. For example, the western slopes of the Altay receive over 1,500 mm of precipitation a year and are heavily covered in Siberian larch; the eastern slopes, on the other hand, may receive less than 200 mm of precipitation and support semidesert shrubby vegetation.

Siberia has many federally protected parks. The Great Arctic Zapovednik, created in 1993 on 4 million ha (including some sections of the Arctic Ocean), is the largest preserve in Russia. It is the Russian version of the Arctic National Wildlife Refuge, protecting the fragile ecosystems of the northern Taymyr coast at 73–75°N latitude. Herds of wild reindeer and dozens of shorebirds and geese make their homes here. Lake Baikal is a World Heritage Site, and also the location of two zapovedniks and three national parks. The Altaysky Zapovednik in the south protects the pristine forests and alpine tundras of the Altay Mountains along the eastern shores of Teletskoe Lake.

Cultural and Historical Features

Culturally, Siberia is about 85% Russian, with substantial Ukrainian and German minorities in the agricultural south and a few dozen indigenous tribal minorities in the north. The Tunguss-Manchu people of the Altaic language family are well represented by the Evenki, living along the eastern tributaries of the Yenisei. These are people adapted to the harsh life of northern forests, who traditionally would be reindeer herders and

FIGURE 27.3. Cattle herding is a common activity in southern Siberia. *Photo:* Author.

hunters. The mysterious Kety people of the middle Yenisei have no ethnic relatives anywhere, and are apparently an indigenous Siberian tribe of uncertain origin. In the south, the Altaitsy and Tuvins are Turkic-speaking pastoralists who live in the mountains.

The region's most important city, Novosibirsk, was dubbed "Siberian Chicago" for its rapid rise from a little village along the newly built Trans-Siberian Railroad in the 1890s to the third largest city in Russia (with over 1.4 million people in 2006). Other important cities include Omsk, Krasnoyarsk, Irkutsk, Barnaul, Novokuznetsk, Kemerovo, and Tomsk (Figure 27.4) ranging in size from 500,000 to slightly over 1 million. Some Siberian cities, such as Biysk (Vignette 27.1), were founded during the 17th and 18th centuries as forts along rivers to facilitate Russian settlement of the distant parts of the expanding Russian Empire. Some have historical city centers (Tomsk and Biysk), while others (Novosibirsk and Kemerovo) do not, because they were developed primarily during the Soviet period. All Siberian big cities, except Norilsk, are located between 49°N (the same latitude as most of the U.S.–Canada border) and 60°N (cf. Anchorage, Akaska). Norilsk, with 213,000 residents, is the largest city above the Arctic Circle in the world, located on top of a massive polymetallic ore deposit at almost 70°N. It is connected to the nearby port, Dudinka, by railroad and highway links; it is also one of the most polluted cities in Russia.

Siberia is certainly the region of the free and the brave. It shares a lot of cultural characteristics with Alaska and the Yukon. The Russian word for a native of Siberia, *sibiryak*, connotes health and reliability. Siberians are known for their resourcefulness, self-sufficiency, and independence. The Russian peasants in Siberia have never known serfdom and live in spacious villages along lush river valleys. The Siberian Cossacks formed a distinct frontier subethnic group of the Russian people and are famous for their hunting and horse-riding skills. Today Siberian culture is known through the crafts, songs, and folklore of the indigenous Siberian tribes (especially the Yakuts and Evenks) on the one hand, and through books and movies about Siberian Russia on the other. The beauty of the wild Siberian nature and the epic struggles of the people sent to conquer the willful land are two sources of inspiration for these books and films. Well worth reading are the works of the Soviet-period writers Viktor Astafyev (Krasnoyarsk), Valentin Rasputin (Irkutsk), and Vasily Shukshin (Srostki, Altaysky Kray). Siberia is also known for its talented sportsmen (e.g., Irina Chashchina from Omsk, a

FIGURE 27.4. Tomsk is the old cultural capital of Siberia. *Photo:* A. Fristad.

Vignette 27.1. Profile of Biysk

Biysk (52°31'N, 85°10'E) is a typical medium-sized industrial city in Siberia. It is located near the confluence of the Biya and Katun Rivers, which form the Ob; it is about 160 km southeast of Barnaul, the capital of Altaysky Kray. Biysk is a city of 220,000 people, about the size of Olympia, Washington. It was established as a fortress under Peter the Great in 1709. The city is the final stop on the Novosibirsk–Altay railroad. Downtown Biysk has a wealth of late-19th-century historical buildings, including merchant houses, warehouses, banks, and stores (Figure 1). Biysk was a center for missionary activities of the Russian Orthodox Church in the mid-19th century, and a beautiful Russian Orthodox cathedral serves as a reminder of this. Biysk was chosen as a location for a number of Soviet chemical factories, including the top-secret "Altay" enterprise and other factories making explosives and solid rocket fuel. The Polyeks factory is one of the largest producers of cotton- and other-fiber-based products, including fire hoses, surgical cotton, varnishes, and enamels. Also, two large pharmaceutical factories (Altayvitaminy and Evalar) are busy with orders. Other factories include a boiler plant, an oleum plant, a fiberglass factory, an electric furnace manufacturer, and a tobacco factory.

In 2007 Biysk was named as one of about 10 *naukograds* (scientific cities) of Russia, which should result in a major infusion of federal cash. The city is home to a large regional university—the Shukshin Biysk Pedagogical State University with about 5,000 students. It also has a foreign-language lyceum, a few technical colleges, a medical college, and a college of economics and law. There is a fine museum of regional studies named after Vitaly Bianki, a well-known Soviet nature writer. An academy of science and arts and a small drama theater provide needed cultural influences. There are three local TV channels and four radio stations. Biysk's future may not be only in the chemical industry: It is the gateway city to the greater Altay, which begins about an hour's drive south along the federal Chuya highway. Itineraries for trips to Teletskoe Lake, Mt. Belukha, and the Katun River gorge all begin in Biysk.

FIGURE 1. An old merchant house in downtown Biysk. *Photo:* Author.

world-class gymnast). Kalinov Most from Novosibirsk and Grazhdanskaya Oborona from Omsk are two well-known Siberian rock bands. On the technical front, the Siberian inventor of the AK-47 automatic rifle, Mikhail Kalashnikov, is from Altaysky Kray; he celebrated his 90th birthday in 2009.

Economics

Siberia is an economic powerhouse, mainly in the extracting industries (coal, timber, gold), as well as in metal smelting (nickel, copper, aluminum). It also generates hydropower; produces specialized military equipment, weapons, and electronics; and manufactures many other industrial products. Tomsk, Novosibirsk, Omsk, and Barnaul have excellent technical universities and dozens of large research institutes. Siberia is particularly attractive for Asian investments: It shares a long border with China, while Japan, South Korea, and Taiwan are also relatively close by.

In the west of the region, Omsk and Tomsk specialize in petroleum refining and petrochemical production. Omsk is also a large military–industrial center. Tomsk has the oldest university in Siberia (established 1880) and is a refined intellectual center. However, it lost out to Novosibirsk when the Trans-Siberian Railroad bypassed Tomsk in favor of better-located Novonikolaevsk (as Novosibirsk was then called) on the Ob in 1896. Novosibirsk is the largest financial and business center in Siberia; it is also its main manufacturing, transportation, and high-tech research hub.

In the center of the region, Kuzbass is the largest coal-producing basin in Russia, accounting for about 70% of the nation's total coal production. Kuzbass coal is low in ash and sulfur content, and about 50% of it is now mined above ground. Nevertheless, the area still leads the country in the number of tragic underground mining accidents, due to frequent violation of safety norms by greedy mine owners. The presence of cheap coal and local iron ore in Shoria and Khakassia permitted the development of an entire agglomeration of cities specializing in coal, steel, plastics, and fertilizer production: Kemerovo, Novokuznetsk, Kiselevsk, Prokopyevsk, and others. Just to the east of Kuzbass, the Kansk-Achinsk coal basin in Krasnoyarsky Kray contains even larger coal deposits. Coal-mining machinery is also produced here.

Located in central Siberia on the Yenisei, Krasnoyarsk is an important center of scientific and military research, including a few cities in the vicinity directly involved in the nuclear weapons production. Krasnoyarsky Kray is the leading producer of hydropower in the country, with over 20,000 megawatts (MW) installed capacity in four giant dams on the Yenisei and its Angara tributary. Because of all this hydropower, aluminum smelters are also located here, taking full advantage of the low electricity rates. Moreover, Krasnoyarsky Kray and Irkutsk Oblast together account for about 20% of all timber harvested in Russia. Major paper and lumber mills are located here, and much timber is also exported as logs to China, Japan, and other countries via the Trans-Siberian Railroad. Krasnoyarsk also has many machine-building enterprises for the coal and transport industries.

In the east of the region, Irkutsk is a historical Siberian city (established 1661), an important university and industrial center, and a gateway to Lake Baikal. Nearby Angarsk is the eastern terminal of a large pipeline and a site of major petroleum refinery. Sayansk (salt, plastics), Angarsk (petrochemicals), and Irkutsk are responsible for a lot of air and water pollution in eastern Siberia. Chita Oblast and Buryat Republic mine copper, molybdenum, tin, and uranium. Over 70% of their agricultural production is beef, lamb, and wool. Few crops are capable of growing in eastern Siberia's severely continental climate, with almost no snow in winter, but with frosts averaging –30°C.

Siberia is located in the very center of Northern Eurasia, with rail connections to east and west via the Trans-Siberian Railroad and the Baikul–Amur Mainline (BAM), and summertime river links to the Arctic Ocean via the Ob and the Yenisei. The BAM at present is running at only 20% capacity; completion of a new tunnel in Chita Oblast will allow the Trans-Siberian's capacity to double. Novosibirsk's international airport, Tolmachevo, has an ambitious program for development as a continental passenger and cargo hub. Lake Baikal is a gem of global significance, but is at present only moderately visited (400,000 people visit Irkutsk Oblast per year, including about 50,000 foreign tourists) because of distance, expensive transportation and food, and lack of tourist infrastructure and advertisement. Baikal also continues to be polluted by the discharge from the Baikalsk paper and pulp mill, located at the southern end of the lake. Despite 40 years of public protests and political declarations, the company remains there and keeps polluting the lake.

Challenges and Opportunities in Siberia

The future of Siberia depends on the ability of the federal and regional governments to stop dramatic population loss, increase investments in social and physical infrastructure for those who still live in the region, and provide additional incentives for people from outside the region to come and settle here. Nearby China is a source of cheap labor, but also a looming demographic threat, at least as perceived by the locals. Japan, South Korea, and Taiwan, on the other hand, are eager suppliers of new technologies and financial backing for future development. Altaysky Kray is one of only four subjects of federation in Russia chosen to have a free gambling zone, and this should attract much investment and attention to the region. It is also the main tourist destination in Siberia, with Lake Baikal in second place.

EXERCISES

1. Find specific examples from recent economic development that support Mikhail Lomonosov's prediction about Siberia.
2. Use a world gazetteer, the Website of Russia's 2002 census, or any other source that lists all major cities in Siberia with their populations. What proportion of these are located right on the Trans-Siberian Railroad? What major cities are exceptions? What are the reasons why these cities far away from this main transportation artery are so large?
3. Using Table 27.1, try to find counterparts to each of Siberia's subjects of federation in your own country, based on similar economic profiles. This may be a small-group class activity.
4. Search for online collections of photo portraits of Siberians (both Russian and non-Russian ethnicities). Share pictures that you like with your classmates, and discuss differences and similarities between these people and anybody you may know. To what extent is the Siberian culture evident in these photos? What are its characteristics?

Further Reading

Bassin, M. (1983). The Russian Geographical Society, the Amur epoch and the Great Siberian Expedition, 1855–63. *Annals of the Association of American Geographers, 73*, 240–256.

Bobrick, B. (1992). *East of the sun: The epic conquest and tragic history of Siberia*. New York: Poseidon Press.

Edwards, M. (2003, June). Siberia's Scythian masters of gold. *National Geographic*, pp. 112–129.

Glendinning, A., & West, P. (2007). Young people's mental health in context: Comparing life in the city and small communities in Siberia. *Social Science and Medicine, 65*, 1180–1192.

Hartshorne, J. (2004, September–October). Saving Baikal. *Russian Life*, pp. 23–29.

Hill, F., & Gaddy, C. G. (2003). *The Siberian curse: How communist planners left Russia out in the cold*. Washington, DC: Brookings Institution Press.

Ivanov, V. I. (2006). Russia's energy future and northeast Asia. *Asia–Pacific Review, 13*, 46–60.

Kane, H. (2002, March–April). Who speaks for Siberia? *Worldwatch Magazine*, pp. 14–23.

Lincoln, B. (1994). *The conquest of a continent: Siberia and the Russians*. New York: Random House.

Matthiesen, P. (1992). *Baikal: The sacred sea of Siberia*. San Francisco: Sierra Club Books.

Mommen, A. (2007). China's hunger for oil: The Russian connection. *Journal of Developing Societies, 23*, 435–467.

Montaigne, F. (1998, June). Russia's iron road. *National Geographic*, pp. 2–33.

Odier, P. (2004). *Some last people: Vanishing tribes of Bhutan, China, Mexico, Mongolia, and Siberia*. Eagle Rock, CA: L'Image Odier.

Scherbakova, A., & Monroe, S. (1995). The Urals and Siberia. In P. R. Pryde (Ed.), *Environmental resources and constraints in the former Soviet republics* (pp. 61–77). Boulder, CO: Westview Press.

Shevyrnogov, A., Vysotskaya, G., Sukhinin, A., Frolikova, O., & Tchernetsky, M. (2008). Results of analysis of human impact on environment using the time series of vegetation satellite images around large industrial centers. *Advances in Space Research, 41*, 36–41.

Thubron, C. (1999). *In Siberia*. New York: HarperCollins.

Wolff, L. (2006). The global perspective of enlightened travelers: Philosophic geography from Siberia to the Pacific Ocean. *European Review of History, 13*, 437–454.

CHAPTER 28

The Far East

THE RUSSIAN PACIFIC

Across the Pacific Ocean from the United States, the giant Far East federal district of Russia is the largest unit in the country (6.2 million km^2). It includes huge Sakha (Yakutiya), which alone is bigger than the European Union (EU) in area, at about 3 million km^2. The Russian Pacific in the narrow sense includes only the units that border the ocean: Chukotka Autonomous Okrug, Kamchatka Kray, Magadan Oblast, Sakhalin Oblast, Khabarovsky Kray, and Primorsky Kray. Inland, Amur Oblast and Jewish Autonomous Oblast are located in the Amur watershed along the Chinese border (Table 8.3, Figure 28.1, and Table 28.1).

The region thus defined has merely 6.7 million residents, giving it a population density of 1.1/km^2—the lowest average density in Russia, and only one-third of Canada's density. To put it another way, this huge region is settled by only about half as many people as live in Moscow. The population is 76% urban. In many ways, the Far East is analogous to northern Canada or Alaska. It has plenty of natural resources (especially timber, fish, and gold) and is a perfect location for Russian naval bases, airfields, missile defense, and more. The Pacific Rim is an emerging powerhouse of the 21th century, and Russia's Far East is part of it. Unfortunately, the region is losing population fast, at a rate of about –0.7% per year. As in the case of Siberia (Chapter 27), this is due to very low fertility and fairly high mortality, coupled with emigration. Many people are now choosing to relocate to the western part of the country, where the climate is warmer and living expenses are lower. The Far East has lost about 1 million people since 1991. Far Eastern cities are the most expensive places in Russia to live. Magadan, Anadyr, and Petropavlovsk lead the country in the price of a minimal shopping basket of goods, at about 8,000 rubles per capita per month required in 2009 (as compared to the national average of 5,000). Basic supplies must be brought in from far away by ship or plane. Proximity to China across the Amur River allows cheaper food supplies to reach the south of the region. It also supplies laborers, but this creates uncertainty regarding the growing political influence of China in the region. Japan competes with China for major economic influence, supplying the majority of used passenger cars, as well as other high-tech consumer goods. It continues to claim the four southernmost Kuril Islands and refuses to sign a peace agreement with Russia, but pragmatically maintains good trade relations. The Kurils and the southern half of Sakhalin were controlled by Japan between 1904 and 1945, but were taken over by Russia as an outcome of World War II. South Korea is a third

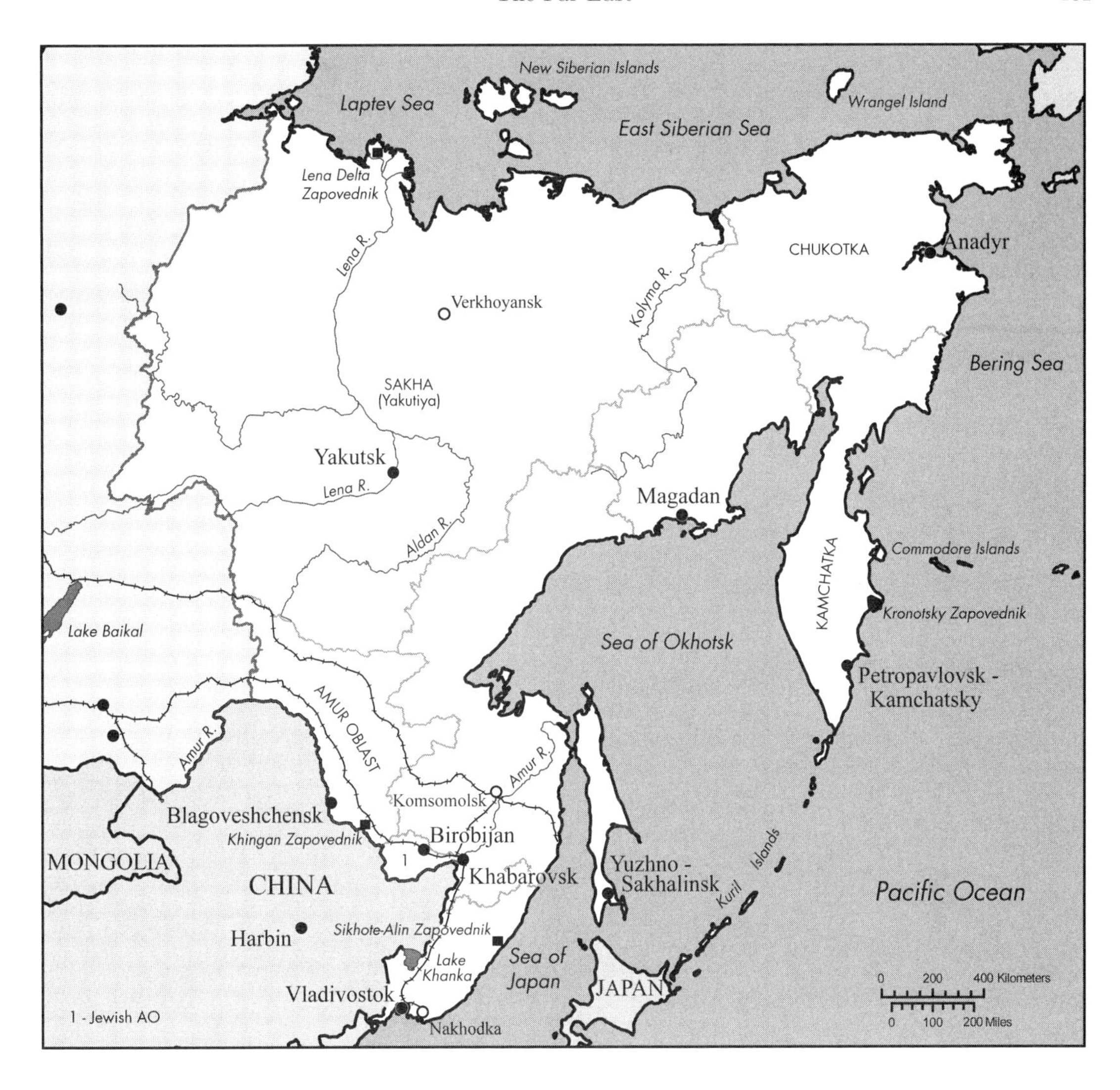

FIGURE 28.1. The Far East federal district of Russia. Map: J. Torguson.

important economic partner of Russia in the region, followed by Taiwan and Vietnam. Contacts with North Korea are now severely restricted along the common border, and the political situation in that country casts its uneasy shadow on the otherwise peaceful region.

Physical Geography

The Far East is more mountainous than the rest of Asiatic Russia. Mt. Pobeda in the Chersky range (source of the Indigirka and Kolyma Rivers) reaches 3,000 m. The Stanovoy range in Khabarovsky Kray and Amur Oblast reaches 2,255 m elevation. The Sikhote-Alin range (home to a few hundred Siberian tigers) north of Vladivostok reaches about 1,700 m. The giants of the region, however, are the volcanoes in Kamchatka (Figure 28.2), about 20 of them active; the highest, Klyuchevskaya, reaches 4,688 m. The Lena River, which drains much of Yakutiya, is the second biggest river in Russia in terms of annual runoff and the third in overall length and basin size. The Amur is the fourth longest river (see Chapter 2, Table 2.2). The Far East accounts for 10% of Russia's coal deposits, 9% of its phosphates, and 8% of its iron ore. The

TABLE 28.1. Subjects of Federation of the Far East Federal District

Subject	Administrative center	Population (2006) (thousands)	% ethnic Russian	GRP/capita (rubles in 2006)	Top products and areas of economic specialization within Russia (% of nation's total production for some)
Sakha Republic (Yakutiya)	Yakutsk	950	41	216,692	Diamonds, gold, metals, coal, timber, furs, reindeer herding, food industry
Amur Oblast	Blagoveshchensk	881	92	103,983	Shipbuilding, agricultural machinery, grains, soy, furs, food industry
Khabarovsky Kray	Khabarovsk	1,412	90	139,271	Timber (7%), cranes (6%), oil refining (5%), TV sets (3%), food, defense
Evreyskaya Autonomous Oblast	Birobidzhan	187	90	96,028	Textiles and clothing, combines, soy, grains, vegetables
Primorsky Kray	Vladivostok	2,020	90	103,769	Seafood canning (17%), coal, timber, tourism, navy, shipping
Sakhalin Oblast	Yuzhno-Sakhalinsk	526	84	310,557	Fishing (15%), seafood canning (9%), petroleum, lumber, shipping
Magadan Oblast	Magadan	172	80	175,620	Reindeer herding, furs, mining equipment, fishing
Kamchatsky Kray	Petropavlovsk	349	81	154,250	Fishing (18%), ship repair, navy, tourism
Chukotsky Autonomous Okrug	Anadyr	51	52	289,856	Reindeer herding, walrus and whale hunting, fishing, furs, coal, tin, gold

Note. Data from the Federal Service of State Statistics, Russian Federation (2009). To convert gross regional product (GRP) to U.S. dollars, divide GRP by 27.19.

size of the huge deposits of oil and natural gas on the Sakhalin shelf is still being determined, but they are likely to be bigger than the Norwegian or U.K. deposits in the North Sea (the petroleum is estimated at 10 billion barrels). The region is heavily forested (46%, or 278 million ha) and accounts for 27% of all Russia's timber reserves. On the other hand, farmland is quite limited (only 4.5 million ha, or 2% of the total). All of the farmland is concentrated in two areas: along the Amur River floodplain and near Lake Khanka in the extreme south.

The Far East has a unique climate: Its northeast is the coldest and most continental place on earth, whereas the southeast is monsoonal (with frequent storms in the late summer and early fall, and relatively mild winters). Verkhoyansk and Oymyakon share the distinction of being the coldest places outside of Antarctica, with the minimum record temperature of –72°C (average annual temperature = –15.1°C) and meager annual precipitation (172 mm), like a desert. Winter frost here lasts 7½ months. Heavy permafrost makes any construction tricky; the cost of 1 m^2 of living space in Chukotka is 10 times that in Vladivostok! That southern city, on the other hand, has an average temperature of +4.2°C and receives 816 mm of precipitation, more than Moscow or St. Petersburg. It has snow for "only" 5 months of the year.

The Far East has many federally protected territories, mainly zapovedniks (there are no national parks in the Far East). Wrangel Island Zapovednik is located off the northeastern Russian coast in the Arctic Ocean. It protects the unique tundra–steppe communities of western Beringia. It is famous for its large population of emperor geese, as well as cormorants, seagulls, eider ducks, murres, puffins, and other cliff-nesting seabirds. Polar bears and walruses are common in and around the island. This was apparently the last place on earth where mammoths went extinct. Lena Delta Zapovednik protects very important wetlands at the mouth of the Lena River. Kronotsky Biosphere Reserve in Kamchatka is home to a famous geyser valley, a number of volcanic craters, lots of brown bears, and wild salmon-filled rivers. Khingan Zapovednik protects two species of cranes in the undisturbed forest–steppe landscapes of the Amur basin on the Chinese border. Sikhote-Alin Biosphere Reserve in Primorsky Kray is home to about half of Russia's Siberian tigers (250) and a rare species of goat-like goral. The reserve encompasses some of the least disturbed and most diverse mixed forests of the Russian Far East (Figure 28.3).

FIGURE 28.2. Kamchatka's Avachinsky volcano. *Photo:* A. Bogolyubov (*www.ecosystema.ru*).

Cultural and Historical Features

Culturally, the Far East region was settled mainly by Russians (88%), with a substantial Ukrainian presence (7%). Most of the Russian and Ukrainian settlers moved to the region relatively recently, during Soviet times. The indigenous groups of the Far East are the most diverse in all of Russia; they include the Turkic-speaking Yakuts in Yakutiya, the Evens of the taiga zone, the Udege and other related tribes in the southeast, the Paleoasiatic Chukchi and Koryaks in the northeast of the region, and the Aleuts and Inuits along the Pacific Coast. The Nivkhs and Aynu are two endangered ethnicities of Sakhalin Island. Many traditional groups are great hunters and fishermen. The Paleoasiatic people of the northeast also hunt walruses, seals, and whales. In more recent times, there has been substantial Chinese and Korean immigration into Primorsky and Khabarovsky Krays and Amur Oblast. The exact numbers are not known, but probably exceed 50,000.

Cultural notables of the Far East include Russian writer and explorer Vladimir K. Arse-

FIGURE 28.3. The Sikhote-Alin range has the highest diversity of trees in Russia. *Photo:* A. Bogolyubov (*www.ecosystema.ru*).

niev (1872–1930), author of *Dersu Uzala* (*Dersu the Trapper*); Yakut writer Platon Oyunsky (1893–1939); Nobel Prize-winning physicist Igor Tamm (1895–1971); Vladivostok-born American actor Yul Brynner; tennis player Igor Kunitsyn; Khabarovsk-born National Hockey League player Alexander Mogilny; and members of the rock band Mumiy Troll.

Economics

The Far East has many important industries, including mining coal and gold, fishing for salmon and crab, cutting timber, pumping petroleum, and making the Su-series of fighter jets (Table 28.1). However, its main importance to Russia is strategic: It is the only region that is open to the vast Pacific Ocean and borders China, Japan, and the United States. Nuclear submarines in Petropavlovsk and Vladivostok, and radar installations on the Kuril Islands, enable Russia to demonstrate its military might to the most important emerging economic powerhouse in the world. The Far East is also a natural trading and tourist gateway to and from Asia. As Chinese and other Asian economies continue to grow, the importance of the few entry points into Russia increases proportionally. The seaports of Nakhodka, Vanino, Vladivostok, and Yuzhno-Sakhalinsk are extremely useful for export–import operations.

With respect to economic development, the southern part of the region along the Trans-Siberian Railroad is more or less contiguously settled. In the north, there are three isolated clusters of development (around Yakutsk, Magadan, and Petropavlovsk), with virtually untouched wilderness in between. The official capital of the federal district is Khabarovsk (established 1858), the second biggest but most centrally located city, which has 580,000 people and is a bustling industrial center on the Amur River. Its diverse industries include machinery, motor, and ship factories; petrochemicals; and wood and food processing. Farther downstream, Komsomolsk-on-Amur (established 1931) builds ships and fighter jets, and has a large ferrous metal factory and a petroleum refinery. Two other important industrial and agricultural centers, Birobidzhan and Blagoveshchensk, are located upstream from Khabarovsk along the Chinese border. Nearby Svobodny has a new space launching pad, which can be used for both military and civilian satellite launches.

Sakhalin Island has as its main focus fishing and petroleum industries. Most of the petroleum extraction is concentrated along the northeastern shore in a few international projects (Sakhalin-1, -2, -3, and so on). In the south, Yuzhno-Sakhalinsk and Kholmsk are diversified seaports. Petropavlovsk and Magadan are mainly fishing ports, with Petropavlovsk also serving as a naval base for the Pacific Fleet and as the only tourist gateway to Kamchatka, full of geysers and grizzlies. The nearby Commodore Islands are a nature reserve supporting healthy populations of sea lions and other marine mammals (Figure 28.4).

Vladivostok, surrounding the Golden Horn Bay, is home to the largest naval base in the region and is the largest city (population 620,000). It is sometimes dubbed "the San Francisco of the Far East," because of its picturesque natural location on the hills around a huge inland bay (Figure 28.5). However, the ubiquitous Soviet concrete apartment blocks make it less beautiful than its American counterpart. It is a major center for machine building, food processing, and the construction industry. It is also a hub for the Russian Academy of Sciences, with many research institutes and universities focusing on oceanic research. Other cities in Primorsky Kray specialize in coal and metal mining, fishing, container shipping, machine building, and military services. There is a proposal to build a new seaport terminal at the triple junction of China, North Korea, and Russia, to facilitate shipping from China's landlocked northeastern provinces.

Yakutiya could be its own country, given its size. In fact, it tried to proclaim independence in the first year of Yeltsin's presidency and proudly renamed itself Sakha, based on the local language. Although it covers over 3 million km^2 of land, it has less than 1 million people and is essentially landlocked. The Lena and Yana Rivers provide access to the Arctic Ocean, but given the present climate the coastal waters remain frozen for over 7 months of the year. As yet, there is no railroad link to Yakutsk from the Baikal–Amur Mainline (BAM); nor is there a year-round passable highway to Magadan. During the winter months, all rivers freeze and turn into icy highways. The old Magadan–Yakutsk highway is being upgraded to become more functional by the end of 2008. Yakutsk is a major industrial center that was established amazingly early by Far Eastern standards, in 1632. Its main industries are food processing, furniture making, and construction materials manufacturing. It is also an important cultural and scientific center. Eastern Yakutiya specializes in gold, tin, and tungsten mining. Reindeer

FIGURE 28.4. The village of Nikolskoe on Bering Island, off the coast of Kamchatka. *Photo:* A. Litsis.

FIGURE 28.5. The city of Vladivostok, viewed from the Golden Horn Bay. *Photo:* A. Osipenko.

herding and horse breeding are indigenous traditions. Western Yakutiya, centered on the city of Mirny, is the largest producer of diamonds in Russia and one of the largest in the world. There are also substantial, but insufficiently explored, petroleum and natural gas deposits. Finally, there is plenty of timber harvesting and hydropower production in the republic, with a large potential for future expansion.

Challenges and Opportunities in the Far East

Future development of the Far East depends on a number of key issues:

- The success of efforts at plugging up the existing demographic hole: reducing emigration, encouraging immigration (both from within Russia and from abroad), and providing incentives for local parents-to-be.
- The impact of the East Siberian pipeline (started in 2006, to be finished by the end of 2008) from Tayshet (Irkutsk Oblast) to Skovorodino (Amur Oblast) and eventually to the Pacific Ocean.
- The impact of Sakhalin's petroleum projects, both positive (cash flow, labor, infrastructure) and negative (potential massive oil spills, uncertain profitability for the local residents).
- Further development of economic ties with Japan and China, as well as lesser economies of the Pacific (South Korea, Taiwan, Vietnam). The region is already being fed primarily from China, because transportation of food from Central Russia has become prohibitively expensive.
- Reorganization of the Russian armed forces, including the Pacific navy fleet.
- Development of tourism along the Trans-Siberian Railroad, as well as ecotourism development in Kamchatka and Primorsky Kray, and at other less exotic and more inland destinations.
- General stability of relations with the United States, another key player in the Pacific. Scientific and technological cooperation between the two countries is already present here, especially in Beringia, a globally significant former land bridge and a great place to conduct research on global climate change between Chukotka and Alaska. This is also the area where an underwater tunnel could connect Eurasia with North America in a futuristic transportation corridor (see Chapter 21).

EXERCISES

1. Use Table 28.1 and additional information from online sources to rank the subjects of federation of the Far East in terms of their attractiveness for foreign investment. Explain your rationale.
2. Compare and contrast the ecotourism potential of Primorye and Kamchatka. Which region, in your opinion, is better positioned for future ecotourism development? Make sure to use concrete statistics to back up your claim.
3. Investigate the impact of the Trans-Siberian Railroad, the BAM railroad, and the new petroleum pipeline on the economy and environment of the region.
4. Propose a program that would boost population of the Far East. What are the key components of your program? To what extent does it balance natural fertility and immigration from other parts of Russia and from abroad?
5. What are the arguments that Japan uses to claim the Kuril Islands? What are the Russian counterarguments? Which side sounds more convincing? Use published articles and treaties in your analysis.

Further Reading

Arseniev, V. K. (1996). *Dersu the trapper.* Kingston, NY: McPherson.

Bassin, M. (1999). *Imperial visions: Nationalist imagination and geographical expansion in the Russian Far East, 1840–1865.* Cambridge, UK: Cambridge University Press.

Blank, S., & Rubinstein, A. (Eds.). (1997). *Imperial decline: Russia's changing role in Asia.* Durham, NC: Duke University Press.

Bloch, A., & Kendall, L. (2004). *The museum at the end of the world: Encounters in the Russian Far East.* Philadelphia: University of Pennsylvania Press.

Bradshaw, M. (2001). *The Russian Far East and Pacific Asia: Unfulfilled potential.* London: Routledge.

Bradshaw, M. (2006). Battle for Sakhalin. *World Today, 62*(11), 18–20.

Chivers, C. J. (2005). An unexpected catch. *Wildlife Conservation, 108*(6), 14.

Devaeva, E. The foreign trade of Russia's Far East. *Far Eastern Affairs, 33*(4), 52–64.

Dienes, L. (2002). Reflections on a geographic dichotomy: Archipelago Russia. *Eurasian Geography and Economics, 43*(6), 443–458.

Hudgins, S. (2003). *The other side of Russia: A slice of life in Siberia and the Russian Far East.* College Station: Texas A&M University Press.

Kotkin, S., & Wolff, D. (Eds.). (1995). *Rediscovering Russia in Asia: Siberia and the Russian Far East.* Armonk, NY: Sharpe.

Linge, G. J. R. (1995). The Kuriles: The geo-political spanner in the geo-economic works. *Australian Geographical Studies, 33*, 116–132.

Newell, J. (2004). *The Russian Far East: A reference guide for conservation and development.* McKinleyville, CA: Daniel & Daniel.

Nigge, K. (1999, March). The Russian realm of Steller's sea eagles. *National Geographic*, pp. 60–72.

Rodgers, A. (1990). *The Soviet Far East: Geographic perspectives on development.* London: Routledge.

Rozman, G. (2008). Strategic thinking about the Russian Far East: A resurgent Russia eyes its future in northeast Asia. *Problems of Post-Communism, 55*, 36–49.

Stephan, J. J. (1996). *Russian Far East: A history.* Stanford, CA: Stanford University Press.

Thornton, J., & Ziegler, C. E. (Eds.). (2002). *Russia's Far East: A region at risk.* Seattle: University of Washington Press.

Walker, B. L. (2007). Mamiya Rinzō and the Japanese exploration of Sakhalin Island: Cartography and empire. *Journal of Historical Geography, 33*(2), 283–313.

CHAPTER 29

The Baltics

Europeysky, Not *Sovetsky*

The three Baltic countries—Estonia, Latvia, and Lithuania—share a history of having forcibly been made part of the Soviet Union in 1940. They were the first among the 15 Soviet republics to proclaim their independence from the U.S.S.R. in 1990, and they gained it the following year. Since then, they have eagerly sought integration into European political and economic structures. In 2004 they were admitted into both the North Atlantic Treaty Organization (NATO) and the European Union (EU). Thus they are culturally, economically, and politically now part of (Western) Europe. They have never joined the Commonwealth of Independent States (CIS) and are now routinely included in European regional texts. Nevertheless, there are reasons why they merit at least a mention in this textbook.

First of all, for better or for worse, these three countries were part of the U.S.S.R. for half a century. The imprint of this period is still visible in their landscapes. There are, for example, remnants of large Soviet enterprises left, including a nuclear station in Lithuania and oil shale factories in Estonia. Another example is the housing: In many places the same old Soviet apartments are still inhabited, with largely unchanged designs. Moreover, many of the people who are now in control in the Baltic states were born and raised during Soviet times and share a collective memory of this period. Surprisingly to Western observers, some Soviet traditions stubbornly persist. The Baltic states also lead the EU in some negative social statistics related to the Soviet past. For example, abortion rates in all three are much higher than is common in Western or Central Europe. On the positive side, the Baltic states have retained some of the social benefits of the Soviet period under different political and economic realities.

Second, long before the Soviet Union existed, the Baltic countries were involved with the Russian Empire—sometimes fighting against it, sometimes collaborating with it, and sometimes part of it. Surrounded by historically powerful Sweden, Germany, Poland, and Russia, they had to be politically aligned with one or two of these great powers over much of their history. Thus the presence of Polish, German, Swedish, and Russian cultural elements today is a unifying pattern in all three states.

Third, the Baltic republics still have large Russian-speaking minorities, ranging from 6% in Lithuania to 35% in Latvia. Russian language and culture remain integral aspects of their society, to an extent unknown in the rest of Europe. Fourth, and significantly, these countries continue to engage with Russia on a deeper economic level than most other European economies because of their close geographic proximity, joint borders, and common trade.

TABLE 29.1. Comparative Characteristics of the Baltic States and Selected Other Countries or Regions

Country	Population (millions)	Ethnic Russian (%)	GDP PPP/capita (2008, U.S. dollars)	GDP growth (2008, %)	Services (2008, % of GDP)	Exports (2008, % of GDP)
Estonia	1.3	26	21,200	–3.0	65	52
Latvia	2.2	30	17,800	–5.0	75	30
Lithuania	3.5	6	17,700	3.2	63	45
Russia	140	80	15,800	6.0	55	28
CIS-12	280	50	10,500	10.3	50	32
EU-27	495	<1	32,300	3.0	71	8
United States	306	<1	47,000	1.3	79	8

Note. Data from CIA World Factbook (2009) and calculations by author. GDP PPP, gross domestic product adjusted for purchssing power parity. CIS-12, the Commonwealth of Independent States up to 2008 (which included Russia and 11 other former Soviet Union republics, but not the Baltics). EU-27, the European Union with 27 members in 2008 (including Estonia, Latvia, and Lithuania). Exports for the EU-27 do not include trade within the EU itself.

The Baltic states are very small in both territory and population. Estonia has merely 1.3 million people, and the biggest, Lithuania, has only 3.4 million—smaller than most American states (Table 29.1). In area (see Chapter 23, Figure 23.1), Latvia and Lithuania are about the size of West Virginia (or Sri Lanka), while Estonia is the size of Vermont and New Hampshire combined (or Denmark). All have rapidly declining populations, with rates of decline comparable to Russia's—about –0.6% per year in Estonia and Latvia, and –0.25% in Lithuania.

Physical Geography

In terms of physical environment, all three countries are coastal lowlands, covered by thick glacial landforms from the last Ice Age (which ended about 12,000 years ago). Sand dunes are prominent along the Baltic Sea coast (Figure 29.1). Estonia controls two large islands in the Baltic Sea, Saaremaa and Hiiumaa. Glacial lakes, marshes, and scattered forests create a picturesque mosaic of habitats farther inland. Lake Peipus (Chudskoe) is shared by Estonia and Russia. This was

FIGURE 29.1. Dunes of the Curonian Spit in the Baltic Sea. *Photo:* A. Bogolyubov (*www.ecosystema.ru*).

the site of the famous battle between the Russians under Prince Alexander Nevsky and the Teutonic Livonian knights, who were defeated on the ice of the lake in 1242. Estonia's highest point is only 300 m above sea level. The Baltic climate is maritime—cool and wet. The average temperature in Tallinn in January is –5°C, slightly below freezing. It is about +17°C in July. Inland Vilnius is slightly warmer. One can swim in the Baltic Sea in July, but even then the water is rarely warmer than +17°C.

Soils are mainly podzolic and are moderately productive. Mixed forests of pine, spruce, aspen, birch, maple, basswood, and oak are common. All three countries are net timber exporters. They also have important fisheries in the Baltic Sea. There are few mineral deposits, except construction stone, oil shales (only in Estonia), and amber (principally in Lithuania). Each country has a few national parks open to the public. The parks have human settlements within their borders. There are some excellent areas for migratory bird watching on the Baltic Coast, especially on the Curonian Spit shared by Kaliningrad Oblast and Lithuania.

Cultural and Historical Features

Culturally, the Estonians are of Finnish–Ugric (Uralic) extraction, while the Latvians and Lithuanians are Baltic peoples, closely related to the Slavs. With respect to religion, Estonians and Latvians were historically Lutherans, having been influenced by the Germans and the Swedes, whereas Lithuanians were traditionally Roman Catholics. Lithuania was the last European country to convert to Christianity, toward the end of the 14th century. Because of a close political alliance with Poland over much of its medieval history, the country has become heavily Roman Catholic. Lithuania also has a higher proportion of churchgoers than more secularized Estonia or Latvia.

Economics

Estonia has the most open and wealthiest economy of the three Baltic states. By 2007 it was already at Portugal's level of gross domestic product (GDP) per capita, and ahead of Slovakia or Hungary (see Table 29.1). Latvia and Lithuania are trailing somewhat behind, but are still ahead of Russia, the wealthiest economy of the CIS, by about 20%. Estonia also leads the other two countries in the percentage of Internet users and mobile phone users per capita. Its economy is the most export oriented of the three as well, with exports accounting for about half of the GDP, and it attracts 42% of all foreign direct investments in the Baltics. Estonia's main industries are electronic equipment manufacturing, wood processing, oil shale energy production, telecommunications (call centers and cell phone parts manufacturing), textiles, shoemaking, and tourism. Furniture, prefabricated parts for log cabins, and wooden toys are some of its important exports.

The most important cultural and economic centers in each republic are their capitals, which are also the biggest cities: Tallinn (population 420,000), Riga (820,000), and Vilnius (580,000). Riga is the most industrialized of the three, while Tallinn is arguably the most picturesque, with its charming medieval German old town (Figure 29.2) and Toompea Castle in the middle. All three attract foreign tourists, especially from Europe (Germany, Sweden, and Finland) and Russia. Other cities in Estonia include the university town of Tartu, located deep inland in the southeast; a fishing port, Pärnu, in the southwest; and an industrial city, Narva, on the border with Russia. Although Finland is the largest trade partner of Estonia, Russia and Latvia are also in the top five. Latvia and Lithuania, for example, receive Estonian food products (processed food and drinks), various chemical products, metals (especially Latvia), machinery and equipment, clothes (Lithuania), and electricity (Latvia).

Latvia and Lithuania have generally similar profiles, with Lithuania being the most industry-heavy (about one-third of its GDP, as opposed to only 20% for the other two). Lithuania and Estonia are electricity exporters, while Latvia has to buy electricity from the other two or from Russia. Agriculture continues to play an important role in all three countries. Dairy farming and diversified cold-climate grain and vegetable production are economically important products as well as

FIGURE 29.2. View of Old Tallinn. *Photo:* G. B. Pedersen (public domain).

culturally traditional. Strong family farms were not fully destroyed even by the years of Communism here and are now very productive, although less so than in the Western European countries with similar climate (Denmark and Germany).

Latvia's GDP is at about the level of Poland's. Its economy was particularly hard hit by the recession of 2008–2009 because of a large external debt exposure: The GDP plunged over 5% in 2008 and 20% in 2009. Latvia has some advantages over Estonia in the long run, however. It has a bigger population and a larger land base. Moreover, Latvia had a well-developed manufacturing economy during the Soviet period; for example, RAF minivans, suburban electric trains, and radio sets were made here for the entire U.S.S.R. Like Estonia, Latvia has a thriving fishing fleet. It also has two major export-oriented seaports at Liepaja and Daugavpils, which handle transshipments of wood, fertilizer, and petroleum products from Russia and Belarus. Latvia has a slightly warmer climate than Estonia and more productive soils, both of which encourage farming (about one-third of the country is arable, as compared to only 16% in Estonia). Dairy farming and food processing are very important. Latvian smoked sardines (sprats) and chocolate candy are well-known delicacies sold all over the world. Latvia also has a well-developed tourism infrastructure focused on the sea beaches (Yurmala) and diverse food industry. Riga is the most cosmopolitan city in the Baltic states, and the Russian language is still very common there. Its magnificent Dom Cathedral was the largest medieval church built in the Baltics. It also has a well-preserved castle (built in 1330), which is now the official residence of the president and also houses a few museums (Figure 29.3).

Lithuania is the largest of the three Baltic states by population, has the most farmland (45%), and is about as wealthy as Latvia. Besides Vilnius—which, unlike Riga and Tallinn, is an inland capital—Kaunas is a big historical city and the ancient capital (Figure 29.4). Before World War II, Vilnius was home to almost 500,000 Jews and a substantial number of Poles. Most of the Jewish population perished in the Nazi concentration camps. Some of the worst perpetrators of the Holocaust in Lithuania were the Lithuanian nationalists, who were more supportive of Hitler than of Stalin. Lithuania continues to expect war reparations from Russia (it asked for $20 billion in 2004), to compensate for the damage done by the Soviet Union during the war. Not surprisingly, such requests meet with an icy-cold reception in Moscow.

FIGURE 29.3. Riga Castle. *Photo:* L. Latj (public domain).

As noted earlier, Lithuania has a larger share of industry in its economy than Estonia and Latvia do, although many factories are still the old Soviet-period ones and are only slowly being upgraded. It is also the only Baltic state to have a Soviet-built nuclear power station, which produces 1,300 megawatts (MW) of electricity at Ignalina. The EU would like to see it decommissioned and replaced with renewable energy sources, but in this age of skyrocketing energy costs, such a decision requires careful deliberation. Lithuania exports some of its surplus electricity to Latvia. Important industries in Lithuania include wood processing, food processing, furniture making, and machine tool manufacturing, as well as production of TVs, refrigerators, small ships, and textiles. Its main seaport, Klaipeda, is an important trading post on the

FIGURE 29.4. Trakai Castle in Kaunas, the old capital of Lithuania. *Photo:* V. Satas.

FIGURE 29.5. A street in Vilnius. *Photo:* V. Satas.

Baltic Sea. Amber jewelry and wooden toys are the local folk crafts. Tourism is also well developed, especially sea-based tourism in Palanga on the coast, and historical tourism in Kaunas and Vilnius (Figure 29.5).

Challenges and Opportunities in the Baltics

The future of the Baltic region is largely in the EU's hands. NATO continues to play a big role as well. All three countries have contributed a few soldiers and some technical support to the American-led coalition forces in Iraq, with tiny Estonia sending 40 troops, and Latvia and Lithuania a little over 100 each. The three countries form a strong pro-American bloc in the current EU, consistently voting for decisions that are in the best interests of the United States. However, because they have very small populations and so are represented by only a few votes in the European Parliament, they cannot exert much influence over European politics. Also, practical matters demand more focus on local and regional than on global affairs. Estonia, for example, trades much more with Russia than it does with the United States, and Russian tourists are more common in Tallinn than American visitors are (although German and Finnish tourists are even more common).

At the same time, all three countries have unresolved issues with their big eastern neighbor. Both Latvia and Estonia have unsettled disputes with Russia along their eastern borders. Meanwhile, Russia demands better treatment of the Russian-speaking minorities in Estonia and Latvia, who are marginalized by the nationalistic governments of those countries. In recent elections in both states, pro-Russian parties enjoyed strong support among voters. Riga's current city council chairman, elected in 2009, is a Russian national advocating closer ties with Russia. Moscow also pushes for recognition of the role played by Baltic ultranationalists in the genocide of the Soviet Jews during World War II, as well as for protection of the Soviet war monuments and graves in the Baltic states. Lithuania enjoys the friendliest relations with Russia, probably because it has the smallest Russian-speaking minority and so has fewer issues to worry about. Also, it shares a border with Russia only in Kaliningrad Oblast. In general, however, relations between Russia and the Baltic states could be much improved. New generations of leaders in all four countries will have to seek improved cooperation and collaboration as a top priority.

EXERCISES

1. Use available economic and social data to compare and contrast Estonia, Latvia, and Lithuania with each other and with the following European countries: Finland, Poland, the Czech Republic, Portugal, and Belarus. Present your findings in class (this can be done as a group exercise, with each group picking one of the Baltic countries to focus on).
2. Evaluate tourism options available to you that would include any one of the Baltic countries as a destination.
3. Have an in-class debate about the merits of a recent proposal by the U.S. government to locate antimissile radar installations in Poland and the Czech Republic. Have a team representing the Baltic states, a team representing Russia, and a team representing the U.S./NATO experts.
4. Use U.S. international trade information (available from *www.trade.gov*) to find out what items are imported to the United States from the Baltic states, and what items are exported from the United States there.
5. Prepare a report about any one of the national parks in Estonia, Latvia, or Lithuania.

Further Reading

Ashborne, A. (1999). *Lithuania: The rebirth of a nation 1991–1994.* Lanham, MD: Rowman & Littlefield.

Berg, E. (2003). Some unintended consequences of geopolitical reasoning in post-Soviet Estonia: Texts and policy streams, maps and cartoons. *Geopolitics, 8*, 101–120.

Ceccato, V. (2008). Expressive crimes in post-socialist states of Estonia, Latvia and Lithuania. *Journal of Scandinavian Studies in Criminology and Crime Prevention, 9*, 2–31.

Clemens, W. C. (2001). *The Baltic transformed: Complexity theory and European security.* Lanham, MD: Rowman & Littlefield.

Feakins, M, & Bialasiewicz, L. (2006). "Trouble in the East": The new entrants and challenges to the European ideal. *Eurasian Geography and Economics, 47*, 647–662.

Galbreath, D. J. (2008). Still "treading air"?: Looking at the post-enlargement challenges to democracy in the Baltic states. *Demokratizatsiya, 16*, 87–96.

Hansen, B., & Heurlen, B. (Eds.). (1998). *The Baltic states in world politics.* New York: St. Martin's Press.

Kolsto, P. (Ed.). (1999). *Nation building and ethnic integration in post-Soviet societies: An investigation of Latvia and Kazakhstan.* Boulder, CO: Westview Press.

Kulu, H. (2004). Determinants of residence and migration in the Soviet Union after World War 2: The immigrant population in Estonia. *Environment and Planning A, 36*, 305–325.

Leetmaa, K., & Tammaru, T. (2007). Suburbanization in countries in transition: Destination of suburbanizers in the Tallinn metropolitan area. *Geografiska Annaler, Series B: Human Geography, 89*, 127–146.

Meleshevich, A. (2006). Geographical patterns of party support in the Baltic states, Russia, and Ukraine. *European Urban and Regional Studies, 13*, 113–129.

Miniotaite, G. (2003). Convergent geography and divergent identities: A decade of transformation in the Baltic states. *Cambridge Review of International Affairs, 16*, 209–222.

O'Connor, K. (2003). *A history of the Baltic states.* Westport, CT: Greenwood Press.

Ostergren, R. C., & Rice, J. G. (2004). *The Europeans: A geography of people, culture, and environment.* New York: Guilford Press.

Pabriks, A., & Purs, A. (2001). *Latvia: The challenges of change.* New York: Routledge.

Peil, T. (2006). Emerging, submerging and persisting ideas: Is there social and cultural geography in Estonia? *Social and Cultural Geography, 7*, 463–492.

Steen, A. (1997). *Between past and future: Elites, democracy, and the state in post-Communist countries: A comparison of Estonia, Latvia, and Lithuania.* Brookfield, VT: Ashgate.

Waitt, G. (2005). Sexual citizenship in Latvia: Geographies of the Latvian closet. *Social and Cultural Geography, 6*, 161–182.

Weeks, T. R. (2006). A multi-ethnic city in transition: Vilnius's stormy decade, 1939–1949. *Eurasian Geography and Economics, 47*, 153–175.

CHAPTER 30

Eastern Europeans

UKRAINE, BELARUS, AND MOLDOVA

The three independent states discussed in this chapter have strong historical links with Russia. The first two are, like Russia, Eastern Slavic nations with Slavic languages and an Orthodox Christian religious tradition; they share much of their history, and were tightly economically integrated during the Soviet period. Moldova has the same Orthodox religion, but is culturally and linguistically Romanian. Nevertheless, it too has at various times been part of either the Russian Empire or the Soviet Union, and shares many cultural and economic features with the other two republics (Table 30.1 and Figure 30.1).

Among the three, Belarus (population 10 million) is by far the most thoroughly integrated into post-Soviet Russian space. Since 1997, there has been a movement toward complete integration of the two countries into a union with a common currency, borders, military, and government. Although it does make political sense, such a union has not yet fully materialized because of the complicated maneuvering involved. The ma-

TABLE 30.1. Comparative Characteristics of Belarus, Ukraine, Moldova, and Selected Other Countries or Regions

Country	Population (millions)	Ethnic Russian (%)	GDP PPP/capita (2008, U.S. dollars)	GDP growth (2008, %)	Services (2008, % of GDP)	Exports (2008, % of GDP)
Belarus	9.6	11	11,800	9.2	50	51
Ukraine	46	17	6,900	2.1	59	35
Moldova	4.3	6	2,500	7.3	61	32
Russia	140	80	15,800	6.0	55	28
CIS-12	280	50	10,500	10.3	50	32
EU-27	495	<1	32,300	3.0	71	8
United States	306	<1	47,000	1.3	79	8

Note. Data from CIA World Factbook (2009) and calculations by author. GDP PPP, gross domestic product adjusted for purchasing power parity. CIS-12, the Commonwealth of Independent States ip to 2008 (which included Russia, Belarus, Ukraine, Moldova, and eight other former Soviet Union republics). EU-27, the European Union with 27 members in 2008. Exports for the EU-27 do not include trade within the EU itself.

FIGURE 30.1. Belarus, Ukraine, and Moldova. Map: J. Torguson.

jority of Belarusians speak Russian as their first language and continue to live close to the land in medium-sized to small towns (Figure 30.2). The economy of Belarus was tightly integrated into the R.S.F.S.R. economy during the Soviet period and remains intertwined with the Russian economy today, so a full union in the future does seem likely. At the same time, however, Belarus is experiencing problems with human rights abuses and is a pariah in Europe because of its highly authoritarian current leader, Alexander Lukashenko. In essence, Belarus remains a closed society and thus is unable to attract significant investments from the West. Nevertheless, its close economic ties with Russia translate into an average income almost twice as high as in neighboring Ukraine.

The much larger Ukraine (population 46 million, with an area bigger than France) is a bilingual country, split about 25% to 75% between Russian and Ukrainian speakers, and is much more independent minded. The Ukrainian and Russian languages diverged only a few centuries ago and are mutually intelligible. The Dnieper River is the main linguistic divide: More people west of it speak Ukrainian, and more people east of it (and also in the south and the Crimea) speak Russian. This prominent division is apparent in

FIGURE 30.2. A typical house in a small Belarus town. Although the country is industrialized, many people continue to live close to the land. *Photo:* P. Miltenoff.

the country's post-Soviet politics, as evidenced by strong support in the 2004 presidential election for the more nationalist Viktor Yushchenko in the west, and the more pro-Russian Viktor Yanukovych in the east and south (Figure 30.3).

Tiny Moldova (population 4 million) is predominantly Romanian-speaking, but has about a 14% Russian- or Ukrainian-speaking minority, as well as 4% Turkic-speaking Gagauz and 2% Bulgarians. It is also home to a secessionist controversy and a self-proclaimed republic (see below). All three countries used to have substantial peasant Jewish settlements (*shtetls*), which were largely destroyed in nationalist pogroms in the early 20th century or by the Nazis during World War II. Urban Jewish communities remain prominent in Kiev, Odessa, Kharkov, Minsk, Kishinev, and other large cities. Although Moldova is 80% Orthodox Christian, it has many other religions, including Baptist, Armenian, and Jewish.

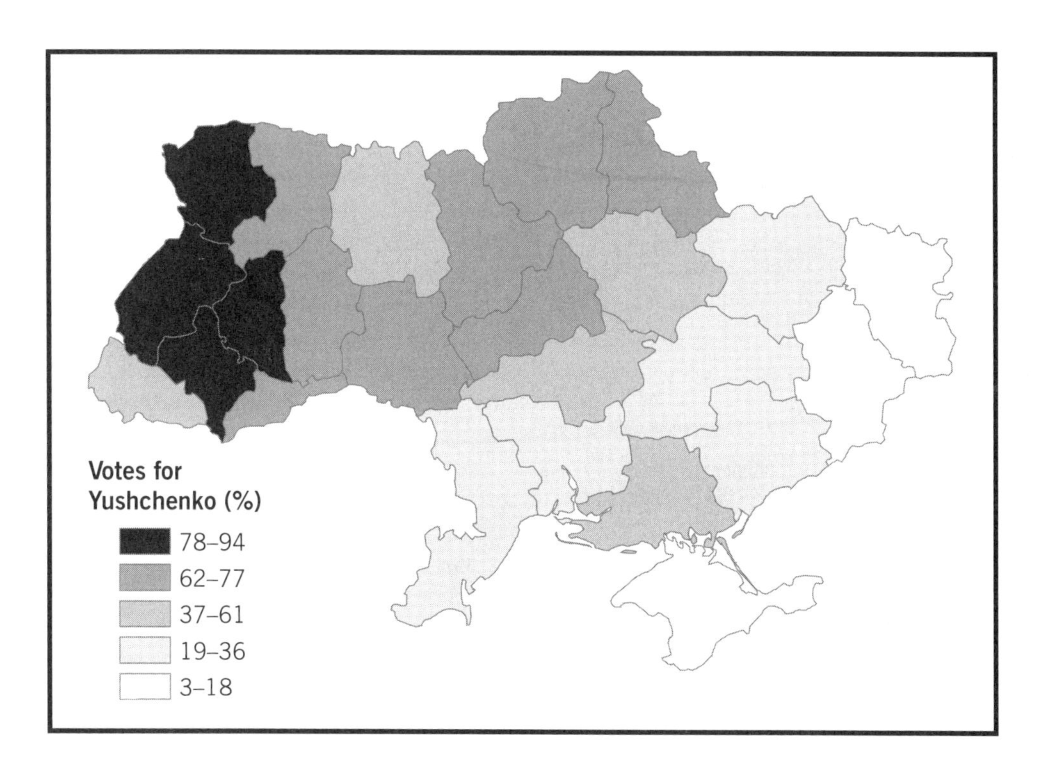

FIGURE 30.3. Results of the 2004 presidential elections in Ukraine show the prominent geographic rift between the nationalistic Ukrainian (north)west and the pro-Russian (south)east of the country. The top two candidates, Viktor Yushchenko and Viktor Yanukovych, received 39.87% and 39.32% of the votes, respectively, in a very close race. Data from the Ukraine Electoral Commission.

Physical Geography

The climate of this section of Northern Eurasia is distinctly European. Although winters here are colder on average than in much of Western Europe, the typical winter temperatures in northern Ukraine are similar to those experienced in the U.S. upper Midwest and not much colder than in Hungary or Slovakia. The average January temperature in Kharkov is only –7°C, and in Odessa –2.5°C. The Carpathians shelter Moldova from the worst outbursts of Arctic air, as do the much smaller Crimean Mountains protecting Yalta along the balmy south Crimean coast. These areas have a Dfa climate, according to the Köppen classification; this is comparable to Ohio or the Balkans. The rest of the region has a Dfb climate, similar to Moscow or Minnesota. Much of Belarus is covered with forested swamps, called *polesye*. Potato, sugar beets, and flax are the most important crops in the republic. Ukraine has some forests in the west, but is primarily a steppe country with well-developed agriculture on the famous chernozems, or black soils. It was the breadbasket of the Soviet Union and remains a globally significant agricultural producer of wheat, corn, sunflowers, hogs, geese, and other commodities. Moldova's agricultural specialties are orchard crops, vegetables, and wine production. Large sections of northern Ukraine and especially eastern Belarus suffered from Chernobyl's radioactive fallout in 1986 and remain out of production today. Moldova suffers from decades of pesticide and fertilizer overuse in its agricultural areas.

Cultural and Historical Features

The three countries are culturally distinct, but with many similarities as well—for example, in the distinctly Eastern European styles of folk songs, dances, dress, architecture, and diet common to all three. For centuries, these were peasant societies with a strong communal village life. Some villages in Ukraine remain huge (over 5,000 residents) as compared to Russian ones, especially along fertile river valleys. Parts of southern Ukraine were settled by the seminomadic Cossacks who had staged raids on the surrounding areas along the Black Sea border and into Poland and Turkey in the 15th–17th centuries. They protected the core Slavic territories from invaders from both the east and west. The history of Ukraine's statehood is a long and convoluted one, but essentially centers on internal struggles between pro-Russian and pro-Polish groups and on its emerging nationalism since the mid-18th century, with perpetually shifting affinities and borders. Areas of western Ukraine have seen hundreds of border adjustments in the past five centuries, with portions of it belonging to Poland, Prussia, Austria–Hungary, Russia, and even Turkey and Sweden for various periods of time. Ukraine in this sense is a classic example of a political transition zone in perpetual search of an identity. Post-Soviet Ukraine remains in the same position today, politically torn between East and West.

Ukrainian culture incorporated elements of Polish high culture during the long period of Polish rule (1400s–1654) and non-Slavic elements from the nomadic steppe cultures (Huns, Polovtsy, Khazars, Tatars) into its primarily Eastern Slavic basis. A few aspects of its culture come from pre-Christian times—for example, the tradition of coloring waxed eggs for Easter (*pysanki*), which used to symbolize the cult of the sun. Ukraine has produced many distinguished cultural figures. Its most famous poet, Taras Shevchenko (1814–1861), and the lesser-known Lesya Ukrainka (1871–1913) were among the first to write in the Ukrainian language. Nikolai Gogol was from Ukraine and publicized many of the local folk tales from his childhood village in the short stories *Evenings in Dikanka*, although he mostly wrote in Russian and lived in St. Petersburg. Joseph Conrad, a Pole, and Sholom Aleichem, a Jew, were born in Ukraine and vividly described it in their stories. There are a few contemporary writers in the West who have Ukrainian origins (e.g., Chuck Palahniuk). The modernist painter Kazimir Malevich was from Ukraine. Two famous Ukrainian composers were Dmitry Bortniansky and Mykola Leontovitch. A survey of Hollywood stars reveals many with Ukrainian or Jewish–Ukrainian roots, including Steven Spielberg, Dustin Hoffman, Lee Strasberg, and Milla Jovovich. There are literally dozens of U.S. and Canadian athletes who have some Ukrainian roots (e.g., the hockey greats Wayne Gretzky and Terry Sawchuk). In the world of science and technology, Igor Sikorsky the inventor of the helicopter, was born in Kiev. Famous mathematicians, including Mikhail Krawtchouk, Viktor Bunyakovsky, and Georgy Voronoi, worked in Ukraine. The list could go on and on. Belarus and Moldova are smaller countries and have proportionally fewer famous names associated with them, but with a careful look one discovers quite a few.

A final cultural fact worth mentioning is that many American and Canadian families with early-20th-century "Russian roots" have ancestors who actually came to North America from one of these three countries (especially Ukraine or Belarus), not from Russia proper. Canada alone received close to 2 million Ukrainian migrants into its prairie provinces by the beginning of the 20th century.

Economics

Although Ukraine is of course by far the biggest of these three countries, its gross domestic product (GDP) per capita ($6,900 in 2008) trails far behind that of Belarus ($11,800). This may come as a surprise, given the poor image Belarus has in the Western media because of its human rights violations (Table 30.1). Economics and politics are separate things, however, and the prosperity of Belarus is determined primarily by the overall integration of its economy into Russia's much larger economy, not by its political virtue. Ukraine and Belarus are both below average in income globally. Moldova, on the other hand, is the poorest economy in Europe; its GDP per capita ($2,500) is similar to that of Mongolia, India, or Nicaragua. This may seem surprisingly low; after all, the country enjoys a favorable climate, rich agricultural soils, and a well-educated workforce, and its cities seem to look better than the income alone suggests. Nevertheless, it has no fossil fuels and lacks sea access for trade (it was deliberately landlocked by the Soviet planners to prevent secession after World War II; the main sea outlet was given to Ukraine), and it suffers from many years of incomplete reforms and an ambivalent stance toward both Russia and Ukraine. Moldova could be strengthened by a closer alliance with ethnically similar Romania, since the latter is now a full European Union (EU) member, but there are enough internal differences between the two to keep them separate.

Belarus is a heavily industrialized economy centered on Minsk (population 1.8 million). Other big and historical cities include Vitebsk, Polotsk, Mogilev, Brest, and Gomel. Machine building accounts for almost one-quarter of the country's GDP. Since the Soviet period, it has been a center of tractor and heavy truck building; electronics manufacturing; and potassium fertilizer, rubber, and plastics production. It has adequate timber and water reserves, but low energy reserves. Belarus is also the main conduit of Russian oil and gas to Europe, and of European food and goods to Russia, via its pipelines and highways. However, the authoritarian government of President Lukashenko hampers free market reforms, punishes political opposition, and creates unpredictability and discomfort in doing business here. As noted earlier, Belarus is a pariah state in Europe, with little to no representation in the affairs of the EU or other important bodies. Many of its politicians are even refused entry to the EU or the United States because of the country's abysmal political rights record with respect to dissidents. However, Belarus does a lot of business with Russia, Ukraine, Germany, Poland, the Netherlands, the United Kingdom, and some Middle Eastern and Asian partners. A recent visit from Venezuela's Hugo Chavez, seeking Belarusian weapons and offering cheap petroleum, showed an interesting new connection across the Atlantic among the "coalition of the unwilling" to accept U.S. global hegemony.

The Ukrainian economy is diverse. Its main strengths are mining of iron ore in Kryvoy Rog, manganese near Nikopol, and coal in Donbass; hydropower production and associated aluminum smelters on the Dnieper River; various types of manufacturing (shipbuilding, train engines, railroad cars, cars, engines, small and large military equipment, electrical and refrigeration equipment); and agriculture. About one-quarter of the industrial output is accounted for by metallurgy, and another quarter by energy production. Dnepropetrovsk and Zaporozhye were major industrial centers of the Soviet military. The largest intercontinental ballistic missiles and some airplanes were assembled here. Today many of the military factories have been converted to civilian uses (furniture, appliances). There are five major hydropower facilities on the Dnieper and five nuclear stations (with a total of 15 reactors, including the remaining 3 in Chernobyl, which were finally shut down in 2003). Food processing accounts for 16% of the total industrial output. Ukraine is, however, deficient in petroleum and natural gas. Some limited production of oil occurs along the Black Sea margins, but most must be imported from Russia. Transshipment of oil and gas from Russia and of gas from Turkmenistan provides two major sources of revenue, but is also the subject of frequent bickering over "fair" prices. Mining and manufacturing in the eastern part of the country (Donetsk and Lugansk Oblasts) produce close to half of all industrial output. The cities in the middle (Kiev, Dnepropetrovsk, Zaporozhye) are intermediate in terms of industrial production. Although Ukraine's primate city, Kiev, is a large city with over 3 million residents, it has relatively little manufacturing for a city of its size (Figure 30.4). However, it plays a major political and cultural role in unifying the divided country and is home to many museums, theaters, hospitals, universities, and schools. It also leads the country in investments and consumption. The hilly but nationalist west is much poorer, and is predominantly agricultural; cities there include Lvov and Ivano-Frankovsk (Figure 30.5).

Tourism is well developed in the Crimea Peninsula, near Odessa, and in the Carpathians. In the Crimea, it is much hampered by high prices, low-quality service, and poor infrastructure as compared to nearby Turkey or Bulgaria. The presence of the Russian Navy in Sevastopol is a source of additional tension (Figure 30.6); a new agreement pushed by Kiev would gradually phase it out of the country by 2014. Still, the Crimea is one of the most beautiful corners of Northern Eurasia.

FIGURE 30.4. Independence Square in Kiev, the site of the "Orange Revolution" in 2004. It is noteworthy that while all the street signs in the square are now in Ukrainian, just a few streets away Russian street and shop names remain in place, as they were during Soviet times. Most Kiev residents can speak Russian or Ukrainian with equal ease, but all government business is now conducted strictly in Ukrainian. *Photo:* J. Lindsey.

FIGURE 30.5. Lvov is a beautiful historical city in western Ukraine. *Photo:* J. Lindsey.

Whereas Ukraine is 68% urban, Moldova is only 46%. About half of its industry is in food processing. Agriculture accounts for 18% of its GDP, as compared to only 9% for Ukraine or Belarus. Many agricultural products of Moldova are expensive perishable crops, such as grapes, berries, and fresh vegetables. Politically motivated bans on Moldovan wine and fruit exports to Russia in 2006 seriously crippled the already fragile economy; Moldova has vacillated in recent years between more pro-Russian and pro-EU stances, and the export ban was the kremlin's punishment for that. The country is also experiencing a shortage of labor: Close to a million Moldovans have left the country for employment in the construction, retail, food, and textile industries of Russia, Ukraine, Turkey, Italy, and France. Ukraine and Moldova are also notorious exporters of sex industry workers for the urban European and Asian markets, especially the United Kingdom, Greece, the Netherlands, and Israel.

FIGURE 30.6. The Russian Navy still maintains a base in Sevastopol, Ukraine. *Photo:* O. Voskresensky.

In addition, Moldova is home to an ongoing political secessionist conflict in the Trans-Dniester region—the only such conflict in Northern Eurasia outside the Caucasus. The unrecognized Trans-Dniester Republic (TDR) lies east of the Dniester River along the Ukrainian border, with Tiraspol as its self-proclaimed capital; it accounted for only 20% of Moldova's population, but almost half of its industrial output, in 1991. This region has long-standing historical ties to the Russian Empire as compared to the part of Moldova (Bessarabia) on the west bank of the Dniester, and is heavily Russian- and Ukrainian-speaking. In early 1992 the TDR sought greater cooperation with Moscow and Kiev than was allowed by the newly independent government in Kishinev. After a few skirmishes, a major war was prevented by the Soviet 14th Army, which was positioned in the republic. Some Russian peacekeepers remain to guard the TDR today on an informal basis, but in general the Russian leadership has chosen to distance itself from the TDR. The Moldovan government is forced to tolerate the status quo, which may or may not last very long. The TDR is not officially recognized by any foreign government, but in the aftermath of Kosovo's declaration of independence in 2008, the TDR is likely to seek similar solutions. It is important to note here that the TDR is not merely a Russian–Ukrainian exclave inside Moldova, as it is sometimes wrongly perceived to be. Instead, a full quarter of its population is Moldovan; however, its leadership continues to favor unity with Russia, not Romania, as the long-term political goal—a very different viewpoint from that of the official government in Kishinev. The TDR also glorifies all things Soviet and in some respects resembles a Soviet theme park, complete with old Soviet slogans and even Soviet vending machines from about 1980 in the streets.

Challenges and Opportunities in Ukraine, Belarus, and Moldova

The countries described in this chapter are on divergent tracks with respect to future development. Belarus seems to be firmly committed to greater political unity with Russia, at the expense of political freedoms at home. Ukraine is more and more openly seeking full membership in NATO and eventually the EU, while remaining a pragmatic trade partner with Russia and other Commonwealth of Independent States (CIS) republics. It has the most established democracy of the three, but is experiencing fierce internal competition between its western and eastern regional elites, so its future is quite uncertain. In the spring of 2010, Victor Yanukovich became the new president of Ukraine, reflecting growing dissatisfaction with the poor state of the Ukrainian economy 6 years after the Orange Revolution. For the moment, at least, it seems that both countries will be moving forward with shelving some of the old disputes. They have already made the decision to extend Russia's lease of the navy port in Sevastopol in exchange for cheaper natural gas prices for Ukraine. Ukraine farther distanced itself from NATO and is likely to start looking for fresh opportunities in reengaging with its northern and eastern neighbor. Nevertheless, other pro-Western or nationalistic politicians, especially the outspoken former Prime Minister, Yulia Timoshenko, remain active and will continue to reshape Ukraine's political future. Moldova is in a severe recession at the moment, and it is hard to know how soon and in what direction the situation will change there. Greater unity with Romania seems likely, but full integration into the latter or the EU either may never happen or may still be 10–15 years away.

EXERCISES

1. Explore the meaning of the word *Ukraina* (Ukraine). What does it suggest about the country's geographic position? What might it be called from a Polish, Russian, or Crimean Tatar perspective? From a Ukrainian nationalist perspective? Have a debate in class, outlining views on Ukraine's history from each of these perspectives.
2. Study famous battlefields of World War II that involved territories in Belarus (Brest, Vitebsk, Minsk) and Ukraine (Kiev, Kursk, Lugansk, Sevastopol, Kerch). In small groups, give in-class presentations comparing the geographic aspects of these battlefield locations.
3. Explore in greater depth the electoral geography of Ukraine today, using online sources and some of the suggested readings at the end of the chapter. Why do you think the rift is so severe?

4. Compare and contrast the Ukraine–Russia relationship with the Canada–U.S. relationship. List similarities and differences (e.g., relative sizes, languages, political systems). Is this a fair comparison? Why or why not? Can it help predict the future development of Ukraine–Russia relations?
5. Investigate the geographic history of the borders of any one of the following historical entities: Galicia, Volhynia, Bukovina, Transcarpathia, Bessarabia, Transnistria. How does the history shape the existing territorial claims in the region?
6. Study the history of recent political developments in Belarus. Is it fair to call the political state under Alexander Lukashenko "neo-Stalinist"? Why or why not?
7. Evaluate available tourism options for the Crimean Peninsula and the Carpathian Mountains.
8. Make an inventory of items you personally own (clothing, electronics, etc.). Are any of those made in either Ukraine or Belarus? What does this suggest about the foreign trade of those countries with your country? You can expand your search by visiting a few department stores in your area.
9. Discuss the pros and cons of Ukraine's joining the EU and NATO in the future.
10. Discuss the pros and cons of Moldova's joining Romania in the future.

Further Reading

Åslund, A., & McFaul, M. (Eds.). (2006). *Revolution in orange: The origins of Ukraine's democratic breakthrough.* Washington, DC: Carnegie Endowment for International Peace.

Banaian, K. (1999). *The Ukrainian economy since independence.* Northampton, MA: Elgar.

Bruchis, M. (1996). *The republic of Moldova: From the collapse of the Soviet empire to the restoration of the Russian empire.* New York: Columbia University Press.

Clem, R. S., & Craumer, P. R. (2008). Orange, blue and white, and blonde: The electoral geography of Ukraine's 2006 and 2007 Rada elections. *Eurasian Geography and Economics, 49*, 127–152.

Foer, J. S. (2002). *Everything is illuminated.* Boston: Houghton Mifflin. (See also the movie based on this novel.)

Garnett, S. W., & Legvold, R. (Eds.). (1999). *Belarus at the crossroad.* Washington, DC: Brookings Institution Press.

Harris, C. D. (1999). New European countries and their minorities. *Geographical Review, 83*, 301–320.

Ioffe, G. (2007). Unfinished nation-building in Belarus and the 2006 presidential election. *Eurasian Geography and Economics, 48*, 37–58.

Jordan, P., & Klemenčić, M. (2003). Transcarpathia—bridgehead or periphery? *Eurasian Geography and Economics, 44*, 497–514.

Kuzio, T., & Wilson, A. (2000). *Ukraine: Perestroika to independence.* New York: Palgrave.

Lieven, A. (1999). *Ukraine and Russia: A fraternal rivalry.* Washington, DC: U.S. Institute of Peace Press.

Magosci, P. R. (2007). *Ukraine: An illustrated history.* Seattle: University of Washington Press.

Mould, R. F. (2000). *Chernobyl record: The definitive history of the Chernobyl catastrophe.* Philadelphia: Institute of Physics.

Reid, A. (1999). *Borderland: A journey through the history of Ukraine.* Boulder, CO: Westview Press.

Rowland, R. H. (2004). National and regional population trends in Ukraine: Results from the most recent census. *Eurasian Geography and Economics, 45*, 491–515.

Swain, A. (2006). Soft capitalism and a hard industry: Virtualism, the "transition industry" and the restructuring of the Ukrainian coal industry. *Transactions of the Institute of British Geographers, 31*, 208–224.

Williams, B. G. (2000). *The Crimean Tatars: The diaspora experience and the forging of a nation.* Boston: Brill.

Wilson, A. (2000). *The Ukrainians' unexpected nation.* New Haven, CT: Yale University Press.

CHAPTER 31

Central Asia

The Heart of Eurasia

In this chapter I focus on the five independent republics of Central Asia, the five "-stans" (*stan* means "state" in Turkic) (Figure 31.1). Collectively known as "Turkestan," they have much in common: an arid physical environment, Turkic languages (except for Tajikistan), Sunni Islam, and Asian cultural traditions. All five are also landlocked, if one discounts access to the inland Caspian Sea for Kazakhstan and Turkmenistan. Whereas Uzbekistan is the biggest of the five by population, Kazakhstan is both the largest by area and the most prosperous (Table 31.1). Kazakhstan is just a little smaller than Argentina and is the ninth largest country worldwide; more than 80% of its territory is steppe or desert. Tajikistan is both the poorest and the smallest by area, while Kyrgyzstan has the smallest population. Unlike the rest of the former Soviet Union (FSU), all of the Central Asian states have growing populations, with the annual growth rates ranging from +0.4% in Kazakhstan to +1.9% in Tajikistan.

Physical Geography

Most of the region has semidesert or desert climate, with high mountains rising in the south and east. The flattest and most desert-like is Turkmenistan, with 80% of its territory occupied by the sands and gravel of the Kara Kum. Uzbekistan is home to the Kyzyl Kum desert, while Kazakhstan has the Moynkum and Saryesik-Atyrau deserts near Lake Balkhash. The highest mountains are the Pamirs in Tajikistan, reaching 7,495 m above sea level, and the second highest are the Tien Shan in Kyrgyzstan, reaching 7,439 m. The amount of arable land varies from 3% in Turkmenistan to 11% in Uzbekistan. The most productive area of the region is the fertile Fergana Valley along the upper Syr Darya River, which is shared by Uzbekistan, Kyrgyzstan, and Tajikistan (Figure 31.2). It is easy to find it on a map; just look for the jigsaw-puzzle-like pattern made by the borders of the three countries where they meet in Fergana. Located along the northern branch of the famous medieval Silk Route, the valley has been a magnet for settlement since antiquity.

All five "-stans" are deficient in water and timber. Water is greatly needed for irrigation and is likely to become increasingly scarce in the course of the 21st century's warming. Timber has to be imported from outside the region, mainly from Russia, although some can be produced domestically in the mountains. Kazakhstan is a mining

FIGURE 31.1. The five independent Central Asian republics. Map: J. Torguson.

TABLE 31.1. Comparative Characteristics of the Central Asian States and Selected Other Countries and Regions

Country	Population (millions)	% ethnic Russian	GDP PPP/capita (2008, U.S. dollars)	GDP growth (2008, %)	Services (2008, % of GDP)	Exports (2008, % of GDP)
Kazakhstan	15	30	11,500	3.0	55	47
Kyrgyzstan	5.4	13	2,100	6.0	49	28
Tajikistan	7.3	1	2,100	7.9	48	44
Turkmenistan	4.9	4	6,100	10	50	28
Uzbekistan	28	6	2,600	8.9	38	36
Russia	140	80	15,800	6.0	55	28
CIS-12	280	50	10,500	10.3	50	32
EU-27	495	<1	32,300	3.0	71	8
United States	306	<1	47,000	1.3	79	8

Note. Data from CIA World Factbook (2009) and calculations by author. GDP PPP, gross domestic product adjusted for purchasing power parity. CIS-12, the Commonwealth of Independent States (which include Russia, the five Central Asian states, and six other former Soviet Union republics). EU-27, the European Union with 27 members in 2008. Exports for the EU-27 do not include trade within the EU itself.

FIGURE 31.2. The Fergana Valley is the most fertile area of Central Asia; it is shared by Uzbekistan, Kyrgyzstan, and Tajikistan. Baking flatbread and cooking a lamb and rice dish called *plov* are two local culinary traditions. *Photo:* C. Burke.

giant, with substantial deposits of coal, oil, natural gas, gold, uranium, manganese, chromium, copper, and other metallic ores. Uzbekistan has gold and limited petroleum and natural gas resources. Turkmenistan has a lot of natural gas and some petroleum as well along the Caspian Sea coast. In contrast, Tajikistan and Kyrgyzstan have few commercially exploitable mineral resources except construction stone. All five countries have plentiful rangelands for sheep, goats, and cattle. Uzbekistan and Turkmenistan are major cotton producers on irrigated lands south of the Aral Sea (Figure 31.3). Wool is a major export for Kyrgyzstan. All five countries have fertile valleys where orchard and vegetable crops can be grown, including peaches, plums, apricots, quince, watermelons, melons, and grapes. Kazakhstan is a large producer of wheat, growing about 2.2% of the world's wheat in 2006. Much of it is grown in the area that was plowed under during Khrushchev's Virgin Lands campaign in the 1950s, around Tselinograd (now the new capital, Astana). Fisheries of the Caspian Sea and Balkhash Lake are locally important. Regrettably, the formerly rich Aral Sea fisheries are almost gone now because the lake is drying (Chapter 5). The steppes of Central Asia are still inhabited by endangered taiga antelopes, wild donkeys, camels, and a few smaller species of game that are trophy-hunted. The elusive snow leopard of the Central Asian mountains is the largest predator in the region now that the Central Asian subspecies of tiger is extinct.

Cultural and Historical Features

Four of the five Central Asian states speak Turkic-based languages of the Altaic family. Kyrgyz and Kazakh are very close, mutually intelligible languages. The Turkmen language is closest to Turkic and Azeri, whereas the Uzbek language is closest to Uygur of northwestern China and also incorporates many Persian words. During Soviet times all countries were forced to use the Cyrillic alphabet, but some are now in the process of changing it either to Latin (Turkmenistan and Uzbekistan) or to Arabic (Tajikistan) script. The Tajik language is Indo-European and is closely related to Farsi of Iran and to the languages of Afghanistan and Pakistan (Pashtu and Urdu). Culturally, the Kazakhs and Kyrgyz were nomadic people traveling with their herds across the vast Kyrgyz steppe. The Uzbeks, Turkmens, and Tajiks led more sedentary farming lifestyles. The cities of Bukhara and Samarkand in Uzbekistan, along the Zeravshan River, were centers of powerful emirates formed in the 18th century as Islamic states. Samarkand's most famous ruler was Tamerlane (Timur), a man who was not afraid to challenge rulers from Turkey to India in the mid-14th century.

Besides the main five ethnicities, there are many others; for example, Karakalpaks, Russians, Ukrainians, Germans, Jews, Chinese, and Koreans live in the region. Outside the region, many Tajiks, Uzbeks, and Turkmens live in

FIGURE 31.3. The Kara Kum canal stretches from the Amu Darya River across Turkmenistan and is used for cotton irrigation. Cotton is grown mainly in Turkmenistan and Uzbekistan and is the biggest cash crop in the region. *Photo:* C. Burke.

Afghanistan, and some Kazakhs live in northwestern China and in the Altay in Russia. More recently, many thousands of Tajiks and Uzbeks have migrated to Russia's big cities in search of employment.

Probably the best-known creative works originating in Central Asia are the medieval poems and philosophical works of Alisher Navoi, who lived in Herat (now in Afghanistan) in the 15th century. He wrote in two languages, Persian and Old Uzbek. The Uzbek government named a city after him in 1958, and many monuments are dedicated to him in the region. An earlier writer, Firdousi (10th century), wrote in Persian, and is well regarded as a father of the literature in Tajikistan and Afghanistan. More recently, the poet and thinker Abay Kunanbaev (1845–1904) greatly influenced Kazakh literature (Figure 31.4). Prior to his period, virtually all Kazakh literary genres were oral tales or songs. He made major efforts to reach out to the society at large with his own poetic and philosophical works and with his translations from other languages (e.g., the poems of Pushkin and Byron). Although he criticized Russian colonial policies, he strongly believed that Russian cultural influences were beneficial for Kazakhstan and would open up the world to the young nation.

FIGURE 31.4. Abay Kunanbaev was the most important Kazakh literary figure of the mid-19th century. This monument is in Semey. *Photo:* Author.

Economics

Kazakhstan

Kazakhstan currently has by far the largest economy in Central Asia, accounting for two-thirds of the regional gross domestic product (GDP) and having a per capita income twice as high as the second highest, Turkmenistan's (Table 31.1). In addition to the natural resources mentioned above, it has a well-developed industrial sector, including machine and tractor building, steel and nonferrous metallurgy, construction, and financial and high-tech services. Kazakhstan also leases the Baykonur launch pad to the Russian space agency. Like Ukraine, it gave up its nuclear arsenal to Russia after the collapse of the Soviet Union, but continues to mine uranium. Russia is Kazakhstan's largest trade partner, providing over one-third of its imports and receiving 12% of its exports. The two countries border each other over 5,000 km. Germany is the biggest trade partner of Kazakhstan in Europe, and China is the biggest in Asia. The country's most discussed development issue is of course Caspian Sea oil, of which Kazakhstan has about two-thirds of the total reserves. Many conflicting estimates exist for the Caspian Sea petroleum reserves (Dekmejian & Simonian, 2001), but a recent independent report from the Energy Watch Group suggests that the total reserves of Kazakhstan are likely to be about 33 billion barrels of oil—about three times as large as those of Azerbaijan, and just a little less than the revised reserves of Kuwait (usually reported at close to 100 billion barrels, but probably only 35 billion barrels).

Kazakhstan's petroleum is produced mainly in Atyrau and Mangistau Oblasts in the western part of the country. The largest onshore field, Tengiz, was discovered in 1979. It is the sixth largest oil field in the world, about the size of Alaska's Prudhoe Bay (9 billion barrels), and is currently developed by an international consortium in which Chevron and ExxonMobil are heavily involved. The even larger offshore Kashagan oil field, discovered in 2000, will require major investments in underwater drilling technology and is not likely to begin producing oil for several years yet. Unlike Russia, Kazakhstan invited major Western oil companies to develop its oil fields at an early stage. Today it is estimated that almost 80% of all petroleum produced in the country is produced by U.S. and European companies, although Russian and Chinese oil companies also participate. A new pipeline from Kazakhstan into China's Xinjiang region was completed in 2009 and can handle 120,000 barrels per day. However, talk about renationalizing some of these assets is making investors nervous (Olcott, 2002).

Kazakhstan's new capital, Astana, has a population of only half a million, with glitzy skyscrapers and a brand-new airport impressing its first-time visitors. It is designed as a city for the modern business and government elite, and styles itself as a Dubai of Central Asia. However, the former capital, Almaty (population 1.2 million), remains the business and banking hub of the country. It also attracts numerous foreign visitors, especially from China. The eastern mining centers of Ekibastuz and Karaganda (coal) and Oskemen (nonferrous metals) remain heavily Russian populated (Figure 31.5), whereas the cities in the rural south (Shimkent and especially Kyzyl-Orda) are over 60% Kazakh populated.

Uzbekistan

Uzbekistan has a much bigger population than Kazakhstan's, but has a smaller economy, mainly because of a lack of economic and political reforms. Although both countries have had authoritarian leaders since the dissolution of the U.S.S.R. (Islam Karimov and Nursultan Nazarbayev, respectively), Kazakhstan has largely followed the Russian model in rapidly privatizing its economy, reforming its banking sector, and making major investments in education and infrastructure. This did not happen in Uzbekistan, where a much larger agricultural sector was particularly hard to reform, and the local corrupt Communist bosses remained unchallenged and unchanged from Soviet times. Other problems in Uzbekistan have included the presence of a militant underground Islamist movement and a lack of direct access to Russia or Western markets. Moreover, Uzbekistan's leading export is not oil, but cotton; its major industry is not machine building, but textiles. It does have limited natural gas supplies, but very little petroleum. In short, it has relatively little

FIGURE 31.5. East Kazakhstan Oblast, centered on Oskemen (Ust-Kamenogorsk), is 48% Kazakh and 45% Russian populated. It is the largest center of nonferrous metallurgy in the country. *Photo:* Author.

to offer to the world and has a long way to go to become another important economic producer in Central Asia. Nevertheless, it is strategically located in the very middle of the region and is the only country to border all four of the others. It is the logical central location for pan-Central Asian functions. Uzbekistan opened its airspace and airfields to the North Atlantic Treaty Organization (NATO) forces bound for Afghanistan in 2002, but more recently has restricted access to these as a backlash against Western demands for more human rights in the country. In fact, Uzbekistan has some of the worst corruption in the world as measured by Transparency International, and it also has one of the most brutal and least transparent judicial systems. In particular, opposition journalists are persecuted and sometimes disappear without a trace.

The capital of Uzbekistan, Tashkent, is a large city of over 3 million. Originally founded in an oasis along the Silk Route, it was greatly expanded during the Soviet period. Tashkent was devastated by an earthquake in 1966, but was quickly rebuilt. It has lavish tree-lined streets, fountains, large government buildings, hospitals, schools, and even its own subway. Tashkent remains the center of Uzbekistan and is the most cosmopolitan city in the country, with many nationalities peacefully living together. It also remains the largest industrial center in the country, with tractor and airplane factories, as well as a number of enterprises making equipment for the cotton industry. Much of the new construction since 1991 has been done to accommodate the modest increase in international business, including office towers, banking centers, plazas, and malls—not only in Tashkent, but in smaller cities as well (Figure 31.6). Daewoo has opened a car factory in Asaka near Andijan in the extreme east of the country. There is significant gold and uranium production in the Kyzyl Kum desert. Historical Samarkand and Bukhara attract thousands of tourists to their World Heritage Sites, which include mosques, religious schools (Figure 31.7), and mausoleums. Samarkand is at least 2,750 years old, which makes it one of the oldest, continuously inhabited cities worldwide.

Kyrgyzstan

Kyrgyzstan is another struggling economy in the region. Although it was the first Central Asian state to launch market reforms and political democratization in the early 1990s, it soon fell out of pace with Kazakhstan and Russia because of internal political tensions. After the 2005 ouster of President Askar Akaev (who had provided the impetus for many of the earlier reforms), the new,

FIGURE 31.6. Tennis courts have sprung up all over Uzbekistan, largely due to President Karimov's fondness for the sport. This complex is located in Termiz, along the border with Afghanistan. *Photo:* C. Burke.

more nationalist President Kurmanbek Bakiyev, had to deal with a declining output, corruption, and lack of foreign investments (Marat, 2006).

As this book was being prepared for publication, a public uprising in April of 2010 ousted President Bakiyev. Initially peaceful protests turned into a bloody revolt when security forces close to the president were ordered to shoot into crowds. The revolt was mainly precipitated by the deepening economic crisis, a lack of economic opportunities, earlier increases in utility costs, and overall public dissatisfaction with the pervasive corruption in the circles close to the former president. Bakiyev fled the country in a Kazakhstan-brokered escape attempt and found asylum in Belarus. A new transitional government composed of diverse opposition figures was formed in Bishkek and a new constitution was drafted. Parliamentary elections are scheduled for the fall of 2010. While the new government has expressed its political neutrality with respect to both the United States and Russia, it is clear that the new regime is likely to prove more pro-Russian than that of its predecessor. At the same time, the United States and China are strategically interested, along with Russia, in Kyrgyzstan's future prosperity and stability.

Exports of gold, tin, and antimony are significant sources of foreign revenue. Traditional exports also include wool and mutton, cotton, tobacco products, and limited uranium and natural gas exports. Kyrgyzstan has surplus hydropower from dams on the Naryn. The biggest station, Toktogul, has a respectable capacity of 1,200 megawatts (MW). The country also exports 75% of the water in its reservoirs to the neighboring states. Its main trade partners are Russia, Kazakhstan, and China. Kyrgyzstan remains one of the most heavily Russian-speaking countries of

FIGURE 31.7. Registan Square in Samarkand is a World Heritage Site. It houses an assembly of three religious schools built in the 15th–17th centuries. *Photo:* S. Newton.

the post-Soviet region, and the Russian language is recognized as a language of intercultural communication; in fact, it enjoys broader recognition than in any non-Russian FSU republic except Kazakhstan and Belarus.

One of the potential bright spots on the generally bleak economic map of Kyrgyzstan is Lake Issyk-Kul, with its associated tourism development. The high mountain lake is one of the largest and purest in Asia (180 x 60 km in area, and over 600 m deep). It never freezes in winter and provides wintering grounds for millions of migratory birds. Backpacking, mountaineering, and horseback tourism are well developed in the mountains around the lake, especially in Karakol. The lake has a few endemic and endangered species of fish. It is mildly saline, however, and its water level is dropping slightly in response to the warming climate.

Kyrgyzstan's capital, Bishkek, is a primate city with over 600,000 residents; it is home to the main government institutions, as well as industrial enterprises, banking, universities, and hospitals. Manas Airport nearby has served as a major logistical air hub for the NATO efforts in Afghanistan. The second biggest city, Osh, is in the south of the country and is poorly connected to the north. It has been inhabited for about 3,000 years, being located along the strategic Ak-Burra River as it enters the Fergana Valley. Osh is a major cotton textile center.

Tajikistan

Tajikistan is the least developed, poorest, and most mountainous country in the FSU. Like its neighbor to the south, Afghanistan, Tajikistan experienced a bitter civil war, although this war lasted for a much shorter period of time (1992–1997) as the progovernmental Kulabis (south central) fought the Garmis (central) and the Gorno-Badakhshanis (southern mountains) over political control. The government in Dushanbe was supported by Russia and Uzbekistan, while the mountainous united opposition had supporters among Islamist movements in Uzbekistan, Afghanistan, and elsewhere. The war ended with the signing of a truce in 1997, but the situation today is not absolutely stable. Besides ethnic Tajiks, about one-quarter of the population consists of Uzbeks (who are very similar to the Tajiks in customs, diet, and dress, but speak a Turkic rather than an Iranian language). The rapid emigration of qualified Russian teachers and engineers since independence has resulted in a dearth of professional workers in the republic. Although officially a secular state, Tajikistan has an increasingly vocal Muslim population divided into Sunnis, Shiites, and a few other sects. Russian military units located along the border with Afghanistan help prevent infiltration of extremists from the south and at least partially maintain internal law and order. There are unresolved border disputes with Uzbekistan and Kyrgyzstan in the Fergana Valley, resulting in frequent border closures among the three. Despite the poor economy, 98% of Tajikistan's population is literate, a legacy of the Soviet period.

Tajikistan's economy is dominated by an aluminum smelter, Talco (aluminum accounts for almost 60% of Tajikistan's export revenues), and by hydropower facilities on the Vaksh River. The Nurek dam is the highest in the world at 300 m, producing 3,000 MW of power (about 50% more than Hoover Dam on the Colorado). The Nurek station alone can supply most of the nation's need for electricity, but new dams are being built with Russian, Iranian, and Chinese involvement. Another major source of income is cotton exports; Tajikistan also grows a lot of wheat. In addition, the country is similar to Armenia and Moldova in the FSU (and to El Salvador and the Dominican Republic in Latin America) in its reliance on remittances sent back home by migrant workers who are employed outside the home country. The remittances are thought to be one of the top three sources of foreign revenue, accounting for about one-third of the country's GDP. A major problem for Tajikistan is the increase in production and transport of opiate drugs from Afghanistan through Tajik territory. Located across the Panj River from Afghanistan, a country that grows 80% of the world's opium, Tajikistan is the logical gateway for traffickers en route to Russia and Europe.

Dushanbe (population 680,000) is the capital of the country and its primate city. It is a center for cotton and silk production; it is also home to the main government institutions, universities, and museums. Tajikistan's biggest geographic liability is the fact that the country is landlocked and very mountainous. Future development of ecotourism in the mountains is possible, however. The country has the majority of the FSU's mountains above 6,000 m; the FSU's longest glacier, Fedchenko, in the central Pamirs; and the FSU's highest summit, Ismail Samoni (formerly Peak Communism) at 7,495 m.

Turkmenistan

Turkmenistan is the most closed society of Central Asia. Its development was severely hampered by 15 years of the autocratic rule of Saparmat Niyazov, who even had his image printed on banknotes. Under Niyazov, Turkmenistan pursued increasingly isolationist policies; for example, instruction in both Russian and English was forbidden at most universities, and few foreigners were allowed into the country. President Niyazov spent much of the country's revenue on extensively renovating cities—particularly the capital, Ashgabat, where lavish palaces and monuments were erected in his honor (Figure 31.8). After Niyazov's death in 2006, his successor, Gurbanguly Berdimuhamedov, began cautiously easing some of the restrictions of the former regime. Turkmenistan's two economic staples are cotton (10th largest producer in the world) and natural gas (5th largest producer). Other exports include wool (including famous wool rugs), vegetable oil, and fruit (Turkmen melons are of legendary quality). Its economy, however, is one of the least privatized in the FSU, with about 70% of all assets still state owned. Russia is one of its leading trading partners, along with Ukraine, Turkey, Iran, Germany, and the United States. Future development is to a large extent tied to planned pipelines for natural gas into Iran and Turkey in the west, and into Afghanistan and Pakistan in the east.

Challenges and Opportunities in Central Asia

The countries described in this chapter are diverse and yet in many ways alike. All of their economies have been recently growing at a rapid rate, attracting much-needed foreign investments, and opening up to the rest of the world. But geographic limitations cannot be ignored: Central Asia remains one of the remotest areas of the world,

FIGURE 31.8. Turkmenbashi, the presidential palace in Ashgabat, is reminiscent of some palaces built in Iraq under Saddam Hussein. *Photo:* C. Burke.

far away from the economic powerhouses of Asia, Europe, or North America, and is entirely landlocked. The future of the region depends on a few key external players—particularly Russia and China, but also Iran, Turkey, and Saudi Arabia, as well as the European Union and the United States. For example, Iran has a natural interest in Turkmenistan, because the two countries share a border, and Iran has some Turkmen population. Turkey is also heavily involved in Turkmenistan. Saudi Arabia is keen to be involved in promoting Sunni Islam in all of Central Asia.

At the moment, Kazakhstan's prospects appear particularly shiny, Sasha Baron Cohen's insinuations in the movie *Borat* notwithstanding. If you travel to the region, visits to the architectural gems of Samarkand and Bukhara in Uzbekistan should be at the top of your list. Central Asia has a wealth of cultural and natural tourism opportunities awaiting exploration in every country, and at present only a fraction of its vast potential is utilized.

EXERCISES

1. Create a table comparing and contrasting the five main ethnicities of Central Asia with respect to their languages, religion, diet, dress, music, main economic activities, and any other characteristics that you think may be appropriate.
2. Investigate recent political developments in the Fergana Valley, Tajikistan, Turkmenistan, and Kyrgyzstan. Do any common underlying themes seem evident in each one of these conflicts? Would you characterize these conflicts as primarily shaped by local forces, or by forces outside the region?
3. Make an inventory of the protected natural areas of any Central Asian republic. (Turkmenistan and Kazakhstan have some of the most famous parks, including Repetek Preserve in the former and Almatinsky Zapovednik in the latter.) What can visitors to these parks see and do? What are the main threats to the ecosystems of the parks?
4. Make a study of the economic, political, and cultural connections between Russia and Kazakhstan, both historical and current. To what extent may Kazakhstan be likened to Canada, and Russia to the United States? Produce a policy statement that argues for or against tighter integration between Kazakhstan and Russia.
5. Which cities of Central Asia may be directly traced to the Silk Route?
6. Use Google Earth to track a segment of the border between Turkmenistan and Iran or between Tajikistan and Afghanistan. What role does topography play in delineating the border?

Further Reading

Alaolmolki, N. (2001). *Life after the Soviet Union: The newly independent republics of the Transcaucasus and Central Asia*. Albany: State University of New York Press.

Dekmejian, R. H., & Simonian, H. H. (2001). *Troubled waters: The geopolitics of the Caspian region*. London: Tauris.

Ebel, R., & Mennon, R. (Eds.). (2000). *Energy and conflict in Central Asia and the Caucasus*. Lanham, MD: Rowman & Littlefield.

Farrant, A. (2006). Mission impossible: The politico-geographical engineering of Soviet Central Asia's republican boundaries. *Central Asian Survey, 25*, 61–75.

Heathershaw, J. (2007). Worlds apart: The making and remaking of geopolitical space in the US–Uzbekistani strategic partnership. *Central Asian Survey, 26*, 123–141.

Marat, E. (2006). *The Tulip Revolution: Kyrgyzstan one year later.* Washington, DC: Jamestown.

Olcott, M. B. (1996). *Central Asia's new states: Independence, foreign policy, and regional security*. Washington, DC: U.S. Institute of Peace Press.

Olcott, M. B. (2002). *Kazakhstan: Unfulfilled promise.* Washington, DC: Carnegie Endowment for International Peace.

Pomfret, R. W. T. (2006). *The Central Asian economies since independence*. Princeton, NJ: Princeton University Press.

Rashid, A. (2002). *The rise of militant Islam in Central Asia*. New Haven, CT: Yale University Press.

Roy, O. (2000). *The new Central Asia: The creation of nations*. London: Tauris.

Rumor, B., & Zhukov, S. (Eds.). (2000). *Central Asia and the new global economy*. Armonk, NY: Sharpe.

Weinthal, E. (2002). *State making and environmental cooperation: Linking domestic and international politics in Central Asia*. Cambridge, MA: MIT Press.

CHAPTER 32

Epilogue

ENGAGING WITH POST-SOVIET NORTHERN EURASIA

In a book titled *Russia 2010*, Yergin and Gustafson (1993) attempted to predict what the country would be like after 15 years of reforms. They envisioned four broad scenarios: (1) "Chaos" (dissolution and an all-out civil war); (2) "Two-Headed Eagle" (the restoration of an authoritarian state, albeit with a capitalist economy); (3) "Russian Bear" (the rise of anti-Western security or military forces, which would run the country along more socialist lines); and (4) "*Chudo*" (i.e., "[Economic] Miracle," in which Russia would become a stable democracy and an increasingly powerful, competitive market economy that was able to export not only natural resources, but high-tech goods as well). Remarkably, many of these predictions have come true. At this writing in 2010, Russia seems to be somewhere between scenarios 2 and 4, with the arguably much worse alternatives (1 and 3) safely avoided. The worst economic times seem to be over. The economy is not growing as fast as expected, but 5–6% annual growth is still much better than what leading Western economies are experiencing nowadays. Russia and some other former Soviet Union (FSU) states are enjoying the windfall of high petroleum and natural gas prices. Politically, the security forces (mainly ex-KGB officers) do in effect rule the country, but they are not as isolationist or nationalist as could be feared. In fact, it is obvious that Russia's current leadership (the Putin–Medvedev team) is very well tuned in to the greater world. Russia is a full participant in the Group of Eight (G8), is working in cooperation with the North American Treaty Organization (NATO) on many issues, is a member of the Council of Europe, takes part in World Trade Organization talks, and remains committed to participation in all other key economic and political global initiatives as a partner of all major world powers. Russia does favor a multipolar world order and, like the United States, is wary of the rise of China. It also does not like to play "second fiddle" to the United States, which is natural for a country of its size and history. In fact, in the past few years Russia has been more and more assertive of its interests not only in other FSU regions (Central Asia or the Caucasus), but increasingly in other parts of the world (Africa, the Middle East, or even Latin America). It also sometimes does disagree with the U.S. or other NATO allies on key issues.

Although the real Russia 2010 does not have independent TV networks or elected regional governors any more, it still remains a multiparty democracy. Its citizens are allowed to read and watch diverse media sources, surf the Internet without censorship, and travel around the world. Its economy is more privatized than those of some European countries, and its companies are not only oil and gas giants, but increasingly competitive small businesses providing consumer goods and some high-tech products that are much needed at home. In short, Russia 2010 is a much better place to be than Russia 1993.

Many aspects of the preceding description apply to other FSU republics. In some (the Baltics), the transition to Western-style democracies and economies is virtually complete, although their nationalistic agendas and overexposure to global financial flows make these countries vulnerable. Other countries range from economically developed but politically oppressive Belarus to politically developed but economically precarious Ukraine and Georgia, with Kazakhstan somewhere in the middle. Although it is impossible to predict the future, I would like to propose some observations on likely developments for Russia and the rest of post-Soviet Northern Eurasia here. First I outline the most likely trajectories for Russia and the other countries; I then provide some ideas about how you, the reader of this book, may *personally* wish to engage with the region.

Russia 2020

In less than 10 years, Russia is likely to emerge as one of the top five economies in the world—bigger than Germany or France, and just a little smaller than Japan by nominal gross domestic product (GDP). It is already considered a pivotal emerging economy, along with the other three so-called BRIC countries (China, India, and Brazil). Russia will continue to dominate Northern Eurasia; it is likely to become an increasingly important transportation corridor between Europe and Asia, capitalizing on its vast railroads and the increasingly ice free Arctic Ocean; and it almost certainly will remain one of only two countries on earth able to challenge the United States militarily (the other one, of course, will be China). More importantly, Russia is likely to become more democratic. If the past is any indication of the future, periods when there is a relative lack of democracy are replaced with periods when there is a drive for more. Good models for authoritarian Russia to follow might be Chile and South Korea. Although much smaller, both emerged as great success stories economically and politically after decades of autocratic military governments. Compared to both, Russia actually has stronger and older traditions of democratic rule and economic prosperity. Russia also continues to have a well-educated population, and will be able to tap into the large diaspora of professionals of Russian descent around the world.

Russia's other strengths, of course, include its plentiful natural resources. These will continue to play an important role, especially its oil, gas, timber, and metals. The size of the land, especially the arable farmland that is at present underutilized, will become an increasingly important asset in the crowded and warmer future world. Russia has an untapped wealth of alternative energy sources. It also has substantial, clean freshwater reserves, including the Siberian rivers, mountain runoff, polar ice caps, and Lake Baikal. Water is emerging as the most critical resource of the 21th century and is likely to become even more important in global affairs by 2020 than petroleum is today. If the amount of climate change that is predicted even in cautious global-warming scenarios takes place, Russia is poised to benefit more than most other countries on earth.

A big weakness of Russia that will become more apparent by 2020 is its inability to wean itself off the petroleum habit, just as its main deposits are beginning to be depleted. Also, Russia's tremendous size and still cold climate will continue to be impediments to development, especially in the most remote regions of the north and in the Far East. Some other major threats are presented by the continued decrease of its population, the rising HIV/AIDS rates, poor air and water quality in the major cities, and the depopulation of large swaths of its agricultural countryside. In addition, Russia must make peace with its southern and eastern neighbors, while at the same time reducing the size of its military and abolishing the obsolete army draft system. It also must reduce its prison population without resorting to executions or release of hardened criminals into its already disorderly cities. Yet another major challenge is pervasive corruption at all levels of its government, especially in the police. Finally, Russia must solve problems in the Caucasus and avoid farther confrontations with either the United States or China. In fact, a strong Russia is vital for enduring global peace, because it is one of very few global players able to be a neutral third party to any developing confrontation between China and the United States.

Those of you who will travel to Moscow in 2020 will still find the beautiful old Kremlin and the cobblestone streets of downtown Moscow charmingly intact. You will also have a chance to board a high-speed bullet train ride to St. Petersburg (2½ hours); observe the city from the top of the highest skyscraper in Europe at 500 m; and eat in the newest all-organic Mama Russia bistro featuring locally grown, carbon-neutral, genetically unmodified, healthy food. Traveling on a bicycle outside the city, you will see wind turbines on the western hills of Istra district and will be able to paraglide in the "Moscow Switzerland" park nearby. Ethnically, the country will retain a Russian majority, but its share will drop to about 75% from its present 80%. The population will probably be smaller than today (about 130 million), but perhaps you will see families with two or three children, rather than only one, strolling in parks. Half of all Russians will speak English when you meet them (only 10% do today). If you are a European or an American, you may not need a visa to enter Russia in 2020, unlike now. If you are a citizen of Kazakhstan or Ukraine, you will be able to travel from your own country to Russia with much greater speed and ease than today; you will not even need to show your passport, and certainly will not be harassed by grim border guards.

One thing that is almost certain to change is that travelers to Russia in 2020 will not be content just to see Moscow, St. Petersburg, and the Golden Ring. There will be hundreds of other opportunities for travel or work in most parts of the country. The most successful regions in the new Russia will no longer be just the oil- and gas-rich provinces, but also the revitalized farming belt, the Black Sea resorts, and central Siberia (including the Altay and the Sayan Mountains). Also thriving will be the Russian Far East, the gateway to the Pacific. Within the FSU, Belarus will have joined Russia, in a full union (the two will have effectively become one country again); Kazakhstan and Ukraine will have free-trade agreements and an open-border policy (similar to what the United States and Canada have today); and almost all other FSU republics will be actively engaged with Russia in many economic projects. The only country that will remain isolated is Georgia, mired in two regional conflicts and unable to fix its internal economic problems on its own, but still unable to join either the Russian sphere of influence or the European Union (EU).

Although the account above may seem a bit too optimistic, I have many reasons to believe it. Geographic research proves that a country's success depends on multiple factors, and Russia seems to have just the right mix of them at the moment to help the country lift itself up and move forward at an unprecedented rate. However, political instability at home, in Asia, or in the Middle East; growing energy costs; and the demographic situation may tip the odds against a prosperous and stable Russia. What about the other Eurasian states?

Future Europeans?

Besides the Baltics, which have already joined the EU and NATO, Ukraine and Georgia are two serious candidates for NATO membership. Moldova may be able to join the EU and/or NATO in 15–20 years, perhaps merging with Romania in the process. However, by 2020 the world may have already moved past NATO. For example, if the U.S. economy continues to deteriorate between 2010 and 2012 under its internal and external debt (which seems possible at this writing), the Europeans will have to build their own defense system to replace the obsolete, U.S.-dominated NATO framework. Also, pragmatically speaking, it is unclear how well the EU itself will fare in the event of a global economic crisis caused by the oil peak, global warming, an emerging flu pandemic, major terrorist attacks, or any other factor(s). The EU is very attractive at the moment because it is prosperous and generous, but this may not remain forever the case. Admission of Turkey into the EU by 2015, which is likely, may change the internal balance of that organization to such an extent that admission of Ukraine or Georgia may become either certain or impossible. The EU countries' indigenous populations are already shrinking and are quickly being replaced by people of very different ethnic and/or cultural backgrounds, with different aspirations and traditions. Their Europe may not be as welcoming to the new members as the current one. Although there is no question in my mind that the "old Europe" will still be around in 2020, it may not be as generous or prosperous as it is today, and thus may well be less inviting to new members.

Are there other countries within the FSU that may become more deeply engaged with Europe in the future? Belarus certainly can and should, provided that its leadership changes. Russia itself can and probably will pursue more pro-European (and perhaps more anti-American) economic policies, with its energy resources. After all, Moscow is a lot closer to Berlin than to Washington, D.C. Armenia is certainly a nation that may very well be admitted, if not to the EU, then perhaps to some strategic economic alliance with the European countries. Culturally, it is the most European of all FSU states in the Caucasus or Central Asia.

The most critical question is this: What will Ukraine do? If Russia is excluded, it is the biggest country in Europe by size and the fifth biggest by population. Ideally, it should find itself in the position of becoming an EU member or associate member while maintaining pragmatic, friendly relations with Russia. The latter is unlikely, however, if Ukraine actively pursues NATO membership. Ukraine, however, is simply too big and too important for the world to allow it to fail. Therefore, it is likely that all major world players will continue to engage with it at various levels, providing necessary political and economic assistance. One hopes that its leadership and its people will be able to figure out the best course for the nation, regardless of pressures from other places.

The Central Asian States

The Central Asian economies at the moment are rapidly developing, which is encouraging. However, they are skewed too sharply toward production of only a handful of commodities (oil in Kazakhstan, cotton in Uzbekistan, gold in Kyrgyzstan, etc.), which is not sustainable in the long run. The biggest opportunities seem to be in developing new resources that are yet untapped, increasing education and health services, and encouraging a new generation of world-savvy political leaders. Of all the republics, Kazakhstan seems to be furthest along in these respects; however, even there the degree of provincialism is quite obvious. The Central Asian countries are among the least engaged players in the new global economy—only marginally better than many African states, and substantially behind most of Asia or Latin America—and are unlikely to change soon.

Some of the biggest uncertainties for the region lie in the ambivalent U.S. and Russian policies toward the region's development, as well as the increasingly powerful interests of China, Iran, Turkey, and Saudi Arabia intersecting in the region. If Iran, for example, wants to encourage Shiite traditions in Tajikistan, the Saudis may prefer to support Sunni movements there. Proximity to Afghanistan does not make any of these countries very safe, and the presence of militant Islamist movements in Uzbekistan and Tajikistan provides an extra layer of complexity in those countries. Nevertheless, the five "-stans" (and Azerbaijan as well) are likely to develop fairly rapidly in the next 5–10 years and will be increasingly visible and accessible from outside the region.

Engaging with the FSU Yourself

It is my sincere hope that after reading this book, you will want to become more engaged with post-Soviet Northern Eurasia. I have taken students on several class trips there and am always amazed at how much joy they experience in their first direct encounters with Moscow, Siberia, or other corners of this vast territory. Remember that you do not need to be a professional *geographer* to do *geography*. Even the simple act of taking a map out and looking at it engages you somewhat with a place; choosing to read a first-hand account of travels in that place will involve you farther. And, of course, you will become even more engaged if you travel through the area, noticing similarities and differences in the physical and cultural landscapes on the way. Below I provide a few concrete ideas about how you personally may experience the FSU:

- Read as much as you can about the countries of Northern Eurasia. The **Further Reading** section at the end of each chapter of this book is a good place to start.
- Use online tools (e.g., Google Earth, the CIA World Factbook, and country-specific Websites) to broaden your knowledge of specific places.
- Watch movies made in the Soviet Union or the FSU republics (see Chapter 15 for specific recommendations).
- Study the Russian language, and perhaps another language from the FSU. (My personal choice would be to try any of the many Turkic languages—e.g., Azeri or Kazakh.)
- Meet immigrants from the FSU in the city where you live or at school where you study. Most American colleges, for example, have Russian-language programs with associated student organizations. Most big cities in North America or Europe have Russian and Ukrainian groceries, restaurants, and bookstores.
- Become a host family for a high school or college exchange student from Northern Eurasia.
- Sign up for a class that goes on a short study-abroad tour of Russia or any other FSU republic.
- Study abroad for a semester or more in any of the many universities in the FSU offering classes to foreigners.
- Join the Peace Corps and live for some time in any of the FSU republics where the Peace Corps is active.
- Travel to any FSU nation as part of a church group or environmental group.
- Buy a commercial tour package, or, better yet, travel on your own in any of the FSU republics. Travel guides will explain how to do this, even on a small budget.
- Consider finding a job in any of the emerging markets of Northern Eurasia. If you know one of the languages and have a good education, you will be very welcome in many positions. You can work for a Western or domestic company, for your government, or even as a freelance translator or tutor.
- Make friends with the locals while you are there. The people of Northern Eurasia are very friendly, although in many places they are not as used to seeing foreigners as in parts of Latin America or Western Europe. Some people I know have made lifelong friends over there, and a few have even found their future spouses.

Above all, no matter what you do, remember that the FSU is waiting for you to explore it. Best wishes to you, and Do Svidaniya!

Index

Page numbers followed by *f* indicate figures and by *t* indicate tables.

About the Author

Mikhail S. Blinnikov, PhD, is Professor of Geography and Graduate Coordinator at St. Cloud State University (SCSU) in St. Cloud, Minnesota. He is a native of Moscow, Russia, and has traveled extensively in the former Soviet Union (FSU). He has done fieldwork on the White, Baltic, and Black Seas; in Central Russia; on the Volga; and in the Crimea, the Caucasus, and the Altay. Besides his work in Russia, he has visited Ukraine, Belarus, Lithuania, Georgia, and Kazakhstan on extended field trips. Dr. Blinnikov's research and publications focus on the late Pleistocene biogeography of grasslands; phytolith analysis; remote sensing/geographic information systems (GIS); protected natural areas and green spaces in and near cities; young-naturalist movements and nongovernmental organizations; and Orthodox religious landscapes. He has worked with the Biodiversity Conservation Center, the Center for Russian Nature Conservation, Nearby Nature, the Nature Conservancy, and the World Wildlife Fund–Russia, among others. He has also worked as an interpreter and translator for a few Russian–American exchanges on environmental projects and written two books in Russian. At SCSU, Dr. Blinnikov has taught classes on the geography of Russia and the FSU; Soviet geography; the geography of Siberia (for SCSU honors); and global geography, GIS, conservation, and biogeography. He has also directed an SCSU study-abroad tour in Moscow and the Altay. He is a member of the Russian, Soviet, and Eastern European specialty group of the Association of American Geographers.